Modeling and Optimization in Science and Technologies

Volume 9

About this Series

The book series *Modeling and Optimization in Science and Technologies (MOST)* publishes basic principles as well as novel theories and methods in the fast-evolving field of modeling and optimization. Topics of interest include, but are not limited to: methods for analysis, design and control of complex systems, networks and machines; methods for analysis, visualization and management of large data sets; use of supercomputers for modeling complex systems; digital signal processing; molecular modeling; and tools and software solutions for different scientific and technological purposes. Special emphasis is given to publications discussing novel theories and practical solutions that, by overcoming the limitations of traditional methods, may successfully address modern scientific challenges, thus promoting scientific and technological progress. The series publishes monographs, contributed volumes and conference proceedings, as well as advanced textbooks. The main targets of the series are graduate students, researchers and professionals working at the forefront of their fields.

More information about this series at http://www.springer.com/series/10577

Junichi Suzuki · Tadashi Nakano
Michael John Moore
Editors

Modeling, Methodologies and Tools for Molecular and Nano-scale Communications

Modeling, Methodologies and Tools

Springer

Editors
Junichi Suzuki
Department of Computer Science
University of Massachusetts
Boston, MA
USA

Michael John Moore
Pennsylvania State University
State College, PA
USA

Tadashi Nakano
Graduate School of Biological Sciences
Osaka University
Osaka
Japan

ISSN 2196-7326 ISSN 2196-7334 (electronic)
Modeling and Optimization in Science and Technologies
ISBN 978-3-319-50686-9 ISBN 978-3-319-50688-3 (eBook)
DOI 10.1007/978-3-319-50688-3

Library of Congress Control Number: 2016959967

Printed on acid-free paper

This Springer imprint is published by Springer Nature
The registered company is Springer International Publishing AG
The registered company address is: Gewerbestrasse 11, 6330 Cham, Switzerland

Preface

Nanoscale communication has emerged as a new research topic in communications engineering. The term "nano" typically refers to the dimensions from a nanometer to hundred nanometers, and nanoscale communication is loosely defined as communication that may occur in that range. Communication engineers have started to investigate two distinct forms of nanoscale communication: molecular communication and electromagnetic-based nanoscale communication. In molecular communication, chemical signals are used for communication among nano- or microscale devices, while in EM-based nanoscale communication, electromagnetic waves are used. As nanoscale communication further develops, it may enable new applications in medicine, environment, and industry. It may also enable the integration of nano- or microscale devices into currently available communication networks to form future communication networks.

The research area of nanoscale communication is rapidly growing as evidenced by an increasing number of papers that have been published in recent years. In editing this book, our goal was to collect, from leading experts of nanoscale communication, comprehensive summary articles with recent research results on nanoscale communication. Toward this goal, this book consists of 24 articles collectively covering

- **Fundamentals of Molecular Communication** (Chapters "Concentration-Encoded Molecular Communication in Nanonetworks. Part 1: Fundamentals, Issues, and Challenges"–"An Architecture of Calcium Signaling for Molecular Communication Based Nano Network"), in which signal processing, communication theory, architectures, and protocols for molecular communication are presented,
- **Molecular Communication in Biology** (Chapters "On Regulation of Neuro-spike Communication for Healthy Brain"–"Quantifying Robustness in Biological Networks Using NS-2"), in which signal propagation in the brain, inter-molecule interactions, gene regulatory networks, and cell communication are studied,

- **Electromagnetic–Based Nano-scale Communication** (Chapters "Fundamentals of Graphene-Enabled Wireless On-Chip Networking"–Nanoscale Communications Based on Fluorescence Resonance Energy Transfer (FRET)"), in which system design and methods of electromagnetic-based nanoscale communication are presented,
- **Nanomaterial and Nanostructure** (Chapters "Ultrasonics—An Effective Non-invasive Tool to Characterize Nanofluids"–"Reliable Design for Crossbar Nano-architectures"), in which nanotechnology for applied purposes such as security and computing is discussed,
- **Medical Applications of Nanoscale Communication** (Chapters "Effect of Aging, Disease Versus Health Conditions in the Design of Nano-communications in Blood Vessels"–"Computational Biosensors: Molecules, Algorithms, and Detection Platforms"), in which disease diagnosis, drug delivery, and biomolecular sensing are presented.

We believe this book serves as a point of references for students and researchers who are interested in this emerging area of nanoscale communication as well as those who are working in the area. Finally, we thank all chapters' authors and reviewers who made this book possible.

Boston, MA, USA — Junichi Suzuki
Osaka, Japan — Tadashi Nakano
State College, PA, USA — Michael John Moore

Contents

Part I
Fundamentals of Molecular Communication

Concentration-Encoded Molecular Communication in Nanonetworks. Part 1: Fundamentals, Issues, and Challenges

Mohammad Upal Mahfuz, Dimitrios Makrakis and Hussein T. Mouftah

Abstract Concentration-encoded molecular communication (CEMC) is a technique in molecular communication (MC) paradigm where information is encoded into the amplitude of the transmission rate of molecules at the transmitting nanomachine (TN) and, correspondingly, the transmitted information is decoded by observing the concentration of information molecules at the receiving nanomachine (RN). In this chapter, we particularly focus on the fundamentals, issues, and challenges of CEMC system towards the realization of molecular nanonetworks. CEMC is a simple encoding approach in MC using a single type of information molecules only and without having to alter the internal structure of molecules, or use distinct molecules. Despite its simplicity, CEMC suffers from several challenges that need to be addressed in detail. Although there exists some literature on MC and nanonetworks in general, in this chapter, we particularly focus on CEMC system and provide a comprehensive overview of the principles, prospects, issues, and challenges of CEMC system.

Keywords Molecular communication · Concentration encoding · Biological nanomachines · Nanonetworks · System design · Nanoscale communication systems

---

This research work was completed while M.U. Mahfuz was with the University of Ottawa, Canada.

M.U. Mahfuz (✉)
Department of Natural and Applied Sciences (Engineering Technology), University of Wisconsin-Green Bay, 2420 Nicolet Drive, Green Bay, Wisconsin 54311, USA
e-mail: mahfuzm@uwgb.edu

D. Makrakis · H.T. Mouftah
School of Electrical Engineering and Computer Science, University of Ottawa, 800 King Edward Ave, Ottawa, ON K1N 6N5, Canada

J. Suzuki et al. (eds.), *Modeling, Methodologies and Tools for Molecular and Nano-scale Communications*, Modeling and Optimization in Science and Technologies 9, DOI 10.1007/978-3-319-50688-3_1

Acronyms

CEMC	Concentration-encoded molecular communication
EM	Electromagnetic
ISI	Intersymbol interference
LRBP	Ligand-receptor binding process
M-AM	Multiple amplitude modulation
MC	Molecular communication
NEMS	Nano-electromechanical systems
OOK	On-off keying
PAM	Pulse amplitude modulation
RN	Receiving nanomachine
TN	Transmitting nanomachine
VAI	Vibrio fischeri Auto-Inducer
VRV	Virtual receive volume

1 Introduction

Molecular communication (MC), in general, requires interdisciplinary knowledge from several technical fields of study ranging from material science through biophysics and computer science to electrical engineering (e.g. communication systems and computer networks). Concentration-encoded molecular communication (CEMC) is a new class of MC paradigm where information in encoded into the amplitude of the transmission rate of molecules at the transmitting nanomachine (TN) and, correspondingly, information is decoded by observing the concentration of information molecules at the receiving nanomachine (RN).

The random walk motion of a particle, for example, a molecule, suspended in a fluid medium e.g. air, water, or blood plasma, is due to the billions of collisions of the particular molecule with the molecules of the surrounding environment within a very small time period, where these collision events take place at the time scale of picoseconds [1]. Although the idea of continuous-time and discrete-time random walk are the same, they are mainly distinguished by the very nature of the step size the particle takes between any two states, and the corresponding time the particle remains in the previous state before it jumps into the next step [2]. While random walk (also known as *Brownian motion*) of particles was first observed in 1827 by Scottish botanist Robert Brown (1773–1858), it was not until 1905 that such motion observed by Robert Brown was explained by Albert Einstein such that it was due to the extremely large number of collisions among particles [3]. For example, due to collisions a small particle named *lysozyme*[1] [4], p. 6, can take 10^{12} steps in one second [5]. As an emerging communication paradigm, MC is being considered as a

[1]Lysozyme is a kind of enzyme [4].

new physical layer option for communicating nanomachines [5]. Random walk of molecules is the main underlying communication principle of diffusion-based CEMC.

Like in other fields of studies in science and engineering, as a remarkable progress in research in the field of communication systems, nanotechnology has enabled the existence of communication systems and networks at the nanoscale. The word "nanotechnology" refers to creating and manipulating the matter at the scale of an atom i.e. at the nanometre scale. On 29 December 1959 the Nobel laureate scientist Dr. Richard Feynman (1918–1988) delivered a talk entitled "*There's Plenty of Room at the Bottom An Invitation to Enter a New Field of Physics*" where he first mentioned that in the near future devices would be miniaturized down to atomic scale [7]. In addition, he pointed out proposals of how to possibly accomplish the miniaturization and prepare devices at the atomic scale. Fifteen years later, in 1974 the term "nanotechnology" was first defined by Japanese scientist Norio Taniguchi [8] to describe the control of materials at the nanometre scale as "*Nano-technology mainly consists of the processing of, separation, consolidation, and deformation of materials by one atom or one molecule.*" Since then the idea of nanotechnology is still new in our society. Nanoscale refers to the dimension in the range from 1[2] to 100 nm [9–11]. When size goes down to nanoscale, physical, chemical, electrical, magnetic, optical, and mechanical properties of materials change and novel properties arise in the matter [12], which can be beneficial to communication engineering if harnessed properly. An example of the novel characteristics that arises at the nanoscale is the melting point of a particle of 5 nm size that may deviate "as much as couple of hundreds of degrees" as compared to the bulk melting point [12]. Another example would be a silicon nanowire of 5 nm diameter whose band gap may increase from 1.1 eV to nearly approximately 3 eV [12].

Nanoscale devices have at least one dimension in the range from 1 to 100 nm [11]. According to available literature, the dimension of 100 nm is significantly important in the sense that under this limit new material properties arise due to the laws of quantum physics [13]. In this chapter, we are interested to see how nanotechnology can benefit the communication engineering discipline. In particular, we discuss CEMC system in detail in this chapter.

The chapter is organized as follows: Sect. 2 discusses nanomachines and nanonetworks in general as well as the applications of CEMC-based nanonetworks. Sections 3 and 4 describe the communication principles of CEMC in detail. Limitations of CEMC are explained in Sect. 5. Research issues and challenges in CEMC are described in Sect. 6. Finally, Sect. 7 concludes the chapter with a summary of the opportunities of CEMC in nanonetworks.

[2] 1 nm is equal to 10^{-9} (i.e. billionth) of a metre. It is approximately 1/80,000 of the typical diameter of a human hair, or 10 times the diameter of a hydrogen atom [13].

2 Nanomachines and Nanonetworks

Nanoscale communication is targeted for nanomachines that can communicate among themselves in order to form "*nanonetworks*" and/or initiate certain functions [10]. A nanomachine is an artificial or naturally (e.g. biologically) occurring entity of nano-scale to micro-scale dimensions that is capable for performing some simple tasks. A nanomachine can be considered as the most basic functional unit at the dimensions equivalent to the atomic and molecular scales. Examples of the tasks that a nanomachine is capable of are simple molecular computations, sensing and detection of molecules, generation of motion, and performing chemical reactions. In the area of molecular nanotechnology, most molecular biological systems are themselves nanomachines, examples of which include molecular motors, namely, *kinesin*, *dynein*, and *myosin* that convert chemical energy in the form of ATP hydrolysis to mechanical work and thus generate motion [14], p. 33. While man-made nanomachines are yet to be manufactured, natural biological nanomachines are already available in the nature and are able to be engineered to perform tasks in a controlled manner [15]. The engineered nanomachines are discussed in a later subsection. In its structure, a nanomachine is assumed to have an arranged set of molecules that are able to perform one or more of the tasks mentioned above. In addition, it is also assumed that a nanomachine is a building block of more complex systems, e.g. nano-robots (a.k.a. nanobots), nano-processors, nano-memories, or nano-clocks [10].

On the other hand, a large number of nanomachines can collaborate with one another and thereby perform comparatively complex tasks in a distributed manner, which would otherwise not be possible by a single nanomachine. The resulting interconnection of a number of nanomachines is known as *nanonetworks* [10]. Collaboration and coordination among nanomachines necessarily require a nanomachine to communicate with other nanomachines, which is the main research topic of *nanoscale communication networks.*

Nanonetworks would be able to enhance the functionality of a single nanomachine in several ways, for instance, by performing complex tasks, enhancing workspace, performing controlled behaviour, and forming networks of nanomachines [10].

2.1 Approaches to Development of Nanomachines

At the present time, three approaches to nanomachine development, namely, *top-down approach*, *bottom-up approach*, and *bio-hybrid approach*, have been considered as the most likely for use in the artificial nanomachine development [10].

The *top-down approach* is based on the downscaling the micro-scale components with the help of available manufacturing techniques, for instance, electron beam lithography [16], micro-contact printing [17]. Examples of nanomachines built by using this principle are the nano-electromechanical systems (NEMS) components

[10, 12]. Top-down approach requires a very sophisticated processing of materials at the nanoscale and thus experiences a high cost in the manufacturing process when the component size approaches the nanoscale dimensions [18]. For example, the cost of photolithography equipment advanced with nanofabrication capability can be as high as $50 million [18]. In addition, when size goes down to nanoscale, the atoms or molecules tend to stick together due to the Van Der Waals attraction force [19], p. 550. Recent developments of nanomachines by following the top-down approach [10, 18, 20] are still in the early stage of research and development.

On the other hand, the *bottom-up approach* to developing nanomachines is based on using molecules as the building block to develop the nanomachines. Recently, a number of nanomachines, e.g. molecular differential gears and pumps, have been designed theoretically by using this approach [10]. Although the manufacturing technique, called *molecular manufacturing*, using controlled arrangements of molecules is yet to become reality, there is a good hope that molecular manufacturing techniques would become matured in the next few decades and thus the development of nanomachines with the *bottom-up approach* would become reasonably possible. And, in addition, it is highly expected that the bottom-up approach will have certain advantages over the top-down approach [21].

Finally, the bio-hybrid approach to developing nanomachines is based on the inspiration that several nanoscale components present in the biological cells can be considered as nanomachines and thus be used and even engineered as nanomachines to perform desired functions in a controlled manner. An example of the bio-hybrid approach to developing nanomachines is an engineered biological cell that has a nucleus (nano-processing unit), endoplasmic reticulum (nano-storage unit), gap junctions (biological transceivers), receptors (nano-sensors), flagellum (nano-actuators), mitochondrion (nano-power unit), and vacuoles (nano-energy scavenger) [10]. The bio-hybrid approach also encourages us to use these nanomachines in the biological cell as models to develop new nanomachines and/or use them as building blocks in order to develop more complex systems known as nano-robots.

2.2 *Expected Features and Functionalities of a Nanomachine*

Although natural nanomachines are already there in nature, artificial man-made nanomachines are yet to be manufactured. However, the expected features and functionalities of nanomachines are given in the following. Nanomachines are expected to be self-contented meaning that they should have a set of instructions to realize the intended task. The set of instructions could either be embedded on their own molecular structure, or stored in another storage from where the nanomachines would be able to read the instructions when required [10]. Self-assembly would allow the nanomachines to be assembled from several parts without external intervention. Self-replication allows a nanomachine to make a copy of it by using the external elements. Locomotion enables a nanomachine to move from one place to another. Communication allows nanomachines to create more complex structures by communicating, cooperating, and collaborating with other nanomachines. Communication among nanomachines also enables decentralization and distributed intelligence, a.k.a. swarm intelligence [13].

While all of the desired features mentioned above can be obtained from the natural nanomachines, e.g. a biological cell, there can be additional features, e.g. multi-tasking and multi-interfacing, that can also be considered as expected features of a nanomachine. Multi-tasking allows a cell to perform multiple tasks, e.g. taking nutrients from the environment, running chemical processes etc., at the same time. Multi-interfacing allows a cell to communicate with several other entities using different communication methods. For example, a cell can simultaneously communicate with ligand-receptor binding process (LRBP), gap junctions, and molecular motors. Therefore, it is expected that a nanomachine should also be able to achieve multi-interfacing ability.

2.3 Architecture of a Nanomachine

Referring back to the expected features of a nanomachine, in order to realize a nanomachine, the generalized architecture of a nanomachine that is somewhat complete in its functionalities should have the following functional units: *control unit* that executes the instructions stored in the instruction storage and that is equivalent to the nucleus of a biological cell [22], *communication unit* that communicates with other nanomachines by sending and receiving messages and is equivalent to the gap junctions and ligand receptors in a biological cell [14, 23], *reproduction unit* that generates a replica of itself and is equivalent to the DNA sequence in a biological cell, *power unit* that powers all the individual components of the nanomachine and is equivalent to the mitochondrion in a biological cell, *sensor and actuator units* that interface the nanomachine with the external environment and are equivalent to the ligand receptors and flagella respectively in a biological cell [10, 13]. It is highly hoped that in the near future completely synthetic nanomachines [21] would become a reality, achieve the expected features of a nanomachine in general, and thus be able to perform the required tasks of a TN and an RN.

2.4 Engineered Natural and Synthetic Nanomachines

The TN and the RN need communication functionality in its structure in order to realize a communication system working. These functionalities include synthesizing, storing, and releasing information molecules [15]. This subsection describes how natural nanomachines can be engineered to perform the desired tasks. It also introduces how artificial nanomachine (and cell-like) structures can be built based upon biologically-friendly materials and techniques. One of the existing techniques to develop synthetic nanomachines is to modify and add communication and logic functionalities in the existing biological nanomachines through genetic engineering [15, 24, 25]. For example, it is found that sender nanomachines can be engineered to synthesize and release information molecules, e.g. acyl homoserine lactone (AHL) molecules, by using the specific metabolic pathways and those receiver nanomachines can be engineered to react to a specific concentration level of information molecules and thus react accordingly by synthesizing reporter proteins [15].

The information molecules transmitted by the TN can diffuse freely in the propagation environment and thus reach the RNs and make the communication possible between them. By using the principles of synthetic biology, more specifically, genetic engineering, it is possible to develop *logic functions, toggle switch,* and *oscillators* by using a biological nanomachine, e.g. a biological cell [15]. On the other hand, it is also possible to build artificial cell-like structures, by using biological materials e.g. lipid bilayer similar to a cell membrane structure, that can enclose some fundamental components of the artificial cell [15]. When lipid bilayer membrane forms a cell-like structure it is then possible to add functional protein inside the vesicles created by lipid bilayer. Research shows that such a cell-like structure is capable of replicating itself and producing sender and receiver functionalities [15].

3 Communication Between Nanomachines in a Nanonetwork

Although a single nanomachine is of very limited capabilities and can handle forces of the order of pico-Newtons (pN), a huge number of nanomachines would form a nanonetwork in order to handle big tasks that require forces of the order of Newtons. The analysis of a unicast communication link between a pair of nanomachines is important in the sense that it is fundamental to the understanding of the communication system in nanonetworks. In nature, it is found that two biological cells can communicate in the concentration-dependent manner using information molecules [15]. When communication at the nanoscale is investigated, available literature suggests that there can be four possible ways of communication, namely, *nano-mechanical*, *nano-acoustic*, *nano-electromagnetic* (EM), and *molecular communication* (MC). In nano-mechanical communication, the TN and the RN are in direct contact with each other and the information is communicated by means of the movements of the nano-mechanical contacts (hinges) between them [23], p. 183. In nano-acoustic communication, the information is communicated by means of an encoding technique that relies on the variations of the pressure waves [23], p. 177. EM-based nanoscale communication technique relies on the modulation of EM waves transmitted by and received from nanomachines [26]. Finally, MC technique uses molecules as the messages in order to encode the information from the TN destined to the RN. In MC paradigm, the TN transmits information molecules in order to communicate information to the RN [6, 27, 28].

MC-based techniques for communicating nanomachines offer the following two main benefits over other communication techniques at the nanoscale: first, in nano-mechanical communication the TN and the RN would need to be in direct contact with each other, which may not always be possible for a pair of communicating nanomachines; MC-based techniques relax this restriction. Second, because of the size and operating principles of acoustic transducers and RF transceivers, it is reported that their respective implementations in integrated circuits at nanoscale and molecular scale are not feasible [10]. While on the other hand, the (bio-inspired) MC transceivers are basically nanomachines that can transmit and receive molecules and many of these nanomachines already exist in the nature, thus the ability to materialize them is already proven. Therefore, of the four communication techniques mentioned above, in the available literature, the MC technique is thought to be the most biologically suitable technique [10] for communication among nano- to micro-scale biological nanomachines. In this chapter, we provide some fundamentals of nanoscale and molecular communication in general and gradually focus on CEMC system in particular. On another note, it is understood that MC is more advantageous [10] compared to other nanoscale communication techniques and several examples of MC can be found in the Mother Nature.

3.1 *Communication at the Nanoscale*

Nanonetworks basically refer to the two branches of research on the communication technologies at the nanoscale: first, dry techniques that involve nanostructures, modified materials, and interconnected devices derived from the downscaling of the existing micro-scale technologies, and second, wet techniques that are derived from the communication mechanisms and components mainly inspired from the biological systems and materials. While both dry and wet techniques represent communication mechanism at the nanoscale, communication takes place in different forms in both of these techniques. This section represents a brief overview of the different techniques and components of communication at the nanoscale.

3.2 *Dry and Wet Techniques*

Dry techniques refer to the communication techniques that are derived from nanoscale techniques that deal with the fabrication of the carbon, silicon, and other inorganic materials and their involvement with communication at the nanoscale. As mentioned earlier, dry techniques offer communication solutions by downscaling the currently available micro-scale technologies and are meant for the building of the devices at the nanoscale. Examples of dry techniques of nanoscale communication include nanowired communication [13], carbon nanotube (CNT)-based communication [14], nanophotonic communication [13], and free space wireless optical communication [29].

Wet techniques, on the other hand, deal with the study of biological systems that are already available in the nature, and use the inspiration from those already existing systems to design a nanoscale communication system capable of offering a basis for realizing nanonetworks. Biological systems mostly operate in the aqueous environment and hence the name of these "wet" communication techniques. MC is a very well-known example of wet techniques, and a new paradigm for communication at the nanoscale. MC is a new physical layer option that is being considered for communication and networking among a huge number of natural and man-made nanomachines [10, 25, 28, 30–32]. In addition, the highly noticeable feature of MC is that MC is a truly interdisciplinary field of research that spans from physics, nanotechnology, chemistry, biotechnology, and communication technologies. MC relies on communicating with the molecules, meaning that the molecules carry the information from the transmitter to the receiver. MC, as a wet communication technique, would be a main focus of this research study. CEMC is concentration-based approach to MC among nanomachines.

3.3 *MC and CEMC*

In general, MC examines how bio-nanomachines exchange information among themselves. A reasonable volume of research works on MC can be found in several of the existing literature [6, 10, 27, 31, 33, 34]. MC is a bio-inspired paradigm of the nanoscale communication and nanonetworking field. Unlike EM wave-based communications, the nanomachines communicate with molecules in MC. A generic MC system consists of a TN, an RN, and a propagation medium between them. The propagation medium is also known as the *channel* [35]. Figure 1 shows a typical diffusion-based MC channel between a TN and an RN. As shown in Fig. 1, the TN is located at the origin (0, 0, 0). The molecules released by the TN undergo ideal (i.e. free) diffusion in three dimensions in the propagation medium, while being governed by random walk-based diffusion process. The molecules probabilistically reach the RN located at a distance r from the TN. As shown in Fig. 1, the TN transmits information molecules at a given (predefined) rate. Examples of information molecules include proteins and ions that contain information to be transmitted [36]. Examples of the propagation media include water, blood plasma, and air [13].

Like TN, RN is a nanomachine that can receive the information molecules at its receptor sensors. Propagation of molecules can take place with either active transportation (e.g. molecular motors) [31, 37] or passive transportation known as *diffusion* [5, 38]. Diffusion is a passive transportation mechanism in which the information molecules propagate spontaneously in the surrounding environment due to thermal effects. In this chapter, as shown in Fig. 2, we consider a diffusion-based propagation channel for the CEMC purposes between a pair of nanomachines in nanonetworks.

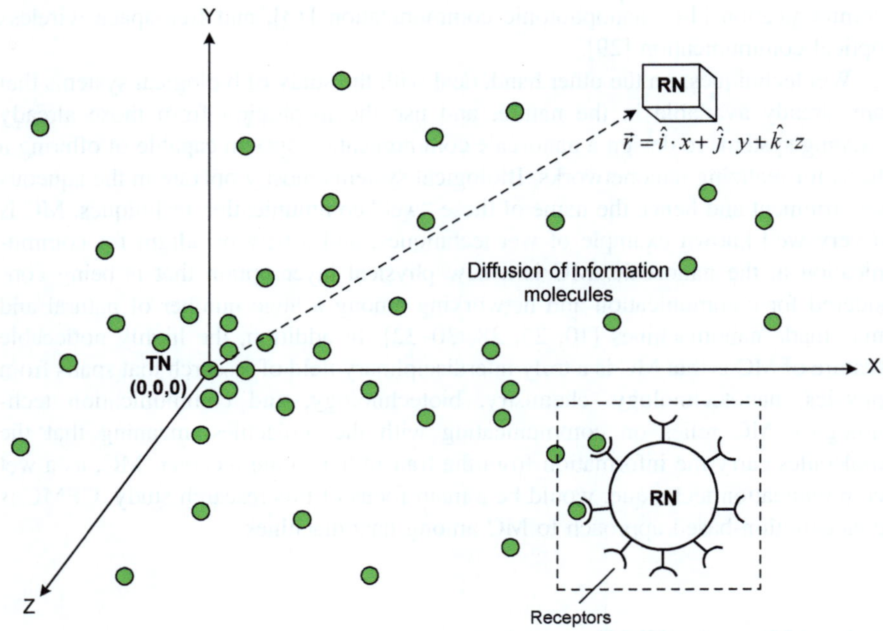

Fig. 1 Ideal (i.e. free) diffusion of information molecules in three dimensions in the unbounded propagation medium. The RN is shown to be in the centre of a small virtual receive volume (VRV) [39]. The receptors of the RN shown in inset bind with a single type of molecules

3.4 CEMC Phases

A TN communicates with an RN by means of five-phase communication mechanism. The first phase is "*encoding*" with which the TN translates the information into the information molecules such that it becomes possible for RN to detect the information. Information may be encoded onto several characteristics of the information molecules, for instance, type of the information molecules used i.e. *distinct molecules* [40], internal structure of the information molecules i.e. *molecular-encoding* [13], or the concentration of information molecules i.e. *concentration-encoding* [30, 41]. *Concentration-encoding* enables information to be encoded in the amplitude of the transmission rate of the molecules by the TN, and correspondingly, by decoding the encoded information by observing the number of information molecules received by the RN per unit volume of the solvent molecules used. Concentration-encoding is relatively simple because it does not require altering the internal structure of the information molecules, nor does it require sending distinct molecules for information communication. This has given us the impetus to concentrate on the CEMC system in this chapter and the next. In this chapter, we focus on concentration-encoding and investigate into various characteristics of a CEMC system between a pair of nanomachines in a fluidic medium.

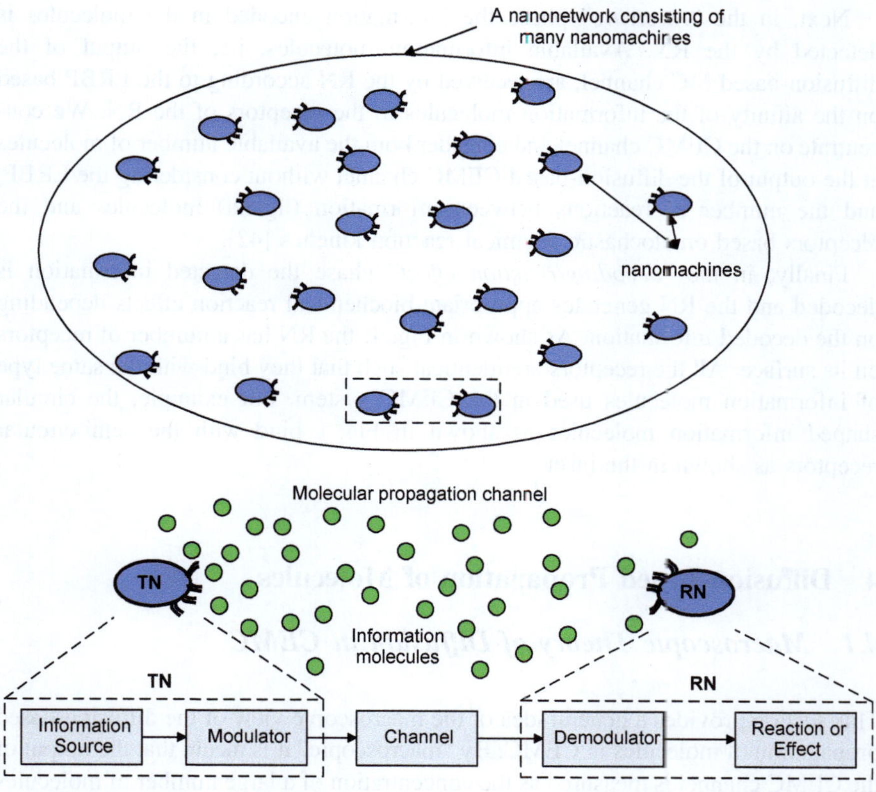

Fig. 2 General view of a nanonetwork (*top*) and a molecular communication channel between a pair of a TN and an RN (*bottom*) shown. The five components of MC system are also shown. The *green* circles indicate the information molecules in the propagation medium

In the second phase of MC, the TN transmits the encoded information molecules in the propagation medium, the communication phase being known as the "*sending*" phase.

The transmitted molecules propagate in the environment and the phase of communication is known as the "*propagation*" phase. In passive transportation, the information molecules thus sent by the TN diffuse passively in the environment due to Brownian motion without using chemical energy. Diffusion of information molecules depends on the size and molecular weight of the molecules, structural characteristics of the propagation medium, e.g. viscosity, flow of the medium, pressure, dispersion, and absolute temperature [13, 14]. However, in this study in order to investigate into the characteristics of communication system only, we have not considered the variations of these factors over the observation time; rather we have considered commonly used values for these factors, e.g. diffusion constants, where necessary.

Next, in the "*receiving*" phase the information encoded in the molecules is detected by the RN. Available information molecules, i.e. the output of the diffusion-based MC channel, are received by the RN according to the LRBP based on the affinity of the information molecules to the receptors of the RN. We concentrate on the CEMC channel and consider both the available number of molecules at the output of the diffusion-based CEMC channel without considering the LRBP, and the number of reactions between information (ligand) molecules and the receptors based on stochastic chemical reaction kinetics [42].

Finally, in the "*decoding/reaction effect*" phase the detected information is decoded and the RN generates appropriate biochemical reaction effects depending on the decoded information. As shown in Fig. 1, the RN has a number of receptors on its surface. All the receptors are identical such that they bind with the same type of information molecules used in the CEMC system. For example, the circular shaped information molecules as shown in Fig. 1 bind with the semi-circular receptors as shown in the inset.

4 Diffusion-Based Propagation of Molecules

4.1 Macroscopic Theory of Diffusion in CEMC

This section provides a general idea of the macroscopic view of the diffusion-based propagation of molecules in CEMC. By "macroscopic" it is meant that the output of the CEMC channel is measured as the concentration of a large number of molecules at the RN. Unlike macroscopic view of the molecular diffusion, we describe the microscopic view of the molecular diffusion in the next section where we consider each molecule individually and we compute the concentration of large number of molecules at the RN by considering each molecule individually based on its probability of availability at the RN.

Due to the probabilistic nature of the number of the molecules available at the RN, the *macroscopic theory* of diffusion [5] explains the available *mean* number of molecules at the RN by providing the average value of the amplitude of concentration signal that is present at the location of the RN. The propagation of information molecules can be explained with two main theories, namely, the *macroscopic theory* and the *microscopic theory* of diffusion [5]. When diffusion is explained in terms of the macroscopic theory, it is assumed that there exist a large number of information molecules in the system and the diffusion process is explained in terms of the average number of the molecules in the system. This subsection provides the details of macroscopic theory of diffusion based on average value of concentration of molecules. The details of diffusion based on microscopic theory will be provided in the next subsection along with channel characteristics and diffusion-based noise model.

Diffusion of molecules in the fluidic medium can be explained by the Fick's first and second laws of diffusion [5, 43]. Fick's laws of diffusion are the differential equations that describe the non-uniform distribution of molecules in the fluidic medium in spatial and temporal domain.

Fick's first law states that net flux of molecules at x is proportional to the slope of the concentration function at x, the constant of proportionality being equal to $-D$. Therefore, Fick's first law can be expressed as the following, meaning that there will not be any particle flux (i.e. $J_x = 0$) if the molecules are uniformly distributed over space, and correspondingly, the distribution of molecules will not change over time if $J_x = 0$.

$$J_x = -D\frac{\partial U}{\partial x} \tag{1}$$

Fick's second law follows from his first law and represents the time rate of change of the concentration of the information molecules at a location provided that the total number of information molecules is conserved, i.e. the information molecules are neither created nor destroyed by the system. Fick's second law can be expressed as below.

$$\frac{\partial U}{\partial t} = -\frac{\partial J_x}{\partial x} \tag{2}$$

Using Eqs. (1) and (2) yields the following.

$$\frac{\partial U}{\partial t} = D\frac{\partial^2 U}{\partial x^2} \tag{3}$$

Fick's second law of diffusion states the time rate of change of concentration of information molecules at x and t and is proportional to the curvature of the concentration function at x and t, the constant of proportionality being equal to D. If the slope of concentration $\partial U/\partial x$ is constant, i.e. $\partial^2 U/\partial x^2 = 0$, then the concentration is stationary: $\partial U/\partial t = 0$, meaning that the number of information molecules diffusing in will be equal to that diffusing out [5].

Extension to Three Dimensions

Referring to Fig. 1, the information molecules undergo ideal (i.e. free) diffusion in the propagation medium in three dimensions according to Fick's laws [5] and the molecules can, therefore, become available to the RN at multiple times. In addition, the TN releases a single type of information molecules; and the TN and the RN are assumed to be synchronized in time [25]. In the three dimensional case, the information molecules propagate independently in three dimensions and so the diffusion equations can be written in three dimensions as well. In general, considering independent diffusion constants in each dimension we can write the equations for net molecular flux and concentrations as below

$$J_x = -D_x \frac{\partial U}{\partial x}, \quad J_y = -D_y \frac{\partial U}{\partial y}, \text{ and } \quad J_z = -D_z \frac{\partial U}{\partial z} \tag{4}$$

However, in homogenous medium it is customary to assume that the diffusion constants in each dimension are equal, i.e. $D_x = D_y = D_z = D$ [5] and so the net flux vector can be written as $\mathbf{J} = -D\,\mathrm{grad}\,U$ and the concentration changes with space and time as

$$\frac{\partial U}{\partial t} = D\nabla^2 U \tag{5}$$

when $U(x, y, z, t)$ is written in short as U by not writing the functional dependence of (x, y, z, t), ∇^2 is the three-dimensional Laplacian operator denoted as $\nabla^2 = \partial^2/\partial x^2 + \partial^2/\partial y^2 + \partial^2/\partial z^2$, and $\vec{r} = \hat{\imath}\cdot x + \hat{\jmath}\cdot y + \hat{k}\cdot z$ is the vector representing the distance between the TN and the RN in the three dimensional space using Cartesian coordinates.

Solution of Fick's Second Law in Three Dimensions

When information molecules are released by the TN in an impulsive manner, i.e. Q_0 molecules are released at the location of the TN $(0, 0, 0)$ at time $t = 0$, the solution to the diffusion equation in three dimensions can found as below [5].

$$U(r, t) = \frac{Q_0}{(4\pi D t)^{3/2}} \exp \frac{-r^2}{4Dt} \tag{6}$$

where $r^2 = x^2 + y^2 + z^2$ and assuming radial symmetry this indicates the mean number of information molecules that become available at the location of the RN.

The concentration of available molecules in molecules per unit volume depends on the location of the RN and the time elapsed after the release of the molecules. For the sake of simplicity, the available concentration $U(r, t)$ of molecules at the location of RN can also be termed as the "*signal intensity*" and any integral of $U(r, t)$ can be termed as the "*signal strength*" [41]. The significance of Eq. (6) is that it indicates the channel impulse response (CIR) of the CEMC channel [35]. By taking the derivative of Eq. (6) it can be shown that the RN located at a distance r will see that the concentration of diffused molecules at r peaks at $t = r^2/6D$ at concentration value of $U = 0.0736 Q_0/r^3$.

When considering diffusion in homogenous medium, we have considered that D does not vary over time and space and that the chosen value of D of small information molecules in water medium is 10^{-6} cm^2/s. The values of D based on other types of information molecules and/or propagation media can also be found in several text books, e.g. [4, 23]. Understanding the solution to the diffusion equation in three-dimensional space we can say that the quantity $U(r, t)$ indicates the available information molecules at the location of the RN.

The propagation of information molecules is affected significantly by the noise generated from and within the propagation environment itself. Examples of such noise generating sources are thermal energy, electric fields, magnetic fields, EM waves, e.g. energy of a photon absorbed by the solvent and solute molecules, and also the molecules that do not take part in MC, as well as other nanomachines that are idle in the system [25]. The quantity $U(r,t)$ denotes the mean number of molecules that are available per unit volume of the solvent molecules at the RN and ready to be received by the RN. As shown in Fig. 1, the RN is assumed to be located at the centre of the small volume VRV and thus the total number of molecules available for reception in the small volume of VRV would represent the amplitude $s(t)$ of the available molecular concentration signal at the RN [41] expressed as below.

$$s(t) = \iiint\limits_{\mathrm{VRV}} U(x,y,z,t)dx\,dy\,dz = \iiint\limits_{\mathrm{VRV}} \frac{Q_0}{(4\pi Dt)^{3/2}} \exp\frac{-(x^2+y^2+z^2)}{4Dt} dx\,dy\,dz \quad (7)$$

Here VRV also denotes the total sensing volume of the RN, *dxdydz* is the differential volume in the VRV, and $r^2 = x^2 + y^2 + z^2$.

4.2 Microscopic Theory of Diffusion in CEMC

In this subsection, propagation of the molecules has been explained from the *microscopic theory* of diffusion, which explains the propagation of a single molecule that is released by the TN. Correspondingly, the probability of getting the emitted molecule at the RN would be investigated. The term "microscopic" refers to the single molecule involved. In the following, the propagation of a single molecule emitted by the TN has been investigated from the view point of two basic principles: first, with the understanding of the Wiener process, and second, with the application of the binomial theorem. However, it is found that both approaches provide the same expression of the "*probability of having a single molecule at the location of the RN,*" which, therefore, proves the correctness of both approaches, especially in the calculation of the additive diffusion noise in the concentration signal available for reception at the RN.

The unbiased random walk in one dimension has been shown in Fig. 3. As a result, the particles remain at (0, 0, 0) on average, and their root mean square (RMS) distance is proportional to the square root of the time. This also yields that the variance of the molecule displacement can be expressed as the following.

$$\mathrm{var}(x(t)) = \langle x^2(t)\rangle - (\langle x(t)\rangle)^2 = 2Dt \quad (8)$$

Here the operator $\langle\cdot\rangle$ denotes the mean of any quantity. Therefore, it can be understood that if the molecule is at $x_i(t)$ at time t then at $(t+\tau)$ the position (displacements) of the molecule in x, y, z dimensions will be as shown below.

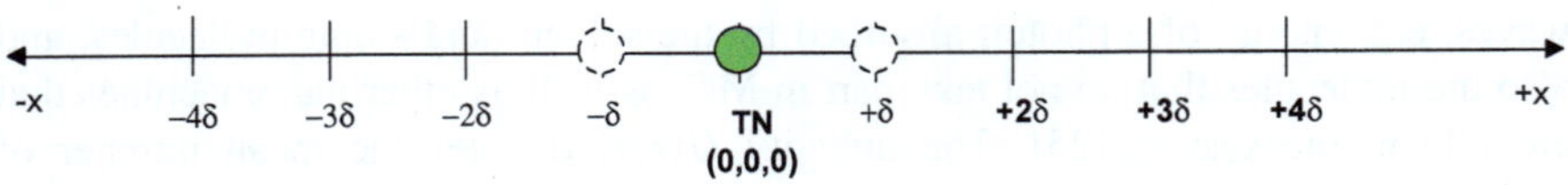

Fig. 3 Molecules released at TN execute one-dimensional random walk. The molecule released at TN $(0,\ 0,\ 0)$ at $t=0$ has equal probability of taking the step to the *right* or to the *left* and move to $x=+\delta$ or $x=-\delta$ respectively at $t=\tau$. The similar phenomenon takes place after every step, thus completing an unbiased random walk in one dimension

$$\begin{aligned} x_i(t+\tau) &= x_i(t) + \sqrt{2D\tau}\mathcal{N}(0,\ 1) \\ y_i(t+\tau) &= y_i(t) + \sqrt{2D\tau}\mathcal{N}(0,\ 1) \\ z_i(t+\tau) &= z_i(t) + \sqrt{2D\tau}\mathcal{N}(0,\ 1) \end{aligned} \tag{9}$$

The diffusion constant D, in cm^2/s unit, can also be defined as $D=\frac{kT}{6\pi\eta\rho}$, where k is the Boltzman constant, T is the absolute temperature, η is the viscosity of the fluid, and ρ is the radius of the information molecules [23].

Explanation with Wiener Process

A molecule starts its journey at time $t=0$ at $(0,\ 0,\ 0)$, i.e. at the location of the TN. The random walk model in one dimension assumes the following rules: each molecule moves to the right or to the left in exactly every τ seconds with a velocity of $\pm v_x$ a distance $\pm v_x\tau$. Although the parameters τ and δ depend on several factors, e.g. the size of the information molecules, the size of the solvent (liquid) molecules, the structure of the liquid, and the absolute temperature, in ideal diffusion-based MC system for the sake of simplicity we tend to consider τ and δ as constants and that probabilities of a molecule going to the left (p_L) and to the right (p_R) are equal, i.e. each probability being equal to $p_L=p_R=0.5$.

Each step is memory-less, meaning that the molecule jumps randomly to right or left forgetting what direction it went in its last leg. Successive steps taken by the molecules are independent and the random walk is unbiased. Each molecule moves independently of all other molecules without interacting with one another, which is a reasonably realistic assumption when the solution is dilute [5]. The propagation medium is assumed to be homogenous for the information molecules. The diffusion constant is assumed to remain unchanged in the entire duration of the observation time.

Each of the Q_0 molecules transmitted by the TN propagates independently in the three-dimensional unbounded propagation environment as per ideal (i.e. free) diffusion. The ideal diffusion of a molecule can be explained as Wiener process [44] whereby the displacement of each molecule in the infinitesimal time τ (i.e. after completing one jump) can be modeled by a Gaussian random variable with a variance of $2D\tau$, where D is the diffusion constant of the information molecules in the propagation medium. Since the mean position of the molecules does not change after each jump and each jump of each molecule is a random process, the diffusion process is statistically independent in each of the three dimensions [5].

As a result, the position of each of the molecules in each of the three dimensions after time t (i.e. after completing n jumps, each taking time τ seconds, meaning that $t = n\tau$) can be given by the following.

$$\begin{aligned} x(t) &= \mathcal{N}(0, \mathrm{var}x(t)) = 0 + \sqrt{2Dt}\mathcal{N}(0,\ 1) \\ y(t) &= \mathcal{N}(0, \mathrm{var}y(t)) = 0 + \sqrt{2Dt}\mathcal{N}(0,\ 1) \\ z(t) &= \mathcal{N}(0, \mathrm{var}z(t)) = 0 + \sqrt{2Dt}\mathcal{N}(0,\ 1) \end{aligned} \tag{10}$$

Considering each of the transmitted molecules separately, the probability density function (pdf) of a single molecule being available at the three-dimensional space $(x(t), y(t), z(t))$ is a three-dimensional *normal* distribution with mean located at $(0, 0, 0)$, i.e. the location of the TN, and variance $2Dt$ in each dimension, and therefore, can be expressed as below.

$$\begin{aligned} P_{XYZ}(x,y,z) &= P_X P_Y P_Z \\ &= \left\{ \frac{1}{\sqrt{2\pi\sigma_X^2}} \exp \frac{-(x(t)-0)^2}{2\sigma_X^2} \right\} \cdot \left\{ \frac{1}{\sqrt{2\pi\sigma_Y^2}} \exp \frac{-(y(t)-0)^2}{2\sigma_Y^2} \right\} \cdot \left\{ \frac{1}{\sqrt{2\pi\sigma_Z^2}} \exp \frac{-(z(t)-0)^2}{2\sigma_Z^2} \right\} \\ &= \frac{1}{(4\pi Dt)^{\frac{3}{2}}} \exp \frac{-(x^2+y^2+z^2)}{4Dt} \\ &= \frac{1}{(4\pi Dt)^{\frac{3}{2}}} \exp \frac{-r^2}{4Dt} \end{aligned} \tag{11}$$

where $\sigma_X^2 = \sigma_Y^2 = \sigma_Z^2 = 2Dt$ and for simplicity, wherever appropriate, we drop from P_{XYZ} the explicit temporal functional reference (t) in this chapter.

Explanation with Binomial Distribution

As mentioned before, the molecule steps to the right with a probability p_R and to the left with a probability p_L, where $p_L = 1 - p_R$. As before, the molecule needs an infinitesimal time τ at each state and then steps to the next state, where after n steps the molecule reaches a distance at time t such that $t = n\tau$. An information molecule steps an enormous number of times (n) every second. As a result, the values of both n and np_R (i.e. the number of steps taken to the right) are finite and large numbers. Correspondingly, when $n \to \infty$, $np_R \to \infty$ is valid and under these conditions a binomial distribution can be approximated to a normal distribution as below [5]

$$P(x)dx = \frac{1}{(4\pi Dt)^{1/2}} \exp\left(-\frac{x^2}{4Dt}\right) dx \tag{12}$$

where $P(x)dx$ is the probability of finding a molecule between x and $(x + dx)$, and

$$P(x,t) = \frac{1}{(4\pi Dt)^{1/2}} \exp\left(-\frac{x^2}{4Dt}\right) \tag{13}$$

is the probability of finding one molecule at a distance x and at time t. The variance of this distribution is $\sigma_x^2 = 2Dt$. Considering ideal diffusion process independently in three dimensions, the probability of finding a molecule in three dimensional space at a distance r where $\vec{r} = \hat{\imath} \cdot x + \hat{\jmath} \cdot y + \hat{k} \cdot z$ and at time t can be expressed as shown below.

$$P(x,y,z,t) = P(x,t)P(y,t)P(z,t) = \frac{1}{(4\pi Dt)^{3/2}} \exp\left(- \frac{x^2+y^2+z^2}{4Dt} \right) \tag{14}$$

Assuming spherical symmetry [5] this can be written as

$$P(\vec{r},t) = \frac{1}{(4\pi Dt)^{3/2}} \exp\left(- \frac{r^2}{4Dt} \right) \text{ where } r^2 = x^2+y^2+z^2. \tag{15}$$

Channel Impulse Response Identification

CIR of the CEMC channel in nanonetworks can be obtained from the well-known time-dependent solution to concentration of diffused substance as governed by the macroscopic theory of diffusion and expressed by Fick's laws [5, 38]. Unlike EM wave-based propagation, molecular propagation should be treated with the particle theory of propagation. For example, the average number of molecules available for reception from a point source per unit of volume can be computed from the solution to Fick's laws of diffusion as the following [38]

$$U(r,t) = \frac{Q_0}{(4\pi Dt)^{3/2}} \exp\left(\frac{-r^2}{4Dt} \right) \tag{16}$$

where the symbols have already been explained earlier. Here D depends on the medium through which the molecules propagate. The average CIR of CEMC channel and its characteristics can be deduced from Eq. (16) [35]. In order to determine the CIR, the CEMC channel is excited by an impulsive emission of Q_0 molecules released at time $t = 0$. The average CIR in response to this can be expressed as Eq. (16) and, in addition,

$$U(r,t) = Q_0 \otimes g(r,t) \text{ where } g(r,\ t) = \frac{1}{(4\pi Dt)^{3/2}} \exp\left(\frac{-r^2}{4Dt} \right) \tag{17}$$

where $g(r,\ t)$ is the CIR of the CEMC channel and the symbol $\otimes$ indicates convolution operation in time domain. In practical scenarios, in the numerical experiments shown in the next chapter, $g(r,\ t)$ is normalized to the total energy (i.e. strength) during the entire observation time as $g(r,\ t)/\int_0^{T_{\text{obs}}} g(r,t)dt$ (see the next chapter, Sect. 2.2).

Extension to Time-Dependent Transmission

In addition to the instantaneous transmission as mentioned above, time-dependent transmission rate, i.e. when the transmission rate at the TN is a function of time, is quite common in nature [45, 46]. According to the available literature, time-dependent transmission is as equally realistic as the instantaneous transmission in case of CEMC found in nature [4, 47, 48]. The output of the CEMC channel in response to the time-dependent transmission signals can be computed by taking the time-domain convolution of the transmission rate with the CIR of the CEMC channel. For example, considering spherical symmetry, when $Q(t)$ is the time-dependent transmission rate of molecules in molecules per second unit, the mean available concentration signal intensity at a distance r and time t can be expressed by taking the integral as below [38].

$$U(r,t) = \int_0^t \frac{Q(\varsigma)}{\{4\pi D(t-\varsigma)\}^{\frac{3}{2}}} e^{\frac{-r^2}{4D(t-\varsigma)}} d\varsigma = Q(t) \otimes g(r,t) \tag{18}$$

where $r^2 = x^2 + y^2 + z^2$ when a Cartesian coordinate system is assumed and ς is the variable of integration and the other symbols are as described earlier.

4.3 Diffusion-Based Noise and Interference

The explanations of the diffusion process from the viewpoints of both the macroscopic theory and the microscopic theory have been discussed in order to build the concept of the diffusion process and the understanding of the diffusion noise and interference. The macroscopic theory of diffusion deals with the mean value of the received molecules considering a large number of propagating molecules, whereas the microscopic theory of diffusion deals with the individual molecules separately. Thus the expression derived in the case of macroscopic model of diffusion constitutes for the mean number of molecules that are available at the location of the RN.

The TN transmits a symbol at the beginning of the time interval known as the symbol interval T_S. Let us consider a system such that the RN will receive a number of molecules during the i-th symbol interval where the TN has transmitted Q_m molecules at the beginning of the i-th symbol interval. At the i-th symbol interval, the RN would receive some of the molecules that were transmitted by the TN at the beginning of the i-th symbol interval, plus some of the molecules that were transmitted by the TN during the previous symbol durations, i.e. from the first symbol duration up to the $(i-1)$-th symbol duration. The former constitutes the desired signal part and the latter constitutes the ISI part. Having known the mean number of molecules available at any time t at the RN, however, the variance of the number of the molecules at the RN needs to be investigated in order to explain the effects of diffusion noise as well as the ISI during the i-th symbol.

The pdf's shown earlier are very significant in the sense that it shows how many of the molecules would be available at the RN when $Q_m(t)$ molecules are transmitted by the TN. As shown in Eq. (22), the mean total number of molecules received in the small volume VRV would represent the amplitude $s(t)$ of the molecular concentration signal available for reception at the RN [41]. However, when a single molecule is transmitted by the TN, the probability that the molecule be received in the VRV can be expressed as below.

$$p = \iiint\limits_{\text{VRV}} P(\vec{r}, t)dx\,dy\,dz = \iiint\limits_{\text{VRV}} \frac{1}{(4\pi Dt)^{3/2}} \exp\left(-\frac{x^2+y^2+z^2}{4Dt}\right) dx\,dy\,dz \tag{19}$$

As mentioned before, in general, in the M-ary CEMC system, the TN sends one of the M different amplitude levels with $Q_m, m = 1, 2, \ldots M$ molecules in the medium for each symbol. Therefore, the probability of having k molecules out of the Q_m molecules during the i-th symbol interval (i.e. whether the k molecules arrive at the RN or not) can be expressed by the binomial distribution function as below.

$$\Pr(k; Q_m, p) = \frac{Q_m!}{k!(Q_m - k)!} p^k (1-p)^{(Q_m - k)} \tag{20}$$

For a reasonably large number of molecules $Q_m \gg 1$ and when p is not close to 1 or 0 and p is finite such that as $n \to \infty$, $np \to \infty$, the binomial distribution in Eq. (20) can be approximated to a normal distribution $\mathcal{N}(\mu, \sigma^2)$ where the mean (μ) and the variance (σ^2) can be expressed as

$$\mu = Q_m p \quad \text{and} \quad \sigma^2 = Q_m p(1-p). \tag{21}$$

As a result, it can be found that the mean of the number of molecules available for reception at the RN is actually the *deterministic* signal (i.e. the mean value) that can be found using the macroscopic theory of diffusion as shown in Sect. 4.1. Therefore, the total number of molecules received, $y(t)$, as a result of diffusion only can be expressed as a normal distributed random variable as below [49].

$$\begin{aligned} &\mathcal{N}(Q_m p, Q_m p(1-p)) \Rightarrow \mathcal{N}(s(t), s(t)(1-p)) \\ &\Rightarrow y(t) = s(t) + n_s(t) \text{ where } s(t) = Q_m p \text{ and} \\ &n_s(t) \sim \mathcal{N}(0, s(t)(1-p)) \Rightarrow n_s(t) \sim \sqrt{s(t)(1-p)}\,\mathcal{N}(0,1) \end{aligned} \tag{22}$$

So, the signaling model can be expressed as below.

$$z_S(t) = s(t) + n_s(t) + n_{\text{ISI}}(t) \tag{23}$$

where $z_S(t)$ denotes the total concentration signal intensity available at the RN and $n_{\text{ISI}}(t)$ denotes the concentration signal intensity contributed by the residual molecules that are present at the RN at time t due to the ISI effects during the current symbol duration. Figure 4 shows the signaling model of CEMC with diffusion noise and ISI.

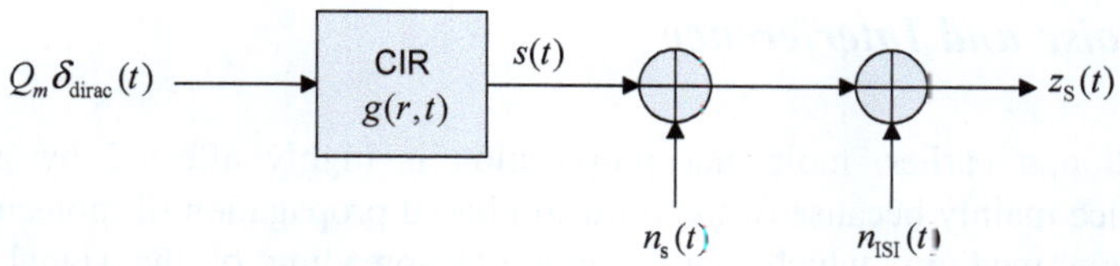

Fig. 4 Signaling model in CEMC with diffusion noise and ISI

5 CEMC Limitations

5.1 *Speed of Communication*

Unlike EM wave-based communication where the EM wave propagates at the speed of light (3 × 10^8 m/s) CEMC is extremely slow, providing the low speed in communication approximately in the range from nm/s to μm/s [15]. In the case of diffusion-based CEMC, the molecules propagate by means of thermal excitation without expending any chemical energy. As a result, due to the random walk motion the molecules face an enormous number of collisions in the path and the effective communication range becomes less. Yet diffusion-based passive transport mechanism in MC provides a unique way to communicate at the nanoscale. As an example of MC, a macro-molecule, e.g. a functional protein, has a complex chemical structure that represents some specific biological functions, e.g. catalysis of a chemical reaction [15]. Therefore, even if the speed is low CEMC provides a promising way of communication at the nanoscale.

5.2 *Communication Ranges*

The communication range is another important issue that limits the effectiveness of CEMC. Unlike EM wave-based communication where the transmitter and the receiver can be as long as hundreds of kilometres away but still communicating reliably enough, the TN and the RN in CEMC should be in the range from nm to μm in order to make the communication reliable [30]. Effective communication range is short because the concentration of molecules experiences attenuation and temporal spreading [35] that make the available concentration of molecules very low in amplitude and distorted in nature when the RN is located far from the TN. It has been found that the closer the TN and the RN are the better the CEMC between them becomes. However, it is also found in biology that biological cells can relay molecular concentration signals to a comparatively longer distance, e.g. in calcium signaling among biological cells, by repeating the processes of diffusion and amplification of the information molecules [33, 50, 51]. Such a mechanism of calcium signaling can possibly be used in increasing the communication range of operation effectively in CEMC.

5.3 Noise and Interference

As mentioned earlier, molecular propagation is highly affected by noise and interference mainly because of the diffusion-based propagation of molecules in the propagation medium, which causes temporal spreading of the signal, and the undesired effects the signal receives from the surrounding environment itself. The former produces ISI while the latter produces undesired molecules and undesired molecular reactions that create noise in the desired signal. In addition, characteristics of propagation medium and that of information molecules also contribute to the noise effects that the signal experiences in the CEMC channel. For example, in case of inhomogeneous medium, propagation effects may not be like those in the homogenous medium [52]. Several research works that deal with noise and interference in diffusion-based CEMC and/or MC systems in general are reported in [49, 53–63].

6 Research Issues and Challenges

6.1 Concentration Attenuation

Due to the nature of the diffusion process itself, the concentration of the molecules available at the location of the RN gets attenuated [30, 35], and sometimes depending on the TN-RN distance, the strength of the concentration is so low that the signal detection may become challenging, if not impossible. Therefore, to increase the signal strength is considered as a challenge in diffusion-based CEMC system. In a particular symbol, the desired strength of concentration becomes less because of the fact that the CIR of the molecular channel gets temporally spread [35]. As a result, it is necessary to choose signal amplification techniques to order to increase the signal level at the RN [64]. Existing research in this regard suggests that signal repeaters can be used in the signal pathway between the TN and the RN for signal amplification purposes [65]. For example, a model of signal repeaters using non-excitable biological cells and inter-cellular Ca^{+2} signaling is presented in [65] that takes into account specified conditions.

6.2 Intersymbol Interference

Figure 5 shows that CIR of the CEMC channel becomes temporally spread [35] such that there is a possibility that depending on the symbol duration the molecules intending for the previous symbols may interfere with the molecules intended for the current symbol and thus cause ISI to the current symbol. Due to the nature of the diffusion process itself the RN finds it difficult to combat the ISI in the current

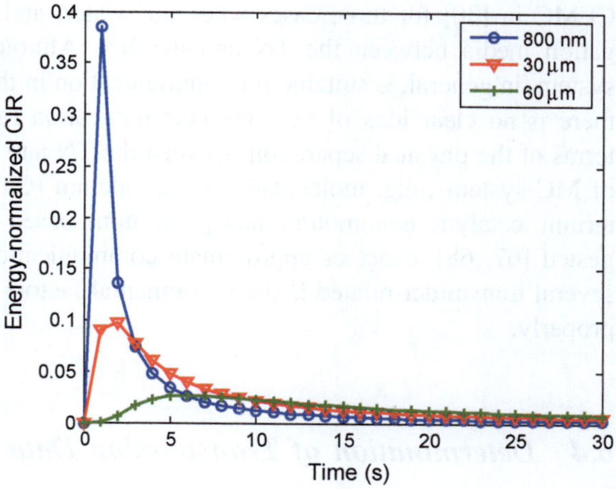

Fig. 5 Energy-normalized CIR when the RN is at 800 nm, 30 and 60 μm from the TN in water medium ($D = 10^{-6}$ cm^2/s) [64]

symbol. Therefore, the effects of ISI vary with the temporal spreading properties of the CEMC channel. As a result, the effects of ISI become worse when TN-RN distance and/or communication data rate increase(s). The impact of the ISI is significant in the signal detection process and reduces the effective communication range for reliable CEMC among nanomachines.

In order to ensure successful detection of the transmitted messages at the receiver, and thereby, to ensure a reliable CEMC, the effects of ISI need to be taken care of properly. As an option to reduce the effects of ISI, a reduced pulse-width scheme has been proposed for pulse-transmitted on-off keying (OOK)-based CEMC system [53]. However, the work in [53] did not investigate completely on the transmission power level adjustment at the TN. On the other hand, an enzyme-based scheme has been proposed in [58], where the diffused enzymes react with the ISI-producing information molecules to form intermediate products through chemical reaction, and therefore, keep the information molecules from the previous symbols away not to cause ISI in the current symbol. Finally, the ISI issue in multiple amplitude modulation (M-AM)-based CEMC is another important research area.

6.3 Determination of Communication Ranges

Determination of the appropriate communication ranges is a challenge in CEMC. This is because future CEMC-based molecular nanonetworks applications would possibly require a suitable range of TN-RN distances in which the nanomachines would like to be reliably communicating with each other. So far, recent studies do not provide the appropriate set of communication ranges for nanomachines in a nanonetwork. However, a few studies exist in this field. For example, CEMC-based communication ranges have been categorized into *short*-, *medium*-, and *long*-range

CEMC in [30] for three cases when air, water, and blood plasma are the propagation media between the TN and the RN. Although it is anticipated that MC system, in general, is suitable for communication in the ranges from nm to μm [66], there is no clear idea of how the communication range would be categorized in terms of the physical separation between the TN and the RN. While different types of MC systems, e.g. molecular motors, calcium (Ca^{+2}) signaling, flagellated bacterium, catalytic nanomotors, and pheromone-based MC systems, have been suggested [67, 68], exact or approximate communication ranges seem to depend on several transmitter-related and environmental factors and so they are not classified properly.

6.4 Determination of Transmission Data Rates

Like all communication systems, in CEMC it is also desired to have a higher data rate between the TN and the RN. However, determination of the most suitable transmission data rate is another challenge in CEMC system. Transmission data rate is related to the duration of the transmitted symbols. In order to ensure reliable CEMC among nanomachines, although the suitable transmission data rates are not clearly known yet, which somewhat depend on a particular system under investigation, it is found in some recent literature that the TN can communicate with the RN at data rates in the range from 0.01 to 1 bps reasonably acceptably by using ASK and FSK-based modulation schemes [69]. Provided that nanomachines are limited in their capabilities, to which data rates a CEMC system would perform the best is still an open question to the research community. In addition, as data rate increases the effects of ISI at the RN increases. As a result, employing higher data rates may not provide reliable CEMC between nanomachines [41, 49].

6.5 Addressing Mechanisms

Addressing is another important issue in CEMC-based nanonetworks. A TN can choose an addressing mechanism with which it can address an RN and thus communicate with it. In a CEMC-based molecular nanonetwork, a TN can address an RN in concentration dependent manner, meaning that the TN can selectively choose one or more of the RNs that it would like to communicate with by varying the number of transmitted molecules [15]. When the concentration of molecules at the RN becomes higher than a given concentration threshold, or lies within a range concentration thresholds, the RN responds to the TN. Therefore, these selectively chosen RNs will only be able to communicate, while the others cannot.

In general, addressing in MC is done by using either specific molecule type or beacon distances [70]. In [70], the TN addresses the RN by using distances of the RN to a number of molecular beacons. The information is carried by a carrier that can measure its distances from the given set of molecular beacons, where each beacon produces a separate concentration gradient using a different type of molecules. In CEMC, addressing based on beacon distance may become challenging when all the beacons would be using the same type of molecules, making the addressing scheme somewhat complicated because the carrier would not be able to distinguish between molecules being received.

6.6 Efficient Signal Detection Schemes

Selection of the most appropriate symbol detection scheme in CEMC is a significant challenge and thus demands a considerable amount of research in molecular nanonetworks. While so far research on signal detection in CEMC systems mainly focused on either sampling-based [49, 71] or strength-based [60, 71, 72] signal detection, it is anticipated that in the future more efficient signal detection schemes need to be investigated, especially the ones with the capability to combat ISI and signal attenuation. Since CEMC relies on a single type of information molecules to be used by all the nanomachines, it can be understood that the detection challenge would be even more severe in CEMC-based molecular nanonetworks, where a very large number of nanomachines located randomly would be communicating with one another simultaneously, where each communication link would need to be reliable and efficient. As a result, it goes without saying that current research would need to focus on efficient detection schemes for CEMC. Effects of diffusion-based noise and LRBP on the new detection schemes as well as their computational complexity would also need to be investigated, which would surely open new avenues for further research in this area.

6.7 Investigations into Channel Coding Schemes

The application of channel codes in CEMC is relatively new and, therefore, needs more research. The recent investigation on applying convolutional codes into pulse amplitude modulation (PAM)-based CEMC scheme found that it is possible to have increased communication ranges by using convolutional coded system [73]. However, a formal evaluation on such system with detailed investigation on the design of the encoder and decoder circuits with biological components and bio-nanomachines is still an open area for research. It is also not clear whether or not traditional communication coding theory would be applicable to CEMC-based molecular nanonetworks, which can create a new paradigm for channel coded CEMC system [74]. Performance of existing and new channel codes in CEMC is an emerging area that needs a significant amount of investigation in detail [73, 75]. In addition, efficient decoding scheme for CEMC system is also considered as a new avenue for research in this field [76].

6.8 System Model and Performance Evaluation

Since the concepts of both MC and CEMC are very new in the area of nanoscale communication networks, the amount of research done in these fields is still at an early age. Several researchers have approached the MC system and the corresponding performance evaluation from various viewpoints; for example, based on ideal (i.e. free) diffusion process [30, 39] and diffusion with molecules removed from the system upon its first hit with or without molecular drift velocity, as reported in [77, 78] and [79, 80] respectively. Since CEMC would possibly be used in a large variety of applications in the future, we believe that there is still room for research in developing the application-specific system model in nanonetworks and hence evaluating the performance of CEMC systems under specific applications. Appropriate performance metrics need to be identified in a particular application scenario and their influence and validity on the analysis and description of the system needs to be studied completely. Efficient modulation schemes suitable for CEMC systems need to be investigated and their communication range-dependent characteristics need to be studied. While most recent studies, e.g. [41, 62, 69], show that amplitude-based and frequency-based modulation techniques can be used for CEMC, it holds a bright prospect to investigate into other new modulation schemes that would be suitable for CEMC as well as their biological relevance to nanomachine functionalities. Research results are coming out on this aspect very recently, and we strongly believe that the same trend would continue through providing new results on CEMC system models and performance evaluations of CEMC systems in particular, which would be able to answer the research questions completely and help the research community to understand the CEMC system with greater details.

6.9 Testbed Development, Simulator, and Experimental Opportunities

Up to now, very few works have been performed on the performance of CEMC system with real testbed and experiments. While theoretical studies provide us with the necessary theoretical results of the CEMC system, experimental studies are extremely important in the sense that they make the analyses complete, with real test data of the CEMC system under investigation. This is why testbeds need to be developed with CEMC testing capability such that real experiments with engineered biological cells [15, 25] and/or artificially created cell-like structures [15] can be performed. We strongly believe that as the research on engineered bio-nanomachines will progress in the near future, the possibilities of real experiments on the performance of CEMC system would be matured enough to provide new results and so help MC engineers to evaluate the CEMC system with more details suitable for nanonetworks applications. On the other hand, some research works have been done on the development of simulators capable of simulating the MC behaviour through computer models [67]. While simulators provide an insight to expected performance of the system, real experiments give completeness to the theoretical knowledge in such cases. A recent review on the experimental opportunities of MC from the viewpoint of molecular biology can be found in [81].

7 Engineered CEMC

Ideal diffusion-based CEMC is in principle a broadcast communication technique where the information molecules undergo random walk motion. As shown earlier, the RN contains a finite number of receptors on its surface and the receptors are specific and can respond to a single type of information molecules only in a concentration-dependent manner, meaning that the RN responds only when a certain number of information molecules is present in a high enough concentration, e.g. above a threshold value or between a band of concentration levels [15]. This gives a provision that the TN can select an RN by sending the information molecules in an appropriate number such that they can be present at the RN with a desired concentration. By controlling the number of transmitted molecules the TN can select the desired RN. For example, when a TN wants to communicate with a nearby RN (with a known concentration band on which the RN responds), the TN can adjust the transmitted number of molecules such that the concentration of molecules at a particular RN nearby only reaches a desired threshold concentration and the concentration of information molecules at distant RNs cannot reach the desired concentration and so they cannot respond to the TN's communication. In addition, different vesicle sizes encapsulating various numbers of information molecules can also be used in order to realize concentration-dependent effects at the RN where it may react with different concentrations of the information molecules carried to it by the vesicles.

Therefore, a TN can adopt an addressing mechanism in a concentration-dependent manner [15]. For instance, it is possible to engineer a bacterium to transmit information molecules called Vibrio fischeri Auto-Inducer (VAI) and the VAI molecules create a concentration gradient in the fluidic medium, the highest of the concentration gradient being located at the location of the TN. Thus it is possible for a TN to communicate selectively with a desired RN by adjusting the number of transmitted molecules.

8 Applications of CEMC and Nanonetworks

Since CEMC is a concentration-based MC paradigm, the application areas of CEMC in particular would be similar to the ones of MC and molecular nanonetworks in general, which would most likely bring about constructive changes to our emerging society through many fruitful applications [10, 28].

Biomedical applications are the most important and major application group that would be benefitted by CEMC-based nanonetworks. For example, it is now already known that with the help of nanotechnology and nanoscale communication it is possible to interact with biological cells, tissues, and organs [15, 82]. In addition, CEMC-based nanonetworks would offer biocompatibility and biostability. The principal applications of CEMC-based nanonetworks in the biomedical field would be in establishing communication links among nanomachines and so in developing immune system support systems, bio-hybrid implant systems, targeted drug delivery systems, health monitoring systems, and nano-systems modified with genetic engineering [10]. CEMC can also help fight against cancer by providing new knowledge in its detection and treatment. For instance, when a nanomachine can detect some special molecules released from a sick (e.g. cancerous) cell, it can announce, through the nanonetworks it belongs to, that there is a sick cell in its vicinity and thus other nanomachines and/or cells nearby become careful about that and can turn on their protective measures against it [82].

In the development of industrial and consumer products, CEMC-based nanonetworks can help with the new materials, processes, and novel quality control techniques. For example, advanced food and water quality control techniques could be developed that are capable of detecting small bacteria and toxic substance in food and water [10]. In addition, advanced fabrics that contain CEMC-based nanonetworks would be able to provide improved functionalities, e.g. antimicrobial textiles [10]. Apart from these, CEMC-based nanonetworks would also be able to provide potential applications in the fields of military, environmental, bio-degradation, and bio-diversity controls [10, 28]. For a detailed account of application areas of MC and nanonetworks, interested readers should refer to [10, 28].

9 Conclusion

This chapter presents an overview of the fundamentals, issues, and challenges of CEMC in view of the bio-inspired communication paradigm and molecular nanonetworks. A brief background on the involvement of nanotechnology into communication system is presented by describing the relationship among nanomachines, nanonetworks, and MC technology. CEMC system is then described in detail by providing with brief descriptions of system components and processes. The CEMC system has been explained from both microscopic and macroscopic theories of molecular diffusion. This chapter also touches upon the possibilities of engineered CEMC systems as well as the limitations involved in CEMC.

The fundamental concepts provided in this chapter should be useful in in-depth understanding of CEMC systems in materializing molecular nanonetworks. In addition, the technical issues and the research challenges addressed in this chapter should provide the CEMC engineers with the required thoughts in order to build a reliable CEMC system for molecular nanonetworks applications. This chapter should serve as a starting point to necessary details of CEMC-based molecular nanonetworks for readers from interdisciplinary fields. With the background knowledge on CEMC presented in this chapter, the next chapter will evaluate the performance of a CEMC system with more details.

Acknowledgements M.U. Mahfuz would like to thank the Natural Sciences and Engineering Research Council of Canada (NSERC) for the financial support in the form of PGS-D scholarship during the years 2010–2013.

References

1. Höfling F, Franosch T (2013) Anomalous transport in the crowded world of biological cells. Rep Prog Phys 76:046602
2. Metzler R, Klafter J (2000) The random walk's guide to anomalous diffusion: a fractional dynamics approach. Phys Rep 339:1–77, 12
3. Einstein A (1905) On the movement of small particles suspended in stationary liquids required by the molecular-kinetic theory of heat. Ann Phys 17:549–560
4. Alberts B, Johnson A, Lewis J, Raff M, Roberts K, Walter P (2008) Molecular biology of the cell. Garland Science, New York
5. Berg HC (1993) Random walks in biology. Princeton University Press, NJ, USA
6. Hiyama S, Moritani Y, Suda T, Egashira R, Enomoto A, Moore M, Nakano T (2005) Molecular communication. In: Proceedings of the NSTI nanotechnology conference
7. Feynman RP (1960) There's Plenty of Room at the Bottom. Eng Sci, Caltech, USA:22–36
8. Taniguchi N (1974) On the basic concept of nanotechnology. In: Proceedings of the international conference on product engineering, Part II. Tokyo, Japan Society of Precision Engineering, pp 18–23
9. Freitas RA (2005) Nanotechnology, nanomedicine and nanosurgery. Int J Surg 3:243–246
10. Akyildiz IF, Brunetti F, Blazquez C (2008) Nanonetworks: a new communication paradigm. Comput Netw J 52:2260–2279
11. Xia Y, Yang P, Sun Y, Wu Y, Mayers B, Gates B, Yin Y, Kim F, Yan H (2003) One-dimensional nanostructures: synthesis, characterization, and applications. Adv Mater 15:353–389
12. Meyyappan M, Li J, Li J, Cassell A (2006) Nanotechnology: an overview and integration with MEMS. In: 19th IEEE international conference on micro electro mechanical systems (MEMS 2006). Istanbul, pp 1–3
13. Lacasa NR (2009) Modeling the molecular communication nanonetworks. M.Sc. thesis, The Universitat Politècnica de Catalunya (UPC), Spain
14. Bush SF (2010) Nanoscale communication networks. Artech House Incorporated
15. Nakano T, Moore M, Enomoto A, Suda T (2011) Molecular communication technology as a biological ICT. In: Sawai H (ed) Biological functions for information and communication technologies. Springer, Berlin Heidelberg, pp 49–86
16. Tseng AA, Chen K, Chen CD, Ma KJ (2003) Electron beam lithography in nanoscale fabrication: recent development. IEEE Trans Electron Packag Manufact 26:141–149
17. Wilbur JL, Kumar A, Biebuyck HA, Kim E, Whitesides GM (1996) Microcontact printing of self-assembled monolayers: applications in microfabrication. Nanotechnol 7:452–457
18. Qin D, Riggs BA (2012) Nanotechnology: a top–down approach
19. Bhushan B (ed) (2004) Springer handbook of nanotechnology. Springer, Berlin; New York
20. Yun YJ, Ah CS, Kim S, Yun WS, Park BC, Ha DH (2007) Manipulation of freestanding au nanogears using an atomic force microscope. Nanotechnol 18:505304

21. Ozin GA, Manners I, Fournier-Bidoz S, Arsenault A (2005) Dream nanomachines. Adv Mater 17:3011–3018
22. Alberts B, Johnson A, Lewis J, Raff M, Roberts K, Walter P (2002) Molecular biology of the cell. Garland Science, New York
23. Freitas RA (1999) Nanomedicine, vol 1: basic capabilities. Landes Bioscience, Austin, TX
24. Nakano T, Moore M (2011) Molecular communication paradigm overview. JNIT: J Next Gener Inf Technol 2:9–16
25. Moore MJ, Enomoto A, Suda T, Nakano T, Okaie Y (2007) Molecular communication: new paradigm for communication among nano-scale biological machines. In: Bidgoli H (ed) The handbook of computer networks. Wiley
26. Akyildiz IF, Jornet JM, Pierobon M (2010) Propagation models for nanocommunication networks. In: Proceedings of the fourth European conference on antennas and propagation (EuCAP), pp 1–5
27. Suda T, Moore M, Nakano T, Egashira R, Enomoto A (2005) Exploratory research on molecular communication between nanomachines. In: Genetic and evolutionary computation conference (GECCO), Late Breaking Papers, 25–29 June, Washington, DC, USA
28. Nakano T, Moore MJ, Wei F, Vasilakos AV, Shuai J (2012) Molecular communication and networking: opportunities and challenges. IEEE Trans Nanobiosci 11:135–148
29. Hranilovic S (2005) Wireless optical communication systems. Springer, New York
30. Mahfuz MU, Makrakis D, Mouftah HT (2010) On the characterization of binary concentration-encoded molecular communication in nanonetworks. Nano Commun Netw J 1:289–300
31. Moritani Y, Hiyama S, Suda T (2007) Molecular communication a biochemically-engineered communication system. In: Frontiers in the convergence of bioscience and information technologies, FBIT 2007, pp 839–844
32. Parcerisa Giné L, Akyildiz IF (2009) Molecular communication options for long range nanonetworks. Comput Netw 53:2753–2766
33. Nakano T, Suda T, Moore M, Egashira R, Enomoto A, Arima, K (2005) Molecular communication for nanomachines using intercellular calcium signaling. In: 2005 5th IEEE conference on nanotechnology, vol. 2, pp 478–481
34. Moritani Y, Hiyama S, Suda T (2006) Molecular communication for health care applications. In: Fourth annual IEEE international conference on pervasive computing and communications workshops (PerCom Workshops 2006), pp 5–553
35. Mahfuz MU, Makrakis D, Mouftah HT (2010) Characterization of molecular communication channel for nanoscale networks. In: Proceedings 3rd international conference on bio-inspired systems and signal processing (BIOSIGNALS-2010). Valencia, Spain, pp 327–332
36. Smith JM (2000) The concept of information in biology. Philos Sci 67:177–194
37. Farsad N, Eckford AW, Hiyama S, Moritani Y (2011) A simple mathematical model for information rate of active transport molecular communication. In: 2011 IEEE conference on computer communications workshops (INFOCOM WKSHPS), pp 473–478
38. Bossert WH, Wilson EO (1963) The analysis of olfactory communication among animals. J Theor Biol 5:443–469
39. Atakan B, Akan OB (2010) Deterministic capacity of information flow in molecular nanonetworks. Nano Commun Netw 1:31–42
40. Eckford AW (2007) Achievable information rates for molecular communication with distinct molecules. In: 2nd Bio-Inspired models of network, information and computing systems, 2007 (Bionetics 2007), pp 313–315
41. Mahfuz MU, Makrakis D, Mouftah HT (2011) A comprehensive study of concentration-encoded unicast molecular communication with binary pulse transmission. In: 2011 11th IEEE conference on nanotechnology (IEEE-NANO), pp 227–232
42. Gillespie DT (2000) The chemical Langevin equation. J Chem Phys 113:297–306
43. Crank J (1975) The mathematics of diffusion. Clarendon Press, Oxford, Eng
44. Durrett R (2010) Probability: Theory and examples, 4th edn. Cambridge University Press

45. Krivan V, Lánský P, Rospars JP (2002) Coding of periodic pulse stimulation in chemoreceptors. BioSyst 67:121–128
46. Rospars J, Křivan V, Lánský P (2000) Perireceptor and receptor events in olfaction. comparison of concentration and flux detectors: a modeling study. Chem Senses 25:293–311
47. Moore MJ, Nakano T (2011) Synchronization of inhibitory molecular spike oscillators. In: BIONETICS-2011
48. Mosharov EV, Sulzer D (2005) Analysis of exocytotic events recorded by amperometry. Nat Methods 2:651–658
49. Mahfuz MU, Makrakis D, Mouftah HT (2013) Sampling based optimum signal detection in concentration-encoded molecular communication receiver architecture and performance. In: Proceedings of the 6th international conference on bio-inspired systems and signal processing (BIOSIGNALS-2013). Barcelona, Spain
50. Nakano T, Liu Jian-Qin (2010) Design and analysis of molecular relay channels: an information theoretic approach. IEEE Trans NanoBiosci 9:213–221
51. Kuran MS, Tugcu T, Edis BO (2012) Calcium signaling: overview and research directions of a molecular communication paradigm. IEEE Wireless Commun 19:20–27
52. van Milligen BP, Bons PD, Carreras BA, Sánchez R (2005) On the applicability of Fick's law to diffusion in inhomogeneous systems. Eur J Phys 26:913–925
53. Mahfuz MU, Makrakis D, Mouftah HT (2011) Characterization of intersymbol interference in concentration-encoded unicast molecular communication. In: 2011 24th Canadian conference on electrical and computer engineering (CCECE), pp 000164–000168
54. Pierobon M, Akyildiz IF (2012) Intersymbol and co-channel interference in diffusion-based molecular communication. In: 2012 IEEE international conference on communications (ICC), pp 6126-6131
55. Kuran MŞ, Yilmaz HB, Tugcu T, Akyildiz IF (2012) Interference effects on modulation techniques in diffusion based nanonetworks. Nano Commun Netw 3:65–73, 201203
56. Kuran MS, Tugcu T (2011) Co-channel interference for communication via diffusion system in molecular communication. In: BIONETICS 2011, York, UK, pp 199–212
57. Kuran MS, Yilmaz HB, Tugcu T, Akyildiz IF (2011) Modulation techniques for communication via diffusion in nanonetworks. In: 2011 IEEE international conference on communications (ICC), pp 1–5
58. Noel A, Cheung KC, Schober R (2013) Improving receiver performance of diffusive molecular communication with enzymes, pp 1–12. arXiv:1305.1926
59. Noel A, Cheung KC, Schober R (2012) Improving diffusion-based molecular communication with unanchored enzymes. In: Proceedings of the 7th international conference on bio-inspired models of network, information, and computing systems (BIONETICS 2012), December
60. Mahfuz MU, Makrakis D, Mouftah HT (2013) A generalized strength-based signal detection model for concentration-encoded molecular communication. In: Proceedings of the 8th international conference on body area networks (BodyNets 2013). Boston, MA, USA (30 Sept.-02 Oct., 2013), pp 461–467
61. Mahfuz MU, Makrakis D, Mouftah HT (2011) On the characteristics of concentration-encoded multi-level amplitude modulated unicast molecular communication. In: 2011 24th Canadian conference on electrical and computer engineering (CCECE), pp 000312–000316
62. Mahfuz MU, Makrakis D, Mouftah H (2010) Spatiotemporal distribution and modulation schemes for concentration-encoded medium-to-long range molecular communication. In: 2010 25th Biennial symposium on communications (QBSC), pp 100–105
63. Moore MJ, Suda T, Oiwa K (2009) Molecular communication: modeling noise effects on information rate. IEEE Trans NanoBiosci 8:169–180
64. Mahfuz MU, Makrakis D, Mouftah HT (2013) Concentration encoded molecular communication: prospects and challenges towards nanoscale networks. In: Proceedings of international conference on engineering, research, innovation and education (ICERIE-2013). Sylhet, Bangladesh pp 508–513

65. Nakano T, Shuai J (2011) Repeater design and modeling for molecular communication networks. In: 2011 IEEE conference on computer communications workshops (INFOCOM WKSHPS), pp 501–506
66. Garralda N, Llatser I, Cabellos-Aparicio A, Pierobon M (2011) Simulation-based evaluation of the diffusion-based physical channel in molecular nanonetworks. In: 2011 IEEE conference on computer communications workshops (INFOCOM WKSHPS), pp 443–448
67. Garralda N, Llatser I, Cabellos-Aparicio A, Alarcón E, Pierobon M (2011) Diffusion-based physical channel identification in molecular nanonetworks. Nano Commun Netw 2:196–204, 201112
68. Llatser I, Pascual I, Garralda N, Cabellos-Aparicio A, Pierobon M, Alarcon E, Sole-Pareta J (2011) Exploring the physical channel of diffusion-based molecular communication by simulation. In: 2011 IEEE global telecommunications conference (GLOBECOM 2011), pp 1–5
69. Mahfuz MU, Makrakis D, Mouftah HT (2011) Transient characterization of concentration-encoded molecular communication with sinusoidal stimulation. In: Proceedings of the 4th international symposium on applied sciences in biomedical and communication technologies (ISABEL'11), Article 14, 6 Pages. Barcelona, Spain
70. Moore MJ, Nakano T (2011) Addressing by beacon distances using molecular communication. Nano Commun Netw J 2:161–173
71. Mahfuz MU, Makrakis D, Mouftah HT (2011) On the detection of binary concentration-encoded unicast molecular communication in nanonetworks. In: Proceedings 4th international conference on bio-inspired systems and signal processing (BIOSIGNALS-2011), 26–29 January. Rome, Italy, pp 446–449
72. Mahfuz MU, Makrakis D, Mouftah HT (2012) Strength based receiver architecture and communication range and rate dependent signal detection characteristics of concentration encoded molecular communication. In: Proceedings BWCCA-2012. Victoria, Canada, pp 28–35
73. Mahfuz MU, Makrakis D, Mouftah HT (2013) Performance analysis of convolutional coding techniques in diffusion-based concentration-encoded PAM molecular communication systems. BioNanoSci 3:270–284
74. Yeh PC, Chen KC, Lee YC, Meng LS, Shih PJ, Ko PY, Lin WA, Lee CH (2012) A new frontier of wireless communication theory: diffusion-based molecular communications. IEEE Wireless Commun 19:28–35
75. ShihP-J, Lee C-H, Yeh P-C (2012) Channel codes for mitigating intersymbol interference in diffusion-based molecular communications. In: 2012 IEEE global communications conference (GLOBECOM), pp 4228–4232
76. Ko P-Y, Lee Y-C, Yeh P-C, Lee C-H, Chen K-C (2012) A new paradigm for channel coding in diffusion-based molecular communications: molecular coding distance function. In: 2012 IEEE global communications conference (GLOBECOM), pp 3748–3753
77. Kadloor S, Adve RS, Eckford AW (2012) Molecular communication using brownian motion with drift. IEEE Trans NanoBiosci 11:89–99
78. Srinivas K, Adve R, Eckford A (2012) Molecular communication in fluid media: the additive inverse gaussian noise channel. IEEE Trans Inf Theory 58:1
79. Nakano T, Okaie Y, Liu Jian-Qin (2012) Channel model and capacity analysis of molecular communication with brownian motion. IEEE Commun Lett 16:797–800
80. Eckford AW (2007) Nanoscale communication with brownian motion. In: 2007. CISS'07. 41st annual conference on information sciences and systems, pp 160–165
81. Balasubramaniam S, Ben-Yehuda S, Pautot S, Jesorka A, Lio P, Koucheryavy Y (2013) A review of experimental opportunities for molecular communication. Nano Commun Netw 4:43–52
82. Mahfuz MU (2012) Nanoscale communication systems and their role in an emerging society. Mini-course lecture slides, University of Ottawa

Concentration-Encoded Molecular Communication in Nanonetworks. Part 2: Performance Evaluation

Mohammad Upal Mahfuz, Dimitrios Makrakis and Hussein T. Mouftah

Abstract As discussed in the previous chapter, concentration-encoded molecular communication (CEMC) is an information encoding approach to molecular communication (MC) where a transmitting nanomachine (TN) encodes information by varying the transmission rate of molecules, and correspondingly, a receiving nanomachine (RN) decodes the transmitted information by observing the concentration of information molecules available at the RN. While the previous chapter basically dealt with the fundamentals, issues, and challenges of CEMC system, the main objective of this chapter is to particularly focus on performance evaluation of CEMC system in detail. Understanding a single CEMC link completely and accurately is of utmost importance in order to fully understand CEMC-based molecular nanonetworks in the emerging biological information and communication technology (bio-ICT) paradigm. Hence this chapter focuses on the performance evaluation of a single-link CEMC system between a pair of nanomachines.

Keywords Molecular communication · Concentration encoding · Performance evaluation · Bit error rate · Biological nanomachines · Nanonetworks · System design · Signal detection

Acronyms

ASK Amplitude-shift keying
BER Bit error rate

This research work was completed while M.U. Mahfuz was with the University of Ottawa, Canada.

M.U. Mahfuz (✉)
Department of Natural and Applied Sciences (Engineering Technology),
University of Wisconsin-Green Bay, 2420 Nicolet Drive, Green Bay, WI 54311, USA
e-mail: mahfuzm@uwgb.edu

D. Makrakis · H.T. Mouftah
School of Electrical Engineering and Computer Science, University of Ottawa,
800 King Edward Ave, Ottawa, ON K1N 6N5, Canada

J. Suzuki et al. (eds.), *Modeling, Methodologies and Tools for Molecular and Nano-scale Communications*, Modeling and Optimization in Science and Technologies 9, DOI 10.1007/978-3-319-50688-3_2

CEMC	Concentration-encoded molecular communication
CIR	Channel impulse response
CME	Chemical master equation
EM	Electromagnetic
FSK	Frequency-shift keying
FPT	First passage time
IM	Impulse modulation
ISI	Intersymbol interference
LRBP	Ligand-receptor binding process
MC	Molecular communication
M-AM	Multilevel amplitude modulation
M-PAM	Multilevel pulse amplitude modulation
OOK	On-off keying
PAM	Pulse amplitude modulation
RN	Receiving nanomachine
RRE	Reaction rate equations
SCK	Stochastic chemical kinetics
TN	Transmitting nanomachine
VAI	Vibrio fischeri Auto-Inducer
VRV	Virtual receiving volume

1 Introduction

Molecular communication (MC) in general requires interdisciplinary knowledge from several technical fields of study ranging from material science through biophysics and computer science to electrical and communication engineering as well as computer networks. While Chap. 1 provides necessary background information on CEMC, this chapter mainly focuses on the performance evaluation of the same. A transmitting nanomachine (TN) releases information molecules in the propagation medium. Solutions to Fick's laws of diffusion provide us with the average concentration of information molecules available at a receiving nanomachine (RN), which is important in the sense that they help us understand diffusion dynamics and communication range- and rate-dependent characteristics of CEMC system in more detail.

In this chapter, we consider a single CEMC link between a pair of nanomachines in aqueous medium. The system model is presented first by presenting the system components and channel impulse response (CIR) characteristics. Suitable signaling schemes in CEMC are described next by detailing transmission and modulation schemes as well as propagation models. CEMC diffusion dynamics are presented next dealing with the average concentration level of available molecules at the RN, without taking into account the noise due to the uncertainty in propagation and reception. Results on diffusion dynamics using average concentration level based on

Fick's laws provide useful insights to identify average performance of a CEMC system, and therefore, are of significant importance in the complete study of CEMC system. Performance of CEMC system by taking into account the randomness due to uncertainty in diffusion-based propagation and reception are also provided, which includes receiver models suitable for CEMC signal detection and analyses in terms of bit error rate (BER) performance and ligand-receptor binding process (LRBP).

Performance evaluation of CEMC system is a wide area open for research at this moment of time. Here in this chapter our focus is on CEMC system based on ideal (i.e. free) diffusion of information molecules. In the simplest form, we assume that propagation medium composes of solvent and information molecules only. Information molecules are of larger size than solvent molecules and collide with solvent molecules randomly and, therefore, become available at the RN in a probabilistic manner. The performance of CEMC system is evaluated by transmitting a random sequence of bits. Communication range varies from several hundreds of nanometres to several tens of micrometres. The effect of intersymbol interference (ISI) is one of the major concerns in CEMC system, which degrades BER performance of the system severely, especially when communication range and/or transmission data rate need(s) to be increased. Effects of ISI have been described in detail. Finally, the findings are summarized with directions to possible future works on the advancements of CEMC system in molecular nanonetworks.

2 System Model

In this section, we describe the system model in general and the characteristics of the CEMC channel. Referring to Fig. 1 in Chap. 1, the assumptions related to diffusion-based CEMC system have been explained in this section. In order to derive the channel characteristics and diffusion-based noise distribution, we first consider an instantaneous transmission of Q_0 molecules at time $t = 0$ at the TN located at (0, 0, 0). Instantaneous transmission is important because it would provide CIR of the CEMC system. When CEMC channel is fully understood, it would be possible to investigate more into other transmission and modulation schemes, e.g. pulse amplitude modulation (PAM), on-off keying (OOK), and frequency-shift keying (FSK) [1]. In this chapter, we also refer to Fig. 1 in the previous chapter and assume the following in the system.

The TN can precisely control the departure time and the number of molecules. The TN is transparent to the released molecules, meaning that the TN does not affect the movement of the released molecules nor does it react with them.

The molecules undergo ideal (i.e. free) diffusion in the propagation medium in three dimensions. This means that in each dimension in space each of the released molecules has equal probability of taking the next step to the right (forward) or to the left (backward) from its previous position [2]. In addition, the molecules propagate infinitely even after the first hitting time at the RN. The RN can sense the number of molecules available for reception at the RN and the molecules that reach the RN are not removed from the system [3]. This ensures free diffusion of molecules according to Fick's laws [2, 4] and, therefore, the molecules can be available to the RN multiple times according to the ideal diffusion phenomenon in three dimensions in an unbounded propagation medium.

The medium between the TN and the RN is three-dimensional, with the transmitter located at the origin $(0, 0, 0)$ and the RN at any other location in the three-dimensional space. The location of the RN can be identified by the vector $\vec{r} = \hat{\imath} \cdot x_r + \hat{\jmath} \cdot y_r + \hat{k} \cdot z_r$ that has the origin as the starting point and the location of the RN as the ending point. Here $\hat{\imath}, \hat{\jmath},$ and $\hat{k}$ denote unit vectors in $x, y,$ and z axes respectively, $r^2 = x_r^2 + y_r^2 + z_r^2$ and r is the Euclidian distance between the TN and the RN.

The TN and the RN are synchronized in time [5, 6]. The TN releases a number of molecules of the same type and each of the molecules propagates independently to the RN meaning that for molecules i and j, the paths $B_i(t)$ and $B_j(t)$ to the RN are independent if $i \neq j$. Following ideal diffusion mechanism based on random walk motion [2], RN decodes information symbols by measuring molecular concentration at its receptor's location [7].

In discrete-time random walk model, referring to Fig. 3 in Chap. 1, the step size δ and the time τ a molecule remains in one state before it moves to the next state are constants [2, 4]. Also, movement of a molecule to a new position is statistically independent of its previous movement. This makes the discrete-time random walk model Markovian. The medium is assumed to be homogenous in nature for the information molecules.

Sensing time of RN is much shorter than passage time of a molecule, regardless of whether it is a first passage time (FPT) or a higher order passage time. FPT is the time duration from the release of a molecule at TN to the time instant when the molecule first hits the receptor of RN [8]. Alternatively, FPT can also be termed as the propagation delay. Higher order passage times indicate scenarios when the molecule reaches the RN multiple times.

RN contains a number of receptors on its surface, see Fig. 1 in the previous chapter. Surrounding the RN we assume a virtual receiving volume (VRV) [3] with the RN located at its centre. Diffused molecules can interact with the receptors and may or may not bind with them according to ligand-receptor binding process (LRBP) that depends on the affinity of information molecules to the receptors on the surface of the RN. In general, we consider available molecules at the location of the RN with and without LRBP.

2.1 System Components

As shown in Fig. 1 in the previous chapter, a generic CEMC system between a pair of nanomachines consists of a TN, an RN, and a propagation medium between them. The propagation medium can also be thought of as "channel" [9] in similarity with the traditional wireless communication systems [10]. TN transmits information molecules at a given rate. Examples of information molecules are proteins and ions that contain information to be transmitted. Examples of propagation medium are water ($D = 10^{-6}$ cm^2/s), air ($D = 0.43$ cm^2/s), and blood plasma ($D = 2.2 \times 10^{-7}$ cm^2/s for simple message and 1×10^{-9} cm^2/s for complex message) [11]. Both TN and RN are artificial devices or modified biological cells [12]. RN can receive information molecules at its receptors.

While in the propagation medium there may exist other molecules that may cause undesired chemical reactions [13] and thus may create noise in the system, we have not considered such cases in our research and so in the system model presented in this chapter we consider that the propagation medium consists of solvent molecules only through which the information molecules diffuse and ultimately reach the RN probabilistically. It should be noted that information molecules (e.g. protein, ion) are different from solvent molecules (e.g. water, air, blood plasma) and of larger size than the solvent molecules.

2.2 *CIR Characteristics: Distance and Temporal Dependence*

The impulse response of CEMC channel needs to be investigated in order to find output concentration of molecules at the location of RN. As shown in the previous chapter, CIR $g(r, t)$ is a function of both time t and TN-RN distance r. Investigating into $g(r, t)$ it is clear that, unlike EM wave-based propagation, modeling CEMC channel cannot be explained in terms of separate distance dependence and temporal dependence. For instance, the exponent part in the expression of $g(r, t)$ is a function of both distance r and time t. In free space, EM waves propagate at the speed of light (3×10^8 m/s). In some cases, wireless channels are realistically assumed to be stationary for short propagation times between sender and receiver. But unlike EM wave propagation, molecular propagation is a very slow process and so spatiotemporal variation of CIR should be investigated in-depth even for short distances [14]. Spatiotemporal variation rather plays a significant role in terms of the analyses of path-loss and output signal. As mentioned earlier, the concentration of molecules at a distance r and at time t, $U(r, t)$, is the intensity [7] of molecular concentration signal at RN. Therefore, any integral of $U(r,t)$ over time would indicate the strength (or, alternatively, energy) [7] of CEMC signal. CIR $g(r, t)$ is normalized to its total energy (i.e. strength) over the entire observation time as $g(r,t)/\int_0^{T_{obs}} g(r,t)dt$ [15] where, in ideal case of diffusion of molecules, $T_{obs} = \infty$; however, in numerical simulations, $T_{obs} = 1$–10 h should be reasonable enough to study the performance of CEMC systems. Figure 1 below shows the energy-normalized CIR $g(r, t)$ at TN-RN distances of 800 nm and 100 μm in water medium. As shown in Fig. 1, CIR becomes temporally spread when r increases causing ISI in signal detection [16–18]. A comparison of the characteristics of CIR of CEMC channel at different TN-RN distances in air medium can be found in [9] whereas the performance of CEMC

Fig. 1 CIR of CEMC channel in water medium

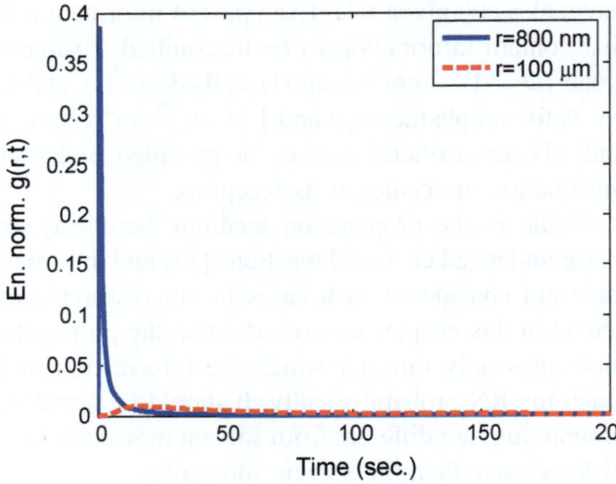

system based on binary pulse transmission scheme in air, water, and blood plasma propagation media can be found in [16]. Figure 1 is similar to Fig. 5 in the previous chapter in the sense that they both show temporal spreading of CIR. However, Fig. 1 shows the temporal spreading of CIR over an extended time scale up to 200 s and an increased communication range of 100 µm.

3 Transmission and Modulation Schemes

In the transmission phase of CEMC, the TN releases molecules in the propagation medium. The transmission rate, or in other words the concentration of molecules at the TN, can be varied in several ways based on the characteristic feature of the concentration signal at the TN. In the following, we focus on three commonly used modulation schemes in CEMC.

3.1 *Impulse Modulation*

In impulse modulation (IM), the TN releases all the molecules as an impulsive manner at the beginning of each symbol. Two different schemes, namely, generalized ASK-based and on-off keying (OOK) schemes can exist in IM. In generalized ASK scheme, as shown in Eq. (1), the TN transmits $Q_1\delta(t)$ and $Q_0\delta(t)$ molecules when it wants to send a bits[1] 1 and 0 respectively [19], where $Q_1 \gg 1, Q_0 \gg 1$, and $\delta(t)$ is the Dirac delta function [20] and $Q_1 > Q_0$ in general.

[1]In binary scheme, each symbol is represented as a bit (1 or 0), while in M-ary scheme, each symbol composes of $\log_2 M$ bits [1].

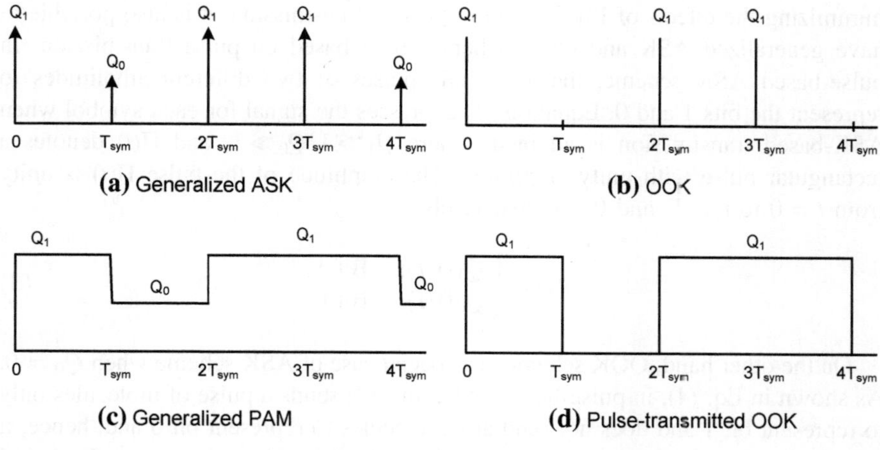

Fig. 2 Generalized ASK and OOK transmission schemes: IM in (**a**) and (**b**), PAM in (**c**) and (**d**). For PAM, $T_p = T_{sym}$ is assumed

$$s(t) = \begin{cases} Q_1\delta(t) & \text{Bit 1} \\ Q_0\delta(t) & \text{Bit 0} \end{cases} \tag{1}$$

However, in OOK scheme, as shown in Eq. (2), the TN sends Q_1 molecules, where $Q_1 \gg 1$, when it wants to send bit 1 and does not send any molecules at all when it wants to send bit 0, meaning that the TN remains apparently in "on" and "off" states while transmitting bits 1 and 0 respectively. Figure 2a, b show ASK and OOK schemes based on IM in CEMC.

$$s(t) = \begin{cases} Q_1\delta(t) & \text{Bit 1} \\ 0 & \text{Bit 0} \end{cases} \tag{2}$$

3.2 *Pulse Amplitude Modulation*

Pulse-based transmission of molecules in CEMC system is also possible. Unlike impulsive transmission, in pulsed transmission, the TN sends a pulse of molecules with a given pulse-width $0 < T_p \leq T_{sym}$ where T_{sym} denotes the symbol interval. A given T_{sym} determines the separation between the starts of two consecutive pulses of molecules representing two different symbols, and so, T_{sym} determines the effective data rate Ω of the CEMC system as $\Omega = 1/T_{sym}$. Like in conventional communication systems, in CEMC it is also desired to have a higher transmission data rate in order to increase the speed of communication, which faces a challenge in providing reliable CEMC with lower BER between nanomachines while minimizing the effects of ISI. Like in impulsive transmission, it is also possible to have generalized ASK and OOK schemes both based on pulse transmission. In pulse-based ASK scheme, the TN sends pulses of two different amplitudes to represent the bits 1 and 0. Equation (3) expresses the signal for each symbol when ASK-based transmission is adopted, where $Q_1 \gg 1, Q_0 \gg 1$, and $\Pi(t)$ denotes a rectangular pulse with unity amplitude. The amplitude of the pulse $\Pi(t)$ is unity from $t = 0$ to $t = T_p$ and 0 everywhere else.

$$s(t) = \begin{cases} Q_1\Pi(t); & \text{Bit 1} \\ Q_0\Pi(t); & \text{Bit 0} \end{cases} \tag{3}$$

On the other hand, OOK scheme is a special case of ASK scheme when $Q_0 = 0$. As shown in Eq. (4), in pulse-based OOK, the TN sends a pulse of molecules only to represent bit 1 and does not send any molecules to represent bit 0 and, hence, it apparently remains "off" to represent bit 0. A detailed account of pulse-based modulation schemes in CEMC can be found in [7, 16–18, 21]. Figure 2c, d show pulse-based transmission schemes in CEMC.

$$s(t) = \begin{cases} Q_1\Pi(t); & \text{Bit 1} \\ 0; & \text{Bit 0} \end{cases} \tag{4}$$

3.3 Multilevel Pulse Amplitude Modulation

By varying the number of molecules transmitted by the TN, it is possible to design a multilevel pulse amplitude modulation (M-PAM) scheme based on generalized ASK. In M-PAM, the TN transmits each symbol by using one of the M different transmitted numbers of molecules, meaning that each symbol can be represented by $\log_2$ M bits being transmitted [17]. It is also possible to implement multilevel amplitude modulation based on IM scheme. Transmitted signals in M-PAM can be expressed as shown in Eq. (5) below.

$$s(t) = Q_m\Pi(t), \text{ where } m = \{1, 2, 3, \ldots, M\} \text{ and } Q_m \gg 1. \tag{5}$$

3.4 Sinusoidal Transmission

While IM, PAM, and M-PAM modulation schemes are all based upon varying the amplitudes of the transmitted number of molecules, it is also possible to vary the rate of change of the sinusoidal variation of molecular transmission rate, thereby making it possible to design frequency-shift keying (FSK) modulation in CEMC [14]. Transmitted signals in sinusoidal-based transmission can be expressed as below.

$$Q(t) = Q_{\text{average}} + Q_{\text{amp}} \sin(2\pi f t) \tag{6}$$

Here Q_{average}, Q_{amp}, and f denote average value, amplitude, and frequency of sinusoidal transmission respectively. Since concentration of molecules can never be a negative number, in Eq. (6), $Q_{\text{average}} \geq Q_{\text{amp}}$. In Eq. (6), different information symbols can be encoded by varying either of Q_{average}, Q_{amp}, f, or any combination of these quantities. Based on sinusoidal signaling, varying Q_{average} and/or Q_{amp} would give multilevel amplitude modulation based on sinusoidal transmission, while varying f would give FSK scheme [16, 22].

4 Fick's Laws Diffusion Dynamics

In CEMC, amplitude and frequency (i.e. the rate of change of amplitude) of the concentration signal are two important quantities of the transmitted signal that can be varied in order to materialize various modulation schemes as shown in Sect. 3. As shown earlier, in IM and PAM, information symbols are represented by the number of transmitted molecules at the beginning of symbol interval by following impulsive and pulsed transmissions respectively. On the other hand, in order to represent different symbols FSK-modulated signaling in CEMC relies on varying the frequency of amplitude variation of information molecules at the TN. In this section, based on Fick's laws of diffusion, we focus on mean concentration of molecules at the location of RN.

4.1 Fick's Laws Concentration Channel

Fick's laws of diffusion have been widely known for a long time [23, 24] and express the mean concentration of information molecules that become available at RN through propagation by the means Brownian motion [4]. In ideal diffusion, it is assumed that the total number of molecules is conserved, i.e. no molecules are created or destroyed, which provides us with the probability of having a molecule available at the RN as shown in Sect. 4 in the previous chapter. Several works that focus on the concentration channel based on Fick's laws are reported in [9, 16, 25].

4.2 Pulse-Transmitted Scheme

4.2.1 Single Pulse Transmission

In pulse-transmitted CEMC system, in order to understand the performance of transmitting a bit sequence completely, it is necessary to first understand the

Fig. 3 Available concentration signal intensity at the RN in response to a pulse transmission of pulse-width of 50 s (i.e. data rate $\Omega = 0.02$ bps) in water medium (adapted from [26])

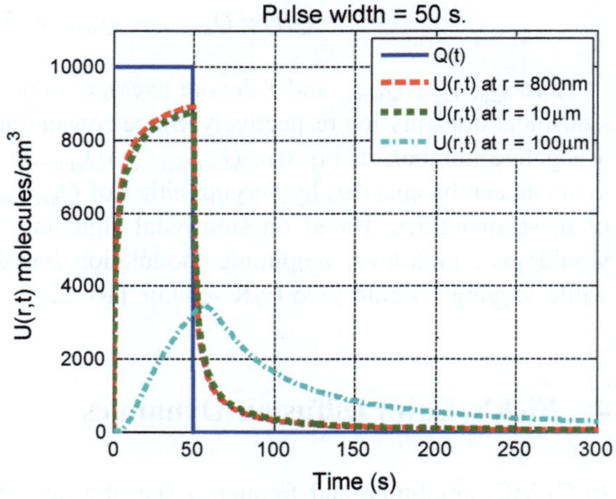

performance of transmitting a single pulse in CEMC channel. Figure 3 shows the output signals at various communication ranges when a single pulse with pulse-width 50 s is transmitted by the TN. As shown in Fig. 3, when communication range increases, signal gets temporally spread and thereby causes an increased level of ISI at the current symbol.

The performance of a pulse transmission can be explained on the basis of *desired signal strength ratio* $S_{(r,f)}$ and *interference strength ratio* $I_{(r,f)}$ expressed as shown in Eq. (7), where E_S, E_U, and E_I are *received desired signal strength*, *received total signal (i.e. desired and interference signals) strength*, and *received interference signal strength* respectively [7].

$$S_{(r,f)} = \frac{E_S}{E_U} = \frac{\int_0^{T_{\text{sym}}} U(r,t)dt}{\int_0^{T_{\text{obs}}} U(r,t)dt} \text{ and}$$
$$I_{(r,f)} = \frac{E_I}{E_U} = \frac{\int_{T_{\text{sym}}}^{T_{\text{obs}}} U(r,t)dt}{\int_0^{T_{obs}} U(r,t)dt} = \frac{(E_U - E_S)}{E_U} = 1 - S_{(r,f)} \quad (7)$$

Here E_S, E_U, and E_I can be found by taking integrals of the output signal $U(r,t)$ over $(0, T_{\text{sym}})$, $(0, T_{\text{obs}})$, and $(T_{\text{sym}}, T_{\text{obs}})$ intervals respectively, as shown in Eq. (7), where T_{sym} and T_{obs} denote symbol duration and observation time respectively.

On the other hand, in case of single pulse transmission, Fig. 4 shows the strength ratio quantities at various communication ranges. When r increases beyond 10 μm, $S_{(r,f)}$ decreases, meaning that the effective desired signal strength decreases. Since at any r and f, $S_{(r,f)} + I_{(r,f)} = 1$, when $S_{(r,f)}$ decreases, $I_{(r,f)}$ increases and vice versa.

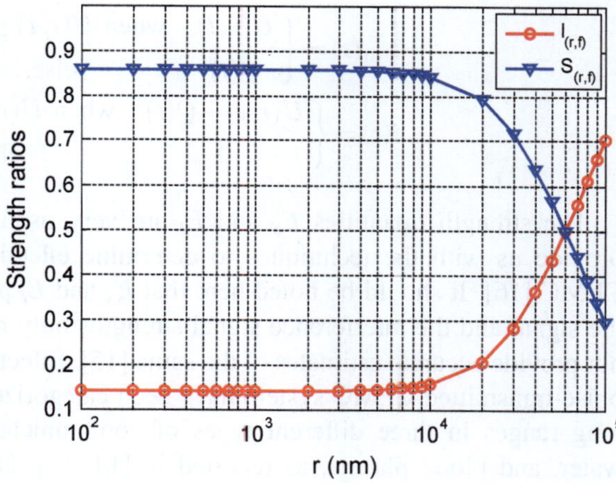

Fig. 4 Strength ratios at various r when pulse-width is 100 s and observation time is 10000 s

A decrease in $S_{(r,f)}$ indicates that more of the output signal strength would be interfering with the detection of the current symbol, and hence an increased level of ISI at the RN.

4.2.2 Bit Sequence Transmission

While the performance of a single pulse transmission provides us with the concepts of desired signal strength and interference signal strength in case of single pulse only, a similar approach can be adopted in pulse-transmitted CEMC scheme in order to send a sequence of bits in the propagation medium. Performance metrics in case of pulse-transmitted OOK-modulated scheme for transmitting a random sequence of bits have been provided in detail in [7]. As shown in Fig. 2d, a TN releases molecules at a fixed rate of Q_1 molecules per second during the entire symbol duration T_{sym} when it wants to send a bit 1 and does not transmit any molecule at all when it wants to send bit 0. Therefore, the entire observation time can be given as $T_{obs} = NT_b$ where N is the total number of bits to be transmitted and T_b is the duration of each bit in binary CEMC and, therefore, $T_b = T_{sym}$ in case of binary CEMC system. In case of transmitting a random sequence of bits based on OCK modulation, the signal strengths E_S and E_I can be expressed as the following [7].

$$E_S = \int_0^{T_{obs}} s_{\text{out}}(t)\,dt \quad \text{and} \quad E_I = \int_0^{T_{obs}} i(t)\,dt \tag{8}$$

Here the signals $s_{\text{out}}(t)$ and $i(t)$ in Eq. (8) can be expressed as follows and incorporate the effects of residual molecules originating from the previous symbols that contribute to desired signal strength at the current symbol [7].

$$s_{\text{out}}(t) = \begin{cases} U(r,t) & \text{when } U(r,t) \leq Q(t) \\ Q(t) & \text{else.} \end{cases}$$
$$i(t) = \begin{cases} U(r,t) - Q(t) & \text{when } U(r,t) > Q(t) \\ 0 & \text{else} \end{cases} \tag{9}$$

The strength quantities E_S and E_I are very useful in the sense that they can provide us with the techniques to determine effective communication ranges in CEMC [16]. It should be noted here that E_S and E_I provide approximate values of the signal and the interference signal strengths only respectively, and therefore, do not provide accurate estimates of the same [15]. Effective communication ranges for pulse-transmitted CEMC system have been categorized among short, medium, and long ranges in three different types of communication environment, e.g. in air, water, and blood plasma, as reported in [14, 16]. Determining effective communication ranges are important when a nanonetwork of a large number of nanomachines is to be materialized. For example, by computing average concentration of molecules at the RN as found using Fick's laws for pulse-transmitted CEMC, effective communication ranges in water medium can be found as <800 nm for short-range, 800 nm up to 10 μm for medium-range, and >10 μm for long range [16]. Similar short, medium, and long ranges in air can also be found as $r < 0.5$ mm, 0.5 mm $\leq r \leq$ 1 cm, and $r >$ 1 cm respectively [16]. Similarly, short, medium, and long ranges in blood plasma can also be found as $r <$ 400 nm, 400 nm $\leq r \leq$ 5 μm, and $r >$ 5 μm respectively [16]. The diffusion constants of information molecules in air, water, and blood plasma media, as considered in the determination of effective communication ranges, are 0.43 cm^2/s, 10^{-6} cm^2/s, and 2.2×10^{-7} cm^2/s respectively [11, 16]. Apart from this, as discussed in the previous chapter, increasing effective communication range is highly desired in CEMC. For example, it is shown in [15] that convolutional coding techniques can be applied to pulse-transmitted CEMC systems in order to increase effective communication range. Signal-related quantities when a random sequence of bits is transmitted are shown in Fig. 5.

4.2.3 Communication Range- and Rate-Dependent Characteristics

Communication range and transmission data rate significantly impact the performance of CEMC system. From CIR point of view, when r increases, CIR becomes temporally spread, meaning that in frequency domain amplitude spectrum squeezes. Therefore, fading characteristics of the signal may become affected. However, for lower data rate signals, larger symbol durations can ensure flat fading characteristics on the transmitted signal [10]. This also ensures that all the frequency components of the transmitted signal experience the same fading characteristics in the propagation channel and so the transmitted signal can retain its shape in temporal domain at the RN. While there still exists an open research question on how to select or optimize symbol duration in MC in general, communication range- and

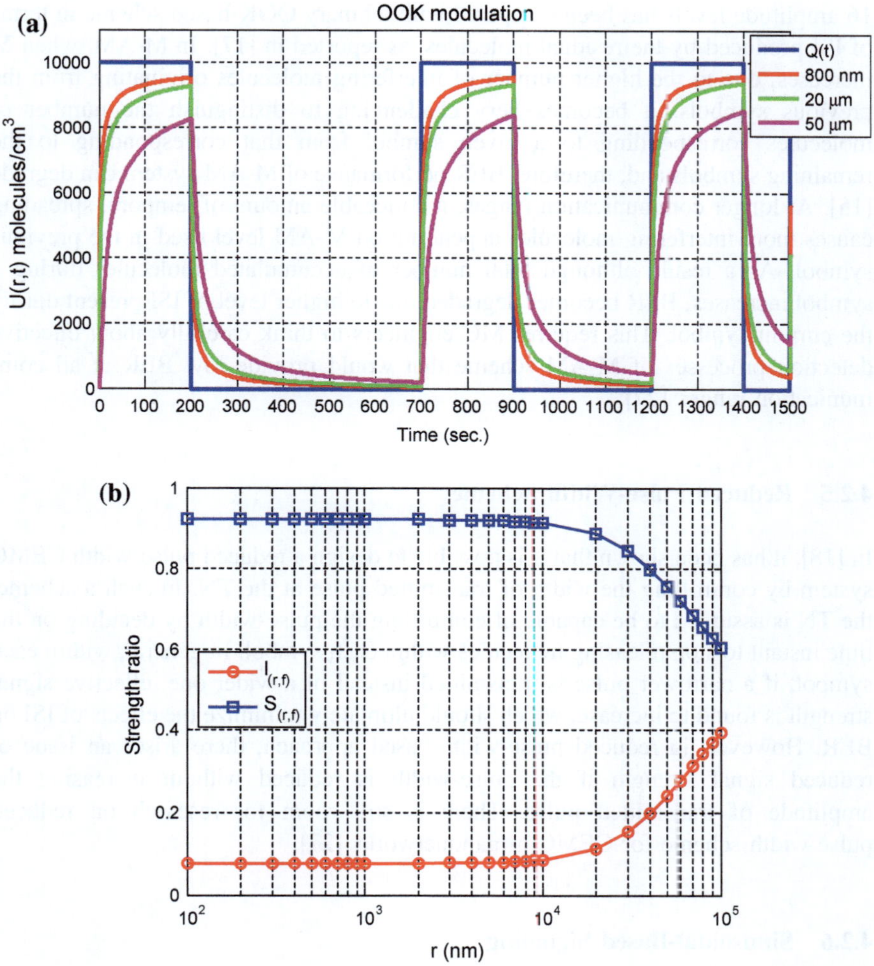

Fig. 5 Transmission of a random sequence of bits in Fick's laws CEMC channel. **a** First 15 bits {110000011000110} of a 100-bit long random sequence of bits is shown where $T_{\text{sym}} = 100$ s. **b** Strength ratios when a 100-bit long random sequence of bits with $T_{\text{sym}} = 100$ s is transmitted by the TN

rate-dependent characteristics of CEMC systems, especially when a random sequence of bits is transmitted, have been explained in [26].

4.2.4 Multilevel PAM Scheme

In pulse-transmitted CEMC system, it is also possible to develop multilevel PAM scheme based on various amplitudes of the transmitted pulses at the TN [17]. For example, a multilevel amplitude modulation (M-AM) scheme with M = 4, 8, and

16 amplitude levels has been compared with a binary OOK-based scheme in terms of ISI produced by the residual molecules, as reported in [17]. In M-AM, when M increases, due to the higher number of interfering molecules originating from the previous symbols, it becomes very challenging to distinguish the number of molecules corresponding to a given symbol from that corresponding to the remaining symbols and, therefore, BER performance of M-AM system can degrade [15]. At longer communication ranges, a noticeable amount of temporal spreading causes more interfering molecules depending on M-AM level used in the previous symbol. As a result, although total number of accumulated molecules during a symbol increases, BER becomes degraded due to higher level of ISI present during the current symbol. This requires MC engineers to think carefully about effective detection processes of M-AM scheme that would provide low BER at all communication ranges [27].

4.2.5 Reduced Pulse-Width Scheme

In [18], it has been shown that it is possible to design a reduced pulse-width CEMC system by controlling the width of transmitted pulse at the TN. In such a scheme, the TN is assumed to be capable of controlling the pulse-width by deciding on the time instant to stop releasing molecules within each symbol. In general, within each symbol, if a narrower pulse is transmitted instead of a wider one, effective signal strength is found to increase, which should ultimately minimize the effects of ISI on BER. However, in reduced pulse-width-based approach, there arises an issue of reduced signal strength if the pulse-width is reduced without increasing the amplitude of transmitted pulse. There is still room for research on reduced pulse-width scheme for CEMC in nanonetworks [27].

4.2.6 Sinusoidal-Based Signaling

Apart from using pulse-transmitted signaling scheme, it is also possible to design a sinusoidal-based signaling scheme in CEMC [22]. Sinusoidal transmission rates can be expressed as shown in Sect. 3.4. The sinusoidal transmission rate as shown in Eq. (6) can be expressed as the following [22].

$$Q'(t) = \begin{cases} 1 + Q'_{\text{amp}} \sin(2\pi f t) & \text{for } t \geq 0 \\ 0 & \text{for } t < 0 \end{cases} \tag{10}$$

Here $Q'(t)$ and Q'_{amp} indicate transmission rate and amplitude of sinusoidal variation respectively, both normalized to Q_{average}, which results in unitless quantities $Q'(t)$ and Q'_{amp} expressed as $Q'(t) = Q(t)/Q_{\text{average}}$ and $Q'_{\text{amp}} = Q_{\text{amp}}/Q_{\text{average}}$ respectively [22]. Correspondingly, $U'(r, t)$ denotes signal intensity in response to $Q'(t)$, meaning that $U'(r, t)$ is equal to $U(r, t)$ normalized to Q_{average}. The zero initial phase assumption in Eq. (6) allows us to investigate into phase errors in

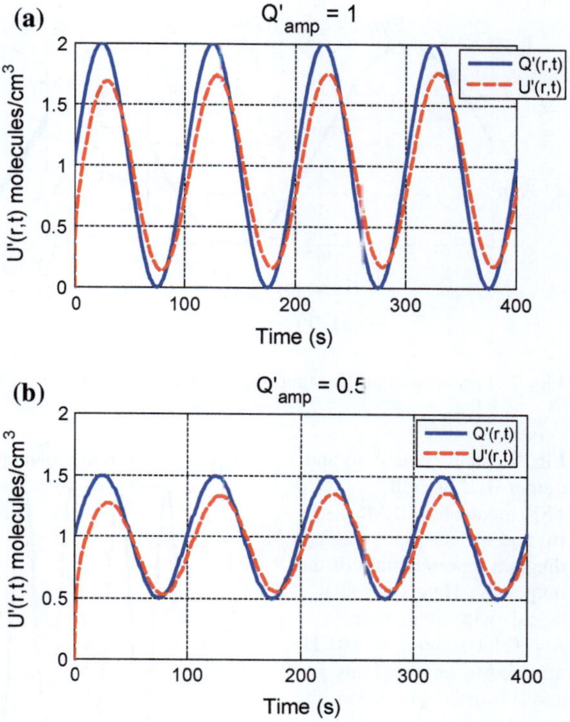

Fig. 6 Output signal intensity $U'(r, t)$ in response to $Q'(t)$ with $f = 0.01$ Hz, $\theta = 0$ rad, and $r = 1$ µm, and **a** $Q'_{amp} = 1$ and **b** $Q'_{amp} = 0.5$. Adapted from [22]

sinusoidal-based signaling in CEMC. Figure 6 shows the variation of signal intensity in response to a sinusoidal transmission rate.

Initial and steady-state phase errors, initial and steady-state amplitude losses, and detection noise margin have been explained in [22] using eye-diagram [28] representations as shown in Fig. 7. Results show that even with zero initial phase $U(r, t)$ suffers from phase errors and amplitude loss that vary over communication ranges and transmission data rates as reported in [22]. Transients of phase errors for a given TN-RN pair would also cause problem in the detection of FSK-modulated signals. Figure 7a, b respectively show the eye diagram representations of input $Q'(t)$ and output $U'(r, t)$ signals at the RN. As shown in Fig. 7b, τ_{ini} and τ_{ss} denote initial and steady state phase errors respectively at the RN [22]. In addition, initial and steady state amplitude losses at the RN can be found as below [22], where the quantities c and d can be found as shown in Fig. 7b.

$$AL_{ini} = 1 - \frac{c}{Q'_{amp}} \quad \text{and} \quad AL_{ss} = 1 - \frac{d}{Q'_{amp}} \tag{11}$$

On the other hand, sinusoidal-based signaling can be used in implementing FSK-modulated CEMC. As shown in Fig. 8, a higher frequency sinusoidal signal

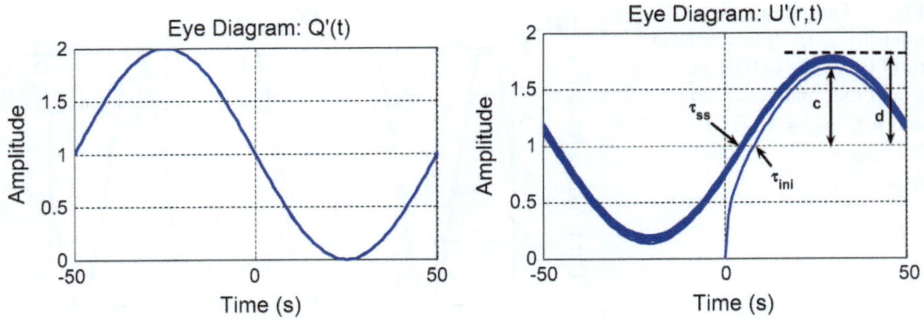

(a) $Q'(t)$ **(b)** $U'(r,t)$

Fig. 7 Representation of input and output signals at $r = 1$ μm using eye diagrams

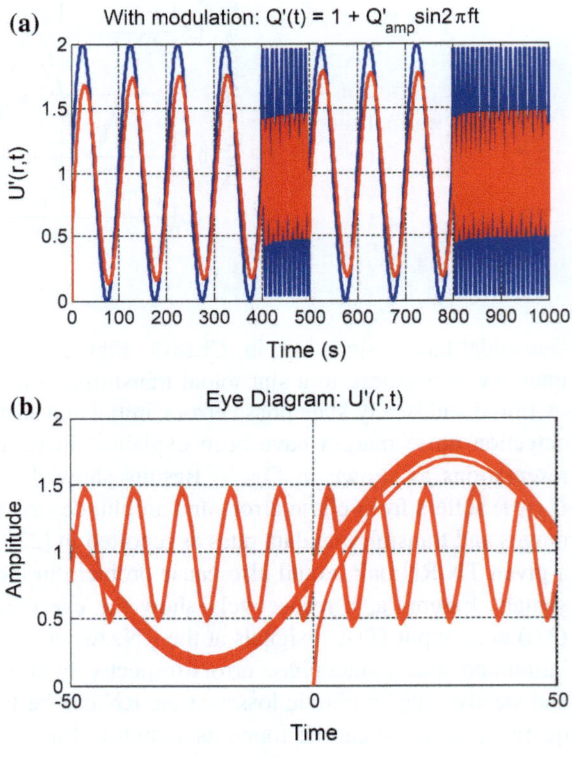

Fig. 8 Input (*blue line*) and output (*red line*) of FSK-modulated CEMC (**a**) and the corresponding eye diagram representation of the output (**b**). Here sinusoidal signals with frequencies $f_1 = 0.1$ Hz and $f_0 = 0.01$ Hz are used to represent bits 1 and 0 respectively, when bit rate $\Omega = 0.01$ bps and $r = 1$ μm

representing bit 1 would experience more amplitude loss at the RN, which would most likely be creating a problem in the detection of output frequency at the RN. In addition to amplitude loss, higher frequency components also experience higher phase errors at the RN [22], which in turn may cause problems in detecting FSK-modulated signals.

5 Signal Detection

5.1 *Stochastic Concentration Channel: Signal and Noise Models*

The exact number of molecules that become available at the RN at a given time instant is a random variable based on the diffusion-based propagation of molecules. A stochastic channel model takes into account the randomness in propagation in form of diffusion-based noise and ISI [19, 29, 30]. With diffusion-based noise and ISI in effect, in generalized ASK-based transmission scheme as shown in Fig. 2a, the output concentration signal $z(t)$ at time instant t at the RN can be expressed as the following [19].

$$z(t) = s_m(t) + n_s(t) + n_{ISI}(t) \tag{12}$$

Here $s_m(t)$ is the deterministic part of the signal available in VRV, $m \in \{0, 1\}$, $n_s(t)$ is the diffusion-based noise present at the RN and can be expressed as a zero-mean normal-distributed random variable with variance $s_m(t)(1-p(t))$, i.e. $n_s(t) \sim \mathcal{N}(0, s_m(t)(1-p(t)))$ [19]. Apart from this, $n_{ISI}(t)$ denotes the residual molecules that cause ISI at the RN and can be expressed as $n_{ISI}(t) \sim \mathcal{N}\left(\mu_{ISI}, \sigma_{ISI}^2\right)$. To incorporate the effects of ISI, μ_{ISI} and σ_{ISI}^2 denote the mean and the variance of ISI-producing molecules at the RN respectively. Details of the stochastic CEMC channel model can be found in [19].

Therefore, binary signal detection model in CEMC can be written formally as below

$$z(t) = \begin{cases} \mathcal{N}\left(s_1(t) + \mu_{ISI}, s_1(t)(1-p(t)) + \sigma_{ISI}^2\right); & \mathrm{H}_1 \\ \mathcal{N}\left(s_0(t) + \mu_{ISI}, s_0(t)(1-p(t)) + \sigma_{ISI}^2\right); & \mathrm{H}_0 \end{cases} \tag{13}$$

where the hypotheses H_1 and H_0 denote the cases when TN sends information bits 1 and 0 respectively; $s_1(t)$ and $s_0(t)$ denote the deterministic mean values of the desired signal at H_1 and H_0 respectively in the absence of diffusion-based noise and ISI, and $p(t)$ is the probability of getting one molecule at the receiver sensing volume VRV.

5.2 *Sampling-Based Detection*

In sampling-based signal detection, the RN samples the concentration signal intensity at one or more temporal instants in each symbol, and correspondingly, takes decision on which symbol being transmitted based on the sample value(s) of concentration [19]. Communication range-dependent characteristics of sampling-based signal detection show that detection performance becomes improved when the number of samples taken in each symbol increases [19], meaning that more number of samples provide more information about the transmitted bit, which ultimately help increase the chance of correct detection of the bit under examination. On the other hand, a larger communication range would experience a higher BER due to ISI-producing molecules originating at the previous symbols. However, the capability of the RN of estimating ISI-producing molecules correctly would also be an important factor in order to improve BER performance in such cases. Detection performance also depends on transmission data rate of the system such that at higher data rates BER performance degrades and vice versa. Details on sampling-based signal detection in CEMC can be found in [19].

5.3 Strength-Based Detection

Unlike sampling-based signal detection, in strength-based signal detection the RN accumulates all the molecules that become available at the RN in each symbol and decides on which symbol being transmitted based on comparing the accumulated number of molecules to a threshold [30]. In strength-based detection, after the RN has sensed the intensity of concentration at regular temporal intervals of t_s seconds, it produces as output the detection variable, which can be expressed as below [30],

$$z_{\mathrm{ED}} = s_{\mathrm{ED}} + n_{\mathrm{ED}}^{\mathrm{Noise}} + n_{\mathrm{ED}}^{\mathrm{ISI}} \tag{14}$$

where s_{ED}, $n_{\mathrm{ED}}^{\mathrm{Noise}}$, and $n_{\mathrm{ED}}^{\mathrm{ISI}}$ denote the strengths of desired (deterministic), diffusion-noise, and ISI signals respectively. Since strength-based signal detection can also be thought as energy-based detection [16, 21], here we use the subscript "$_{\mathrm{ED}}$" to denote the quantities related to strength-based signal detection. Therefore, in binary strength-based detection in CEMC, the detection problem can be expressed as below [30], where hypotheses H_1 and H_0 denote transmissions of bits 1 and 0 respectively.

$$z_{\mathrm{ED}} = \begin{cases} \mathcal{N}\left(s_{\mathrm{ED}}^{(1)} + \mu_{\mathrm{ISI(ED)}}, \sigma_{\mathrm{S(ED)}}^{2(1)} + \sigma_{\mathrm{ISI(ED)}}^{2}\right); & \mathrm{H}_1 \\ \mathcal{N}\left(s_{\mathrm{ED}}^{(0)} + \mu_{\mathrm{ISI(ED)}}, \sigma_{\mathrm{S(ED)}}^{2(0)} + \sigma_{\mathrm{ISI(ED)}}^{2}\right); & \mathrm{H}_0 \end{cases} \tag{15}$$

Here $s_{\mathrm{ED}}^{(m)}$ and $\sigma_{\mathrm{S(ED)}}^{2(m)}$, $m \in \{0, 1\}$, denote the deterministic mean signal strength and the variance of diffusion-based noise strength respectively when H_m is true. The quantities $\mu_{\mathrm{ISI(ED)}}$ and $\sigma_{\mathrm{ISI(ED)}}^{2}$ denote the mean and the variance of signal strength arising from ISI-producing molecules. More details of strength-based signal detection in CEMC can be found in [30].

6 Ligand-Receptor Binding

The molecules that become available at RN interact with the receptors located on its surface and based on their interaction the molecules may or may not bind with the receptors. The binding of information molecules with the receptors is commonly known as *ligand-receptor binding*. As mentioned earlier, in CEMC the information molecules are of single type, so are the receptors on the surface of the RN. In general, ligand-receptor binding can be explained by two main approaches, namely, by reaction rate equation (RRE) and by stochastic chemical kinetics (SCK), as described next.

6.1 *Reaction Rate Equations*

Ligand-receptor binding based on RRE relies on binding and release reactions taking place between information molecules and receptors, where these binding and release reactions are assumed to be taking place only according to the deterministic rates k_+ and k_- respectively [31]. In fact, RREs are a set of coupled ordinary differential equations that explain the temporal evolution of an output (e.g. bound receptors) of a chemical reaction as a deterministic process. In this model, the positions of the receptors and the positions as well as the velocities of the information molecules need to be known accurately in order to derive the temporal evolution of reactions taking place between the information molecules and the receptors. Since it is quite impossible to track the positions and velocities of all the information molecules during a time interval, the temporal evolution of reactions is apparently a non-deterministic process [3], for which the second approach, which is based on SCK, is explained next.

6.2 *Stochastic Chemical Kinetics*

In SCK approach, the information molecules and the receptors are assumed to form a "well-stirred" system [31], meaning that the locations of the receptors and the information molecules are assumed to be randomly distributed inside the VRV. This allows LRBP to be explained only with the populations of the chemical species, excluding the needs to have the positions of the receptors and the positions as well as the velocities of the individual information molecules to be known as in RRE-based approach. In SCK approach, the number of reactions or the population of each chemical species cannot be known deterministically as in the RRE-based approach, rather probabilistically by using chemical master equation (CME) [3, 31]. In CEMC, realistic channel models have been proposed by incorporating SCK-based approach where the number of reactions that take place within some time duration is shown to be a Poisson-distributed random variable with rate depending on the propensity function and the time duration [3]. In SCK, propensity function is known as a function that, when multiplied by a time duration, indicates the probability that a reaction would take place between an information molecule and a receptor [3]. Apart from this, the concept of SCK-based approach has been used to design a receiver model for CEMC based on pulse transmission with OOK modulation [26], where the RN is found to develop an optimum receiver based on the average number of molecules that are available for reception at the RN at any symbol incorporating the effects of ISI [26, 32].

7 Conclusion

In this chapter, a thorough investigation has been made into the performance of a diffusion-based CEMC channel where concentration-based approach has been adopted to encode information. CEMC is quite common among biological nanomachines, and therefore, it holds a bright prospect as an information encoding approach in molecular nanonetworks in the future generations of information technology. While MC is gradually maturing and, as a result, progressive works have started to appear in the research community lately, this chapter would be able to provide a good amount of insight to the performance of CEMC system in general. Notable communication features of CEMC, e.g. communication range and transmission data rate, have been investigated from the solutions of Fick's laws that provide average concentration of information molecules at the RN. On the other hand, the roles of diffusion-noise, ISI, and ligand-receptor binding have also been investigated. However, there are several research areas that demand future works in this field to include the following. Optimum signal detectors with sampling-based, strength-based, and SCK schemes for a variety of modulation formats need to be well investigated in order to ensure reliable CEMC for nanonetworks. In addition, the ability of the RN to sense the concentration signal intensity in a more intelligent fashion and correspondingly a formal evaluation of the actual implementation of the optimum signal detector with biological nanomachines should demand more research to be done in this field in the future. On the other hand, intelligent techniques to improve BER at the RN and to increase effective communication range between a pair of nanomachines in a nanonetwork are two areas that are worth investigating in detail in the future. We strongly believe that the investigations into the performance of CEMC system shown in this chapter would encourage interested readers towards new research on CEMC-based molecular nanonetworks and further the knowledge of CEMC in nanonetworks applications in general.

Acknowledgements M. U. Mahfuz would like to thank the Natural Sciences and Engineering Research Council of Canada (NSERC) for the financial support in the form of PGS-D scholarship during the years 2010–2013.

References

1. Haykin S (2000) Communication systems. Wiley
2. Berg HC (1993) Random walks in biology. Princeton University Press, NJ, USA
3. Atakan B, Akan OB (2010) Deterministic capacity of information flow in molecular nanonetworks. Nano Commun Netw 1:31–42, 201003
4. Bossert WH, Wilson EO (1963) The analysis of olfactory communication among animals. J Theor Biol 5:443–469
5. Moore M, Suda T, Oiwa K (2009) Molecular communication: modeling noise effects on information rate. IEEE Trans NanoBiosci 8:169–180
6. Moore MJ, Nakano T (2011) Synchronization of inhibitory molecular spike oscillators. In: BIONETICS-2011
7. Mahfuz MU, Makrakis D, Mouftah HT (2011) A comprehensive study of concentration-encoded unicast molecular communication with binary pulse transmission. In: 2011 11th IEEE conference on nanotechnology (IEEE-NANO), pp 227–232
8. Eckford AW (2007) Nanoscale communication with brownian motion. In: 41st annual conference on information sciences and systems, 2007. CISS'07, pp 160–165
9. Mahfuz MU, Makrakis D, Mouftah HT (2010) Characterization of molecular communication channel for nanoscale networks. In: Proceedings 3rd international conference on bio-inspired systems and signal processing (BIOSIGNALS-2010), Valencia, Spain, pp 327–332
10. Rappaport TS (2002) Wireless communications: principles and practice. Prentice Hall PTR, Upper Saddle River, N.J
11. Lacasa NR (2009) Modeling the molecular communication nanonetworks, MSc thesis, The Universitat Politècnica de Catalunya (UPC), Spain
12. Nakano T, Moore M, Enomoto A, Suda T (2011) Molecular communication technology as a biological ICT. In: Sawai H (ed) Biological functions for information and communication technologies. Springer, Berlin, pp 49–86
13. Moore MJ, Enomoto A, Suda T, Nakano T, Okaie Y (2007) Molecular communication: new paradigm for communication among nano-scale biological machines. In: Bidgoli H (ed) The handbook of computer networks. Wiley
14. Mahfuz MU, Makrakis D, Mouftah H (2010) Spatiotemporal distribution and modulation schemes for concentration-encoded medium-to-long range molecular communication. In: 2010 25th biennial symposium on communications (QBSC), pp 100–105
15. Mahfuz MU, Makrakis D, Mouftah HT (2013) Performance analysis of convolutional coding techniques in diffusion-based concentration-encoded PAM molecular communication systems. BioNanoScience 3:270–284 (Springer)
16. Mahfuz MU, Makrakis D, Mouftah HT (2010) On the characterization of binary concentration-encoded molecular communication in nanonetworks. Nano Commun Netw J 1:289–300 (Elsevier)
17. Mahfuz MU, Makrakis D, Mouftah HT (2011) On the characteristics of concentration-encoded multi-level amplitude modulated unicast molecular communication. In: 2011 24th Canadian conference on electrical and computer engineering (CCECE), pp 000312–000316
18. Mahfuz MU, Makrakis D, Mouftah HT (2011) Characterization of intersymbol interference in concentration-encoded unicast molecular communication. In: 2011 24th Canadian conference on electrical and computer engineering (CCECE), pp 000164–000168
19. Mahfuz MU, Makrakis D, Mouftah HT (213) Sampling based optimum signal detection in concentration-encoded molecular communication receiver architecture and performance. In: Proceedings of 6th international conference on bio-inspired systems and signal processing (BIOSIGNALS-2013), Barcelona, Spain
20. Haykin S (2002) Signals and systems. Wiley, New York
21. Mahfuz MU, Makrakis D, Mouftah HT (2011) On the detection of binary concentration-encoded unicast molecular communication in nanonetworks. In: Proceedings

4th international conference on bio-inspired systems and signal processing (BIOSIGNALS-2011), 26–29 Jan 2011. Rome, Italy, pp 446–449
22. Mahfuz MU, Makrakis D, Mouftah HT (2011) Transient characterization of concentration-encoded molecular communication with sinusoidal stimulation. In: Proceedings of the 4th international symposium on applied sciences in biomedical and communication technologies (ISABEL'11), Article 14, 6 p. Barcelona, Spain
23. Einstein A (1905) On the movement of small particles suspended in stationary liquids required by the molecular-kinetic theory of heat. Ann Phys 17:549–560
24. Philibert J (2006) One and a half century of diffusion: fick, einstein, before and beyond. Diffus Fundam 4:6.1–6.19
25. ShahMohammadian H, Messier GG, Magierowski S (2012) Optimum receiver for molecule shift keying modulation in diffusion-based molecular communication channels. Nano Commun Netw 3:183–195, 201209
26. Mahfuz MU, Makrakis D, Mouftah HT (2012) Strength based receiver architecture and communication range and rate dependent signal detection characteristics of concentration encoded molecular communication. In: Proceedings of BWCCA-2012, Victoria, Canada, pp 28–35
27. Mahfuz MU, Makrakis D, Mouftah HT (2013) Concentration encoded molecular communication: prospects and challenges towards nanoscale networks. In: Proceedings of international conference on engineering, research, innovation and education (ICERIE-2013), Sylhet, Bangladesh, pp 508–513
28. Lee EA, Messerschmitt DG (1994) Digital communication. Kluwer Academic Publishers, USA
29. Pierobon M, Akyildiz IF (2011) Diffusion-based noise analysis for molecular communication in nanonetworks. IEEE Trans Signal Process 59:2532–2547
30. Mahfuz MU, Makrakis D, Mouftah HT (2013) A generalized strength-based signal detection model for concentration-encoded molecular communication. In: Proceedings 8th international conference on body area networks (BodyNets 2013), Boston, MA, USA (30 Sept–02 Oct 2013), pp 461–467
31. Pierobon M, Akyildiz IF (2011) Noise analysis in ligand-binding reception for molecular communication in nanonetworks. IEEE Trans Signal Process 59:4168–4182
32. Mahfuz MU, Makrakis D, Mouftah HT (2014) Strength-based optimum signal detection in concentration-encoded pulse-transmitted OOK molecular communication with stochastic ligand-receptor binding. Simul Model Pract Theor 42:189–209 (Elsevier)

Physical Channel Model for Molecular Communications

Humaun Kabir and Kyung Sup Kwak

Abstract This chapter comes up with a brief overview of molecular communication models and modulation techniques by reviewing current research works found in the literature. The chapter also provides with an analysis of molecular communication in free diffusion-based molecular communication channel. In this model, the trasmitter nanomachine releases messenger molecules, the molecules diffuse through the channel, and the receiver nanomachine counts the received molecules to decode the information. We consider free diffusion of molecules where no additional force is required. Such a channel is referred to as the diffusion channel and can be modeled by using Ficks law of diffusion. Diffusion coefficient describes the tendency of propagation of the messenger molecules through the medium. Analysis shows that, channel memory offers a significant impact on performance.

1 Introduction

Nanomachines are the blessings of nanotechnology built from individual molecule or arranged set of molecules. Communication between nanomachines can be built through mechanical, acoustic, electromagnetic, and chemical or molecular means [1]. Nanonetwork is the interconnection of nanomachines anticipated to perform collaborative task which an individual nanomachine is unable to do. We are promised with a handful offers by nanonetworks not limited to lab-on-a-chip, health monitoring, drug delivery, regenerative medicine, environment monitoring, waste/population control, pattern and structure formation etc. [2]. However, conventional communication technologies are found inapt for nanonetworks mainly due

H. Kabir (✉)
Inha University, Incheon 402-751, Korea
e-mail: hakim2021@yahoo.com

K.S. Kwak
UWB Wireless Communications Research Center, School of Information and Communication Engineering, Inha University, Incheon 402-751, Korea
e-mail: kskwak@inha.ac.kr

J. Suzuki et al. (eds.), *Modeling, Methodologies and Tools for Molecular and Nano-scale Communications*, Modeling and Optimization in Science and Technologies 9, DOI 10.1007/978-3-319-50688-3_3

Table 1 Electromagnetic/optical communication and molecular communication [4]

	Electromagnetic/optical communication	Molecular communication
Information carrier	Electromagnetic waves, electrical/optical signals	Molecular/Chemical signals
Media	Space, cables	Aqueous
Speed	Speed of light (3×10^8 m/s)	Extremely slow ($n \sim \mu$m/s)
Range	Long distance ($\sim$ km)	Short distance (nm $\sim$ m)
Information	Texts, audio, videos	Chemical reactions, states
Other features	Reliable, high energy consumption	Unreliable, biocompatible and energy efficient

to the size and power consumption of the components in operation [3]. Molecular communication appears to be a promising approach for nanonetworks. First nanonetwork models are stimulated by molecular communications observed in biological systems. Small-scale devices built from biological materials, posing a high degree of energy efficiency and biocompatibility, are capable of interacting with biological molecules and cells in nano to micrometer scale [2]. However, communication based on molecular signal differs from traditional communications. A summary of the features of molecular communication and classical electromagnetic communication is presented in Table 1. Seemingly, unlike electromagnetic or other traditional communications, molecules serve as information carriers in molecular communication. The number of molecules in transmit and received signals can be considered as the amplitudes of the signals. Therefore, existing information and communication theories will likely not be applicable directly. For example, the terms such as encoding, modulation, demodulaltion, transmission power, signal-to-noise-ratio (SNR), bandwidth, inter-channel interference (ICI), inter-symbol interference (ISI), noise, channel memory etc. are supposed to be treated in dissimilar fashion in molecular communication.

2 How Molecular Communication Works?

2.1 Encoding

In molecular communication, information are encoded onto molecules instead of electromagnetic or acoustic waves. Individual molecule properties or molecule ensemble properties are used for encoding. The information may be encoded based on the three-dimensional structure of the information molecule (e.g. a specific structure of molecule), or on the specific composition of the information molecules (e.g. DNA sequence), or on the concentration of information molecules (i.e. the number of information molecules per unit volume) [2].

2.2 *Modulation*

2.2.1 On Off Keying (OOK)

Analogous to classical electromagnetic communication, on-off keying is the process where the transmitter either releases molecules or remains silent over a symbol time period [5]. This is a binary modulation scheme. Decision for choosing 1 or 0 is based upon a threshold number of received molecules.

2.2.2 Concentration Shift Keying (CSK)

Concentration Shift Keying is analogous to amplitude shift keying as in classical electromagnetic communication. In this modulation scheme, information is encoded according to the number of information molecules per unit volume i.e., concentration [6]. Decision for choosing 1 or 0 is based on a threshold of the concentration of received molecules. In binary concentration shift keying, two concentration levels are used and we need one threshold for decision. The scheme can be increased to higher levels depending upon the number of concentration levels. For example, in quadrature concentration shift keying, four concentration levels are used and we need three thresholds for decision. Figure 1 shows the diagram for QCSK modulation where the concentration levels c_0, c_1, c_2, c_3 are used for the symbols $s_0 = (0, 0)$, $s_1 = (0, 1)$, $s_2 = (1, 0)$, $s_3 = (1, 1)$, respectively.

2.2.3 Molecule Shift Keying (MoSK)

Bearing a resemblance to frequency shift keying, in this modulation scheme, information is encoded onto different types of molecules. In binary molecular shift keying, two types of molecules are used [6]. Like concentration shift keying, this scheme can also be increased to higher levels depending upon the types of molecules. For example, in quadrature molecular shift keying, four types of molecules are used for encoding. Figure 2 shows the diagram for QMoSK modulation where n_1, n_2, n_3, n_4 molecules are used for the symbols $s_0 = (0, 0)$, $s_1 = (0, 1)$, $s_2 = (1, 0)$, $s_3 = (1, 1)$, respectively and z_1, z_2, z_3, z_4 are corresponding threshold number of received molecules.

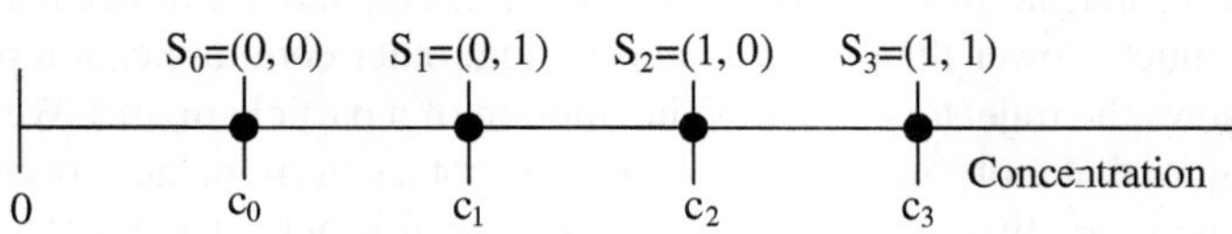

Fig. 1 QCSK modulation

Fig. 2 QMoSK modulation

Symbol	Type of molecules	Number of molecules	Threshold for decision
S_0=(0, 0)	t_1	n_1	z_1
S_1=(0, 1)	t_2	n_2	z_2
S_2=(1, 0)	t_3	n_3	z_3
S_3=(1, 1)	t_4	n_4	z_4

2.2.4 Isomer-Based Ratio Shift Keying (IRSK)

IRSK encodes the information based on the ratios of different types of messenger molecules and this type of modulation scheme has high, theoretically infinite, modulation order [7].

Other modulation schemes include in-sequence molecule delivery, time of dispersal of molecules, etc.

2.3 Transmission

Transmitter nanomachine releases the messenger molecules in the environment either by unbinding the messenger molecules from the sender nanomachine or by opening a gate that allows the information molecules to diffuse away [2].

2.4 Signal Propagation

The propagation of messenger molecules from the transmitter to the receiver nanomachine is governed by Brownian motion and is affected by two parameters: drift velocity and the diffusion coefficient [8]. Information molecules in molecular communication propagate through fluid medium (such as blood or water) whereas electromagnetic waves propagate through free space or wire. Apparently, molecular communication is much slower than electromagnetic and other communication processes. Figure 3 shows the trajectory of Brownian motion of a particle in 20 s. We can infer from the figure that molecular communication is not a deterministic process. Unlike electromagnetic or other communication processes, it is not certain that a molecule will reach a certain distance at a certain time. Molecular signal is also a discrete signal. Even a single molecule can be a signal in molecular communication. The

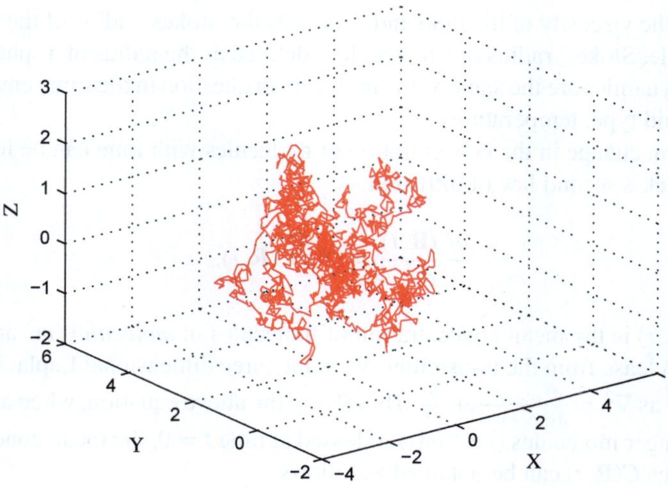

Fig. 3 Brownian motion of one particle in 20 s

diffusion of molecules can be interpreted by using Wiener process in which, the displacement of each molecule during an infinitesimal interval of time dt is modeled with a zero-mean Gaussian distribution with variance $2Ddt$ [9]:

$$\mathbf{R}(t+dt) = \mathbf{R}(t) + \sqrt{2Ddt}\mathbf{v}(t) \quad (1)$$

where $\mathbf{R}(t) = \hat{i}x(t) + \hat{j}y(t) + \hat{k}z(t)$ is the position vector of a molecule with unit vector $(\hat{i}, \hat{j}, \hat{k})$ at time t and $\mathbf{v}(t)$ is a vector of random variables. $\mathbf{v}(t)$ has a multivariate Normal distribution $\mathbf{v}(t) \sim \mathcal{N}(\mathbf{0}, \mathbf{I})$, where $\mathbf{0}$ and $\mathbf{I}$ are the null and identity matrices. Diffusion coefficient D, measured in m^2/s, describes the tendency of propagation of the messenger molecules through the medium. It can be written as

$$D = \frac{k_B T}{b} \quad (2)$$

where k_B is the Boltzmann constant in JK^{-1}, T is the absolute temperature of the environment in Kelvin, and b is the drag constant of the propagating molecule inside the given fluid. The constant b is affected by the comparative sizes between the propagating molecule (S_m) and the molecules of the fluid (S_f) [10]. If the propagating molecule's size is similar to the size of the molecules of the fluid then the fluid can be considered as a continuum [11]. Based on these two different conditions, the constant b is calculated as

$$b = \begin{cases} 4\eta\zeta_s & : S_m \approx S_f \\ 6\eta\zeta_s & : S_m \gg S_f; \end{cases} \quad (3)$$

where η is the viscosity of the fluid and ζ_s defines the Stokes' radius of the propagating molecule. Stokes' radius of a molecule is defined as the radius of a sphere whose diffusion dynamics are the same as the molecule in question in the same environment (such as fluid type, temperature).

The mean change in the concentration of molecules with time can be formalized by using Fick's second law of diffusion:

$$\frac{\partial C(\mathbf{R},t)}{\partial t} = D\nabla^2 C(\mathbf{R},t), \tag{4}$$

where $C(\mathbf{R},t)$ is the mean concentration of molecules in $molecules/m^3$ and t is the time after release from the transmitter, ∇^2 is the three dimensional Laplacian operator denoted as $\nabla^2 = \frac{\partial^2}{\partial x^2} + \frac{\partial^2}{\partial y^2} + \frac{\partial^2}{\partial z^2}$. By solving the above equation, when an impulse of n messenger molecules (i.e., $n\delta(t)$) released at time $t = 0$, the mean concentration of molecules $C(\mathbf{R},t)$ can be obtained as follows:

$$C(\mathbf{R},t) = \frac{n}{\left(\sqrt{4\pi Dt}\right)^3} \cdot \exp\left(-\frac{|\mathbf{R}-\mathbf{R}_{tx}|^2}{4Dt}\right) \tag{5}$$

where $\mathbf{R}_{tx} = \hat{i}x_{tx} + \hat{j}y_{tx} + \hat{k}z_{tx}$ is the position of the transmitter nanomachine. Figure 4 shows the graph for normalized concentration at different diffusion coefficients. We notice from the figure that, the higher the value of the diffusion coefficient, the quicker the normalized concentration reaches its peak.

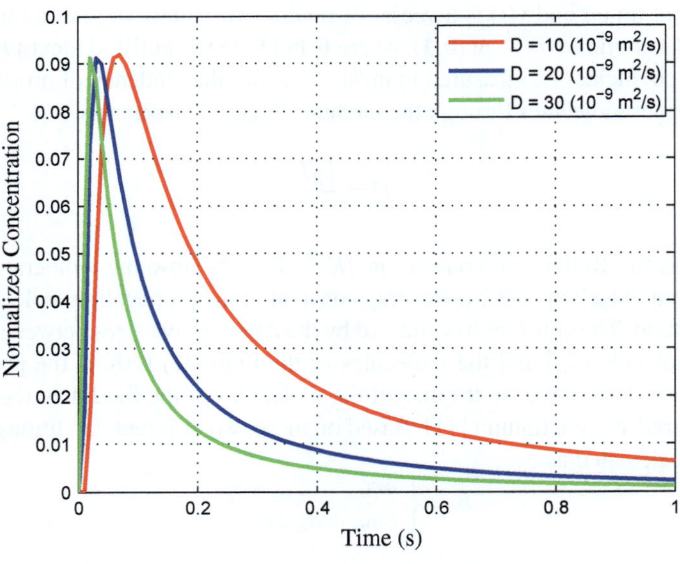

Fig. 4 Normalized concentration ($r = 2 \times 10^{-9}$ m)

The first passage time T of a molecule at a certain distance r with a certain diffusion coefficient D is a random variable. The probability density function of first passage time T is [12]

$$f_T(t) = \begin{cases} 0 & : t = 0 \\ \frac{r}{\sqrt{4\pi D t^3}} \cdot \exp(-\frac{r^2}{4Dt}) & : t > 0 \end{cases} \quad (6)$$

where $\mathbf{r} = \mathbf{R} - \mathbf{R}_{tx}$.

Using Eq. 6, we find the probability that a molecule reaches a receiver nanomachine within $(\tau, \tau + T_s)$ as

$$p(t) = \int_{\tau}^{\tau+T_s} f_T(t)dt, \quad (7)$$

where τ is a given transmission time defined as the delay and T_s is the time slot.

2.5 *Receiving and Decoding*

Upon capturing the incoming molecules, the receiver bio-nanomachines biochemically react with the molecules to decode the information [2]. Receptors, capable of binding to a specific type of information molecules, can be used for capturing. Channels can also be used for capturing (e.g., gap-junction channels) [4]. For decoding, the receiver may produce new molecules, perform simple task, or produce another signal [4].

2.6 *Noise*

Other source(s) may emit the same molecules used to encode the information messages [3] or it can be originated from an undesired reaction occurring between information molecules and other molecules present in the medium. As occurs in traditional communication systems, noise can be overlapped with molecular signals such as concentration level of molecules. If there is a drift, the time an information molecule requires to reach the receiver nanomachine follows an inverse Gaussian distribution [13, 14]. This leads to inverse Gaussian noise in time modulation.

2.7 *Inter-symbol Interference (ISI)*

Inter-symbol interference (ISI) may occur due to the delay of molecules in the medium [13]. The molecules used to convey information (be the information encoded in the number of molecules emitted, in their concentration, or in the time instant of release etc.) can interact with the transmission medium [15]. The molecules emitted from other source(s) can also be overlapped with intended molecular signals.

2.8 Channel Memory

Some molecules stay longer in the medium and reach the receiver after a delay. The transmitter may release the molecule corresponding to the 'next' symbol while the 'previous' molecules are still propagating. These delaying molecules arrive out of order introducing memory to the channel [13]. Following phenomena should also be taken into account:

- Information-carrying molecules may react with other molecules present in the medium and get absorbed.
- Molecules may replicate by stimulating the generation of new molecules.
- Spontaneous emission may take place so that molecules may get generated within the medium without any external intervention.

2.9 Delay

Not all the molecules reach the receiver within a specific time slot. Rather, molecules move around in the medium and reach the receiver after a delay. Since the propagation of molecules is Brownian in nature, the molecules may reach the receiver nanomachine after a long time. These delaying molecules may impinge on the later transmissions contributing to ISI or ICI.

3 Molecular Communication Model

Concepts of molecular nanonetworks are inspired by molecular communications observed in biological systems. Various modes and mechanisms of molecular communications found within and between cells are common and ubiquitous methods by which biological nanomachines communicate. Quorum sensing, calcium signaling etc. are few examples of biological nanonetworks that appear in intra-cell, inter-cell and intra-species communication. Modes and mechanisms of molecular communication are categorized based on how signal molecules propagate through the environment. Molecular signals may simply diffuse through the medium or directionally propagate by consuming chemical energy. The former type is categorized as the passive transport-based molecular communication, and the later as active transport-based molecular communication [4].

3.1 *Passive Transport-Based Molecular Communication*

Passive transport is the simplest method of propagating signal molecules within a cell and between cells in which, no additional force is required to propagate the molecules [2]. Signal molecules diffuse in all available directions in passive transport-based molecular communication and the process is suitable for highly dynamic and unpredictable environment. However, passive transport requires large number of messenger molecules and it is suitable for smaller size of molecules. Passive transport can be categorized as:

3.1.1 Free Diffusion-Based Molecular Communication

Free diffusion appears to be the most fundamental mechanism that molecular communication relies on to propagate a molecule. Figure 5 shows a free diffusion-based molecular communication model. Signal molecules (e.g., proteins and peptides) are released by the transmitter cells into extracellular environment and the molecules diffuse through the medium freely. Neighboring cells capture the signal molecules by using protein receptors which results in activation of chemical reaction such as increasing metabolism or transcription of cellular proteins. In biology, several examples of this class of molecular communication are observed. One illustration is intracellular metabolites propagating between cells. Another example is DNA binding molecules (e.g., repressors) that propagate over a DNA segment to search for a binding site [17]. Quorum sensing, a communication mechanism of bacterial cells, is referred to as a free diffusion-based molecular communication. In this process, bacteria coordinate certain behaviors such as biofilm formation, virulence, and antibiotic resistance, based on the local density of the population of bacteria [18]. They constitutively produce and secrete certain signaling molecules such as autoinducers (e.g., acyl homoserine lactone) or pheromones. They also have a receptor that can specifically detect the signaling molecule (inducer). Molecular communications of this class have been studied extensively in [5, 19–21].

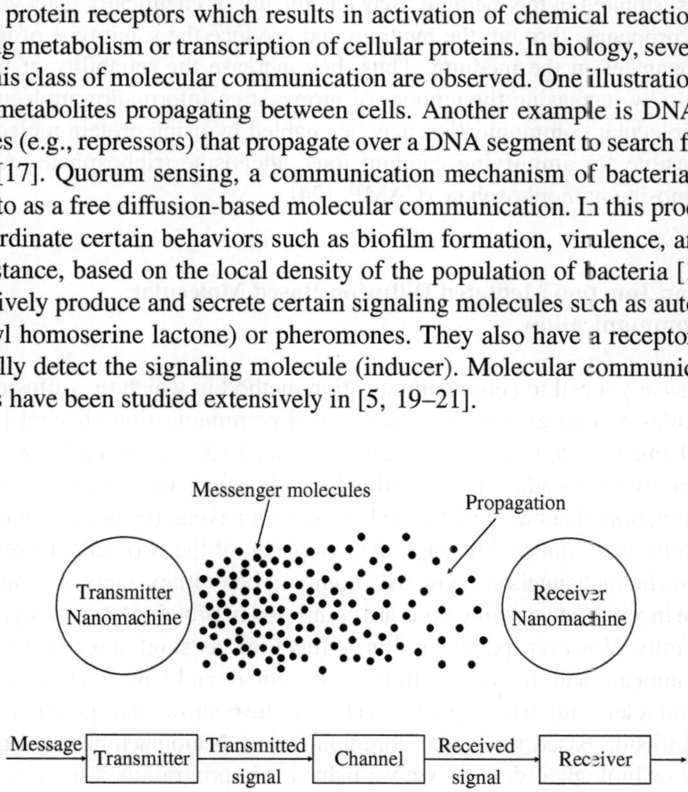

Fig. 5 Free diffusion-based molecular communication [16]

3.1.2 Molecular Diffusion with Drift

In this model, molecules undergo Brownian motion with drift. The propagation of molecules depends upon the drift velocity and diffusion coefficient. Information molecules may have drift velocity during release. The molecules may also undergo directional drift which continuously propagates molecules in the direction of drift [8, 13]. If there is a drift from transmitter to receiver, the first arrival time of the particles has the Inverse Gaussian distribution, leading to the additive Inverse Gaussian (IG) Noise channel [13]. This type of molecular communication is observed in human body. Cells in the body secrete hormonal substances which circulate with the flow of the blood stream and propagate to the distant target cells distributed throughout the body [2]. The process may also represent the active mode of molecular communication by which motor proteins carry and directionally propagate molecules from a sender bio-nanomachine to a receiver bio-nanomachine [22, 23].

3.1.3 Molecular Diffusion with Reactions by Amplifiers

Amplifiers, situated in the medium, may chemically or chemically react with molecules that propagate through the medium and produce the same type of molecules which propagates in the medium. Thus they increase the reliability of molecular propagation by increasing the number of propagating information molecules. This class of molecular communication may be enabled by using protein molecules that are responsible for amplifying calcium ions, adenosine triphosphate (ATP), and cyclic adenosine monophosphate (CAMP) [24].

3.1.4 Gap Junction Mediated Diffusion-Based Molecular Communication

This is basically a cell to cell communication method in which the diffusion of signal molecules can be guided through cell-cell communication channel [24]. Gap junction channels, analogous to the plasmodesmata that join plant cells, are physical channels between two adjacent cells allowing only connected cells to communicate. One gap junction channel is composed of two connexons (or hemichannels). Two adjacent cells are connected through the cytoplasm of the two cells. In vertebrates, gap junction hemichannels are primarily homo- or hetero-hexamers of connexin proteins while invertebrate gap junctions are composed of proteins from the hypothetical innexin family. However, pannexin family functions as single-membrane channels that communicate with the extracellular environment and have been shown to pass calcium and adenosine triphosphate (ATP). Figure 6 shows gap junction mediated reaction-diffusion based molecular communication. Nanomachines are engineered organisms or biological devices whose behavior is programmed to achieve application specific goals, and chemically communicate over a cell-cell communication

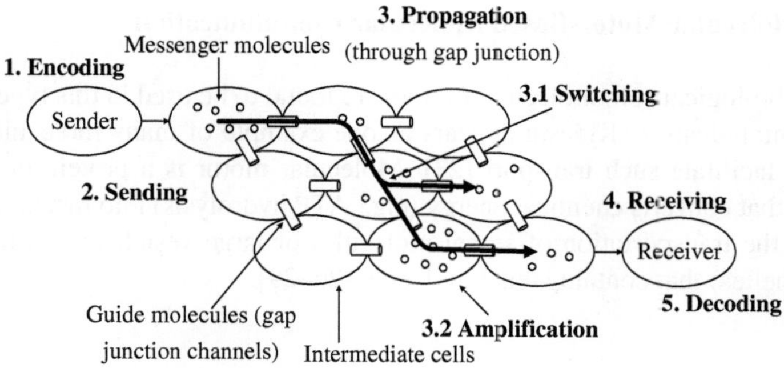

Fig. 6 Gap junction mediated reaction-diffusion based molecular communication [4]

medium. To support molecular communication between nanomachines, the communication medium may provide various networking mechanisms and services such as signal amplification and switching [25].

3.1.5 Reaction-Diffusion-Based Molecular Communication

Diffusion of signal molecules may biochemically react to form a different mode of communication letting the propagation of impulses [4]. Quick variation of the concentration of the signal molecules results in an impulse that propagates through the environment. For example, endoplasmic reticulum (ER) in a cell gathers and stores calcium ions. The cell, if stimulated, releases the stored calcium ions from the ER and the calcium ions diffuse to adjacent cells through cell-cell junction channels. A chain reaction of calcium stimulation occurs because the diffused calcium stimulates the adjacent cells. A short impulse of calcium is created, since the cell pumps calcium within the cell back into the ER and suppresses further stimulation. Likewise, neurons also produce ion impulses that propagate over the length of the neuron.

3.2 Active Transport-Based Molecular Communication

In active transport mechanism, signal molecules are directionally transported to specific locations and can propagate over longer distances as compared to diffusion-based passive transport. The mechanism is suitable for propagating large signal molecules and requires fewer molecules. However, the process requires a regular supply of messenger molecules in order to overcome the chemical reactions between the messenger molecules and the molecules in the environment. Active transport can be categorized as:

3.2.1 Molecular Motor-Based Molecular Communication

Within a biological cell, molecular motors are found to be used in this type of molecular communication. Kinesin appears as one example of many molecular motors that may facilitate such transport [23]. Molecular motor is a protein or a protein complex that converts chemical energy (e.g., ATP hydrolysis) into mechanical work allowing the transportation of signal molecules or large vesicles (e.g., liposomes, cell organelles) that contain signal molecule [26–29].

3.2.2 Bacterial Motor-Based Molecular Communication

Based on chemical concentration in the medium, bacteria can move directionally or it can also exchange DNA through the process of conjugation [4]. In this process, two types of bacteria transfer Deoxyribonucleic acid (DNA) chromosome through a pilus (i.e., a projection from the sender bacterium to the receiver bacterium forming a bridge for transmitting DNA). Sender bacterium here is with an F-plasmid (i.e., a genetic sequence that enhances the transfer of genetic information) while a receiver bacterium is without F-plasmid. The receiver bacterium, in this process, acquires DNA that produces some useful cellular functionality (e.g., protein production, antibiotic resistance). The receiver may also release pheromones creating a chemical gradient that leads a sender bacterium toward a receiver bacterium. Figure 7 shows bacterial motor-based molecular communication.

3.3 *Energy Model*

3.3.1 Energy Model via Diffusion

Energy budget of the transmitting and receiving units is one of the factors that limits over the performance of the communication system. In this model, a system is considered in which each unit is able to produce and store energy [11]. Some of the produced energy is used by routine activities of the unit and the rest is available for communication purposes.

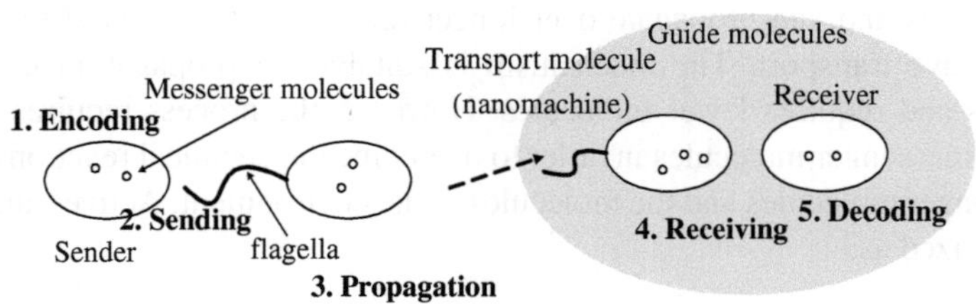

Fig. 7 Bacterial motor-based molecular communication [4]

3.3.2 Forster Resonance Energy Transfer (FRET) Model

Forster resonance energy transfer (FRET) is a nonradiative energy transfer process between fluorescent molecules based on the dipole-dipole interactions of molecules. Energy is transferred rapidly from a donor to an acceptor molecule in a close proximity such as 0–10 nm without radiation of a photon [30].

4 Free Diffusion Molecular Communication Channel

4.1 Capacity Performance

Channel capacity, one of the key measures of the channel, is the upper bound of the information rate that can be reliably transmitted over a communication channel. We calculate the capacity of a diffusion-based communication system. The channel is said to have memory from one previous symbol. Such a channel is termed as first-order memory channel [31]. We consider OOK modulation scheme where the transmitter nanomachine either releases molecules or remains silent over a symbol period. The emitted molecules diffuse through the medium and arrive at the receiver. The receiver counts the number of molecules in order to retrieve the information. The decision is made on the basis of a threshold number of received molecules. No equalizer to mitigate inter-symbol interference (ISI) is considered in this model because the performance depends only on the number of received molecules. Let, n_0 molecules are released for symbol $x = 1$, and 0 molecules for $x = 0$. Transmission time is divided into equal sized slots in which a single symbol (1 or 0) can be sent. The receiving time slots are chosen such a way that they are equal to the symbol durations and denoted by T_s. $n(r, T_s)$ is the number of molecules received at a distance r at time T_s. The decision for choosing between 1 and 0 is denoted by z, where z is a designated number of molecules (threshold). If $n(r, T_s) \geq z$ then we say that 1 is received, and if $n(r, T_s) < z$ then we say that 0 is received. The probability of a molecule hitting the receiver is denoted by $P_{hit}(r, T_s)$. Under the assumption that each molecule propagates independently, if n_0 molecules are sent within a symbol duration, the number of molecules received within the current symbol duration $n(r, T_s) \sim Binomial(n_0, P_{hit}(r, T_s))$ is a random variable and follows a binomial distribution. The first hitting time is τ, as we call delay, T_s is in the time interval from τ to $\tau + T_s$. The number of molecules $n(r, T_s)$ received is the summation of molecules from current and previous symbols. We notice that the current symbol duration falls in the time interval from T_s to $2T_s$ of the previous symbol. If n_p is the number of left over molecules belong to the previous symbol, then $n_p \sim Binomial(n_0, P_{hit}(r, T_p))$, where T_p is in the time interval from T_s to $2T_s$. If the first hitting time is τ, then T_p is in the time interval from $\tau + T_s$ to $\tau + 2T_s$. The number of molecules received within two symbol durations [11]

$$N_{hit} \sim Binomial(n_0, P_{hit}(r, T_s)) + Binomial(n_0, P_{hit}(r, T_p)) \tag{8}$$

Clearly, N_{hit} is nothing but the number of molecules from current symbol received in the time interval from 0 to $2T_s$ or from τ to $\tau + 2T_s$.

$$N_{hit} \sim Binomial(n_0, P_{hit}(r, 2T_s)) \tag{9}$$

According to the probabilistic theory, a Binomial distribution ($Binomial(n, p)$) can be approximated with a normal distribution $\mathcal{N} \sim (np, np(1-p))$, when p is not close to one or zero and np is large enough. Therefore Eq. 9, can be approximated as

$$N_{hit} \sim \mathcal{N}(n_0 P_{hit}(r, 2T_s), n_0 P_{hit}(r, 2T_s)[1 - P_{hit}(r, 2T_s)]) \tag{10}$$

There are four such different cases for decoding the received symbol. From Eq. 10, we obtain $N_{hit} \sim \mathcal{N}(n_0 P_2, n_0 P_2[1 - P_2])$, where $P_2 = P_{hit}(r, 2T_s)$. Probability that 1 is received when current and previous symbols are 1

$$P^{11} = P(N_{hit} \geq z) \approx \mathcal{Q}\left(\frac{z - n_0 P_2}{\sqrt{n_0 P_2[1 - P_2]}}\right) \tag{11}$$

where $\mathcal{Q}(.)$ is the tail probability of the Gaussian probability distribution function. The first and second superscripts in P^{11} denote current and previous symbol molecules respectively. Similarly

$$P^{10} = P(N_{hit} \geq z) \approx \mathcal{Q}\left(\frac{z - n_0 P_1}{\sqrt{n_0 P_1[1 - P_1]}}\right) \tag{12}$$

where $P_1 = P_{hit}(r, T_s)$. Thus, the probability of error p^1 when the symbol 1 is sent is

$$p^1 = 1 - \frac{1}{4}\left(erfc\left(\frac{z - n_0 P_1}{\sqrt{2n_0 P_1[1 - P_1]}}\right) + erfc\left(\frac{z - n_0 P_2}{\sqrt{2n_0 P_2[1 - P_2]}}\right)\right) \tag{13}$$

Similarly, the probability of error p^0 when the symbol 0 is sent is

$$p^0 = \frac{1}{4} erfc\left(\frac{z - n_0(P_2 - P_1)}{\sqrt{2n_0(P_1[1 - P_1] + P_2[1 - P_2])}}\right) \tag{14}$$

Equations 13 and 14 represent the cross-over probabilities for transmit symbols 1 and 0 respectively. Capacity is defined as

$$C = \max_z I(X; Y), \tag{15}$$

with mutual information $I(X; Y) = I(Y; X)$ and $I(Y; X) = H(Y) - H(Y|X)$. $H(Y)$ is entropy of the output y and $H(Y|X)$ is conditional entropy of y. In diffusion process, any number of molecules k at distance r at time t ($t = T_s$ or $t = 2T_s$) is $k(r,t) = n_0 \cdot erfc\left(\frac{r}{\sqrt{4Dt}}\right)$ and the probability of finding k molecules at distance r at time t ($t = T_s$ or $t = 2T_s$), $P(k) = \binom{n_0}{k} p^k q^{n_0-k} = \frac{n_0!}{k!(n_0-k)!} p^k q^{n_0-k}$, where $q = (1-p)$ and p stands for the success for a binomial distribution. The probability of first passage time t, as we defined as delay, is less than or equal to τ, $p = P(t < \tau) = erfc\left(\frac{r}{\sqrt{4D\tau}}\right)$, where τ is maximum delay for a certain diffusion coefficient D and distance r. Apparently, this probability p characterizes the channel. On the basis of this probability p, we can find P_1 and P_2. Figure 8 shows the capacity as a function of normalized diffusion length. The term $2\sqrt{Dt}$ is called the diffusion length which provides a measure of how far the concentration has propagated by diffusion in time t. The diffusion length is normalized to the distance between the transmitter and receiver. Capacity increases as the normalized diffusion length increases. However, the capacity was found to start decreasing after some normalized diffusion length.

4.2 *Bit Error Rate (BER) Performance*

4.2.1 Memoryless Channel

Let us consider that n_0 molecules are released for the symbol $x = 1$. The number of received molecules $n_x(r, T_s)$ is given by [32]

$$n_x(r, T_s) \sim Binomial(n_0, p(r, T_s)) \tag{16}$$

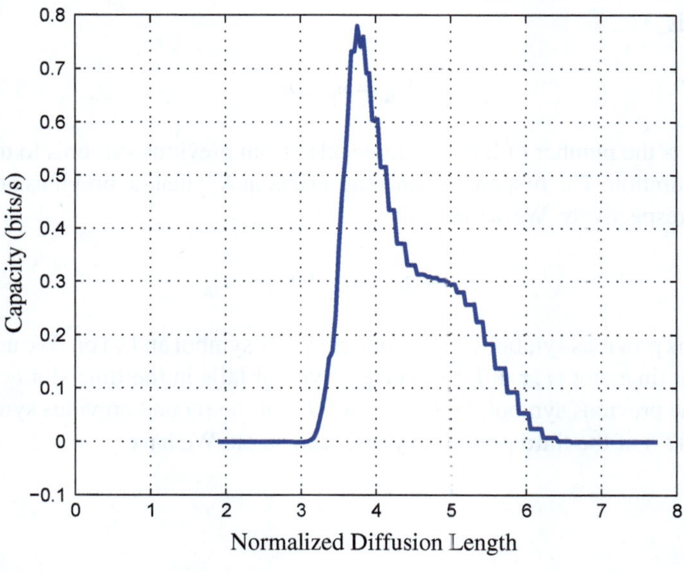

Fig. 8 Capacity as a function of normalized diffusion length [31]

where T_s is in the time interval from τ to $\tau + T_s$. Letting $p = p(r, T_s)$ and $q = 1 - p$, the probability of error for the symbol $x = 1$ is

$$p^1_{memoryless}(n_x < z) = 1 - \sum_{k=z}^{n_0} \binom{n_0}{k} p^k q^{n_0-k} \tag{17}$$

For the symbol $x = 0$, no molecule is sent. The receiver will not receive any molecule and there is no error in this case. Therefore,

$$p^0_{memoryless}(n_x < z) = 0 \tag{18}$$

So, BER for memoryless channel is

$$P_{e,memoryless} = \frac{1}{2}\left(1 - \sum_{k=z}^{n_0} \binom{n_0}{k} p^k q^{n_0-k}\right) \tag{19}$$

4.2.2 Memory Channel

In molecular communication, some molecules stay longer in the medium and impinge on later transmissions [33]. The received molecules are an amalgamation of molecules from current and previous symbols. Let us consider fourth-order memory channel for a test case i.e., the effect of four previous symbols over current symbol is considered. When received, for every transmit symbol $x \in \{0, 1\}$, one of $2^5 = 32$ cases may occur with equal probability; 16 cases for $x = 0$ and 16 cases for $x = 1$ as is shown in Table 2. The number of received molecules N_{px} for a channel with memory is

$$N_{px} = n_p + n_x, \tag{20}$$

where n_p is the number of left over molecules from previous symbols to the current symbol duration. The first and second subscripts in N_{px} denote previous and current symbols respectively. We write n_p as

$$n_p = n_{x_1} + n_{x_2} + n_{x_3} + n_{x_4} \tag{21}$$

where x_1 is previous symbol, x_2 is second previous symbol and so on. We notice from Fig. 9 that time slot $(\tau, \tau + T_s)$ of current symbol falls in the time slot $(\tau + T_s, \tau + 2T_s)$ of the previous symbol, $(\tau + 2T_s, \tau + 3T_s)$ of the second previous symbol, $(\tau + 3T_s, \tau + 4T_s)$ of the third previous symbol and so on. We have

Table 2 Symbols that contribute to the received molecules for transmit symbol x

Transmit symbol x	Cases	Symbols contributing to received molecules
0	1	00000
	2	10000
	...	...
	14	10110
	15	01110
	16	11110
1	17	00001
	18	10001
	...	...
	30	10111
	31	01111
	32	11111

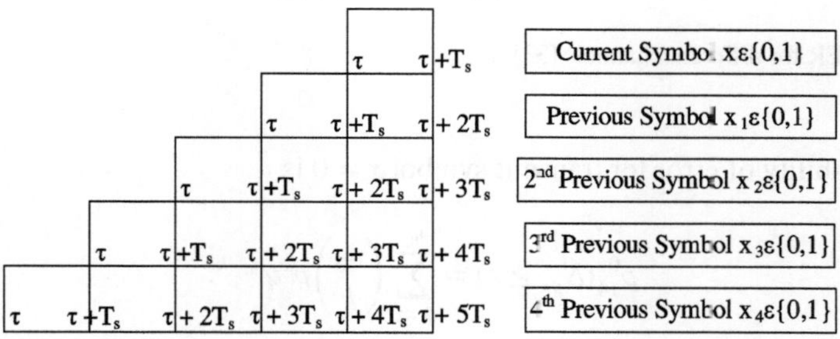

Fig. 9 Receiving history of transmitted molecules [33]

$$\begin{aligned}
n_x &\sim Binomial(n_0, p(r, T_s)) \\
n_{x_1} &\sim Binomial(n_0, p(r, T_1)) \\
n_{x_2} &\sim Binomial(n_0, p(r, T_2)) \\
n_{x_3} &\sim Binomial(n_0, p(r, T_3)) \\
n_{x_4} &\sim Binomial(n_0, p(r, T_4)),
\end{aligned} \tag{22}$$

where T_1 is the time interval between $\tau + T_s$ and $\tau + 2T_s$, T_2 is between $\tau + 2T_s$ and $\tau + 3T_s$ and so on. Let us consider a case when the transmit symbol is $x = 0$. The symbols that contribute to the received molecules are 10110 (case 14 in Table 2). The rightmost digit is the current symbol. Every digit represents the contributor of molecules from that particular symbol. From Eqs. 20 and 21, the number of received molecules is

$$N_{px} = n_{x_1} + n_{x_2} + n_{x_4} \tag{23}$$

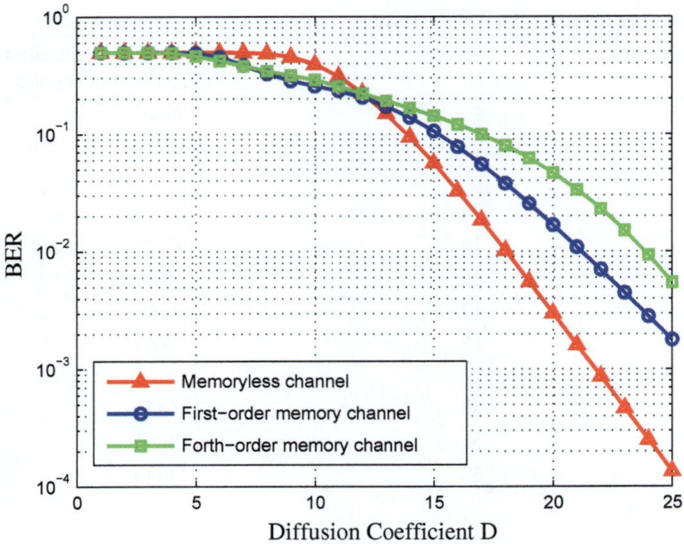

Fig. 10 BER for different channels [33]

The probability of error for transmit symbol $x = 0$ is

$$p_{14}^{0}(N_{px} \geq z) = \sum_{k=z}^{n_0} \binom{n_0}{k} p^k q^{n_0-k}. \tag{24}$$

Similarly, for the case of transmit symbol $x = 1$ (case 30 in Table 2), the probability of error is

$$p_{30}^{1}(N_{px} < z) = 1 - \sum_{k=z}^{n_0} \binom{n_0}{k} p^k q^{n_0-k} \tag{25}$$

where $N_{px} = n_x + n_{x_1} + n_{x_2} + n_{x_4}$. If p^0 and p^1 are the probabilities of errors for $x = 0$ and $x = 1$, respectively, then

$$p^0 = \frac{1}{16} \sum_{i=1}^{16} p_i^0(N_{px} \geq z), \tag{26}$$

$$p^1 = \frac{1}{16} \sum_{i=17}^{32} p_i^1(N_{px} < z). \tag{27}$$

Table 3 Parameters

Parameters	Memoryless channel	First-order memory channel	Fourth-order memory channel
n_0	100	100	400
z	35	40	200

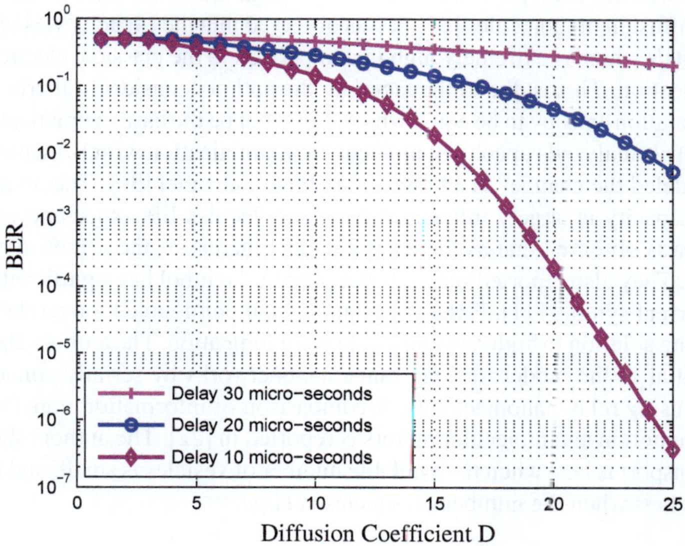

Fig. 11 BER for fourth-order memory channel [33]

Therefore, BER for a channel with memory is

$$P_{e,memory} = \frac{1}{2}(p^1 + p^0). \tag{28}$$

Figure 10 compares BER performances in various channels when the delay is 20 μs and the distance between the transmitter and receiver is 20 μm. Other parameters are shown in the Table 3. We observe that the performance degrades as the order of the channel memory increases. BER performances in fourth-order memory channel for different delays are reported in Fig. 11. As shown in the figure, performance degrades as the delay increases.

5 Related Works

BER is the main performance measure of a communication system. Capacity and delay profile etc. characterize the channel. However, very few papers have been found in literature dealing with BER. Symbol error rate (SER) with respect to drift velocity has been studied in Inverse Gaussian (IG) noise channel [13]. They reported that SER decreases as the drift velocity increases. Leeson, et al. studied forward error correction for molecular communications [34]. They calculated BER versus molecules per bit. The authors reported that, upto a certain limit, performance improves as the number of molecules per bit increases. Many authors have worked with channel capacity. A mathematical expression for the capacity in molecular communication nanonetworks has been provided when the propagation of the information relies on the free diffusion of molecules [35]. It was observed that the capacity was of the order of 10^{36} bits/s which is extremely high in comparison to the classical electromagnetic communication. This is due to the fact that, the authors considered thermal entropy and information entropy to be equal which might not be the case in practical scenario. Nakano et al. analyzed capacity of molecular communication with Brownian motion. They modeled the channel as a time slotted binary channel [36]. The authors found that the capacity is largely influenced by the molecules life expectancy. In energy model it was observed that as the energy budget increases, the achievable data rate increases. They also reported that only the previous symbol has significant ISI effect on the current symbol [11]. FRET-based molecular communication model stands as a promising solution to high rate nanoscale communication. The authors also showed the potential of the model for long-range nanonetworks by serially connecting the channels using relay nanomachines. A comparison of information transfer between Brownian motion and molecular motors is reported in [22]. The authors showed that active transport is best when the available number of vesicles is small, and Brownian motion is best when the number of vesicles is large.

6 Conclusion

We introduce molecular communication models and modulation techniques. A free diffusion-based molecular communication channel has also been analyzed where the first passage time of a molecule has been used as delay. Capacity was found to increase with normalized diffusion length. However, after a certain limit, it starts decreasing. Calculation shows that, BER performance degrades as the memory order increases. BER performance was also found to degrade as the delay increases. Enormous research works have been found analyzing the capacity and BER with respect to fluid velocity, number of molecules per bit, transmission distance, diffusion coefficient etc. However, unlike classical electromagnetic communication, a unique quality measure like signal-to-noise-ratio (SNR) is yet to be defined in molecular communication. Coding is yet to be studied exclusively. Transmitter power and transmitter/receiver diversity should also be studied. No mentionable work is found to deal with multiplexing, equalizer, and filtering etc.

Acknowledgements This work was supported by the National Research Foundation of Korea (NRF) grant funded by the Korean Government (MEST) (No. 2010-0018116).

References

1. Freitas RA (1999) Nanomedicine, vol 1: Basic capabilities. Landes Bioscience Georgetown, TX (1999)
2. Nakano T, Moore MJ, Wei F, Vasilakos AV, Shuai J (2012) Molecular communication and networking: opportunities and challenges. IEEE Trans Nanobiosci 11:135–148
3. Akyildiz IF, Brunetti F, Blázquez C (2008) Nanonetworks: a new communication paradigm. Comput. Netw. 52:2260–2279
4. Sawai H (2011) Biological functions for information and communication technologies: theory and inspiration. Springer
5. Moore MJ, Suda T, Oiwa K (2009) Molecular communication: modeling noise effects on information rate. IEEE Trans Nanobiosci 8:169–180
6. Kuran M, Yilmaz HB, Tugcu T, Akyildiz IF (2011) Modulation techniques for communication via diffusion in nanonetworks. In: 2011 IEEE International Conference on Communications (ICC), pp 1–5
7. Kim NR, Chae CB (2013) Novel modulation techniques using isomers as messenger molecules for nano communication networks via diffusion. IEEE J Sel Areas Commun 31:847–856
8. Kadloor S, Adve R (2009) A framework to study the molecular communication system. In: Proceedings of 18th International Conference on Computer Communication Networks, pp 1–6
9. ShahMohammadian H, Messier GG, Magierowski S (2012) Optimum receiver for molecule shift keying modulation in diffusion-based molecular communication channels. Nano Commun Netw 3:183–195
10. Tyrrell H, Harris K (1984) Diffusion in liquids: a theoretical and experimental study. Butterworth-Heinemann
11. Kuran M, Yilmaz HB, Tugcu T, Zerman B (2010) Energy model for communication via diffusion in nanonetworks. Nano Commun Netw 1:86–95
12. Redner S (2001) A guide to first-passage processes. Cambridge University Press
13. Srinivas KV, Eckford AW, Adve RS (2012) Molecular communication in fluid media: the additive inverse Gaussian noise channel. IEEE Trans Inf Theory 58:4678–4692
14. Khormuji MN (2011) On the capacity of molecular communication over the AIGN channel. In: 2011 45th annual Conference on Information Sciences and Systems (CISS), pp 1–4
15. Miorandi D (2011) A stochastic model for molecular communications. Nano Commun Netw 2:205–212
16. Einolghozati A, Sardari M, Beirami A, Fekri F (2011) Capacity of discrete molecular diffusion channels. In: 2011 IEEE international symposium on Information Theory Proceedings (ISIT), pp 723–727
17. Berg HC (1993) Random walks in biology. Princeton University Press
18. De Kievit TR, Iglewski BH (2000) Bacterial quorumsensing in pathogenic relationships. Infecti Immun 68:4839–4849
19. Eckford AW (2007) Nanoscale communication with brownian motion. In: Proceedings of 41st annual conference on information sciences and systems, pp 160–165
20. Eckford AW (2007) Achievable information rates for molecular communication with distinct molecules. In: Proceedings of workshop computer communications from biological systems: theory and applications, pp 313–315
21. Okaie Y, Nakano T (2012) Nanomachine placement strategies for detecting Brownian molecules in nanonetworks. In: Proceedings of IEEE Wireless Communication Networking Conference (WCNC), pp 1755–1759
22. Eckford AW, Farsad N, Hiyama S, Moritani Y (2010) Microchannel molecular communication with nanoscale carriers: Brownian motion versus active transport. In: Proceedings of IEEE international conference on nanotechnologies, pp 854–858
23. Eckford AW (2009) Timing information rates for active transport molecular communication. In: Nano-Networks, pp 24–28
24. Alberts B, Bray D, Lewis J, Raff M, Roberts K, Watson JD (1994) Molecular biology of the cell (1994) Garland. New York, pp 139–194

25. Nakano T, Suda T, Koujin T, Haraguchi T, Hiraoka Y (2007) Molecular communication through gap junction channels: system design, experiments and modeling. In: Proceedings of International Conference on Bio-Inspired Models of Network, Information and Computing Systems, BIONETICS, pp 139–146 (2007)
26. Hiyama H, Moritani Y, Suda T (2009) A biochemically engineered molecular communication system. Nano Networks, pp 85–94
27. Oiwa K, Sakakibara H (2005) Recent progress in dynein structure and mechanism. Current Opin Cell Biol 17:98–103
28. Shima T, Kon T, Imamula K, Ohkura R, Sutoh K (2006) Two modes of microtubule sliding driven by cytoplasmic dynein. Proc Natl Acad Sci 103:17736–17740
29. Toba S, Oiwa K (2006) Swing or embrace. New aspects of motility inspired by dynein structure in situ. Bioforum Eur 10:14–16
30. Kuscu M, Akan OB (2012) A physical channel model and analysis for nanoscale molecular communication with Forster Resonance Energy Transfer (FRET). IEEE Trans Nanotechnol 11:200–207
31. Kabir MH, Kwak KS (2013) Molecular nanonetwork channel model. Electron Lett 49:1285–1287
32. Socolofsky SA, Jirka GH (2005) CVEN 489-501: Special Topics in Mixing and Transport Processes in the Environment. In: EngineeringLectures. 5th edn., vol 3136. Coastal and Ocean Engineering Division, Texas A&M University, MS, p 77843
33. Kabir MH, Kwak KS (2014) Effect of memory on BER in molecular communication. Electron Lett 50:71–72
34. Leeson MS, Higgins MD (2012) Forward error correction for molecular communications. Nano Commun Netw 3:161–167
35. Pierobon M, Akyildiz IF (2011) Information capacity of diffusion-based molecular communication in nanonetworks. In: INFOCOM, 2011 Proc IEEE, pp 506–510
36. Nakano T, Okaie Y, Jian-Qin L (2012) Channel model and capacity analysis of molecular communication with Brownian motion. IEEE Commun Lett 16:797–800

Modulation in Molecular Communications: A Look on Methodologies

Ecehan Berk Pehlivanoglu, Bige Deniz Unluturk and Ozgur Baris Akan

Abstract Nanonetworking is a recently proposed paradigm that aims to achieve collaboration between nanomachines to carry out complex tasks. Molecular communications has been the most vibrant area of research for nanonetworking, mostly because of its feasibility and existence of communication schemes similar to molecular communications in nature. In molecular communications, two nanomachines communicate with each other via propagation of molecules from the transmitter to the receiver nanomachines through the medium they reside in. How and where to encode the message, i.e. modulation, plays a key role in molecular communications since it greatly affects the communication performance at nanoscale. To this end, in this paper, we examine the landscape of modulation in molecular communications, categorize the modulation schemes in molecular communications by methodology and discuss how convenient they are in terms of synchronization requirements in a nanoscale environment and their biocompatibility for applications inside human body.

1 Introduction

Field of nanotechnology envisages development of very small scale entities, called nanomachines, that are able to carry out tasks such as communication, sensing and computation. Capabilities of a nanomachine on its own is very limited. Hence, to achieve complex tasks, making nanomachines collaborate efficiently is a key design objective. Nanonetworking aims to accomplish this objective by defining a set of rules for communication and collaboration between a collection of nanomachines, so

E.B. Pehlivanoglu (✉) · B.D. Unluturk · O.B. Akan
NWCL Koc University, Istanbul, Turkey
e-mail: epehlivanoglu@ku.edu.tr

B.D. Unluturk
e-mail: bunluturk@ku.edu.tr

O.B. Akan
e-mail: akan@ku.edu.tr

J. Suzuki et al. (eds.), *Modeling, Methodologies and Tools for Molecular and Nano-scale Communications*, Modeling and Optimization in Science and Technologies 9, DOI 10.1007/978-3-319-50688-3_4

as to achieve bigger tasks. The capabilities that numerous nanomachines can achieve via nanonetworking have been exemplified in [1]: For instance, chemical nanosensors will be able to perform complex drug delivery in human body via collaboration. In other scenarios, in cases of deployment over large areas, nanomachines will be able to interact with each other via either multihop or broadcast communication mechanisms.

Networking and collaboration of nanomachines to achieve complex tasks inherently requires solid communication schemes. Communication between nanomachines is a difficult task requiring multifaceted research efforts. The challenges ahead of nanonetworking can be briefly listed as follows:

1. Downsizing of classical communication circuitry to nanoscale is not possible, hence design of nanomachines that are able to carry out communication functionalities is required.
2. Mobility of both the communicating nanomachines and the messengers they use for communication are highly dependent on the medium they reside in and the application they are designed for, bringing a number of channel and communication models to study.
3. Some nanocommunication and nanonetworking applications are designed to operate in living organisms, in which case both the nanomachines and their communication mechanism has to be biocompatible.

Various communication types, namely; acoustic, electomagnetic, nanomechanical and molecular communications have been proposed for nanonetworking. Acoustic and electromagnetic communications types are challenging to be realized until entities that can carry out acoustic or electromagnetic operations at nanoscale can be developed. Nanomechanical communications, on the other hand, require direct contact between communicating parties and is not suitable for distant applications [3]. In this context, molecular communications becomes a viable method for nanocommunications, backed by the fact that similar mechanisms are already present in many living organisms and environments.

In molecular communications, two or more nanomachines are envisaged to communicate by the help of messenger molecule exchanges [4]. Transmitter nanomachine can encode information, i.e. carry out modulation in molecular communication, via various techniques. The messenger molecules, via which the nanomachines communicate, then are released to the medium according to the preferred modulation technique. The messenger molecules propagate through the medium in which the nanomachines reside, from the transmitter nanomachine towards the receiver nanomachine. These molecule(s) eventually arrive at the receiver nanomachine, and the message is decoded at this nanomachine accordingly. A basic molecular communication system is depicted in Fig. 1. The comparison of molecular communications and traditional communications for communication entities and properties is given in Table 1 [1, 4–7].

Molecular communication faces different challenges at different networking layers. Routing needs reconsideration given the aqueous medium and the slow propagation characteristics of molecular communications. How to encode messages into

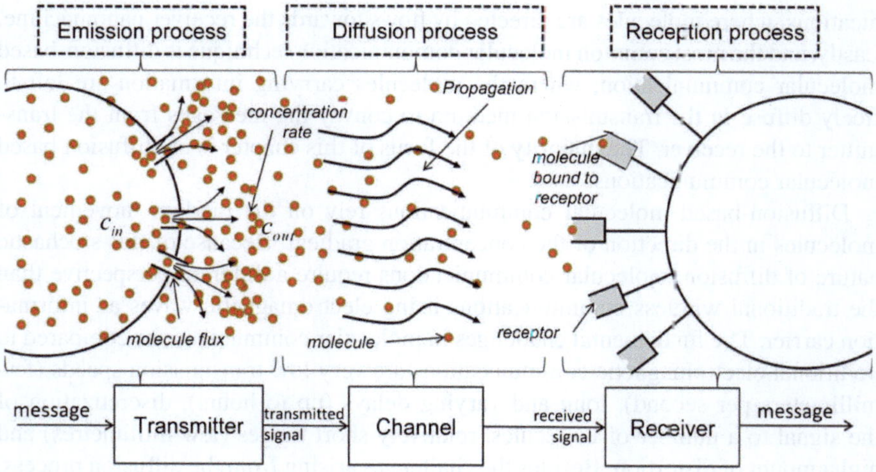

Fig. 1 Diffusion based molecular communication [2]

Table 1 Comparison of traditional and molecular communications [1, 4–7]

	Traditional communications	Molecular communications
Communicating parties	Electronic devices	Nanomachines
Signal types	Electrical or optical	Chemical
Propagation speed	Speed of light	Very slow
Propagation range	On the order of *m–km*	On the order of *nm–μm*
Communication medium	Air and/or wire	Aqueous

molecules and address different nanomachines present significant challenges. Synchronization, multiple access into the channel, channel capacity and interference are just a few other areas that need consideration in molecular communication, since results from traditional communications are not directly applicable [7]. In this chapter we will focus on a very important physical layer issue; namely, modulation; and review the proposed modulation techniques for molecular communications according to their categorization, and discuss their strengths and open issues.

2 Motivation for Modulation Research in Molecular Communications

Molecular communications can be broadly categorized under three main groups according to how molecules propagate in the medium. The first category is walkway-based or active transport molecular communications, where molecules follow predefined paths to the receiver. The second category is the flow based molecular communications, where molecules are directed by flows towards the receiver nanomachine. Lastly, and the most common molecular communication technique is diffusion-based molecular communication, where the molecules carrying information are left to freely diffuse in the transmission medium to convey the messages from the transmitter to the receiver. The majority of the focus of this chapter is on diffusion based molecular communications.

Diffusion-based molecular communications rely on the random movement of molecules in the direction of the concentration gradient. Because of this stochastic nature of diffusion, molecular communications require a different perspective than the traditional wireless communications using electromagnetic waves as information carrier. The fundamental challenges in molecular communications compared to traditional electromagnetic communications are very low transmission speeds (few millimeters per second), long and varying delays (up to hours), discretization of the signal to a number of molecules, relatively short ranges (few millimetres) and high amount of distortion. Besides the challenges arising from the diffusion process, the restricted information processing and memory capabilities of nanomachines also limits the complexity of design solutions for molecular communications.

Modulation of the information in molecular communications is central to many of these challenges. Given the low transmission speeds and long delays, having an efficient modulation scheme for molecular communications is highly necessary. To this end, many modulation scheme have been proposed. The research efforts in this field can be broadly categorized under four main groups:

1. The first group of modulation schemes is called concentration based modulation. In such schemes, the information message is encoded into the concentration of a certain type of molecule within the aqueous medium. When the sender nanomachine intends to send a digital 1, it releases a certain number of that molecule such that the concentration of this molecule is higher than a predefined threshold. When these molecules propagate to the receiver, hopefully their concentration will still be above this threshold, and the receiver nanomachine will correctly decode the bit as 1.
2. The second group comprises molecule type based modulation schemes. In a molecule type based modulation, the information is encoded on the bits of the molecule to be released into the medium. Examples of such messenger molecules are isomers, where the H and F atoms can be considered to represent bits 0 and 1, respectively.
3. The third group corresponds to the molecule release time or molecule release order based modulation schemes. In such schemes, the information is encoded into the release time or order of the messenger molecules.
4. The last group are hybrid modulation schemes, where two or more of the modulation schemes pertaining to the above groups are employed at the same time to increase diversity.

Considering the challenges of molecular communications, it is crucial to understand for which types of applications each of these modulation groups would be most suitable, in terms of achieved communication rate. Besides the rate, need for synchronization could be another important factor for choosing one modulation type over another, since synchronizing the communicating entities may not always be possible. Furthermore, for applications requiring different communication ranges, different modulation schemes are proposed to increase the capacity. For molecular communication, the range of a nanomachine can be short, medium or long which are on the scale of *nm–μm*, *μm–mm*, *mm–m*, respectively. A short range molecular communication takes place inside of a cell or in between two neighbour cells. Quorum sensing in a bacteria population can be considered as medium range molecular communication where the bacteria release a specific substance whose concentration gives the information on the size of the population. For long range molecular communication, the hormonal communication between organs meters apart is a good example.

Applications of molecular communications inside human body would also require the messenger molecules to be biocompatible, since otherwise they would cause biological complexities for the functions of the human body. Given these perspectives, it is necessary to understand what the current landscape in modulation schemes of molecular communications is, for which applications what kind of modulation type would be most suitable and what are the open issues and remaining challenges towards realization of more efficient molecular communications.

3 Modulation in Molecular Communications

In this section, different types of modulation techniques for diffusion-based molecular communication are presented. Some of the techniques have been implemented such as the TEC modulation in bacteria [8] and On-Off Keying modulation in [9]. However, since molecular communication field is still in its early stages, there is not any system which is able to implement and compare all the modulation techniques presented here. This categorized list point outs how the modulation techniques for electromagnetic communication can be adapted to molecular communication and how the peculiarities of molecular communication offer new perspectives for modulation.

3.1 Concentration Based Modulation

In this part, we will explain the different modulation techniques which basically changes the shape of a molecular concentration wave. Similar to electromagnetic waves, both analog and digital modulation schemes can be constructed to encode information on molecular concentration waves.

3.1.1 Amplitude and Frequency Modulation

The idea of using analog modulation in molecular communication is proposed in [10] based on the inherent amplitude modulation (AM) and frequency modulation (FM) of intercellular calcium waves propagating through the gap junctions between cells [11]. Calcium ions are the messenger molecules of intercell communication systems existing in living organisms. By the diffusion of intercellular calcium waves (ICW), groups of cells coordinate with each other and regulate the cellular activities such as chemical secretion, fertilization, neural signalling, contraction and death [11].

One of the regulations of ICWs takes place in immune system cells, which generate high or low concentration Ca^{2+} waves causing different reactions in cells. A naive B lymphocyte (a type of immune system cell) having a contact with the antigen (a substance which provokes an immune system response) for the first time creates a high concentration Ca^{2+} wave inducing reproduction of the cell. However, a B lymphocyte which has encountered the same antigen before creates a low concentration Ca^{2+} wave blocking the reproduction. This process illustrates an AM signalling for diffusion based molecular communication where the same immune system's cells (B lymphocytes) using the same messenger molecule (Ca^{2+} waves) can control two different processes by adjusting the concentration level of Ca^{2+} waves [11].

Another regulation mechanism of ICW's uses the frequency of Ca^{2+} waves instead of the amplitude. The frequency of the regular oscillations in Ca^{2+} concentration can be varied by the concentration of an input signal in order to control the rate of different cellular processes such as fluid secretion by salivary glands or glycogen metabolism in liver cells. These processes corresponds to FM signalling for diffusion based molecular communications.

A nanoscale communication system using ICWs can be formed via a group of neigbouring cells capable of propagating ICWs. A source cell encodes the information on the Ca^{2+} wave using either AM or FM modulation and transmits the wave to the nearest intermediary cell in the channel from the source cell to the destination cell. Each intermediary cell in the channel propagates the wave to its neighbour until it reaches the destination cell where the modulated signal is decoded [12].

As calcium signalling is the main intercellular communication mechanism in biological cells [13], it is definitely a biocompatible modulation technique. Furthermore the transmitter, receiver and intermediary cells are inherently synchronized for the propagation of continuous calcium ion waves in living organisms. Thus, while designing a molecular communication system interacting with living organisms via calcium ion waves, this system must be synchronized to cellular processes in that living organism.

3.1.2 On-Off Keying

On-Off Keying is the basic digital modulation scheme used in diffusion based molecular communications. In this scheme, time is divided into time slots for transmission. At the beginning of each time slot or during the whole time slot, a predefined number of molecules is released from the transmitter to the communication medium to represent bit 1, and no molecule is released to represent bit 0. The released molecules diffuse in medium stochastically and a portion of the released molecules arrive to the vicinity of the receiver. The receiver is capable of sensing the number of molecules in its vicinity and decides whether a 0 or 1 has been transmitted according to a concentration threshold.

On-Off Keying is mostly considered to characterize the diffusion based molecular communication channel and to calculate its capacity. In [14], On-Off Keying is used to provide an information theoretical approach for molecular communications taking into account the binding of the molecules to the receiver and considering an exponential decay for molecule concentrations. Molecules are assumed to be released from the transmitter at a fixed concentration level during the whole time slot corresponding to a square pulse. Then, the mutual information between the point-like transmitter and the point-like receiver is analysed. Being one of the first capacity investigations for molecular communication, [14] provides valuable insight. Nevertheless, [14] models the random propagation of molecules as an exponential decay with time, which is a coarse assumption that was later replaced by Brownian Motion assumption in most of the studies.

Nakano et al. [15] elaborates the capacity analysis by considering both the Brownian Motion propagation and the stability of the molecule in the environment. First, a naive modulation scheme is considered where only one molecule is released to represent bit 1 at the beginning of the time slot and no molecule is transmitted to represent bit 0 corresponding to impulse signal. The receiver decides that a 1 was sent if it receives a molecule in the current time slot. Otherwise, it decides that a 0 was sent. Since the probability that one molecule reaches the receiver in the same time slot while following a Brownian Motion is really low, this scheme yields low capacity. In order to achieve higher capacities, an extended modulation scheme is proposed. A high number of the copies of the same molecule are transmitted to represent bit 1 to increase the probability of reception at the receiver and different number of molecules are considered as the threshold for decoding. By the help of this extended scheme, when the slot time, the number of released molecules for bit 1 and the reception threshold are properly adjusted, capacities in the order of 0.03 bits per channel use can be achieved.

For On-Off Keying, several different pulse shapes are investigated in [16] other than the square pulse considered in [14] and the impulse in [15]. A square, a cosine and a Gaussian pulse having the same time slot duration and total energy are compared. Although it is observed that the Gaussian pulse has the lowest distortion and the maximum received peak level, the pulse shape does not have a significant effect on the total received energy. In addition, it is shown in [16] that to achieve higher bandwidth for longer ranges, the pulse duration should be the minimum possible duration. Considering all of these features, an optimal pulse shape may be chosen according to the decoding scheme.

For all of the modulation schemes proposed for On-Off Keying, synchronizing the time slots of transmitters and receivers is a major issue. Taking into account the low complexity of nanomachines, keeping the time for each communicating party is a challenge. Furthermore, because of the long and random propagation delays there will always be an inconsistency between the clocks of transmitter and receiver even if they do some efforts to synchronize via the diffusion channel.

3.1.3 Concentration Shift Keying

Concentration Shift Keying (CSK) is analogous to Amplitude Shift Keying (ASK) in traditional electromagnetic communications. Here, the concentration of the molecules are the amplitudes of the signals for short and medium range molecular communication. CSK, first proposed in [17], is based on the same assumptions as On-Off Keying. Again, the information sequence is spread over the time slots using one symbol per time slot. However this time, there are more than 2 binary symbols and one symbol represents more than one bit. The concentration level of the transmitted pulse is varied according to the groups of sequential bits.

To represent n bits, 2^n different molecular concentration levels must be generated at the transmitter and $2^n - 1$ threshold levels must be considered at the receiver as shown in Fig. 2. Because of the diffusion dynamics, molecules may not arrive to the receiver in the time slot that they are released, causing incorrect decoding of the current symbol. If they reach the receiver in the following time slots, they cause Inter Symbol Interference (ISI) for the next symbols. Hence, although increasing concentration levels improves bit per symbol rate, it also rises symbol errors because of ISI.

Transmitted molecules exhibit Brownian Motion with reception probability depending on the distance between the transmitter and receiver nanomachines and the duration of the time slots. Receiver nanomachines sense the number of molecules arriving to it, i.e., the concentration level, which is modelled with a Binomial distribution. The noise in this scheme arises from the residual molecules from previous time slots and the molecules present in the environment sent by other information sources and it is modelled as Additive White Gaussian Noise. To investigate the effect of CSK modulation on capacity, Kuran et al. [17] compare two cases where $n = 1$ (Binary CSK) and $n = 2$ (Quadruple CSK). For higher SNR values both cases

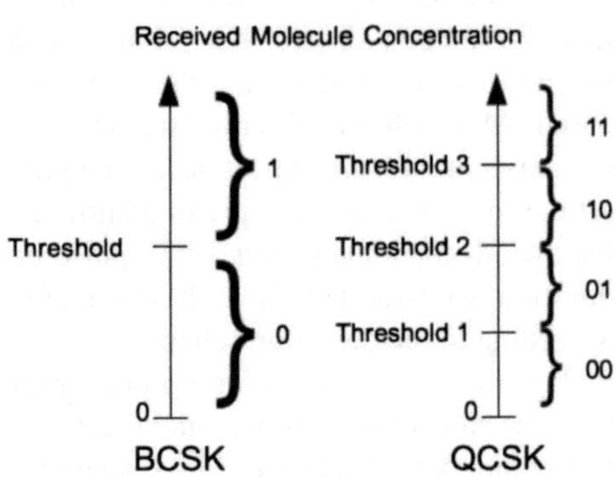

Fig. 2 Detection by thresholds for binary CSK and quadruple CSK

achieve the theoretical limits. However when the noise levels are significant, Binary CSK outperforms Quadruple CSK since for QCSK case the concentration intervals between the thresholds are smaller so it is more vulnerable to noise.

3.1.4 Pass Band Modulation (M-AM and FSK)

For the analog modulation schemes, we covered AM and FM and for the digital modulation schemes we covered base-band modulation techniques such as On-Off Keying and Concentration Shift Keying. In this part, we consider pass band modulation where the amplitude and the frequency of the molecular concentration oscillations are discretely modulated to obtain multilevel amplitude modulation (M-AM) and Frequency Shift Keying (FSK) for medium-to-long range molecular communication.

In [18], binary amplitude modulation and binary frequency shift keying in pass band is considered where the time is slotted. The propagation of messenger molecules are modelled as Brownian Motion, the messenger molecules are assumed to be pheromones propagating long distances in the air. The ligand-binding process of the molecules to the receiver is taken into account. The signal distortion is analysed by simulating the transmitted and the received signals. Furthermore, different detection schemes for M-AM case are proposed. Sampling based detection and energy detection schemes are investigated where in the former the receiver nanomachine senses the concentration of the messenger molecules only for a short period of time and in the latter the receiver counts all of the molecules during the current time slot corresponding to the energy of the signal. In [18], the capacity of the proposed modulation schemes is not considered but this work stands as an exemplary modulation scheme for long range molecular communication.

The synchronization is an essential part of this modulation scheme. The transmitter and the receiver must both know perfectly the operation frequency and the start and end times of the time slots which is a difficult task for nanomachines with limited capabilities.

The pass band modulation schemes can be designed to be biocompatible if existing pheromone molecules are chosen as messenger molecules as in [18]. In this case, for the biocompatibility analysis, the pheromone concentration levels must be adjusted so that it does not affect the biological processes in existing organisms depending on that specific pheromones or the molecular communication system must be isolated from those organisms.

3.2 Molecule Type Based Modulation

In concentration based modulation, the modulation of the signal is obtained by varying the amplitude or frequency of the messenger molecules' concentration while using the same type of molecule for all different symbols. To improve the capacity of molecular channels, molecular diversity can be used as an additional degree of freedom by constructing a molecule type based modulation scheme.

Encoding the information on the type of the messenger molecules is introduced in [19], where the modulation scheme exploits the vast number of different pheromone molecules present in the nature. In one hand, the molecule diversity can be used to create separate non-interfering channels by using different type of molecules for different channels defined as Molecule Division Multiple Access (MDMA) [19] similar to FDMA or TDMA in traditional electromagnetic communications. On the other hand, different types of molecules can be used in the same channel to boost the capacity of the channel defined as Molecule Shift Keying (MoSK) [17] similar to using orthogonal symbols in traditional electromagnetic communications.

To transfer n bits of information, MoSK utilizes 2^n different types of molecules to represent each combination of 2^n different n bit sequences. The time is divided in equal sized time slots and a large number of one type of these 2^n different molecules are released from the transmitter at the beginning of each time slot. If the concentration of one type of molecule exceeds the threshold in the current time slot, n bits represented by that molecule type is decoded. If the concentration of more than one molecule is larger than the threshold or none of the molecule type's concentration exceeds the threshold, an error occurs. Also, the residual molecules from the previous time slots cause ISI [17].

In [17], hydrofluorocarbons are introduced as candidate messenger molecules for MoSK for which n can reach up to 10^9 [20]. However, n should be small enough that the nanomachines can process the molecules and the molecules can propagate in the medium. For example, for intra-body applications, hydrocarbons molecules with n larger than 10^6 cannot propagate in the blood [21].

For hydrocarbon molecules, by attaching H or F atoms to a long carbon chain representing bits 0 and 1, information can be encoded onto molecules with similar physical and chemical properties. In this way, by altering n atoms of a molecule, 2^n different symbols can be produced. An illustrative constellation diagram for Quadruple MoSK is shown in Fig. 3. In [17], it is shown that the capacity of diffusion based molecular channels using Binary MoSK ($n = 1$) may be improved by using a higher

Fig. 3 Constellation diagram for QMoSK [20]

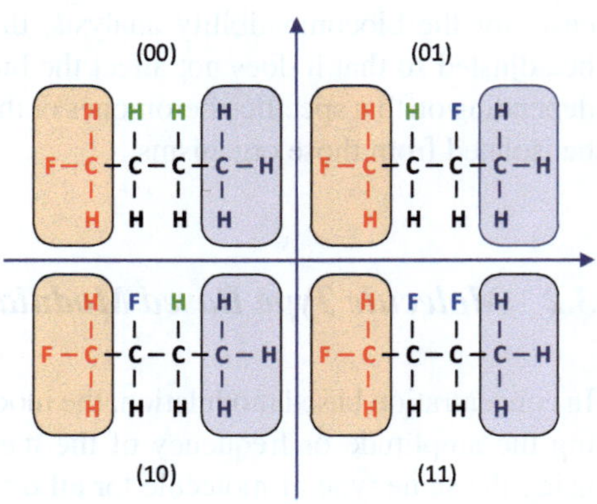

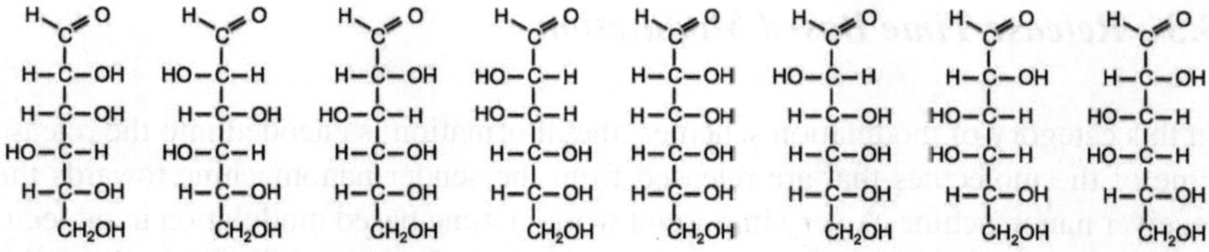

Fig. 4 Isomers of aldohexoses

order modulation such as Quadruple MoSK ($n = 2$) due to the orthogonality of symbols. Thus, as the modulation depth increases the capacity improves. Nevertheless, while the modulation depth increases, the messenger molecules become larger and start to diffuse more slowly reducing the reception probability. In [20], the optimal modulation depth and the conditions yielding it are investigated. By tuning the time slot duration and the concentration level according to the properties of the messenger molecules and the communication medium, the capacity is optimized.

Another technique for molecule type based modulation is using isomer molecules as symbols. In [22], isomers are proposed as messenger molecules since isomers contain same number and type of molecules which facilitates the synthesis of molecules for the transmitter. Specifically, [22] analyses the use isomers of aldohexoses illustrated in Fig. 4, which are monosaccharides with the chemical formula $C_6H_{12}O_6$, for 32-MoSK to increase achievable rate. Using aldohexose isomers is also a biocompatible technique since monosaccharides are naturally found in human body. On the contrary, for hydrofluorocarbons proposed in [17], a toxicity analysis is required to choose the biocompatible sets of messenger molecules since some of the hydrocarbons may be toxic to human body or flammable.

When compared to CSK, MoSK technique is more noise tolerant than CSK since when the modulation depth in CSK increases, the detection thresholds for the concentration of the messenger molecule get closer and errors occur more easily. However, in MoSK, detection depends on a single concentration threshold of the molecule type in use for that symbol.

The challenge in molecule type based modulation is the difficulty of producing, storing and receiving multiple different molecules in transmitter and receiver nanomachines which has limited source and functionality. A developed nanomachine capable of handling more than one type of molecules is required. Besides, every molecule representing a symbol has different physical and chemical properties, thus the propagation of the molecule and the interaction of the molecule with the environment varies favoring some symbols over the others. In this context, some molecules may even be toxic for in-body applications, therefore additional research is needed to ensure biocompatibility in such schemes. Moreover, similar to the concentration based modulation, synchronization of time slots poses an important challenge.

3.3 Release Time Based Modulation

In this category of modulation schemes, the information is encoded into the release time of the molecules that are released from the sender nanomachine towards the receiver nanomachine. A very important work on time based modulation in molecular communications is [8]. Authors have made a significant contribution to the molecular communication research not only in the sense of the modulation considered, but also because they used genetically engineered bacteria in an experimental molecular communication setting. This second point is important in the sense to verify that such bacteria can be used as nanomachines, and since they can be made biocompatible via genetic engineering, biocompatibility at least for the nanomachines in such a setting has been proven to be possible.

Apart from biocompatibility, another result from [8] is that authors employed time based modulation. Before arriving to the time based modulation, however, authors first showed On Off Keying in bacteria. The setup considered in this paper is as follows: The sender has access to a bacterial population, which, when stimulated, releases molecular signals. These molecular signals propagate towards the bacterial population that resides at the receiver site. When these signals are interpreted by the bacteria population at the receiver side, these bacteria respond with fluorescence, which is then interpreted by the receiver circuitry. As a consequence, via bacteria populations, one is able to convey information from sender to receiver.

Krishnaswamy et al. [8] have shown that the bit duration has to be significantly large ($\approx$450 min long) in OOK. This inherently lowers the data rate performance of OOK in bacteria based molecular communications. To this end, *Time Elapse Communications* (TEC) have been proposed. In TEC, the information is encoded in the time interval between two consecutive signals. In this scheme, it is assumed that both the sender and the receiver have the same clock rate f_c, although synchronization of clocks between these two entities is not required. The number of generated molecular signals is always two in this scheme: The first molecular signal is the start signal, and the second molecular signal is the stop signal. The information is encoded in the time between these molecules. Assuming that the information to be sent is represented by v, the start and stop signals are separated by a time period of $\frac{v}{f_c}$, i.e. the bits in v is encoded by the number of $\frac{v}{f_c}$ clock cycles. This modulation scheme is illustrated in Fig. 5.

Another point is that the stop signal can be arranged to be the start signal of the next transmission, further increasing rate. To have a better data rate in TEC with respect to OOK, the clock rate f_c has to be higher than the inverse of the achievable minimum bit duration in OOK, which is denoted as t_b. Although the above illustration is depicted for an error-free communication channel, the molecular start and stop signals can be degraded due to noise, causing errors in the bits encoded in time. Despite this, Krishnaswamy et al. [8] proved that an improvement of approximately an order in data rate is achievable by employing TEC rather than OOK in bacteria based molecular communications.

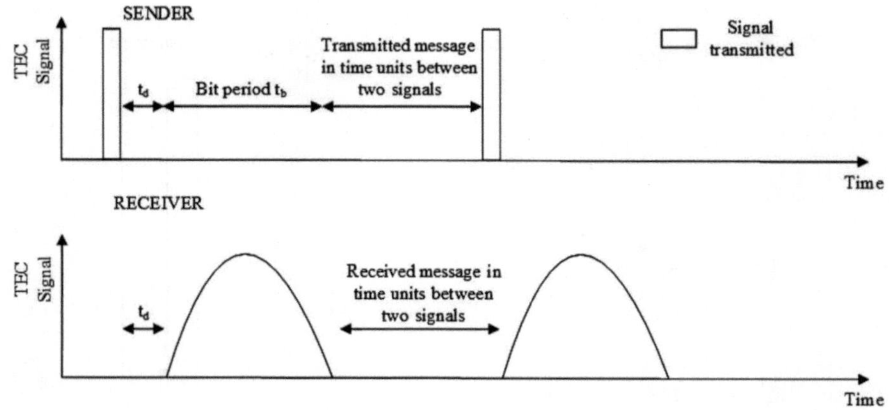

Fig. 5 Illustration of TEC signaling at the sender and receiver sides (Adapted from [8]). Each transmission is started by a start molecular signal, ended by a stop molecular signal, information bits are encoded in the time period between these two signals

Hsieh et al. [23] is another work on time based modulation of information in molecular communications. However, the authors study neither starts with time based modulations, nor stops at that point. The study begins with the consideration of how molecule type based modulation works, then consider how would a release time based modulation scheme could be implemented, and then the authors combine the molecule type based and the release time based modulation schemes to construct a mixed type modulation. This last combined scheme will be presented in Sect. 3.4. For the release time based modulation, authors consider the following model: The two molecule transmission are separated in time by a Bernoulli random variable S, i.e. $P(S = s_0) = \pi_0$ and $P(S = s_1) = 1 - \pi_0$. We assume that M_0 is encoded in case of an s_0 separation in time, and M_1 is encoded in case of an s_1 separation in time. Then, consecutive molecules arrive at the receiver nanomachine, and the information is decided based on the time separation between consecutive molecules, which would be modified until these molecules arrive at the receiver according to the mobility model assumed. The biggest advantage of this model is that it is asynchronous, hence does not require any synchronization nor does it require exact same clocks at the sender and the receiver. Only the receiver is expected to count the time difference between two consecutive molecule arrivals. The model indeed surpasses the need for synchronization or same clocks, however could be prone to error propagation: When a molecule arrives later than expected, not only it would be decoded in error, but also it would affect the timing considerations for the upcoming molecules and may cause them to be decoded in error as well.

Another modulation scheme that relies not on the release time of molecules but on the release order of molecules is presented in [24]. In the described modulation scheme, the release order of different types of molecules determines what message is being transmitted. For instance, let us assume that molecules of type A and B are available. Then, release order of $A - B$ can encode bit 0, while $B - A$ order would

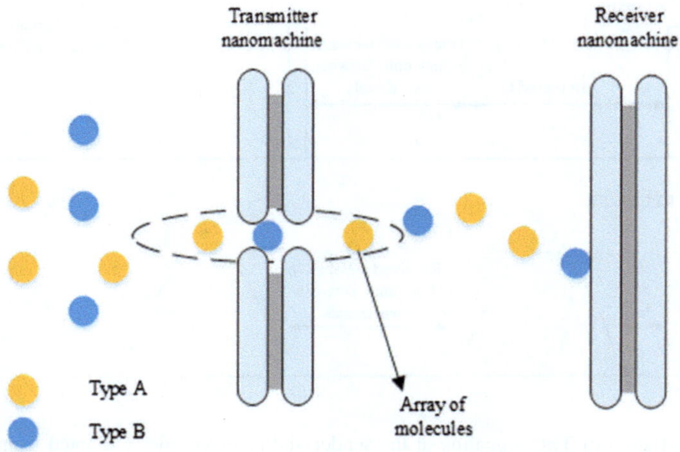

Fig. 6 Illustration of molecular array modulation, when only two molecule types are considered. Accordingly, there are two orders considered in this scheme: $A - B$ or $B - A$, each mapping to a different bit

encode bit 1. An illustration of this scheme is given in Fig. 6. Only the order of these molecules are changed, while the inter transmission time for two different symbols T is kept the same. In this modulation scheme, there are two sources of errors:

- For instance, molecules may be sent in order of $A - B$, but they may arrive to the receiver in the order $B - A$.
- Molecules of the previous bit can interfere with the current bit, causing ISI.

The second source of error can be made insignificant by adjusting the inter transmission time for two symbols T. For the first source of error, one has to increase the inter transmission time for two molecules in a symbol, denoted as t_e. However, in [24] it is shown that as the receiver to transmitter distance increases, the probability of correct decoding at the receiver (i.e. $A - B$ received when $A - B$ sent, $B - A$ received when $B - A$ sent) converges to 0.5, i.e. the correct transmission becomes totally random since the medium characteristics become the dominant factor on the order of the molecules in a symbol. We can deduce that, although the modulation scheme brings great advantages for molecular communication in terms of need for synchronization and achieved capacity, it is not suitable for long range molecular communication networks.

3.4 Mixed Type Modulation

Recently, new modulation schemes that employ more than one of these preceding techniques (i.e. concentration based, molecule type based or molecule release time based) have been proposed in the literature.

Arjmandi et al. [25] have proposed Molecular Concentration Shift Keying (MOCSK). In Concentration Shift Keying (CSK), to transmit b bits at a time, there has to be 2^b different concentration rates. However, since the molecule is the same for all transmissions in CSK, previous transmission can affect the current transmission and cause the decoding to be false. MOCSK brings MOSK idea to the CSK picture. Assuming there exists two molecule types A_1 and A_2, the MOCSK transmitter carries out CSK with A_1 in odd slots, and A_2 in even slots. This alternating molecule type in consecutive slots reduces error probability in decoding since now the receiver also differentiates molecules based on the slot it is currently in. MOCSK, as CSK and many other MOSK schemes, requires synchronization between the sender and the receiver nanomachines.

On the other hand, the study in [23] combines time based modulation and molecule type based modulation, i.e. MOSK. Different molecule types are used to encode different bits of information. Moreover, the inter transmission durations are assumed a simple Bernouilli model. In that sense, for instance, when the time separation between arrivals of two molecules is lower than a threshold, the receiver can decide on bit 0, and when it is higher than this threshold, it can decide on bit 1. Besides this time modulation, the receiver can also make use of the data encoded on the molecules used. Analysis in [23] suggest that encoding information in inter transmission durations improves the capacity of the regular MOSK channels. Nevertheless, error propagations can occur in the information encoded, especially those encoded on the inter transmission durations, since one outlier in terms of delay would affect the upcoming transmissions too.

Diffusion in different fluids such as water, air or blood has different characteristics. Despite the modulation schemes presented in this section are valid for all types of fluid, the parameters of the modulation schemes such as time slot duration or amplitude f the concentrations should be adjusted in order to accommodate the peculiarities of the propagation medium.

3.5 Modulation and Nanomachines in Molecular Communication

All the modulation techniques presented in this section requires complex machinery at nanoscale. There are three approaches for building nanomachines in the literature, namely, the top-down approach, the bottom-up approach and the bio-hybrid approach [1].

The top-down approach suggests miniaturization of current microelectromechanic systems down to nanoscale. However, at nanoscale level, the direct downscaling is hindered by the atomic interactions. Even though nanomicroelectromechanical parts can be fabricated, producing batteries which can support these nanoscale parts is still a challenge.

The bottom-up approach constitutes of building up nanomachines by chemical interactions of individual molecules. With today's technology it is possible to fabricate molecular components such as molecular gears and molecular switches. Nevertheless, the assembly of these components is not straightforward.

The third approach is the bio-hybrid approach suggesting using biological entities which inherently behaves as nanomachines. By tearing apart the useful parts of the cells or reprogramming them as in synthetic biology, a complete nanomachine capable of manipulating molecular communication signals can be obtained. As of now, this approach is the most promising approach since cells already constitute complete nanomachines communicating with each other via molecules.

The modulation techniques discussed in this chapter can be implemented by fabricating artificial nanomachines with the top-down or bottom-up approach which will result in producing electrical circuits at atomic level. Furthermore, by using the bio-hybrid approach, the modification of natural cells may lead us to achieve biocompatible nanomachines.

4 Conclusions

Molecular communication paves the way for the next-generation communication networks at nanoscale. Even though it suffers low data rates and long delays, it carries the potential of directly interacting with nature whose main communication tool consists of molecules. Very diverse biological and biomedical applications can be envisioned from water pollution control to cancer diagnosis/treatment using molecular communication which are delay-tolerant.

In this chapter, several modulation techniques for molecular communication addressing different challenges of nanoscale environment are presented. All of these techniques aspire to increase molecular communication channels' rate and capacity which are very low because of the slow and stochastic nature of diffusion. However, while designing modulation schemes for better performances, the peculiarities of molecular communication such as the limited functionality and memory of nanomachines, biocompatibility, difficulty of synchronization among nanomachines, and mobility should be taken care of. Considering all of these challenges, the most feasible modulation schemes are categorized in four, namely, concentration based modulation, molecule type based modulation, time/order based modulation and mixed modulation.

As the molecular communication signals are transferred with molecule concentration levels, the most instinctive modulation technique is concentration based modulation where the amplitude of concentration signals are varied according to the message to be conveyed. All of the concentration based modulation schemes are prone to errors because of the accumulation of molecules from previous transmissions and digital concentration based modulation techniques suffer severe ISI as the modulation depth increases. The biocompatibility of these techniques depends on the molecule type preferred as messenger molecule. If the messenger molecules are chosen from the molecules which do not interact with biological processes in their communication environment, they may be classified as biocompatible. However, due to the slotted time structure of these techniques, synchronization among communicating parties stands as a major issue.

Another way of modulation for molecular communication channels relies on the usage of different molecules for different symbols. In this case, ISI effects are lowered in expense of the complexity of synthesizing, storing multiple molecules at the transmitter side and the need for multiple receptors and the complexity of decoding at the receiver side. All the proposed molecule type based modulation schemes in the literature takes into account the biocompatibility by intelligently choosing messenger molecule sets. As in the concentration based modulation, molecule type based modulation requires also tight synchronization of transmitters and receivers due to the slotted time assumption.

To address the synchronization problem of the first two techniques, time/order based modulation schemes are introduced. Despite severe ISI, by cleverly encoding the information not on the concentration level of the molecular signal but onto the timing of the signal or the order of signals, we can bypass the need of synchronization. Nevertheless, the disadvantage of these techniques compared to concentration based techniques is the propagation of error over symbols.

The last category of modulation techniques constitutes the techniques combining some of the techniques from the first three category in order to address all of the challenges of molecular communication. These mixed type techniques aim at increasing the rate and capacity, reducing error rates and avoid synchronization issues. Mixed type techniques outperforms the previous techniques in all terms.

All of these four category of modulation are designed for short-to-medium range molecular communication. Designing modulation schemes for long range molecular communication remains as a research challenge.

Eventhough theoretically numerous modulation schemes can be suggested, the practically of each of them in the nanoscale communication environment is still an important question. The determination of which techniques lies in the limits of operation of actual nanomachines and which techniques operate more robustly in the communication environment that might even be a living organisms is the next step in modulation research for molecular communication.

Table 2 Research efforts in modulation in molecular communications

Category	Sub category	Examples	ISI level	Bio compatibility	No need for sync.
Concentration based	On off keying	[14]	Mid	N/A	–
	FM and AM signaling	[10]	N/A	✓	–
	CSK	[17]	High	N/A	–
	Pass band modulation	[18]	High	N/A	–
Type based	Pheromones	[19]	Low	✓	–
	Hydrofluorocarbons	[17]	Low	✓	–
	Aldohexoses	[22]	Low	✓	–
Time/order based	Time elapse	[8]	Mid	✓	✓
	Inter transmission coded	[23]	High	–	✓
	Molecular array	[24]	High	–	✓
Mixed type	MOCSK	[25]	Low	–	–
	Time + MOSK	[23]	Mid	–	✓

Table 2 summarizes the research efforts on modulation in molecular communications, which have also been summarized in Sect. 3.

References

1. Akyildiz IF, Brunetti F, Blazquez C (2008) Nanonetworks: a new communication paradigm. Comput Netw 52(12):2260–2279
2. Akyildiz IF, Jornet JM, Pierobon M (2010) Propagation models for nanocommunication networks. In: Proceedings of EUCAP 2010, Fourth European conference on antennas and propagation (invited paper), Barcelona, Spain, April 2010
3. Guney A, Atakan B, Akan OB (2012) Mobile ad hoc nanonetworks with collision-based molecular communication. IEEE Trans Mob Comput 11(3):353–366
4. Suda T, Moore M, Nakano T, Egashira R, Enomoto A, Hiyama S, Moritani Y (2005) Exploratory research on molecular communication between nanomachines. In: Genetic and evolutionary computation conference (GECCO), Late breaking papers
5. Hiyama S, Moritani Y (2010) Molecular communication: harnessing biochemical materials to engineer biomimetic communication systems. Nano Commun Netw 1(1):20–30
6. Hiyama S, Moritani Y, Suda T, Egashira R, Enomoto A, Moore M, Nakano T (2006) Molecular communication. J Inst Electron Inf Commun Eng 89(2):162
7. Nakano T, Moore MJ, Wei F, Vasilakos AV, Shuai J (2012) Molecular communication and networking: opportunities and challenges. IEEE Trans NanoBiosci 11(2):135–148
8. Krishnaswamy B, Henegar C, Bardill JP, Russakow D, Holst GL, Hammer BK, Forest CR, Sivakumar R (2013) Time-elapse communication: bacterial communication on a microfluidic chip. IEEE Trans Commun 61(12):5139–5151

9. Farsad N, Guo W, Eckford AW (2013) Tabletop molecular communication: text messages through chemical signals. PloS One 8(12):e82935
10. Nakano T, Suda T, Moore M, Egashira R, Enomoto A, Arima K (2005) Molecular communication for nanomachines using intercellular calcium signaling. In: Proceedings of IEEE conference on nanotechnology (IEEE-NANO 2005), Nagoya, Japan, July 2005
11. Berridge MJ (1997) The AM and FM of calcium signalling. Nature 386:759–760
12. Kuran MS, Tugcu T, Edis BO (2012) Calcium signaling: overview and research directions of a molecular communication paradigm. Wirel Commun IEEE 19(5):20–27
13. Scemes E, Giaume C (2006) Astrocyte calcium waves: what they are and what they do. Glia 54(7):71625
14. Atakan B, Akan OB (2007) An information theoretical approach for molecular communication. In: Proceedings of IEEE/ACM BIONETICS 2007, Budapest, Hungary, December, 2007
15. Nakano T, Okaie Y, Liu J-Q (2012) Channel model and capacity analysis of molecular communication with Brownian motion. IEEE Commun Lett 16(6):797–800
16. Garralda N, Llatser I, Cabellos-Aparicio A, Pierobon M (2011) Simulation-based evaluation of the diffusion-based physical channel in molecular nanonetworks. In: 2011 IEEE conference on computer communications workshops (INFOCOM WKSHPS). IEEE
17. Kuran MS, Yilmaz HB, Tugcu T, Akyildiz IF (2011) Modulation techniques for communication via diffusion in nanonetworks. In: 2011 IEEE international conference on communications (ICC). IEEE
18. Mahfuz MU, Makrakis D, Mouftah H (2010) Spatiotemporal distribution and modulation schemes for concentration-encoded medium-to-long range molecular communication. In: 2010 25th biennial symposium on communications (QBSC). IEEE
19. Gine LP, Akyildiz IF (2009) Molecular communication options for long range nanonetworks. Comput Netw 53(16):2753–2766
20. Unluturk BD, Pehlivanoglu EB, Akan OB (2013) Molecular channel model with multiple bit carrying molecules. In: 2013 first international Black Sea conference on communications and networking (BlackSeaCom). IEEE
21. Freitas RA (1999) Nanomedicine, vol. I: basic capabilities. Landes Bioscience, Georgetown, TX
22. Kim N-R, Chae C-B (2012) Novel modulation techniques using isomers as messenger molecules for molecular communication via diffusion. In: 2012 IEEE international conference on communications (ICC). IEEE
23. Hsieh Y-P, Shih P-J, Lee Y-C, Yeh P-C, Chen K-C (2012) An asynchronous communication scheme for molecular communication. In: 2012 IEEE international conference on communications (ICC). IEEE
24. Atakan B, Galmes S, Akan OB (2012) Nanoscale communication with molecular arrays in nanonetworks. IEEE Trans NanoBiosci 11(2):149–160
25. Arjmandi H, Gohari A, Kenari MN, Bateni F (2013) Diffusion-based nanonetworking: a new modulation technique and performance analysis. IEEE Commun Lett 17(4):645–648

Modulation Techniques for Molecular Communication via Diffusion

H. Birkan Yilmaz, Na-Rae Kim and Chan-Byoung Chae

Abstract Molecular Communication via Diffusion (MCvD) is an effective and energy efficient method for transmitting information in nanonetworks. In this chapter, we focus on the modulation techniques in a diffusion-based communication system. We mainly assume the first hitting process for the reception of the signal and it affects the design of the modulation techniques. As observed in the nature, whenever an information carrying molecule hits to the receiver it is removed from the environment. These information molecules are called messenger molecules and can be of many types of chemical compounds such as DNA fragments, proteins, peptides or specifically formed molecules. Information is modulated on one or more physical properties of these molecules or the release timing. In this chapter, we mention four novel modulation techniques, i.e., concentration, frequency, molecular-type, and timing-based modulations for MCvD in a single transmitter and single receiver environment. We also exemplify a systematic realization for molecular-ratio-based modulation using isomers as messenger molecules for MCvD. Next, we compare the pros and cons of the modulation techniques for an absorbing receiver that are studied in the literature. Knowing the workings and the properties of these modulation techniques enables us to use them in combination whenever it is possible.

H.B. Yilmaz
Yonsei Institute of Convergence Technology, Yonsei University, Seoul, Korea
e-mail: birkan.yilmaz@yonsei.ac.kr

N.-R. Kim · C.-B. Chae (✉)
School of Integrated Technology, Yonsei University, Seoul, Korea
e-mail: cbchae@yonsei.ac.kr

N.-R. Kim
e-mail: nrkim@yonsei.ac.kr

J. Suzuki et al. (eds.), *Modeling, Methodologies and Tools for Molecular and Nano-scale Communications*, Modeling and Optimization in Science and Technologies 9, DOI 10.1007/978-3-319-50688-3_5

1 Introduction

Operating in the nano-scale is expected to require high cooperation among multiple devices to make an impact on the macro scale. Nanonetworking is to interconnect several nanoscale machines that has molecular communication as one of the most promising mechanisms [1, 9, 20, 21]. Molecular Communication via Diffusion (MCvD) which is a new paradigm of nano communication networks enables transmitting desired information using biological molecules. On its most basic definition, MCvD deals with communication systems between transmitter-receiver couples; exchanging information via a shared medium using molecules that represent the data. MCvD absent of the need for infrastructural deployment, can be utilized for both intra- and inter-cellular communication. The molecules, called messenger/information molecules, function both as a information and a carrier. These molecules can be of many types of chemical compounds such as DNA fragments, proteins, peptides or specifically formed molecules. Generally, the information is modulated on any physical property or the release timing of the messenger molecules with certain method, which physically travel to a receiver side. Therefore, the modulation techniques are required to properly modulate the desired information.

In this chapter, we mention four novel modulation techniques, i.e., concentration, frequency, molecular-type, and timing-based modulations for MCvD in a single transmitter and single receiver environment [14, 16, 24]. We also investigate molecular-ratio-based modulation using isomers as messenger molecules for MCvD [11]. There is a literature proposing modulation techniques using concentration and types of messenger molecules only conceptually using hydrocarbon molecules. On the other hand, isomers are molecules composed of the same number and types of atoms, and have advantages in terms of complexity and systematic analysis that can be applied into practical systems.

Next, we compare the pros and cons of the aforementioned modulation techniques. For example, the molecular-type-based modulation would have higher transmitter complexity since the transmitter synthesizes multiple types of molecules while only one for the concentration-based method. Nevertheless, it can be more robust with distortion materials in propagating medium than other techniques. There exists modulation techniques those are mutually complementary to each other, the combination of several methods can be chosen depending on system conditions. It is also possible to analyze the achievable data rate performances in specific scenarios.

To make nano communication feasible in practice, proper modulation techniques are one of the important issues to be investigated further. Thus, this chapter will cover the previous work regarding a variety of modulation techniques, and even analyze the system with more complicated and practical situations.

2 Molecular Communication via Diffusion

The messenger molecules are the information particles in molecular scale. In this scale, the movement of particles inside a fluid is modeled by Brownian motion or diffusion process. The motion is governed by the combined forces applied to the messenger molecule by the molecules of the liquid due to thermal energy. In nature, whenever a messenger molecule's body coincides with the body of the receiver, the molecule is received and removed from the environment. Therefore, after that point that molecule cannot move further and constitutes the signal just once. This process is named first passage or hitting process and we are concerned with the probability that a diffusing particle first reaches a specified site or sites at a specified time [23]. If we are dealing with first hitting process, considering the concentration formulation, $C(r, t)$, is not the correct way of finding the hitting time histogram. Using $C(r, t)$ as the channel transfer function, implicitly assumes that the receiver node does not affect the MCvD system, which means molecules are freely moving in the environment and inside the receiver, also passing through the receiver boundary without any resistance. Hence, using first hitting process is more realistic compared to just using propagation process. Some works in the literature consider first hitting and derive channel functions which are inverse Gaussian [6, 24, 25].

The MCvD system for distance d is depicted in Fig. 1, where the radii of transmitter and receiver nodes are denoted by r_{TN} and r_{RN}, respectively. In this communication system, information is sent using a sequence of symbols which are spread over sequential time slots with one symbol in each slot. The symbol sent by the transmitter is called the "intended symbol", and the demodulated symbol at the receiver is called the "received symbol". An MCvD system has five main processes: encoding, emission, propagation, absorption, and decoding [13]. We mainly focus on emission, propagation, and absorption processes in this chapter.

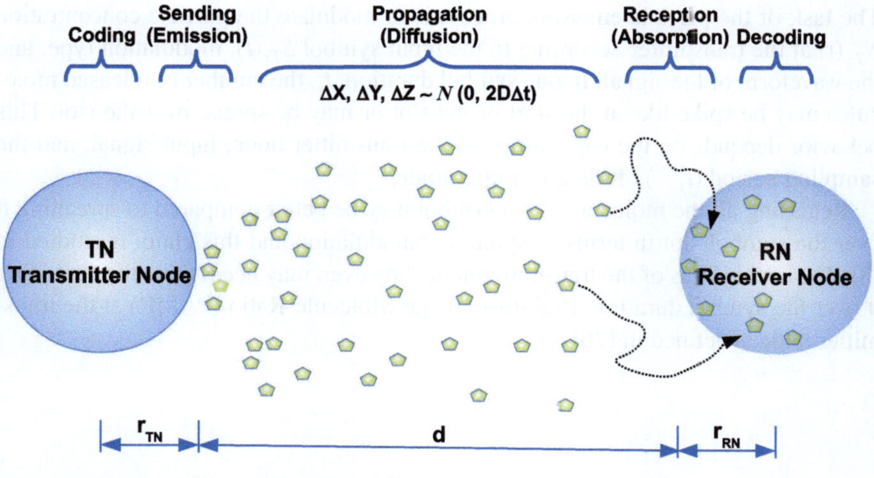

Fig. 1 MCvD system model and processes in a 3-D environment

2.1 Propagation Process

In an n-dimensional space, the total displacement, $\mathbf{r}$ in one time step can be found as

$$\mathbf{r} = (\Delta x_1, \ldots, \Delta x_n) \tag{1}$$

where Δx_i is the displacement at the ith dimension. Movement at each dimension in one time step is modeled independently and follows a Gaussian distribution

$$\Delta x_i \sim \mathcal{N}(0, 2D\Delta t) \quad \forall i \in \{1, \ldots, n\} \tag{2}$$

where Δt and D are the step time and the diffusion coefficient, respectively. Molecules propagate in the environment according to these dynamics. We model the transmitter, the receiver and the messenger molecules as spherical bodies. Our model ignores collisions between the messenger molecules, as done in the literature for simplicity [19]. If the transmitter node is not a point source, then it is a reflecting boundary for messenger molecules, that makes the closed form solution more harder since it changes the symmetry for the solution. Therefore, we simulate the first hitting process by moving messenger molecules according to (2) and reflecting or removing the particles that hit to transmitter or receiver, respectively. While wandering in the environment, a single molecule has a certain hitting probability at the receiver.

2.2 Emission and Reception Processes

The task of the particle emission process is to modulate the particle concentration $N_{Tx}(t)$ at the transmitter according to the input symbol $S_{Tx}(t)$, modulation type, and the waveform of the signal. In one symbol duration, t_s, the number of released molecules may be spike like at the start of the slot or may be spread over the slot. This behavior depends on the capabilities of the transmitter node, input signal, and the sampling period (t_s^{smp}) of the transmitter node.

Releasing all the molecules for a symbol may be better compared to spreading it over the symbol slot in terms of symbol demodulation and this claim is studied in [8]. The capabilities of the transmitter node, however, may necessitate the spreading it over the symbol duration. Peak-to-Average-Molecule-Ratio (PAMR) at the transmitter node is defined in [26] as

$$\mathrm{PAMR}_{Tx} = \frac{\max N_{Tx}(t)}{\mathrm{avg}\, N_{Tx}(t)}. \tag{3}$$

PAMR is a similar concept to Peak-to-Average-Power-Ratio (PAPR) in an OFDM system. If the transmitter node has less space for storing the synthesized messenger molecules, it would necessitate sending the synthesized messenger molecules before the storage areas are full. Hence, depending on the capabilities of the transmitter node, sending one peak at the start of the symbol duration may be possible or not. Having high PAMR value may violate the transmitter node constraints due to capabilities. Similarly, PAMR at the receiver side can be defined and determines the capabilities/constraints of the RN.

On the other hand, whenever a messenger molecule hits the receiver it is directly absorbed and removed from the environment. Demodulation takes place during the reception process. Demodulating the information is achieved by detecting and classifying one or more physical properties of the arriving messenger molecules on which the information is modulated.

Since the molecules propagating through the environment exhibit Brownian motion, a single molecule has a certain hitting probability at the receiver. This probability, $P_{hit}(d, t_s)$, depends on the distance between the transmitter and the receiver, and the symbol duration. Modeling the arrival process as a binomial distribution gives us the model for the received number of molecules, N_{Rx}.

$$N_{Rx} \sim B(N_{Tx}, P_{hit}(d, t_s)) \tag{4}$$

In (4), $B(n, p)$ stands for binomial distribution with parameters n and p. Instead of binomial modeling Gaussian approximation can also be used when N_{Tx} is large enough. Poisson modeling, however, can only be considered when $P_{hit}(d, t_s)$ is very small. Poisson modeling is better than Gaussian modeling when the number of arriving molecules is small enough that Gaussian model's left tail causes error.

3 Modulation Techniques for MCvD

Symbols can be modulated over various "messenger molecule emission properties" at the sender, e.g., concentration, frequency, phase, molecule type, to form a signal. Some of the modulation techniques utilizes carrier waves, however, some do not. Using carrier waves makes the handling of Inter Symbol Interference (ISI) more complex. In this section, we classify four main modulation techniques and elaborate the details of them. Most of the modulation techniques mentioned in this section are studied thoroughly in [12, 26] and some of the results are given to supplement the theory.

3.1 Amplitude-Based Modulations

The concentration of the sent/received messenger molecules is used as the amplitude of the signal in Concentration Shift Keying (CSK), hence it represents data as variations in the amplitude of a carrier wave or pulse. CSK is analogous to Amplitude Shift Keying (ASK) in classical communication. Representing b bits requires 2^b different amplitudes. If the carrier wave is a single point pulse and we use presence and absence of the carrier wave to indicate binary "1" and "0", respectively, it is called Molecular Concentration On-off Keying (MC-OOK) which is a special type of Binary CSK (BCSK). BCSK symbols may also be modulated on a square wave. If there are four amplitude levels for two bit symbols then it is called Quadrature CSK (QCSK). These modulations are depicted in Fig. 2. In all cases, one of the symbols is represented by no emission for reducing the energy consumption and the mean energy consumption is same for all modulations in Fig. 2. Therefore, in MC-OOK that uses a pulse, the transmitter node emits much more molecules just at the start of the symbol slot. MC-OOK requirement in terms of PAMR_{Tx} is higher compared to other modulations using square or cosine wave as a carrier. MC-OOK is imple-

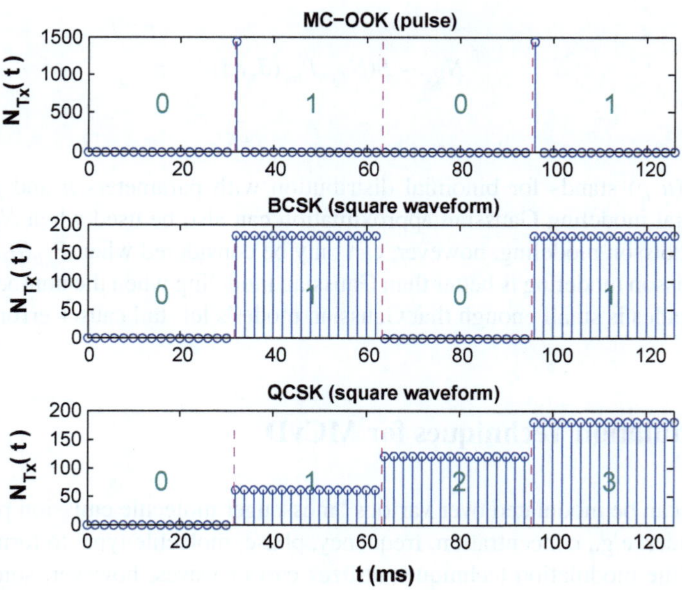

Fig. 2 $N_{Tx}(t)$ for amplitude-based modulations

mented for the tabletop testbed that is introduced in [5]. Because communication is performed through chemical signals, and a limited amount of signaling chemical can be stored in a container at the receiver, the modulation and demodulation scheme should minimize the amount of chemical used. Therefore, authors used zero emission for symbol "0" to reduce the energy cost as mentioned before.

If the data is modulated on the amplitude variations, a thresholding is employed for detecting symbols. For example, demodulating MC-OOK or BCSK symbols requires one threshold. The receiver decodes the intended symbol as "1" if the number of messenger molecules arriving at the receiver during a time slot exceeds a threshold τ, "0" otherwise. The MCvD system using CSK technique can be affected adversely from ISI which can be caused by the surplus molecules from previous symbols. Due to diffusion dynamics, some messenger molecules may arrive after their intended time slot. These molecules may cause the receiver to decode the next intended symbol incorrectly. It is shown in [14] that in the MCvD system, only the previous symbol has a significant ISI effect over the current symbol if the symbol duration is selected appropriately. The severity of this ISI related error depends on the symbol duration, distance, and diffusion channel parameters [15].

In [22], authors proposed Zebra-CSK to overcome ISI problem via utilizing inhibitor molecules. The proposed Zebra-CSK selectively suppresses ISI causing molecules. Numerical results show that Zebra-CSK not only enhances capacity of the molecular channel but also reduces symbol error probability observed at the receiver nanomachine. In [10], authors proposed four methods for a receiver in the molecular communication to recover the transmitted information distorted by both ISI and noise. They introduce sequence detection methods based on maximum a posteriori (MAP) and maximum likelihood (ML) criterions, a linear equalizer based on minimum mean-square error (MMSE) criterion, and a decision-feedback equalizer (DFE) which is a nonlinear equalizer. The results show that using these methods significantly increases the channel capacity in the molecular communication.

3.2 *Molecule Type-Based Modulations*

Molecule type is used to represent information in Molecular Shift Keying (MoSK). Representing b bits requires 2^b different molecules with similar properties. The transmitter releases one of these molecules based on the current intended symbol. The receiver decodes the intended symbol based on the type and the concentration of the molecule received during a time slot. If the concentration of only one molecule type exceeds the threshold τ at the receiver, the symbol is decoded based on the bit sequence corresponding to this molecule type. Another decoding strategy may decide the intended symbol depending on the majority logic. For the first strategy, an error is assumed if the concentration of any molecule types does not exceed the threshold or the concentration of more than one molecule type exceeds the threshold.

Inspired by [7], hydrofluorocarbons can be used as the messenger molecule structure for systematically designing 2^b different molecules for b bit logical information

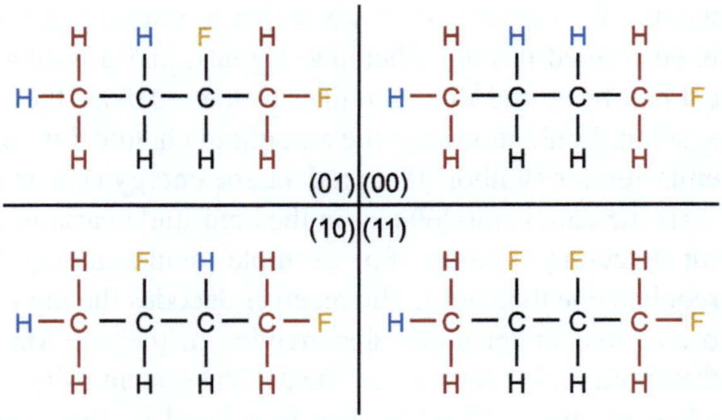

Fig. 3 Constellation of QMoSK using hydrofluorocarbon based messenger molecules

representation. Based on the message to be transmitted, a special messenger molecule is synthesized using three parts: header, trailer, and the chemical bit element. A single header and a single trailer are present in each molecule representing the start and the end of the message. For each bit of information, a chemical bit element is synthesized. A chemical bit element has two forms: one for representing "0" and another one for "1". All of these parts are linked to each other using chemical bonds to form a single messenger molecule. In Fig. 3, we depict a 2-bit constellation realization of this modulation technique called Quadruple MoSK (QMoSK). These molecules are flammable, hence the usage areas may be limited. This constellation is just a hypothetical one and can be done via more appropriate messenger molecules such as isomers [11].

Similar to the CSK technique, the surplus molecules from the previous symbols also cause ISI when MoSK technique is used. However, MoSK is less susceptible to ISI effects than the CSK technique when the bits per symbol is greater than 1. In this case, a single threshold is used for MoSK whereas $2^b - 1$ thresholds are required for CSK. This advantage of the MoSK technique, however, comes at the cost of the requirement for complex molecular mechanisms at both the transmitter and the receiver for messenger molecule synthesis and decoding purposes, respectively. Also, a corruption in such a messenger molecule may cause some or all of the information in the symbol to get lost. This information corruption must be taken into account, since a corruption may still represent some information albeit not the one sent by the transmitter.

In [3], authors combined MoSK and CSK to overcome the ISI problem of CSK. The proposed scheme utilizes two types of molecules and uses them in an alternating fashion. Therefore, ISI on the CSK modulated symbols is reduced significantly. This is due to the fact that the decoding of the current symbol in the proposed scheme does not encounter propagation of error, as the decoding of the current symbol does not depend on the previously transmitted and decoded symbols.

In order to evaluate the performance of different modulation techniques, we consider a channel model as depicted in Fig. 4. Channel model for binary and quadruple modulations are depicted with correct and incorrect demodulations. Modeling

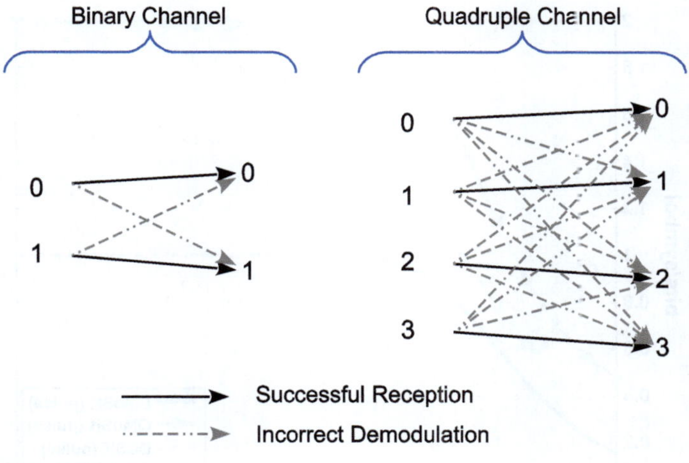

Fig. 4 Channel models

arrival process as in (4) enables us to derive the successful reception and incorrect demodulation probabilities.

Using these probabilities, we can find the mutual information, $I(X;Y)$, given the values for thresholds, distance, and the probability of hit during the current and next symbol durations [2]. By selecting ideal threshold values, the channel capacity (C) can be calculated using the maximum of mutual information as in (5), where b stands for the number of bits per symbol.

$$\begin{aligned} C &= \max_{\tau} I(X,Y) \\ &= \max_{\tau} \sum_{Y=0}^{2^b-1} \sum_{X=0}^{2^b-1} \mathbf{P}_{X,Y}(x,y) \log_2 \frac{\mathbf{P}_{X,Y}(x,y)}{\mathbf{P}_X(x)\mathbf{P}_Y(y)} \end{aligned} \tag{5}$$

Using $P_{hit}(d, t_s)$, arrival model, and channel model, the channel capacities of BCSK, BMoSK, QCSK, and QMoSK implementations with pulse carriers under various Signal-to-Noise Ratio (SNR) values are evaluated. SNR is defined as the square of the reciprocal of the coefficient of variation of the received signal. According to the results given in Fig. 5, all modulation techniques attain their theoretical channel capacity limits when the SNR level is high. As SNR decreases, in case of the binary implementations, BCSK with pulse carrier (a.k.a. MC-OOK) offers more robustness compared to BMoSK. Same amount of noise is applied to both molecule types in BMoSK, since the noise in the channel is AWGN. Thus, BMoSK is more affected by the noise than BCSK.

In case of quadruple implementations (QCSK and QMoSK), this trend changes and QMoSK exhibits higher noise tolerance than QCSK. This behavior is due to the number of threshold values used in these quadruple implementations. Since QMoSK uses a single threshold value, the channel capacity can be kept high by choosing a

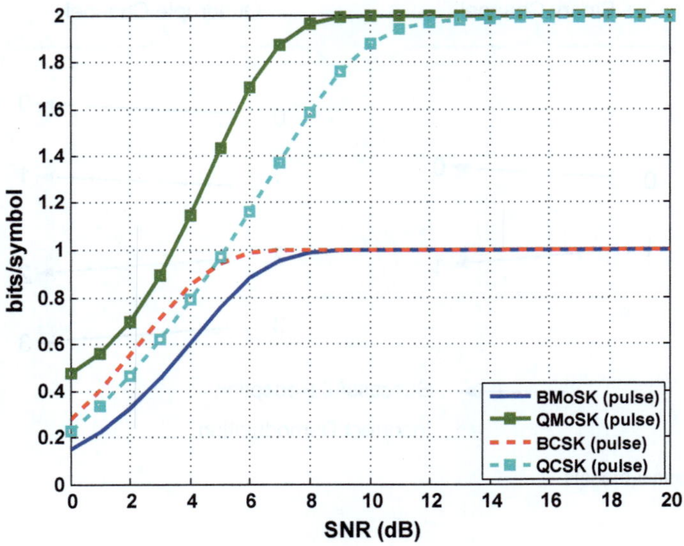

Fig. 5 Channel capacity of different modulation techniques ($N_{Tx} = 1500$ molecules, $D = 79.4\,\mu\text{m}^2/\text{s}$, $d = 16\,\mu\text{m}$, $t_s = 5.9\,\text{s}$, $r_{TN} = r_{RN} = 10\,\mu\text{m}$)

suitable threshold value even when the noise level is high. On the other hand, finding suitable threshold levels to keep the same channel capacity becomes harder as the noise level increases in QCSK.

3.3 Frequency-Based Modulations

Representing the data through discrete frequency changes of a carrier wave of number of messenger molecules is called Molecular Frequency Shift Keying (MFSK). In this modulation scheme, number of molecules sent is not constant during the symbol duration and changes according to cosine wave. Since we cannot send negative number of molecules, we have to shift the amplitude in y-axis up.

If the carrier wave is a cosine wave and there are two frequencies, then it is called Binary MFSK (BMFSK). Similarly, if there are four frequencies for two bit symbols, then it is called Quadrature MFSK (QMFSK). These modulations are illustrated in Fig. 6.

General formula for $N_{Tx}^{s_i}(t)$ with MFSK modulations on cosine waveform is as follows

$$N_{Tx}^{s_i}(t) = \mu_a + \mu_a * \cos(2\pi f_i t) \quad i = 0, \ldots, m \tag{6}$$

where s_i, f_i and μ_a are symbol i, frequency for s_i, and mean amplitude, respectively. If the data is modulated on the frequency variations, the received symbol at the receiver

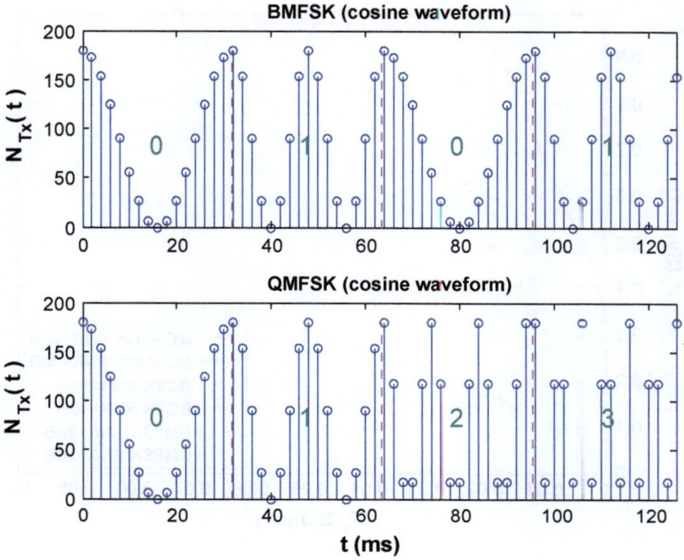

Fig. 6 $N_{Tx}(t)$ for frequency-based modulations

side can be detected via correlating the received signal with the symbol waveforms after synchronizing the signals. You can also look at the received signal in the frequency domain and correlate to the symbol frequency signatures. The MCvD system using MFSK-based modulation is affected less from the ISI, however, CSK-based symbols are easier to detect.

For analyzing symbol detection and false alarm probabilities we compare Receiver Operating Characteristics (ROC) curves. In Fig. 7, ROC curves for symbol "1" of binary modulations are depicted for 3 and 9 dB SNR values. $P_d(\text{Sym} = 1)$ represents the probability of demodulating the received symbol as "1" when the transmitted symbol is actually the symbol "1". Similarly, $P_f(\text{Sym} = 1)$ corresponds to the probability of demodulating the received symbol as "1" when the transmitted symbol is actually the symbol "0", hence it can be seen as false alarm of saying demodulated result is the symbol "1". In Fig. 7, MC-OOK stands for BCSK modulated on pulse at the start of a symbol slot and BCSK stands for BCSK modulated on square wave. Energies spent at a time slot are equal, hence PAMR_{Tx} value of MC-OOK is higher than other modulation techniques.

In Fig. 7, for a given false alarm constraint, achievable detection probability can be compared. For $P_f(\text{Sym} = 1) = 0.05$, we have nearly 0.3, 0.98, and 1 as a detection probability for symbol "1" for 3 dB case for BMFSK, BCSK, and MC-OOK, respectively. Increasing SNR results in better detection probability. Hence, with the same false alarm constraint we have 0.72, 1, and 1 as a detection probability for symbol "1" for 9 dB case for the same order of modulations. MC-OOK has the best performance

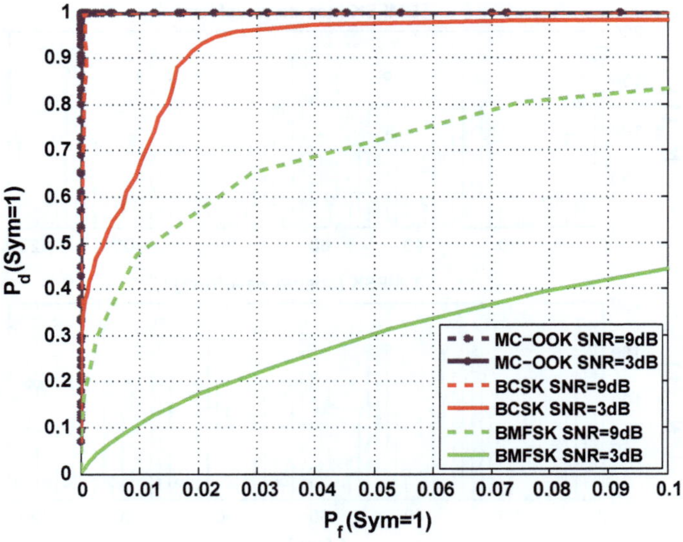

Fig. 7 ROC curves for symbol "1" of binary modulations ($\mu_a = 90$ molecules, $D = 79.4\,\mu\text{m}^2/\text{s}$, $d = 1\,\mu\text{m}$, $t_s = 0.032\,\text{s}$, $t_s^{smp} = 0.002\,\text{s}$, $r_{\text{TN}} = r_{\text{RN}} = 10\,\mu\text{m}$)

in terms of $P_d(\text{Sym} = 1)$ with given $P_f(\text{Sym} = 1)$ constraint. BCSK has comparable performance for 9 dB SNR while 3 dB case can also be satisfactory for some cases.

In Fig. 8, Symbol Error Rate (SER) plots of amplitude and frequency based modulations without ISI filtering are depicted. Without ISI filtering, even BCSK with a square carrier has an error floor. In high SNR regime, BMFSK performs better than BCSK. MC-OOK performs better than all of the modulations compared.

In Fig. 9, SER plots of amplitude and frequency based modulations with ISI filtering are depicted. ISI filter increases the performance of amplitude-based modulations significantly. BCSK with a square carrier wave has an error floor due to ISI and it is eliminated effectively when an ISI filter is applied. MC-OOK has the best performance compared to the BCSK modulated on a square wave and BMFSK modulated on a cosine wave. MC-OOK, however, has the highest PAMR_{Tx} value (32 times higher compared to other modulations for given parameters) while the BCSK, when ISI filter is applied, has an acceptable SER and PAMR_{Tx} value.

In Table 1, we summarize the performances and the capabilities of the modulation schemes. In terms of SER, MC-OOK performs better than the other schemes, however, in terms of PAMR_{Tx} it is the far worse than the others. These results and the summary table can help to the system designers in nanonetworking domain for research and implementation issues.

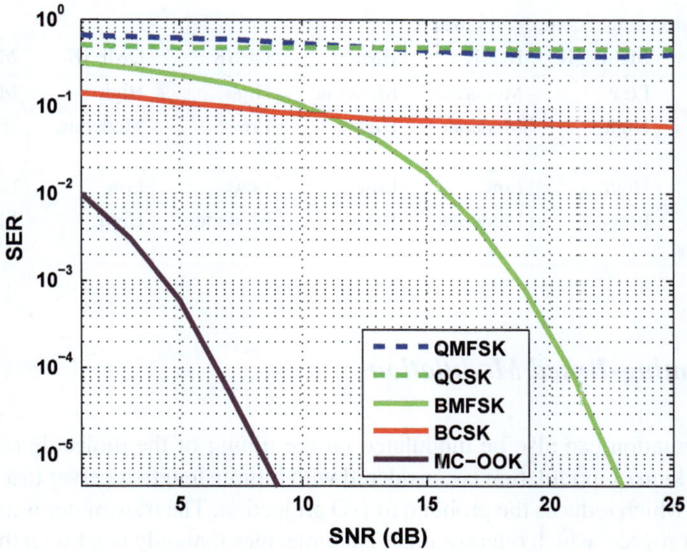

Fig. 8 SER plots of modulation techniques without ISI filtering ($\mu_a = 90$ molecules, $D = 79.4\,\mu\text{m}^2/\text{s}$, $d = 1\,\mu\text{m}$, $t_s = 0.032\,\text{s}$, $t_s^{smp} = 0.002\,\text{s}$, $r_{\text{TN}} = r_{\text{RN}} = 10\,\mu\text{m}$)

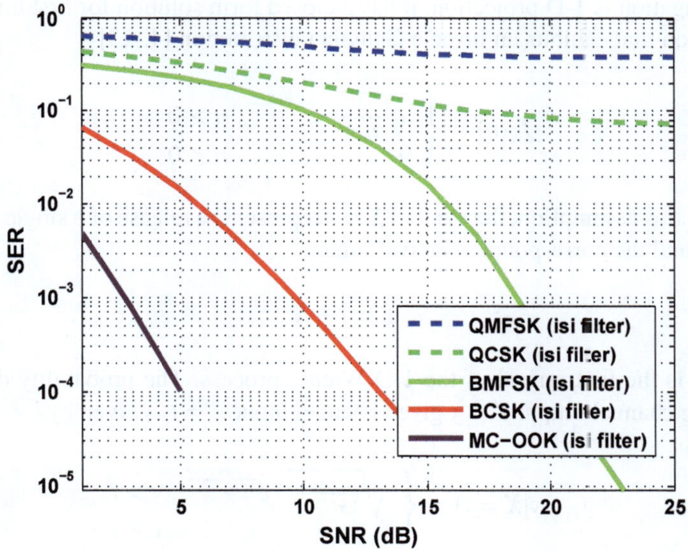

Fig. 9 SER plots of modulation techniques with ISI filtering ($\mu_a = 90$ molecules, $D = 79.4\,\mu\text{m}^2/\text{s}$, $d = 1\,\mu\text{m}$, $t_s = 0.032\,\text{s}$, $t_s^{smp} = 0.002\,\text{s}$, $r_{\text{TN}} = r_{\text{RN}} = 10\,\mu\text{m}$)

Table 1 Modulation comparison matrix

Metric	MC-OOK	BCSK	BMFSK	QCSK	QMFSK	MoSK
SER	Low	Moderate	Moderate	High	High	Moderate
ISI resilience	High	Moderate	High	Low	Moderate	High
$PAMR_{Tx}$	High	Low	Low	Low	Low	Low
Node complexity	Low	Moderate	High	Moderate	High	Moderate

3.4 Timing-Based Modulation

The information can also be modulated on the timing of the molecule releases. In [24], this kind of modulation is considered with infinite length receiver in a 2-D environment, which reduces the problem to 1-D projection. The transmitter is assumed to be a point source, which releases identical molecules that only react with the absorbing boundary.

The transmitter modulates the information on the time of release. The transmitter does not affect the propagation of the molecules once they are released. Since the propagation is 1-D projection, it has a closed form solution for first hitting time distribution that exhibits an inverse Gaussian distribution [4, 23]

$$f_N(n) = \begin{cases} \sqrt{\frac{\lambda}{2\pi n^3}}\ e^{-\frac{\lambda(n-\mu)^2}{2\mu^2 n}}, & n > 0 \\ 0, & n \leq 0 \end{cases} \tag{7}$$

Under this assumption, the symbol $X = t$ represents a release of a single molecule at time t and the corresponding arrival time is

$$Y = t + N \tag{8}$$

where N is the first arrival of the 1-D Wiener process. The probability density of observing channel output $Y = y$ given channel input $X = t$ is given by

$$f_{Y|X}(y|X=t) = \begin{cases} \sqrt{\frac{\lambda}{2\pi(y-t)^3}}\ e^{-\frac{\lambda(y-t-\mu)^2}{2\mu^2(y-t)}}, & y > t \\ 0, & y \leq t \end{cases} \tag{9}$$

As an example of an MCvD system using timing modulation of m symbols, the message alphabet would be $\chi = \{t_1, t_2, \ldots, t_m\}$. The receiver computes an estimate $\hat{X}$ of the transmitted message from Y. The transmission is successful when $\hat{X} = t_i$ and the transmitted message is t_i.

Considering these distributions and maximum likelihood estimator as a receiver yields;

$$\hat{X}_{ML} = y + \frac{\mu^2}{\lambda}\left(\frac{3}{2} - \sqrt{\frac{9}{4} + \frac{\lambda^2}{\mu^2}}\right) \tag{10}$$

where λ and μ are the parameters of the inverse Gaussian distribution. This estimation at the receiver side is valid for infinite length symbol duration. Repeated channel use can also be considered by taking ISI into account. Amplitude-based modulations with appropriate symbol duration can easily be coupled with the timing-based modulations.

3.5 *Realization with Isomers*

In Sect. 3.2, hydrocarbons are used as messenger molecules only to conceptually explain the MoSK technique. The molecules, however, are known to be highly flammable, which makes them less feasible in practical systems, especially for in-vivo applications. Therefore, it is another important thing to be considered in molecular communication systems to choose the proper type of messenger molecules. The authors in [11] propose three isomer-based modulation techniques to use in practical applications, i.e., isomer-based concentration shift keying (ICSK), isomer-based molecule shift keying (IMoSK), and isomer-based ratio shift keying (IRSK).

An important thing when designing messenger molecules is that they have to be non-toxic to the human body. For several reasons, isomers could be potential candidates for messenger molecules which are composed of the same number and types of atoms [17]. First of all, this method can reduce the burden to the transmitter nanomachine that synthesizes messenger molecules since isomers consist of the same type of atoms. Moreover, they have the same physical properties that makes more systematic analysis. Especially in MoSK technique, it is possible to apply the same diffusion coefficient value in spite of using different messenger molecules (or isomers) to represent different information. Here, aldohexoses (i.e., aldose forms of hexoses) are mostly deployed for numerical analysis. Note that other aldoses family (e.g., pentoses, tetroses, or trioses) can also be utilized depending on the required modulation order.

Aldohexoses are monosaccharides that have the chemical formula of $C_6H_{12}O_6$. They have four chiral (not superposable on the mirror images) carbon atoms, which give them 16 (=2^4) stereoisomers [18]. Figure 10 illustrates eight kinds of D-form aldohexeses, and enantiomers (mirror images as shown in Fig. 11) of each molecule compose L-form aldohexoses. Therefore, aldohexoses have 16 different structures. When each isomer is dissolved in an aqueous solution, however, each D- and L-type aldohexoses change the structure into two kinds of cyclic molecules. These are named as α- and β-forms as described in Fig. 12. Thus, by deploying hexoses group into the system, we consider 32 different isomers in total, and three modulation techniques can be used using them as described in Figs. 13, 14 and 15.

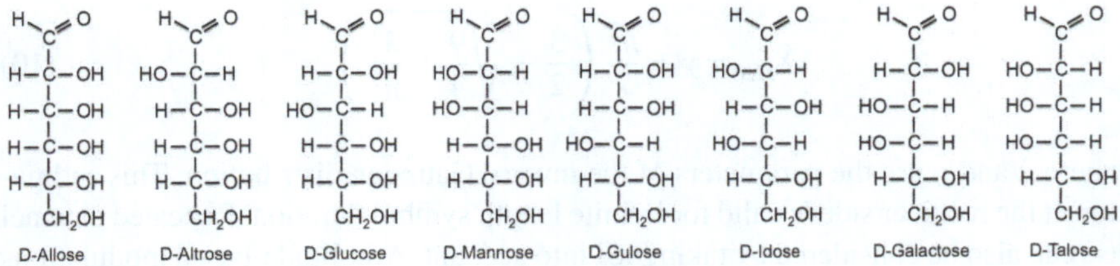

Fig. 10 D-form isomers of aldohexoses

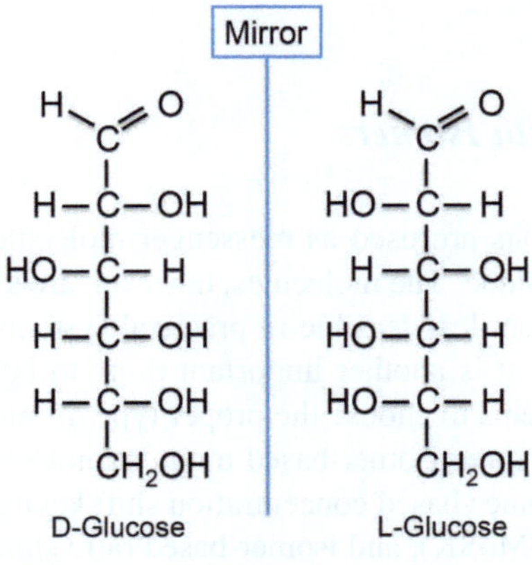

Fig. 11 L- and D-form of glucose that are enantiomers

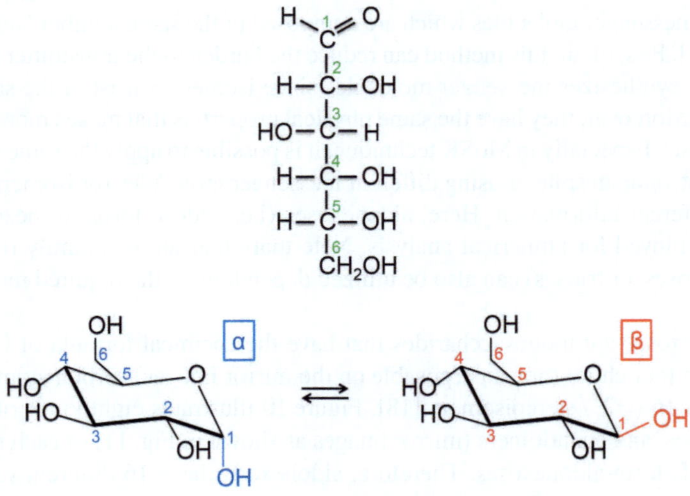

Fig. 12 D-glucose and its α- and β-form isomers in an aqueous solution

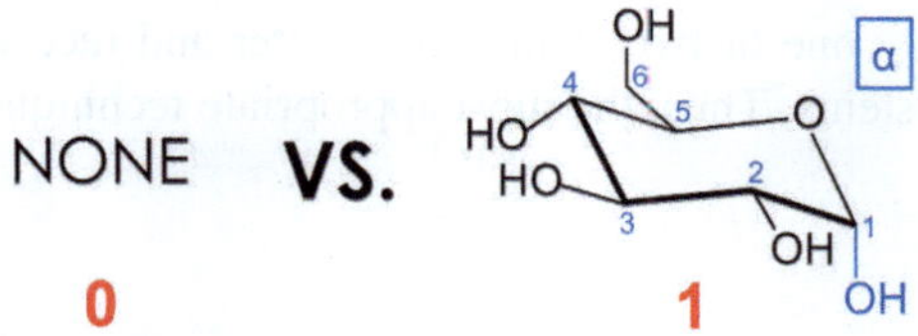

Fig. 13 Conceptual figure of ICSK

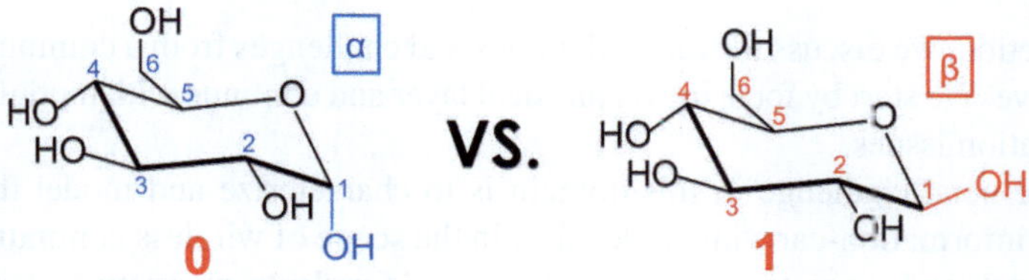

Fig. 14 Conceptual figure of IMoSK

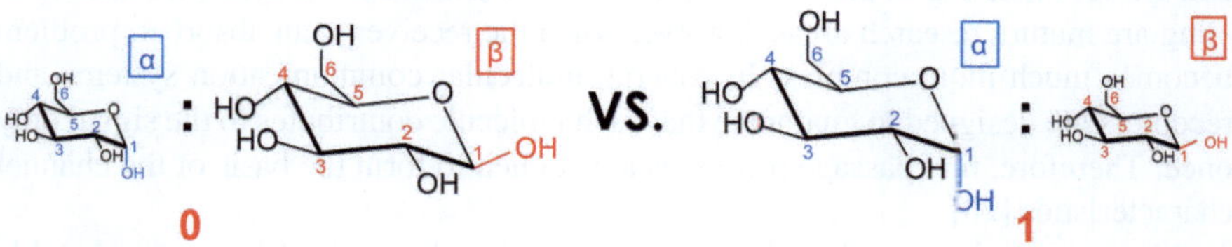

Fig. 15 Conceptual figure of IRSK

First, for realization of concentration-based modulation using isomers, or ICSK, one of the hexose family can be utilized with different concentrations, and basically, the concept is the same as CSK technique. Next, for IMoSK, we can choose one among the several aforementioned isomer sets (e.g., hexoses, pentoses, tetroses, or trioses); a modulation order can be determined by the sets used. For instance, hexoses have a modulation order of up to 32, and trioses of 4 (i.e., we consider 4 isomers for trioses). It has an advantage over MoSK in terms of system complexity since only one diffusion coefficient value is used. Lastly, a new type of modulation technique is proposed using isomers, i.e., ratio-based modulation or IRSK, which encodes information based on the ratios of two isomers. It is basically similar to ICSK except that IRSK uses two, or more, types of isomers. Thus, it can be more robust when the medium has some distortion molecules that have a negative effect to messenger molecules.

Each modulation has pros and cons, and one can be chosen considering system conditions. For example, achievable rates for IMoSK systems can be better for IMoSK systems unless it is binary case. This is because ICSK and IRSK systems have more thresholds, which means shorter minimum distance (i.e., concentration difference) between symbols. The IMoSK system, however, has to generate and detect multiple kinds of messenger molecules whereas ICSK and IRSK systems need to generate only one or two. Thus, transmitter and receiver complexities are higher in IMoSK systems. Thus, the most appropriate technique can be found with different situations.

4 Research Challenges

In this section, we discuss the research issues and challenges from a communications perspective. We start by focusing on physical layer and continue with modulation and demodulation issues.

Most crucial challenge in this domain is to characterize and model the propagation of information-carrying molecules. In the scope of wireless communications, many models and characterizations are developed for electromagnetic waves. To predict and understand the channel response, it is necessary to develop the propagation and arrival modeling in nanonetworking domain. The diffusion dynamics and modeling are mature research areas, however when the receiver is an absorber, problem becomes much more complex. In general, molecular communication systems and receptors are designed to guarantee that each molecule contributes to the signal only once. Therefore, first passage processes are studied to form the basis of the channel characteristics [23].

After physical channel problems are studied and practiced, more predictable and robust modulation techniques can be proposed and implemented. Some of the modulation techniques are introduced in this chapter, however there are unrealistic assumptions including perfect modulation, demodulation, and synchronization. Nanonetworking-enabled nodes are assumed to be capable of synthesizing the required molecules and to be able to store as much as required. There may be constraints on the transmitter receiver pairs for more realistic scenarios [26]. These assumptions should be relaxed in the future in order to make these modulation schemes more feasible.

Another issue in molecular communication domain is the interference. Sources of interference may differ, however due to channel characteristics, ISI is at a crucial point. Therefore, ISI mitigation techniques compatible with nanonetworking-enabled nodes should be designed and implemented. Arrival of diffusing particles spans a huge duration, however on the other hand we are trying to shorten the symbol duration. Therefore, ISI becomes an important problem to be solved. Considering the simplicity and the expected outcomes of the nanomachines, one can deduce that these nonomachines should cooperate and communicate. Their collaborative actions are expected, therefore multi user interference should also be considered.

Other research challenges for modulation issues can be listed as; channel modeling, formulating channel capacity, hardware and device designs, bio-compatibility for in vivo applications, and analyzing the effects of receptor heterogeneity.

5 Conclusions

In this chapter, four types of modulation techniques are introduced for MCvD systems. First, molecular concentration, or amplitude, can be utilized to make CSK or MC-OOK systems using one kind of messenger molecule. When two or more types of messenger molecules are available, MoSK systems are one of possible candidates that encodes information into the type of messenger molecules. Also, frequency of concentration wave can represent information which makes MFSK systems. On the other hand, timing-based method does not utilize the chemical or physical properties of messenger molecules. It encodes information into the release timing of messenger molecules at the transmitter side, instead.

Moreover, the modulation techniques can be realized using isomers as messenger molecules. Isomers have several advantages such as transmitter/receiver complexity and analysis complexity since they have the same physical properties with different structures. Different isomer sets can be chosen depending on the required modulation order or SNR. To sum up, all the techniques explained in this chapter have both advantages and disadvantages, and a proper method has to be applied based on the system conditions and requirements.

References

1. Akyildiz IF, Brunetti F, Blázquez C (2008) Nanonetworks: a new communication paradigm. Comput Netw 52(12):2260–2279
2. Alfano G, Miorandi D (2006) On information transmission among nanomachines. In: Proceedings of international conference on nano-networks and workshops, (NanoNet). IEEE, pp 1–5
3. Arjmandi H, Gohari A, Kenari M, Bateni F (2013) Diffusion-based nanonetworking: a new modulation technique and performance analysis. IEEE Commun Lett 17(4):645–648
4. Chhikara RS, Folks JL (1989) The inverse Gaussian distribution: theory, methodology, and applications, vol 95. CRC Press
5. Farsad N, Guo W, Eckford AW (2013) Tabletop molecular communication: text messages through chemical signals. PloS One 8(12):e82,935
6. Farsad N, Kim NR, Eckford AW, Chae CB (2014) Channel and noise models for nonlinear molecular communication systems. Accepted to IEEE J Sel Areas Commun
7. Freitas RA (1999) Nanomedicine, volume I: basic capabilities. Landes Bioscience Georgetown, TX
8. Garralda N, Llatser I, Cabellos-Aparicio A, Pierobon M (2011) Simulation-based evaluation of the diffusion-based physical channel in molecular nanonetworks. In: Proceedings of IEEE conference on computer communications workshops (INFOCOM WKSHPS). IEEE, pp 443–448
9. Hiyama S, Moritani Y, Suda T, Egashira R, Enomoto A, Moore M, Nakano T (2006) Molecular communication. J Inst Electr Inf Commun Eng 89(2):162
10. Kilinc D, Akan OB (2013) Receiver design for molecular communication. IEEE J Sel Areas Commun 31(12):705–714
11. Kim NR, Chae CB (2013) Novel modulation techniques using isomers as messenger molecules for nano communication networks via diffusion. IEEE J Sel Areas Commun 31(12):847–856
12. Kuran MS, Yilmaz HB, Tugcu T, Akyildiz IF (2011) Modulation techniques for communication via diffusion in nanonetworks. In: Proceedings of IEEE international conference on communications (ICC). IEEE, pp 1–5

13. Kuran MS, Yilmaz HB, Tugcu T, Akyildiz IF (2012) Interference effects on modulation techniques in diffusion based nanonetworks. Elsevier Nano Commun Netw 3(1):65–73
14. Kuran MS, Yilmaz HB, Tugcu T, Ozerman B (2010) Energy model for communication via diffusion in nanonetworks. Elsevier Nano Commun Netw 1(2):86–95
15. Lin WA, Lee YC, Yeh PC, Lee Ch (2012) Signal detection and ISI cancellation for quantity-based amplitude modulation in diffusion-based molecular communications. In: Proceedings of global communications conference (GLOBECOM). IEEE, pp 4362–4367
16. Mahfuz MU, Makrakis D, Mouftah HT (2010) On the characterization of binary concentration-encoded molecular communication in nanonetworks. Elsevier Nano Commun Netw 1(4):289–300
17. McKee T, Mcfee JR (2009) Biochemisty. Oxford University Press
18. McNaught AD, Wilkinson A (1997) Compendium of chemical terminology. IUPAC
19. Moore MJ, Suda T, Oiwa K (2009) Molecular communication: modeling noise effects on information rate. IEEE Trans NanoBiosci 8(2):169–180
20. Nakano T, Suda T, Koujin T, Haraguchi T, Hiraoka Y (2007) Molecular communication through gap junction channels: System design, experiments and modeling. In: Proceedings of IEEE bio-inspired models of network, information and computing systems (Bionetics), Budapest, Hungary, pp 139–146
21. Nakano T, Suda T, Moore M, Egashira R (2005) Molecular communication for nanomachines using intercellular calcium signalling. In: Proceedings of IEEE conference on nanotechnology, pp 478–481
22. Pudasaini S, Shin S, Kwak KS (2014) Robust modulation technique for diffusion-based molecular communications in nanonetworks. arXiv preprint arXiv:1401.3938
23. Redner S (2001) A guide to first-passage processes. Cambridge University Press
24. Srinivas KV, Eckford AW, Adve RS (2012) Molecular communication in fluid media: the additive inverse Gaussian noise channel. IEEE Trans. Inf. Theory 58(7):4678–4692
25. Yilmaz HB, Heren AC, Tugcu T, Chae CB (2014) Three-dimensional channel characteristics for molecular communications with an absorbing receiver. IEEE Commun. Lett. 18(6):929–932
26. Yilmaz HB, Kim NR, Chae CB (2014) Effect of ISI mitigation on modulation techniques in communication via diffusion. In: Proceedings of ACM international conference on nanoscale computing and communication (ACM NanoCom)

The Use of Coding and Protocols Within Molecular Communication Systems

Mark S. Leeson, Matthew D. Higgins, Chenyao Bai, Yi Lu, Xiayang Wang and Ruixiao Yu

Abstract This chapter focuses upon the use of coding and protocols within diffusion based molecular communication systems, laying the groundwork for future development in test bed implementations. The chapter starts with an introduction that briefly discusses coding and protocols used in traditional communication systems. Following this, details of the molecular channel are given, including the energy consumption constraints and a relevant mathematical framework. This discussion then leads onto potential encoding and decoding technologies. Next, original results on the use of Hamming codes in molecular communication systems are presented with a quantitative comparison against an uncoded molecular system. The impact of specific design parameters such as the number of molecules, energy, and transmission distance on the bit error rate (BER) is considered. Finally, a protocol, based upon the use of an acknowledgement (ACK) packet is presented as a further advancement to the field that the reader may wish to consider when designing future systems.

1 Introduction

Today, research on nano-communications can be broadly divided into the following domains [1]: nano-mechanical, acoustic, electromagnetic, and molecular. In the first of these, information exchange between transmitter and receiver is implemented through mechanical contact such as hard junctions between linked devices. In acoustic communication the transmitted information is encoded as ultrasonic waves. This method relies upon the ability to implement ultrasonic transducers which are capable of sensing the rapid variations of pressure and emitting acoustic signals [1] at the nano-scale. Electromagnetic communication transmits information

---

M.S. Leeson (✉) · C. Bai · X. Wang · R. Yu
School of Engineering, University of Warwick, Coventry CV4 7AL, UK
e-mail: mark.leeson@warwick.ac.uk

M.D. Higgins · Y. Lu
WMG, University of Warwick, Coventry CV4 7AL, UK

J. Suzuki et al. (eds.), *Modeling, Methodologies and Tools for Molecular and Nano-scale Communications*, Modeling and Optimization in Science and Technologies 9, DOI 10.1007/978-3-319-50688-3_6

through modulated electromagnetic waves which can propagate with minimal losses either along wires or through an air or fluidic medium. However, it is known to be a challenge to integrate the transceivers at the nano-scale due to the relatively complexity and furthermore, assuming that integration was possible, the output power of the nano-transceiver would be insufficient to guarantee a bidirectional communication channel. As a result, it has been postulated that electromagnetic communication could be used to transmit information from a micro-device to a nano-device, but not from nano-machines to micro-machines, nor among nano-machines [1]. This is still an open research area. Finally, in molecular communications the transmitted information is encoded using molecules. This new and interdisciplinary field combines nano-, bio-, and traditional communication technologies and concepts [2]. Thus, the performance of molecular communication systems is affected by the physical laws governing these fields. In molecular communication, the molecular transmitters are able to release specific molecules in response to an internal command. Similarly, the receivers have the ability to react to specific molecules and thus the local concentration of the molecules at the transmitter and the receiver may be used to understand the molecular bit transmitter sent. Generally, it is common to use existing biological nano-techniques, with design traits found in traditional communications systems to construct the molecular communication system and this idea has typically been proven successful. It is this mind-set that this chapter capitalises upon.

Error detection and correction techniques that enable the reliable delivery of information over unreliable channels are essentially ubiquitous to all communications systems found in our daily lives. Such coding techniques are generally based upon the addition of redundant bits into the source coding representations, followed by the transfer of these bits into the channel where noise is present. Upon reception, an appropriate decoding method is used whereby the redundant bits can be used to check whether or not the information is corrupted [3]. Depending upon the type of source coding, the error may be correctable and the original information appropriately restored. After a thorough analysis of the molecular channel in Sects. 2 and 3 of this chapter, it will be shown in Sect. 4 how this traditional element of communication theory, but more specifically, the use of Hamming codes, can be applied to the molecular channel.

In the same way that error correction techniques are commonplace in all modern systems, so too are a form of protocol. The most well-known examples are the Transmission Control Protocol (TCP) and Internet Protocol (IP), with their v4 and v6 incarnations [4]. In Sect. 5 of this chapter, the implementation of a relatively simple protocol, namely Stop and Wait Automatic reQuest (SW-ARQ) [5, 6] is described, which is also compatible with the work on error correction in Sect. 4.

The use of coding and protocol techniques is still in its infancy at the nano-scale with many outstanding challenges. For example, all the functionalities, including encoding and decoding, or the retransmission mechanics, require energy capabilities which are likely to be one of the limits of any nanomachine that is also constrained mechanically. Moreover, transmission and device speeds are extremely slow in comparison with conventional data networks and devices. In addition, whilst the conceptual design of the communication layer model can be inspired from existing models, the actual implementation of a full protocol stack is also dictated by the availability and capability of each synthetic components used to implement each layer of the stack.

This chapter aims to highlight to the reader how the use of both error correction and protocol can make a significant contribution to the design of any molecular communications system.

2 Analysis of the Diffusive Medium

Modelling molecular propagation in the channel medium is one of the key challenges in predicting molecular communication system performance [1]. The information molecules can be chosen to be proteins, protein complexes, peptides, DNA sequences or other molecular structures [7]. The motion of information molecules is governed by the forces produced by the constant random thermal motion of the molecules within the fluid. To simplify the communication system, the medium is considered to be of extremely large dimensions compared to the size of the information molecules. In addition, collisions between the information molecules are neglected. For this diffusion based molecular communication system, the transmitted information is represented by a sequence of symbols which are distributed over sequential and consecutive time slots with one symbol in each slot. The *intended symbol* refers to the symbol sent by the transmitter and the *received symbol* represents the symbol received by the receiver. The information is encoded by concentration with binary representation. Specifically, if the number of information molecules arriving at the receiver in a given time slot exceeds a threshold τ, the symbol is interpreted as a "1". Otherwise, it will be interpreted as a "0". However, errors may be caused by Intersymbol Interference (ISI), which is an unavoidable consequence of both wired and wireless communication systems and is known to have adverse effects in communication systems, particularly when the system is stochastic [8]. The ISI effect is related to the properties of the medium used, the distance of the symbol propagation and the selection of the threshold value. In the diffusion communication system here, some information molecules may arrive at the receiver after the current time slot according to the diffusion dynamics, which will lead to the possible incorrect decoding of the received symbol in the next time slot.

Here, a three dimensional (3D) diffusion based communication system is to be considered as shown in Fig. 1. The information molecule is at a distance r from the centre of the receiver which has a radius of $R = 5\ \mu\text{m}$ [7].

The diffusion coefficient is taken to be $= 79.4\ \mu\text{m}^2\text{s}^{-1}$, which is a conservative value for insulin in water at human body temperature [7]. The escape probability $S(r, t)$ in the 3D diffusion medium can be described with the following backward difference equation at a given time t [9]:

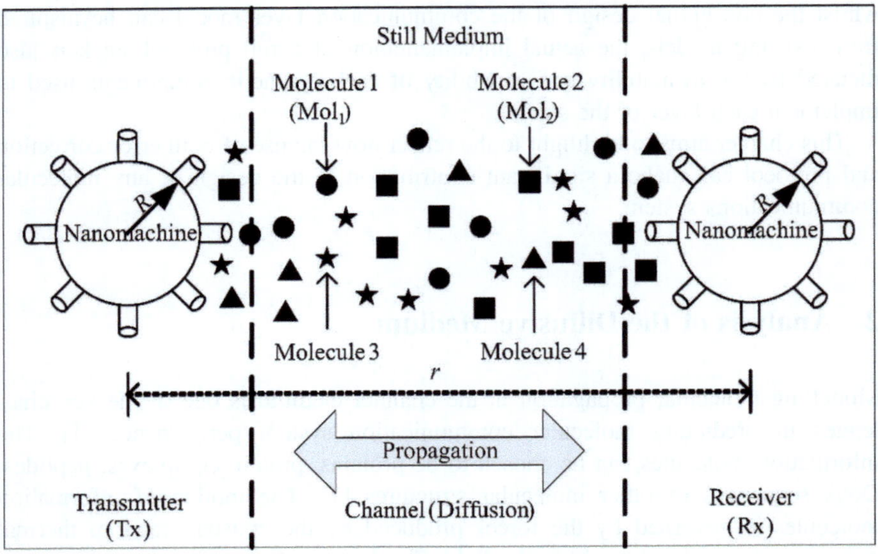

Fig. 1 Model of the 3D diffusion based communication used herein

$$\frac{\partial S(r,t)}{\partial t} = D\nabla^2 S(r,t) \tag{1}$$

For this communication model, the capture probability, rather than the escape probability, is more important. By solving Eq. (1), the capture probability $P_{\text{hit}}(r,t)$ can be calculated as:

$$P_{\text{hit}}(r,t) = 1 - S(r,t) = \frac{R}{r}\text{erfc}\left\{\frac{d}{2\sqrt{Dt}}\right\} \tag{2}$$

where erfc$(\cdot)$ is the complementary error function, and $d = r - R$, the distance between the information molecule and the boundary of the receiver.

Figure 2 illustrates that for a given distance, the capture probability increases with the increasing of time, which means that as time increases, more and more molecules will arrive at the receiver. In addition, for a given period of time, the capture probability decreases when the distance between the information molecule and the receiver increases.

To extract the hit time probability, which refers to the probability that an information molecule arrives at the receiver at a particular time t, Eq. (2) is differentiated with respect to time, obtaining the hit time distribution:

$$h(t) = \frac{R}{r}\frac{d}{2\sqrt{\pi D}}\frac{1}{t^{3/2}}\exp\left(-\frac{d^2}{4Dt}\right) \tag{3}$$

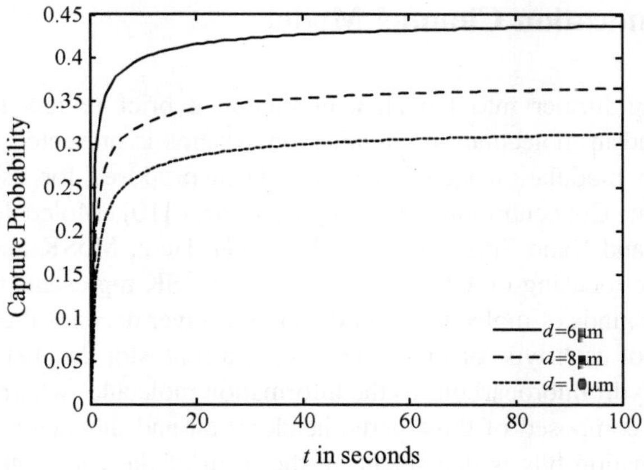

Fig. 2 Capture probabilities versus time for different transmission distances, $d = 6\ \mu m$, $d = 8\ \mu m$ and $d = 10\ \mu m$

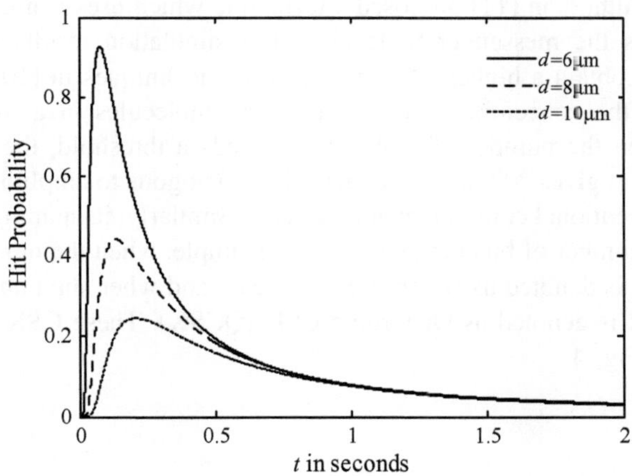

Fig. 3 Hit time distribution plot for different distances $d = 6\ \mu m$, $d = 8\ \mu m$ and $d = 10\ \mu m$

The capture probability shows the probability of successful transmission of the information symbols in a given period of time. To get a better understanding of the probability of success at a specific time, the hit time probability distribution, which encapsulates the response of the molecular channel to the release of an impulse of molecules, is investigated. In Fig. 3, for a given distance, it can be seen that this increases sharply in an extremely short period, it then decreases rapidly and approaches zero as time progresses. This is because that during the whole propagation process, most of the information molecules arrive in a short time whereas a few molecules arrive after a very long period of time, exhibiting a long diffusive tail leading to ISI. In addition, a longer time will be taken for most of the molecules arriving at the receiver when the distance between the molecule and the receiver is larger.

3 Communication Channel Model

Before moving further into the channel model, a brief review the modulation techniques used in molecular communication systems is presented.

Three main modulation techniques have been proposed for use in molecular communication: Concentration Shift Keying (CSK) [10], Molecule Shift Keying (MoSK) [10] and Ratio Shift Keying (RSK) [11]. Here, MoSK and RSK will be outlined before focusing on CSK in more detail. MoSK represents the information using different kinds of molecules, and then the receiver decodes them based on the concentration of each type of molecules during a time slot. In [12], an example is shown using hydrofluorocarbons as the information molecules where the messenger molecules are composed of three parts: header, trail and the chemical bit element, and the information bits is dependent on the form of the chemical element. RSK encodes the information based on the ratios of messenger molecules. Theoretically, it has an infinite modulation order, and needs just two types of molecules in the simplest system; it can also be developed for use in more complex systems. In addition, the authors in [11] proposed a technique which uses isomers, also known as hexoses, as the messenger molecules. The simulation results show that this technique can obtain a higher SNR gain than the techniques in [10, 13].

CSK uses the concentration of the received molecules in a time slot as the criterion. When the number of molecules exceeds a threshold, the receiver gives "1", otherwise it gives "0". It can be viewed as analogous to amplitude shift keying (ASK) in conventional communication systems. Similarly, its naming convention is based on the number of bits per symbol. For example, when the number of bits per symbol is 1, it is denoted as Binary CSK (BCSK) and when the number of bits per symbol is 2, it is denoted as Quadruple CSK (QCSK). These CSK techniques are illustrated in Fig. 4.

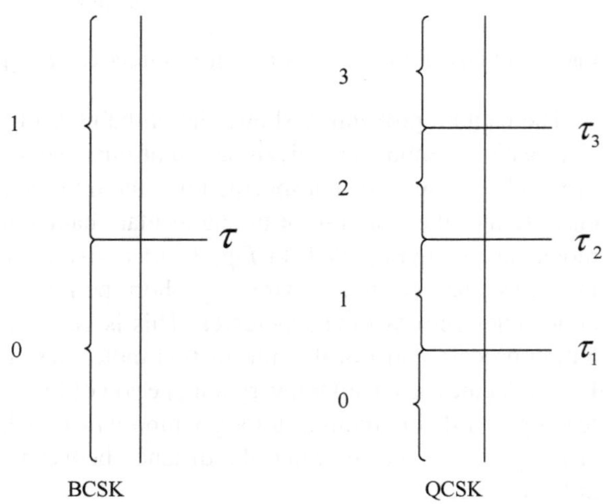

Fig. 4 CSK techniques for 1 and 2 bits per symbol

CSK is susceptible to ISI, and the severity of the effect is dependent on the threshold values and also on the number of bits per symbol. As the number of bits per symbol increases then the impact of ISI errors will be on an increasing number of bits simultaneously.

An outline of the modelling of the communication channel used to analyse the performance of the communication method will now be presented. To effectively represent the transmitted symbols, the propagation time is divided into time slots, also called symbol durations, which have the equal length. Only one symbol propagates in a single time slot the length of which is denoted by t_s. As shown in Sect. 2, there is a probability that the molecule will hit the receiver in a given time slot that depends upon the distance between the transmitter and receiver, the symbol duration, the environment and the chosen type of molecule [7]

The communication channel is a Binomial one, where each molecule arrives at the receiver or does not. It has been stated above that the previous bits can have an influence on the current bit due to ISI. Here, only the previous one time slot will be taken into consideration since this has been shown to be a reasonable approximation [7]. Thus the number of molecules received in a time slot which is denoted by N_{hit} is made up of the molecules sent at the start of the current time slot N_c and the start of the previous symbol duration N_{pre}. N_c and N_{pre} are both Binomial distributions which can be denoted as:

$$N_c = Binomial(N, P_{hit}(d, t_s)) \tag{4}$$

$$N_{pre} = Binomial(N, P_{hit}(d, 2t_s)) - Binomial(N, P_{hit}(d, t_s)). \tag{5}$$

Since no closed form is known for the probability mass function of the difference between two dissimilar Binomial distributions, a Gaussian approximation [7] is employed. Given that the one-bit information of the current intended symbol is S_C and that of the previous time slot is S_{Pre}, the current received symbol depends on both of them. For this binary channel model, both S_{Pre} and S_C can be taken as zero or one as shown in Fig. 5.

As may be seen in Fig. 5, there are four different cases for the binary channel model which are denoted by bit pairs {00, 01, 10, 11} for received symbol decoding, according to the different values of S_C and S_{Pre}. The probability represents the probability of success to receive the current intended symbol in the current

Fig. 5 The binary channel model

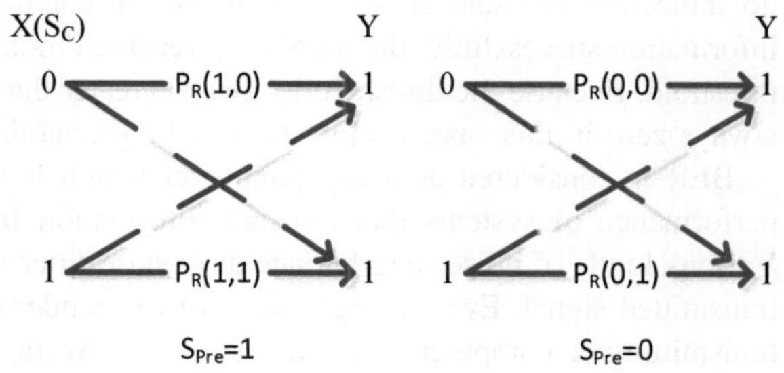

time slot, where S_{Pre} is the one-bit information represented by the previous intended symbol and c is that of the current one. The different four cases are displayed below.

Case {11}: Both the one-bit information represented by the previous intended symbol and current one are "1". In this case, the probability of successfully receiving the current symbol "1" increases because some of the molecules sent at the start of previous time slot will also receive at the current symbol duration. Thus, this is often considered as a favourable case as the ISI adds to the "1". Assuming that $P_1 = P_{hit}(d, t_s)$ and $P_2 = P_{hit}(d, 2t_s)$, the probability of success of this case can be described as:

$$P_{R(1,1)} = Q\left(\frac{\tau - nP_2}{\sqrt{n[P_2(1-P_2) + 2P_1(1-P_1)]}}\right) \tag{6}$$

where $Q(\cdot)$ is the usual tail probability Q-function.

Case {10}: No molecules are sent at the start of the current time slot which means that the received symbol should be "0" if it is correctly decoded. In this case, the molecules overflowing from the previous time slot have a negative influence on the successful decoding of the current intended symbol. It is considered to be an unfavorable case due to ISI. For successful decoding, the received molecules should not exceed the threshold such that:

$$P_{R(1,0)} = 1 - Q\left(\frac{\tau - n(P_2 - P_1)}{\sqrt{n[P_2(1-P_2) + P_1(1-P_1)]}}\right) \tag{7}$$

Case {01}: It means that the number of molecules received in the current time slot is only dependent on the molecules sent at the start of the current time slot, without overflowing molecules from the previous symbol duration. The information will be decoded successfully when the number of received molecules exceeds the threshold. The probability of success can be written as:

$$P_{R(0,1)} = \sum_{k=\tau}^{n} \binom{n}{k} P_1^k (1-P_1)^{n-k} = I_{p_1}(\tau, n-\tau+1) \tag{8}$$

where $I_{p_1}(\tau, n-\tau+1)$ is the regularized incomplete beta function.

Case {00}: There are no overflowing molecules from the previous time slot and no molecules are sent at the start of the current time slot. Thus, to decode the information successfully, the number of received molecules should smaller than the threshold. Because the threshold is always greater than zero, the received symbol is always zero in this case, leading to a 100% probability of success.

BER is considered as a key parameter which is often employed to assess the performance of systems that transmit information from one position to another. Various kinds of noise, interference and phase jitter may cause degradation of the transmitted signal. Even a single bit error can render the program inoperable when transmitting a computer program. The total average BER can be obtained by

Table 1 Values for 60% of molecules to arrive

Distance (μm)	t_s (s) in [7]	t_s (s) based on (2)
1	0.03	0.023
2	0.11	0.092
4	0.4	0.37
8	1.54	1.47
16	5.9	5.86
32	22.01	23.45

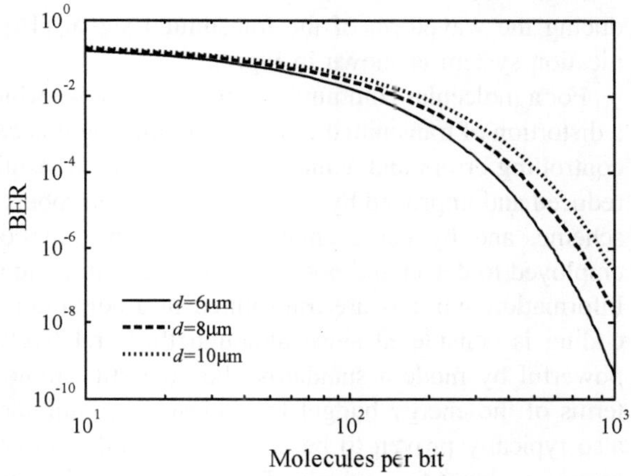

Fig. 6 BER versus molecules per bit

calculating the average BER of all the four states stated above. In this chapter, the symbol duration t_s is chosen as the time before 60% of the molecules arrive at the receiver. A comparison with the results from the literature based on Monte Carle simulation for several values of t_s is given in Table 1. It may be observed that the agreement is excellent and the gain in calculation time is considerable.

From Table 1, it is known that the value of t_s is set according to the transmission distance. Thus, for a fixed transmission distance d, the BER is a function of the threshold τ and the number of molecules per bit N, which is set in the range of 10–1000 here. Also, for each value of N, the threshold must be a specific value such that $\tau \in (1, N)$. Thus, the minimum bit error rate can be obtained for each N, which is the number of molecules per bit.

The optimized BER versus molecules per bit for different transmitted distance when the receiver radius is 5 μm is shown in Fig. 6, which shows that for a fixed distance d, the BER decreases with the increasing of number of molecules per bit N.

This demonstrates that as one might expect, the communication system has a better performance with a superior BER if large numbers of molecules are sent during one time slot. In addition, for a chosen value of N molecules sent in the time slot, the BER increases as the distance between the transmitter and receiver increases as a consequence of by the random property of the diffusion process. Thus, to achieve a better performance of the communication system, it is better to have a smaller propagation distance d and a larger number of transmitted molecules per bit.

4 Error Correction Coding

Coding is considered as a trade-off between reduced throughput and a reduction in the BER because of the redundant bits. Error correction coding covers different families of codes. Some of the traditional codes are shown in Fig. 7, which is far from being exhaustive. At the highest level, codes can be divided into two classes: those that can only detect errors and those can correct errors as well.

Digital communication systems transmit random quantities rather than reproducing the waveform of the transmitted signal [14]. A typical Shannon communication system is shown in Fig. 8.

For a molecular communication system, noise which results from ISI, may cause a distortion of transmitted information, making it necessary to use coding theory for controlling errors and achieving reliable data transmission. The BER can often be reduced and improved by choosing a slow and robust modulation scheme or coding scheme, and by using channel error correction coding schemes which can be employed to detect and possibly correct a certain number of errors that occur when information symbols are transmitted in a communication system. Here Hamming coding is considered since although these relatively simple block codes are not powerful by modern standards, they are still employed, and are very efficient in terms of the energy budget for nanoscale communications. Hamming codes have also typically proven to be a simple and efficient code over a channel where the errors are burst-free widely dispersed. Hamming codes are denoted in the form (n, k), where $n = 2^m - 1$, is the coded length output for a the number of parity check bits m, and $k = 2^m - m - 1$ is the number of date bits per block. The minimum distance d_m of this kind of block code is 3, which means that only one error can be corrected in each block. To compare the coded and uncoded transmission performance, the BER for the Hamming coded operation should be decided which is not an easy task for coding systems in general. However, for binary linear block codes in this case, an approximate BER expression is [3]:

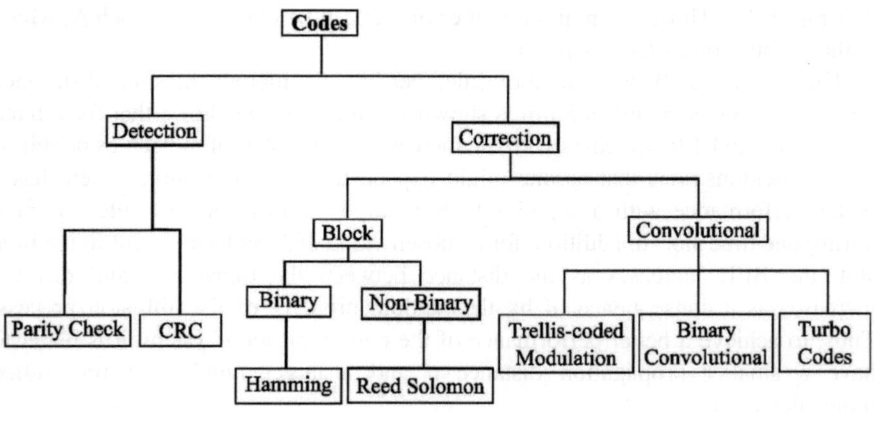

Fig. 7 Error detection and correction code classes

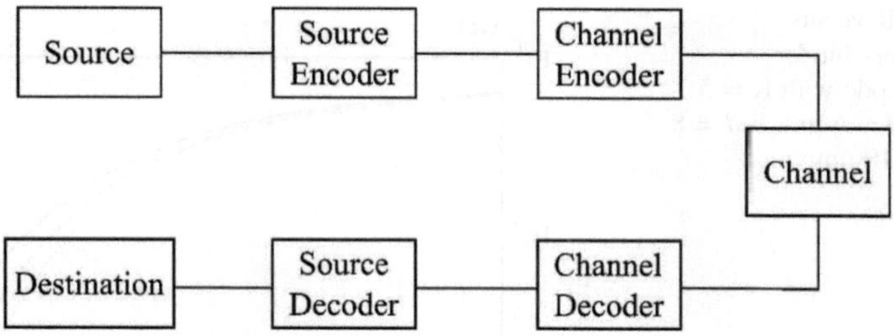

Fig. 8 Typical blocks of the Shannon communication system

$$P_{ec} = \frac{1}{n} \sum_{j=t+1}^{n} j \binom{n}{j} p^j (1-p)^{n-j} \tag{9}$$

where n is the length of code word, $t = (d_m - 1)/2$ is the maximum number of errors that the code can correct, and p is the probability of one bit error. The bit error probability p can be set in different ways [15]. In the case of molecular communications, there are four different probabilities of errors depending on the data pattern because of ISI, which has been stated in the previous section. With the calculation of the optimized BER in Sect. 3, the channel can be considered as a binary symmetric channel (BSC), which means that only bit flip is taken into consideration. Here, the bit flip probability is set as the values of the optimized BER.

Figures 9a through (c) show the BER results as a function of molecules per bit for uncoded, (7, 4) and (15, 11) Hamming-coded transmission by using (9) over distances of 1 μm, 4 μm and 8 μm respectively.

It is clearly seen that the system performance improves when the information sequence is Hamming coded. Generally speaking, to evaluate the performance of the system, the coding gain in dB is often assessed at a specific value of BER for a specific code. This coding gain can be directly obtained by finding out the ratio of molecules needed because there is approximately a linear relationship between the transmission energy and the number of molecules per bit. It can be derived that for a chosen BER level, here 10^{-9}, the coding gain for a (15, 11) Hamming code is 1.71 dB while for a (7, 4) Hamming code, it is 0.89 dB. In addition, an extra distance of 2 μm can be obtained by employing a (15, 11) code and 1 μm for a (7, 4) code. This indicates that the Hamming code is useful to extend the transmission range and achieve a coding gain, which can improve the system performance.

A coded system can give a better performance than an uncoded one but this improvement comes at the extra energy cost associated with the coding and decoding method. The energy saving (or loss) ΔE is defined as:

$$\Delta E = 2450(N_{uncoded} - N_{coded}) - E_{encode} - E_{decode} \tag{10}$$

In (10), $N_{uncoded}$ and N_{coded} are the number of molecules needed to achieve a given BER for un-coding and coding system, E_{encode} and E_{decode} are represent the

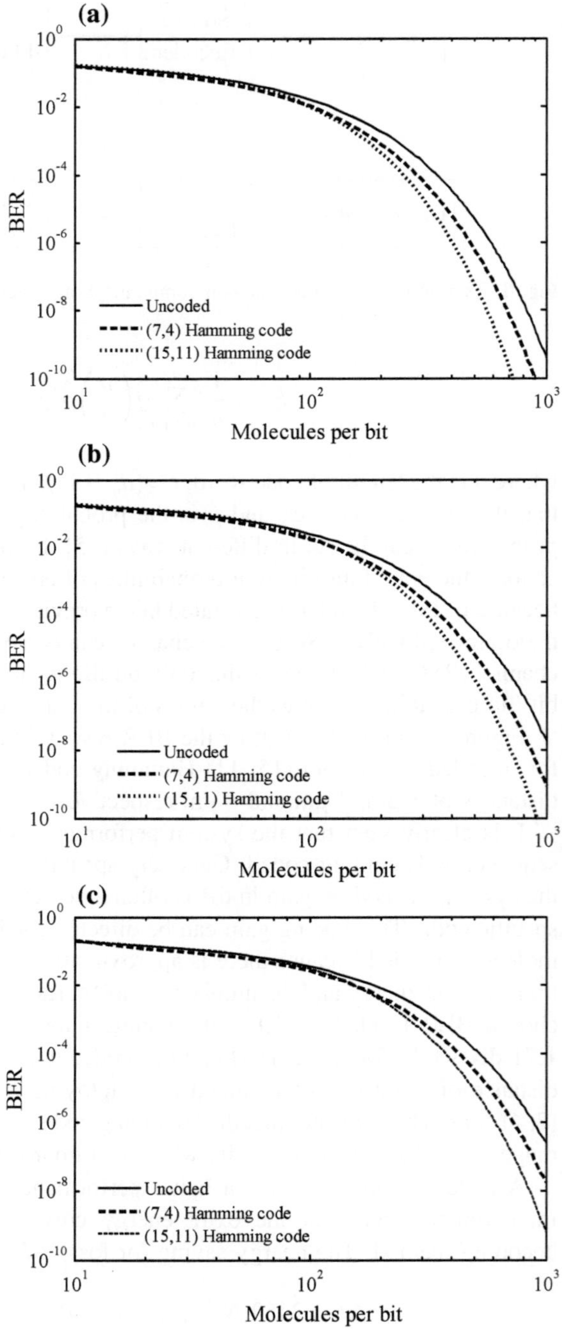

Fig. 9 BER versus molecules per bit for Hamming code with R = 5 μm and **a** d = 6 μm, **b** d = 8 μm, **c** d = 10 μm

energy required to encode and decode, $2450\,K_BT$ is the coat of synthesizing a molecule, where K_B is the Boltzmanm constant. It is also assumed that the system is operating at an absolute temperate of $T = 300$ K. Here K_BT is used as an unit when measuring the energy. It is easy to see that when $\Delta E \geq 0$, it is worth to apply hamming code into the system.

Adenosine triphosphate (ATP) is a nucleoside triphosphate used in cells, it is often called the "molecular unit of currency" of energy transfer between cells in living organisms [16]. Encoding and decoding methods can be developed by using some logic circuits, an sound argument for every transmission through a NAND gate to cost 1 ATP is given by Sauro and Kholodenko [17, 18] and NANDs are universal in that one can construct all other gates from them via the principles of Boolean algebra. So here the NOT gates, AND gates, XOR gates and shift register units cost one, two, five and four ATPs respectively. This provides a basis for obtaining the total energy cost for the coding and decoding systems.

Figure 10a, b show the encoder and decoder for Hamming codes [19]:

The energy consumption for Hamming codes with different orders, m are given by:

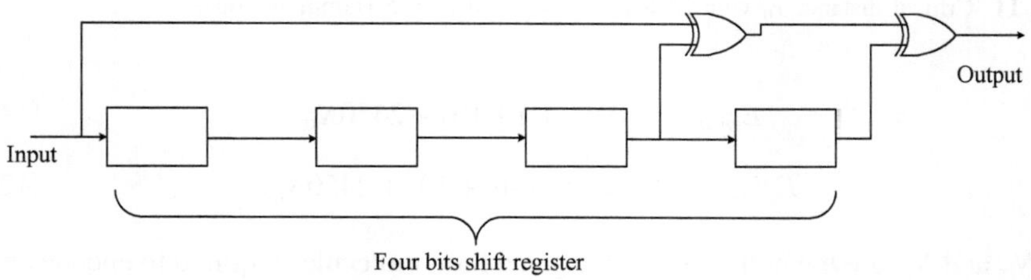

(a) Non-systematic encoder.

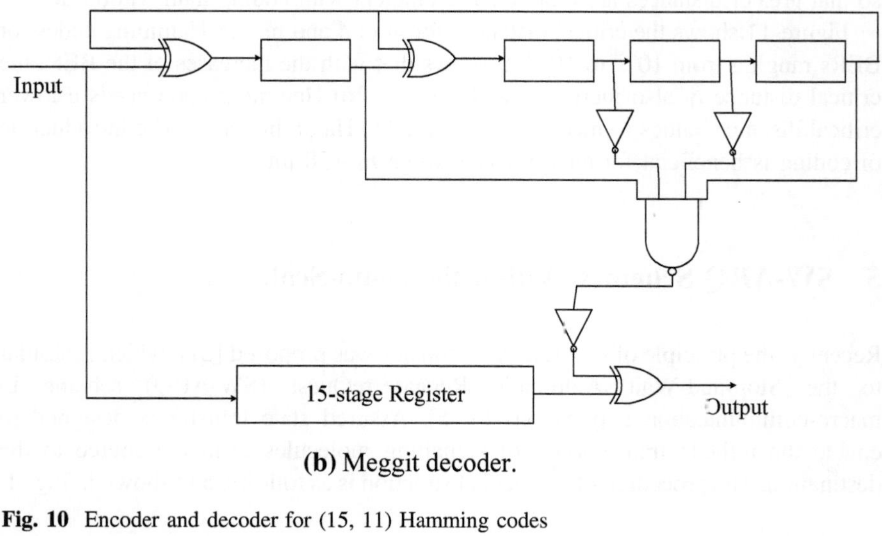

(b) Meggit decoder.

Fig. 10 Encoder and decoder for (15, 11) Hamming codes

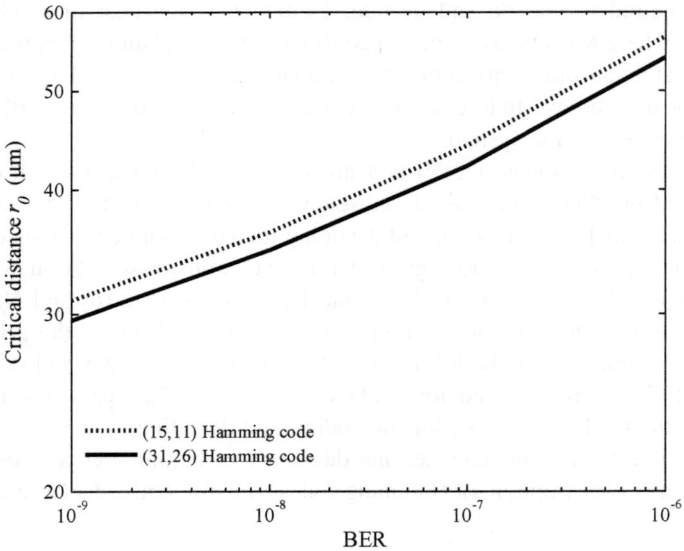

Fig. 11 Critical distance r_0 with BER for m = 4 and m = 5 Hamming codes

$$E_{encode} = 20N_{tx}(4m + 10) + 2450N_{tx} \tag{11}$$

$$E_{decode} = 20N_{rx}(4m + 4n + 19) + 2450N_{rx} \tag{12}$$

N_{tx} and N_{rx} are the numbers of code generation molecules required to encode and decode the data. Assuming that $N_{tx} = N_{rx} = 300$ and one ATP reaction is about $20\,K_BT$ at 300 K. The critical distance [20] is defined as the distance at which $\Delta E = 0$ so that greater distances are more energy efficient with coding than without it.

Figure 11 shows the critical distances for m = 4 and m = 5 Hamming codes for BERs ranging from 10^{-9} to 10^{-6}. It shows that with the increases of the BER, the critical distance r_0 also increases, and the (31, 26) Hamming code needs a lower critical distance values to meet ΔE than (15, 11) Hamming code. The introduction of coding is beneficial for energy usage when $r_0 \geq 30$ μm.

5 SW-ARQ Schemes Within the Nano-Scale

Recently, the principle of *assured state transfer* was proposed [21], which is similar to the Stop-and-Wait Automatic Repeat reQuest (SW-ARQ) scheme in macro-communication applications [5, 6]. Assured state transfer is designed to enable the reliable transmission of signalling molecules from the source to the destination. The procedure of protocol abstraction is as follows, and shown in Fig. 1.

Firstly, the message molecules (represented as Mol_1) are emitted by the transmitting (Tx) nano-machine and propagate to the receiving (Rx) nano-machine. Upon reception, the Mol_1 molecules stimulate Rx to release another kind of molecules, known as an ACKnowledgement or ACK (represented as Mol_2). When Mol_2 is received by Tx, they stimulate the nano-machine to stop releasing Mol_1. In this way, this protocol abstraction guarantees that no more Mol_1 is transmitted to Rx, thus attempting to make each state more reliable. Further work on this protocol abstraction has been carried [22], where simulations on a molecular based reliable communication protocol with nano-logic computation were discussed. In this section, the abstraction is enhanced, proposing two schemes, and combine it with a realistic physical end-to-end model.

The system considered in this chapter, shown in Fig. 1, consists of two nanomachines, the transmitter, Tx, and the receiver, Rx, which communicate via transmission molecules. It is assumed that the nanomachines behave in a way that is analogous to biological entities and that they communicate via a natural ligand-receptor binding mechanism [23]. It is further assumed that, as per the work in [24], that each nano-machine can emit several specific types of molecules which can be recognised by the other nano-machine. For Tx to communicate a message, the total number of bits are broken up into $(n - i)$ bits per packet and then ordered into the packed structure shown in Fig. 12a with i index bits and that the receiver transmits an ACK with structure shown in Fig. 12b. For a message with an index, Ix, ranging from 0 to $2^i - 1$, two different kinds of molecules are required: one for the packet with the index Ix and one for the corresponding ACK with index Ix such that each molecule does not interfere with each other. This is otherwise known as pheromone diversity [13]. It is finally assumed the molecules are chemically inactive so they will not react with other molecules in the medium and that they are easily eliminated after decoding or after a fixed period of time. The medium is assumed to be a diffusing channel [18], where the molecules have a capture probability given by (2).

Two schemes are considered, one of which is very similar to macro-scale SW-ARQ and one that employs multiple acknowledgements.

(1) Scheme 1

As shown in Fig. 13a, the Tx transmits a packet and waits for the ACK from the Rx. If the waiting time is longer than a pre-defined limit, t_{out}, the Tx retransmits the packet. When receiving the packet, the Rx sends back an ACK and waits for the next packet from the Tx. If the waiting time is longer than the same pre-defined limit, the Rx retransmits the ACK.

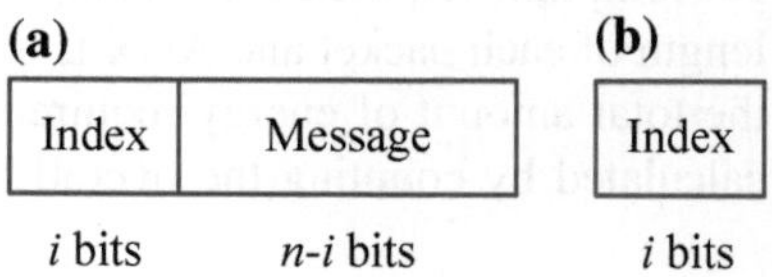

Fig. 12 The structure of: **a** the transmission packet; **b** the ACK

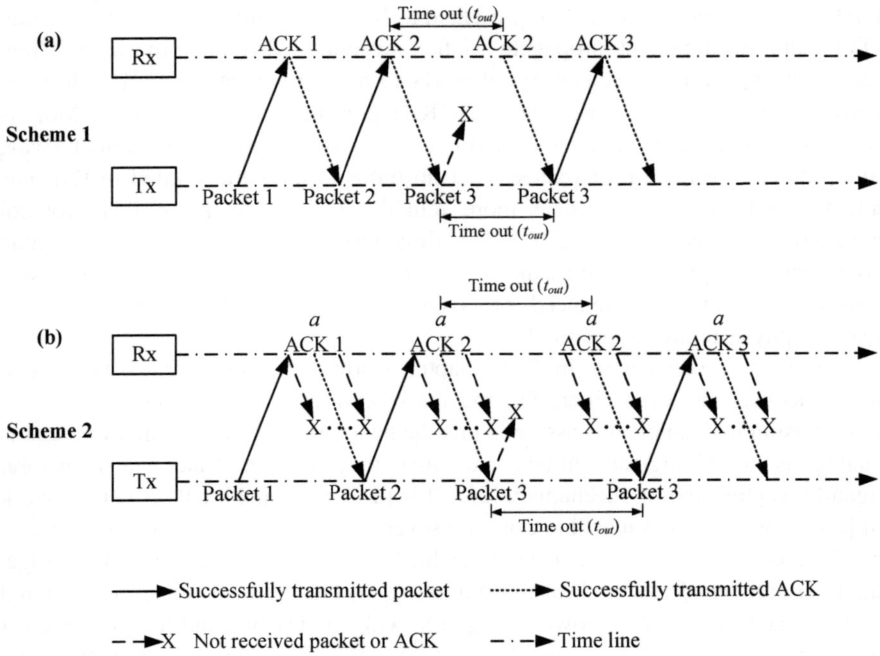

Fig. 13 The two proposed SW-ARQ Schemes

(2) Scheme 2

This scheme, as shown in Fig. 13b is based on the traditional SW-ARQ scheme. But when the Rx receives a packet, it sends back a copies of the corresponding ACK simultaneously. As soon as at least one of the $\boldsymbol{a}$ ACKs is received by the Tx, the next packet will be transmitted. Based on the structures shown in Fig. 12, a length ratio, *Ratio,* can be defined as:

$$Ratio = \frac{\text{The length of the packets}}{\text{The length of the ACKs}} = \frac{n}{i}, \tag{13}$$

which implies that when *Ratio* is large (that the length of the ACKs is significantly smaller than those of packets), when a packet does arrive at Rx, but the corresponding ACK is lost, this scheme will use less energy compared to the Scheme 1.

The first figure of merit in evaluating these transmission schemes is energy consumption required for a complete transmission. Setting $n = 17$ and $i = 2$, the length of each packet and ACK is $l_{pac} = 17$ and $l_{ack} = 2$ (in bits) respectively. Thus the total amount of energy required for a successful complete transmission can be calculated by counting the overall bits transmitted by both Tx and Rx as:

$$\text{Energy} = l_{pac} \times N_{pac} + l_{ack} \times N_{ack} = 17 \times N_{pac} + 2 \times N_{ack} \quad (14)$$

where N_{pac} is the number of packets required and N_{ack} is the number of ACKs required assuming that each bit requires the same amount of energy. It can be noted that the unit of energy in this case is thus normalised to bits, thus allowing the reader to substitute their own energy model given by their own Tx-Rx nano-machine. The second figure of merit is the time for completion of a successful transmission under a given scheme and is calculated though simulation. It should also be noted, that due to this work being simulation based, each of the subsequent results is the average of 5000 retrials.

Considering the Scheme 2, varying a in the interval between 1 and 10, and setting t_{out} = 15 s, it can be seen from Fig. 14 that the scheme 2 greatly reduces the time cost compared to the scheme 1 (a = 1). It can also be seen that this scheme agrees with the known relationship of a 'longer distance requires a longer transmission time', but what is interesting is that for a given r, for an increasing a, the improvement differential (the gain between a and (a − 1)) decreases, implying an optimum value for a given application under this scheme. Furthermore, as can be seen in Fig. 14, using the Scheme 2, more energy is required to perform a successful transmission for any fixed a due to the capture probability decreasing with larger r values. However, what is not so intuitive is that, by increasing a, the probability of a successful transmission is increased, lowering the number of re-transmitted packets. Increasing a though does require more energy, so if a is too high, the cost of transmitting the excess ACKs (or the gain is having more ACKs) may outweigh the energy gain in requiring a lower number of transmission packets. This leads to an interesting trade-off where, as per Fig. 14, it can be seen that the energy cost is actually lowest when a = 3 (Fig. 15).

Referring to Figs. 16 and 17, t_{out} is varied to confirm that as it is reduced, the time to complete the transmission reduces (as the wait is on average shorter), and

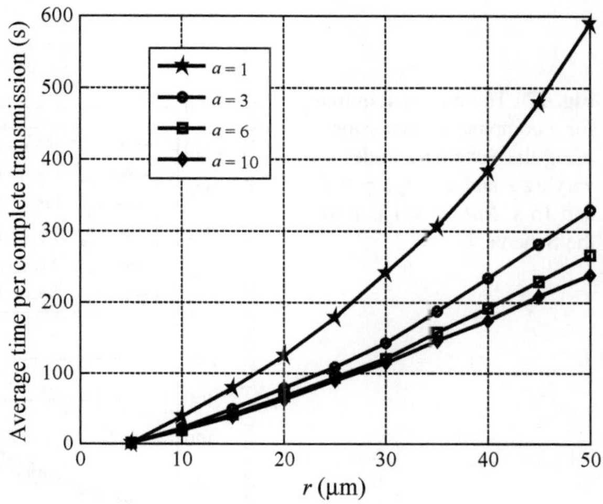

Fig. 14 The time required for a complete transmission using the Scheme 2 under varying r and with t_{out} = 15 s. *Note* a = 1 is also Scheme 1

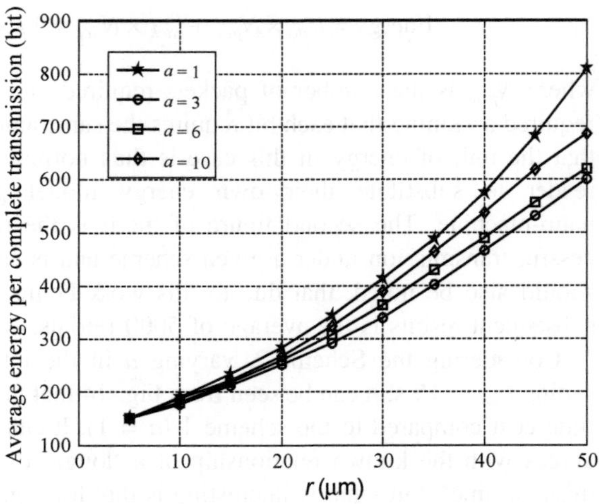

Fig. 15 The energy required for a complete transmission using Scheme 2 under varying r and with $t_{out} = 15$ s. *Note* $a = 1$ is also the Scheme 1

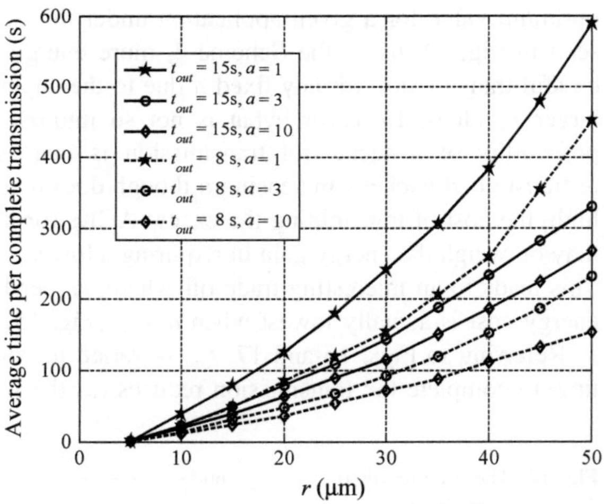

Fig. 16 The time required for a complete transmission using the Scheme 2 under varying r and a at $t_{out} = 8$ s and 15 s. *Note* $a = 1$ is also the Scheme 1

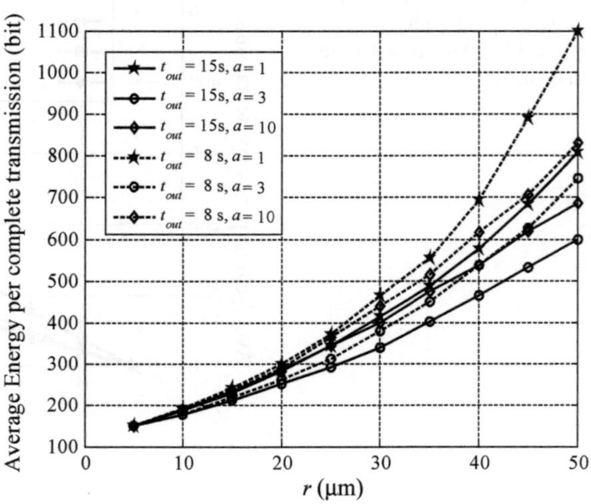

Fig. 17 The energy required for a complete transmission using the Scheme 2 under varying r and a at $t_{out} = 8$ s and 15 s. *Note* $a = 1$ is also the Scheme 1

uses more energy (as more re-transmissions are made). However, this is again somewhat counter intuitive, it can also be seen that there may be a benefit in both time and energy to by carefully selecting t_{out} and a. This is an interesting result as for example, with $r = 50$ μm, and for $t_{out} = 8$ s and $a = 3$, the energy requirement is 747 bits and the time cost is 232 s. For $t_{out} = 15$ s and $a = 3$, the energy consumption is 600 bits, and the time cost is 329 s. That is, with the t_{out} increasing from 8 to 15 s, the energy consumption reduces 20% and the time cost increases 41%. Depending upon the scenario and system requirements, these parameters are available for the system designer to further refine.

Thus, it can stated, based upon the simulation results for the SW-ARQ schemes that Scheme 1 is the best choice for adjacent communications. Access to a lightly higher energy budget will mean that Scheme 2 will provide better performance than scheme 1 for longer communication distances.

6 Summary

This chapter has presented an overview of the essential elements needed to describe analysis and design molecular communication systems employing coding and protocols. After a brief introduction to codes and protocols, and the relevant state of the art, the key differences between macro-scale and nano-scale communications have been outlined. Following this, the diffusive medium and its impact on molecular communications were described leading to quantitative performance illustrations. The modulation techniques used in diffusion based molecular communication system were then introduced. One of them (CSK) formed the basis for more detailed analysis and performance predictions. Next, it was shown that Hamming codes are useful to extend the transmission range and achieve a coding gain, which can improve the system performance. In addition, the results also take energy budget into consideration, thus determining that the critical distance at which coding become beneficial in energy terms is approximately 30 μm. The final section presents the implementation of two SW-ARQ schemes for nano-communication networks, each of which has its own superiority and drawbacks. For short distance, near-adjacent communications ($r \leq 10$ μm) standard SW-ARQ performs well. However, use of multiple acknowledgments is more effective as the distance increases consuming only 50–80% of the energy needed for standard SW-ARQ by the time 50 μm is reached. System parameters need to be carefully selected to achieve the optimal performance and designers should choose the appropriate scheme dependent on the application. Overall, the fundamental aspects germane to the utilization of coding a protocol for molecular communications have been presented with novel insights into these areas. Thus, coding and protocols within molecular communications open up new opportunities and as nano-communications research develops, and it can be expected that new ways of coding and protocols within this exciting paradigm will be investigated.

References

1. Akyildiz IF, Brunetti F, Blázquez C (2008) Nanonetworks: a new communication paradigm. Comput Netw 52:2260–2279
2. Moore M, Enomoto A, Nakano T, Egashira R, Suda T, Kayasuga A et al (2006) A design of a molecular communication system for nanomachines using molecular motors. In: IEEE international conference on pervasive computing and communications workshops, pp 554–559
3. Bernard S (1988) Digital communications: fundamentals and applications. Prentice-Hall
4. Kurose JF (2005) Computer networking: a top-down approach featuring the Internet. Pearson
5. Moeneclaey M, Bruneel H (1984) Efficient ARQ scheme for high error rate channels. Electron Lett 20:986–987
6. De Munnynck M, Lootens A, Wittevrongel S, Bruneel H (2002) Transmitter buffer behaviour of stop-and-wait ARQ schemes with repeated transmissions. In: IEE proceedings in communications, pp 13–17
7. Kuran MŞ, Yilmaz HB, Tugcu T, Özerman B (2010) Energy model for communication via diffusion in nanonetworks. Nano Commun Netw 1:86–95
8. Leeson MS (2000) Performance analysis of direct detection spectrally sliced receivers using Fabry-Perot filters. J Lightwave Technol 18:13–25
9. Ziff RM, Majumdar SN, Comtet A (2009) Capture of particles undergoing discrete random walks. J Chem Phys 130. 27 Mar 2009
10. Kuran MS, Yilmaz HB, Tugcu T, Akyildiz IF (2011) Modulation techniques for communication via diffusion in nanonetworks. In: IEEE International Conference on Communications (ICC), pp 1–5
11. Kim N-R, Chae C-B (2012) Novel modulation techniques using isomers as messenger molecules for molecular communication via diffusion. In: IEEE International Conference on Communications (ICC), pp 6146–6150
12. Freitas RA (1999) Nanomedicine, volume I: basic capabilities. Landes Bioscience
13. Giné LP, Akyildiz IF (2009) Molecular communication options for long range nanonetworks. Comput Netw 53:2753–2766
14. Shannon CE (1948) A mathematical theory of communication. Bell Syst Tech J 27:379–423
15. Yan H, Choe HS, Nam S, Hu Y, Das S, Klemic JF et al (2011) Programmable nanowire circuits for nanoprocessors. Nature 470:240–244
16. Knowles JR (1980) Enzyme-catalyzed phosphoryl transfer reactions. Annu Rev Biochem 49:877–919
17. Sauro HM, Kholodenko BN (2004) Quantitative analysis of signaling networks. Prog Biophys Mol Biol 86:5–43
18. Leeson MS, Higgins MD (2012) Forward error correction for molecular communications. Nano Commun Networks 3:161–167
19. Blahut RE (2003) Algebraic codes for data transmission. Cambridge University Press
20. Howard SL, Schlegel C, Iniewski K (2006) Error control coding in low-power wireless sensor networks: when is ECC energy-efficient? EURASIP J Wireless Commun Networking 2006
21. Akyildiz IF, Fekri F, Sivakumar R, Forest CR, Hammer BK (2012) Monaco: fundamentals of molecular nano-communication networks. IEEE Trans Wireless Commun 19:12–18
22. Walsh F, Balasubramaniam S, Botvich D, Donnelly W, Sergeyev S (2007) Development of molecular based communication protocols for nanomachines. Presented at the proceedings of the 2nd international conference on nano-networks, Catania, Italy
23. Krivan V, Lánský P, Rospars JP (2002) Coding of periodic pulse stimulation in chemoreceptors. BioSystems 67:121–128
24. Atakan B, Akan O (2007) An information theoretical approach for molecular communication. In: Bio-inspired models of network, information and computing systems, pp 33–40

Understanding Communication via Diffusion: Simulation Design and Intricacies

Bilal Acar, Ali Akkaya, Gaye Genc, H. Birkan Yilmaz, M. Şükrü Kuran and Tuna Tugcu

Abstract Understanding Communication via Diffusion (CvD) is key to molecular communications research since it dominates the movement at the nano-scale. The researcher needs to properly understand the random diffusion of the molecules for the analysis of a molecular communication system. This chapter aims explaining the dynamics of diffusion from a communication engineer's perspective as well as providing useful hints for an effective simulation design by discussing some key intricacies. The chapter starts with a brief survey of simulators for molecular communications, followed by the basics of the simulation of Brownian motion and CvD. Several intricacies are addressed to help the researcher in simulation design, such as the number of replications required in terms of movement and bit sequence. We utilize this information further by discussing the design of more complex CvD systems such as tunnel-based approach that utilizes destroyer molecules and distributed simulator design based on HLA. Introduction of more complex CvD systems provides significant improvements in data rate and communications in general, bridging the gap between human-scale and nano-scale systems and enabling nanonetworking as a viable technology.

B. Acar · A. Akkaya · G. Genc · H.B. Yilmaz · T. Tugcu (✉)
NETLAB, Department of Computer Engineering, Bogazici University,
34342 Istanbul, Turkey
e-mail: tugcu@boun.edu.tr

B. Acar
e-mail: bilal.acar.a@gmail.com

A. Akkaya
e-mail: ali.akkaya@boun.edu.tr

G. Genc
e-mail: gaye.genc@boun.edu.tr

H.B. Yilmaz
School of Integrated Technology, Yonsei University, Seoul, South Korea
e-mail: birkan.yilmaz@boun.edu.tr

M. Şükrü Kuran
Abdullah Gul University, Kayseri, Turkey
e-mail: sukru.kuran@agu.edu.tr

J. Suzuki et al. (eds.), *Modeling, Methodologies and Tools for Molecular and Nano-scale Communications*, Modeling and Optimization in Science and Technologies 9, DOI 10.1007/978-3-319-50688-3_7

1 Introduction

Communication via Diffusion (CvD) is one of the key communication mechanisms in the context of Nanonetworks that is proposed by the information and communication technology literature. As the most basic definition, CvD deals with communication systems between transmitter-receiver couples; exchanging information via a shared medium using molecules that represent the data. The information is usually regarded as being encoded onto the quantity of the molecules. Thus, the receiver decodes the information based on the number of molecules it receives within a predefined time period (symbol duration). Regardless of the specific subsystem that is being considered, the medium is assumed to be uncontrolled, or partially controlled, and the molecules propagate through it following the prevalent diffusion dynamics in this miniscule scale. These dynamics define a very unique physical layer unlike the conventional wired and wireless communication media, which brings new challenges specific for this communication system.

Diffusion, as a movement model for small particles, has been studied extensively in the 19th and early 20th century by scientists like Thomas Graham and Adolf Fick. Diffusion focuses on capturing the general behavior of a huge number of diffusing small particles in a medium (e.g., how a drop of dye diffuses through a body of water). However, this diffusing behavior is actually the macroscopic result of some basic movement that is conducted in a microscopic scale, called the Brownian motion. In other words, the movement of individual small particles can be modelled by Brownian motion dynamics.

According to the current literature on diffusion, several system parameters of diffusing particles in several topologies can be evaluated mathematically such as the average arrival time of diffusing particles from a single point source to a given spherical boundary in a 3-D space or the upper bound on the probability of hitting to a spherical body when particles are released again from a given single point source. Nonetheless, when the topology of the system becomes more complex the analytical solution to these parameters become harder to evaluate. For example, in a slightly more complex setting consisting of two spherical bodies, the upper bound of particles hitting to a sphere when released from another sphere is not an easy task and has not been solved analytically yet. On the other hand, in such complex topologies, these system parameters can be evaluated via simulation using Brownian motion dynamics in a short period of time. Therefore, simulation is a crucial method for analyzing the potential performance metrics of a given CvD system.

The research activities on molecular communication utilizes simulations to verify and analyze the performance of proposed models. Due to the different channel characteristics of the fluid environment and the carrier wave properties, current simulation tools can not be used directly for nanonetworking. Simulation of molecular communication requires modeling the new communication paradigm that comprises different options for transmission, propagation, and reception. It should consider possible architectural design options and performance evaluation of molecular communication networks. Since molecular communication involves the modeling of large number of nano-scale objects, scalability of the simulation tool is another important concern. Current simulation tools that are developed for traditional communication models are not suitable to be used as they are for simulation of molecular communication. Extension of current simulation tools or development of new tools are necessary to support research groups working on molecular communication. Another challenge for research groups is to generate a simulation execution plan. The parameters used for the simulations should be sufficient for statistically meaningful results, while they should also be optimal for minimizing the simulation execution time.

In the literature, besides the use of simulation as a research tool, there are several works that have been performed specifically on nano-scale simulation design. In [17], the need for simulation is mentioned and simulation requirements of molecular motor based communication network is briefly defined. In [20], a simulator for 3-D Brownian motion is proposed. The simulator is capable of modeling nano particles under various configurable circumstances to simulate molecule diffusion and reception. The novelty of the proposed model is a dual time step approach to cope with the run time complexity of high number of particles. When the particle is far from the target, the movement is simulated in large time steps, and when it is closer to the target, smaller time steps are used. In [12], the authors introduce a C++ and Tcl based simulation framework developed on top of the commonly used NS-2 discrete event simulator targeted for networking research. It implements diffusive molecular communication in 3-D space using a reaction-diffusion algorithm. The diffusion algorithm is based on the multi-particle lattice gas automata algorithm in which the exact location of particles are not tracked but the medium is divided into lattice slides. Numerical analysis of the presented scenarios are used for the verification of the simulation framework, along with performance evaluation. N3Sim [2] is a Java based simulation tool for diffusion based molecular communication. It enables the evaluation of molecular networks performance in 2-D and in 3-D space for specific scenarios. It uses Brownian motion and considers particle inertia and collisions among particles. The sensing of the local concentration is used for the reception model [15]. In [9], a simulation platform for modeling information exchange at the nano-scale is introduced. A Java based software library is created using object oriented concepts. Elastic collision among molecules and receptor-based reception mechanism are implemented. The model is defined to be generic and can be used for different communication options. A case study is used to demonstrate the features of the simulation tool. In [4], a distributed architecture for molecular communication is proposed. The architecture is built on top of High Level Architecture (HLA), and provides interoperable, re-usable, and scalable design options for simulation of molecullar communication paradigm.

Both the simulator design, and the simulation execution plan are important steps of molecular communication research process. Prior to each research project, a research team should consider the design issues for the simulator selection or development, and also carefully design the simulation execution plan. These two issues greatly influence the time to conclude any results out of simulation execution outputs, and affect the evaluation of alternative options for the proposed model. Flexibility, re-usability, interoperability, and scalability should be considered during simulator design process. Based on the research project needs, an existing simulator can be utilized, or a new simulator can be developed. For the simulation execution plan, researchers should concentrate on the number of replications for Brownian motion randomization and the number of different input sequences used. These parameters affect the simulation time considerably, hence they need to be optimized to minimize simulation time, while still resulting in statistically meaningful results. This chapter aims to guide the reader on the simulator design issues and simulation execution plan. This will help the reader to better plan and execute the simulation step of molecular communication research activities, which in general dominates the overall research plan.

2 Simulation Design of CvD

2.1 Simulation of Brownian Motion

Molecules are free to move in a fluid environment; thus, they move in a random fashion. We study the nature of this random motion within two perspectives: Macroscopic and microscopic views. First, we focus on the macroscopic theory.

The macroscopic theory of diffusion can be developed from two simple and basic assumptions. The first of these is that a substance will move down its concentration gradient. Steeper gradient results in more movement of the material. If the relation between gradient and flux is linear, then in one dimension we have what is known as Fick's first law

$$J = -D\frac{\partial C(x,t)}{\partial x} \tag{1}$$

where x is the position, $C(x, t)$ is the concentration at that point, and D is the diffusion constant. The variable J is the flux, and is defined as the amount of material passing across the point at x (or through a unit area perpendicular to the direction of flow) per unit time. The minus sign means that the flow is in the direction of decreasing concentration.

In a small element of length dx, the flux into the element from the left is different from the flux out of the element from the right. The difference between the two fluxes $J(x)$ and $J(x + \Delta x)$ determines how much material accumulates within the region bounded by x and $x + dx$ in a time interval Δt

$$(J(x + \Delta x) - J(x))\Delta t = -\Delta C \Delta x. \tag{2}$$

After rearranging and converting into derivative form, we get Fick's second law.

$$\frac{\partial C(x,t)}{\partial t} = D\frac{\partial^2 C(x,t)}{\partial x^2} \quad (3)$$

Equation 3 is for the one dimensional case. In three dimensions, the spatial derivative is replaced by the gradient, and combining with the second law we get

$$\frac{\partial C(x,t)}{\partial t} = D\nabla^2 C(x,t) \quad (4)$$

where ∇^2 is the Laplacian operator.

If we just consider the diffusion process starting from origin, the concentration at cite x and time t is given by

$$C(x,t) = \frac{1}{(4\pi Dt)^{m/2}} e^{-||x||^2/4Dt} \quad (5)$$

where m and D are the dimension of the environment and the diffusion coefficient, respectively [18]. The value of D depends on the temperature of the environment, viscosity of the fluid, and the Stokes' radius of the molecule [21].

The microscopic theory of diffusion is utilized for simulating the motion of diffusing particles. Brownian motion, which can be seen as a discrete case of the diffusion process, is simulated by the help of a *good* random number generator. (Based on our experience, we strongly encourage the use of a random number generator derived from Mersenne Twister). For the simulation process, we do not consider the collisions between particles for the sake of simplicity. In the one-dimensional space, the displacement of a single particle in unit time is a random variable ΔX, which follows a normal distribution with zero mean and σ^2 variance

$$\Delta X \sim \mathcal{N}(0, \sigma^2) \quad (6)$$

where $\sigma = \sqrt{2D\Delta t}$, and D is the diffusion coefficient that describes the tendency of the propagating molecules to diffuse through the fluid [6]. As an alternative option for simulating the Brownian motion, one may select the direction randomly and move same amount. Both schemes are equivalent when the Δt is small. When the direction is selected randomly and a fixed length movement is used, the movement becomes correlated. Hence, having normally distributed step lengths have some advantages to simulate the continuous diffusion process.

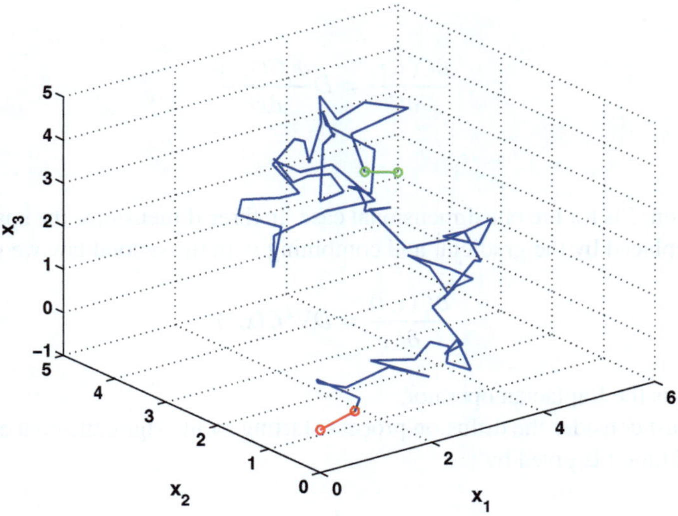

Fig. 1 75 ms trace of a diffusing molecule

If the particles propagate through a three dimensional environment, this movement can be modeled as three independent displacements (one for each dimension) [19] and the total displacement, $\vec{r}$, in one time step can be found as

$$\vec{r} = (\Delta x, \Delta y, \Delta z). \tag{7}$$

Particles (molecules in this scope) are assumed to have spherical bodies. Properties of the molecule and the environment determine the diffusion coefficient, hence the movement dynamics. Equation 6, suggests generating Gaussian random numbers with the given parameters for each dimension and at each step independently.

If you just want to simulate a particle movement starting from the origin without replication, you just create the movement at each time step. In each time step, it is normally distributed and can be considered as the accumulation of normally distributed random variables. You can find a 75 ms trace of a molecule depicted in Fig. 1. Trace of a diffusing molecule is produced by the following MATLAB code. The molecule starts diffusing from the origin $(0, 0, 0)$. First and the last steps are marked.

```
function [ trace ] = ...
    diffusion_sim3D_pointSource_singleMolecule(...
    D,...
    sim_time  ...
    )

% D  :Diffusion coefficient in micro meter^2 / seconds
% sim_time  :Duration of the simulation in seconds
```

```
% Time step is 1 ms
delta_t = 0.001;
sim_step_cnt = floor(sim_time / delta_t);

% Standard deviation of step size N(0,sigma)
sigma = (2*D*delta_t)^0.5;

steps = normrnd (0, sigma, sim_step_cnt, 3);

% Trace is just the cumulative sum of steps
trace = cumsum(steps);

end
```

If you want to simulate the diffusion of many particles released from the origin, and want to find the concentration at a distance depending on the time, you can approximate it with the following MATLAB code. For more exact results you should modify the code and choose Δt as small as possible. In this code snippet, it is assumed that counting the molecules passing from the predetermined distance gives the concentration at that distance if you choose Δt small enough.

```
function [ time_line ] = ...
   diffusion_sim3D_pointSource_timeHistogram(...
   r,   ... Distance in micro meters
   D,... Diffusion coefficient in micro meter^2 / seconds
   numMolecules,  ... Number of released molecules
   sim_time,  ... Duration of the simulation in seconds
   replication  ... Replication count
   )
%%%%%%%%%%%%%%%%%%%%%%%%%%%%%%%%%%%%%%%%%%%%%%%%%%%%%%%%
% This function simulates simple diffusion process
% and finds the C(r,t) estimate at a distance
% No Receiver/Reception is assumed
%%%%%%%%%%%%%%%%%%%%%%%%%%%%%%%%%%%%%%%%%%%%%%%%%%%%%%%%
delta_t = 0.001;
sim_step_cnt = floor(sim_time / delta_t);

% Standard deviation of step size N(0,sigma)
sigma = (2*D*delta_t)^0.5;
mol_position1 = zeros (numMolecules, 3);

time_line = zeros (1, sim_step_cnt);
for i=1:replication
  for t=1:sim_step_cnt
    rem_mol_cnt = size(mol_position1,1);

    % propagate the molecules via diffusion
    mol_displace = normrnd (0, sigma, rem_mol_cnt, 3);
    mol_position2 =  mol_position1 + mol_displace;
```

```
        % evaluate the previous/current distance
        mol_rPrev = sum( mol_position1.^2, 2 ).^0.5;
        mol_rCurr = sum( mol_position2.^2, 2 ).^0.5;

        % find the molecules passing from the r
        r_mask_1 = mol_rPrev < r & mol_rCurr > r;
        r_mask_2 = mol_rPrev > r & mol_rCurr < r;

        % record the time_line
        time_line(t) = time_line(t) + ...
                       nnz(r_mask_1) + nnz(r_mask_2);

        mol_position1 = mol_position2;
      end
    end

    time_line = time_line / replication;

    end
```

For the reception process, we should consider small changes in the simulator code depending on the reception process and the environment. To simulate a CvD system, one needs to define the reception process.

2.1.1 Simulation of CvD

If we consider the CvD system, we should also consider the reception process at the receiver side. In nature reception process is done via ligand-binding and the molecule hitting to the receiver is absorbed and removed from the environment. If we are considering the absorption process at the receiver side, considering the $C(x, t)$ formulation is not the correct way for finding the hitting time histogram.

Therefore we need to consider the absorption process and if we consider the case when there is an absorber, the analytical solution gets complicated for higher dimensions, and adding reflectors in the environment makes it even tougher. One can find the time independent absorption probability in the long run via utilizing the symmetry and the image method. However, in communication theory we need the time distribution of particle absorption we can not have an infinite length symbol duration to reach steady state. While dealing with absorbers, the dynamics of the process changes and is named as the First Passage Process (FPP) [18].

In the 1-D and 2-D environments, diffusing particles hit the receiver in the long run with probability 1 (recurrent process). However, when we consider the 3-D environment, there is a nonzero surviving probability for a diffusing particle [18]. In the 1-D environment, the first hitting probability is

$$f_{hit}(r_0, t) = \frac{r_0}{\sqrt{4\pi D t^3}} e^{-r_0^2/4Dt} \quad (8)$$

where r_0 is the distance to the absorber point. First hitting probability in the 1-D environment is inversely proportional in $t^{3/2}$. In the 1-D environment, we have the closed form solution for first hitting probability function. However for the 2-D or 3-D environments, even with a symmetrical receiver, the closed form solution is a hard surface integration or differential equation problem. Hence, we simulate the diffusion channel and the reception process. If we want to simulate the diffusion with an absorber receiver, we define the reception as removing the hitting molecule from the environment. Hence, we move the molecules in the 3-D environment according to diffusion dynamics, remove and record the molecules when they hit the receiver.

For simulating the basic scenario of a point source transmitter and an adsorbing spherical receiver, we can extend the basic diffusion of multiple particles. Here we give a basic MATLAB code for particles being released from a point source at the origin $(0, 0, 0)$ with a receiver of radius r_{rcv} located at a distance d. Any molecule coinciding with the receiver volume is regarded as a received molecule and removed from the environment.

```
function [ rcv_molecule_histogram ] = ...
   diffusion_sim3D_pointSource_sphereRcv_timeHistogram(...
   d,   ... Source-receiver separation in micro meters
   r_rcv ... Radius of the receiver body
   D,... Diffusion coefficient in micro meter^2 / seconds
   numMolecules,   ... Number of released molecules
   sim_time,    ... Duration of the simulation in seconds
   )
%%%%%%%%%%%%%%%%%%%%%%%%%%%%%%%%%%%%%%%%%%%%%%%%%%%%%%%%%
% This function simulates a simple CvD mechanism using
% the diffusion process for an adsorbing point source and
% sphere receiver and finds the histogram of the number
% of received molecules
%%%%%%%%%%%%%%%%%%%%%%%%%%%%%%%%%%%%%%%%%%%%%%%%%%%%%%%%%
delta_t = 0.001;
sim_step_cnt = floor(sim_time / delta_t);

% Standard deviation of step size N(0,sigma)
sigma = (2*D*delta_t)^0.5;

% Molecule release point is at (0, 0, 0)
mol_position1 = zeros(numMolecules, 3);

% Location of the center of the receiver body
center_rcv = [d + r_rcv, 0, 0];

rcv_molecule_histogram = zeros (1, sim_step_cnt);
for t=1:sim_step_cnt
rem_mol_cnt = size(mol_position1, 1);

% propagate the molecules via diffusion
mol_displace = normrnd(0, sigma, rem_mol_cnt, 3);
mol_position2 =  mol_position1 + mol_displace;
```

```
% calculate the pairwise difference between the location of
% each molecule and the center of the receiver
pairwiseDiff = bsxfun(@minus, mol_position2, center_rcv);

% euclidean distance of molecules to receiver center
dist_2_rcv = sqrt(sum(pairwiseDiff.^2, 2));

% find the received molecules
rcv_mask = dist_2_rcv < r_rcv;

% record the histogram
rcv_molecule_histogram(t) = nnz(rcv_mask);

% remove received molecules from environment
mol_position1 = mol_position2(~rcv_mask, :, :);
end

end
```

Although the code itself is very simple, when simulating a complete communication environment, we may need to perform tens of thousands of molecule releases, each release representing bits or symbols. When this happens, the number of particles roaming free in the communication medium gets very large, and various tricks are needed for a shorter simulation duration.

The first trick is to constrain the size of the communication medium by creating a very big bounding region around the communicting pair. This way, if any molecules reach very far, we can eliminate and remove those molecules from the environment since the probability that they will make it back to the receiver is very low. By the nature of the diffusion process, the longer the communication is carried on, the farther apart the molecules released early on will go. A very large constraining region will have a very small effect on the simulation performance and a very small region constraint will keep molecules from taking the so-called 'scenic route' and reaching the receiver. Therefore, the size of the constraining region should be chosen carefully.

Another improvement can be made by choosing the simulation time step Δt adaptively depending on the closeness to the receiver. The molecules are moved at every time step, hence they have discrete movements and we just consider the final locations to decide whether it is absorbed or not. Therefore, some of the molecule movements, those actually hit to the receiver, can be falsely categorized due to last location being outside the receiver. To resolve this issue, whether implementing the test for line segment (path) and receiver sphere intersection should be considered or choosing Δt as small as possible is encouraged. Choosing Δt as a small increment may increase the run time drastically, hence we also encourage to use region adaptive Δt values. If a molecule is close to the receiver its time step should be small, on the contrary if a molecule is far away from the receiver its time step may be chosen a larger value and it waits for other molecules.

Another notable improvement can be made is by placing the receiver's center at the origin and releasing the molecules from the point $(-(d+r_{rcv}), 0, 0)$. This way, the line of code where we calculate the pairwise difference between the location of each molecule and the center of the receiver will become obsolete. Moreover, instead of calculating the Euclidean distance of the molecules and the receiver's center, we may use the Euclidean squared distance for comparison. This way, we will not need to perform a square root operation in every simulation step, but will detect the received molecules by comparing the sum of squares of the molecule locations with r_{rcv}^2.

This basic point source transmitter—spherical receiver simulation can also be easily converted to one having a spherical transmitter. In this case, it is crucial to implement the characteristics of the transmitter. If the transmitter is also adsorbing, then the reception mechanism should also be implemented at the transmitter side. If the transmitter is not adsorbing, then the molecules trying to diffuse towards the inside of the transmitter should be blocked by the transmitting body. There are several ways of implementing transmitter blockage. The first and easiest way is to roll back the movement of any molecules that end up inside the transmitter body at the end of a simulation step. These molecules can be thought of as staying still for a single simulation step. Another way is to re-draw the molecule displacement from the normal distribution. However, this choice of blockage simulation will result in an immense number of trials at the start of molecule release since the molecules are most likely to end up inside the transmitter early on in the simulation. One other choice is flipping the sign of the molecule displacement vector to move the molecule to the opposite direction, therefore preventing it from stepping into the transmitter body.

2.1.2 Deciding Replications

One of the most crucial factors in running simulations is the number of replications (runs) with different random seeds in order to rule out the effect of random number generation. If we consider the actual implementation of a system as a population and the simulation runs as the samples from that population, according to the Central Limit Theorem, we must run the simulations at least with 30 different random seeds for each of the different random aspects of the simulation. However, if the sample sample size is less than 30, the Central Limit Theorem will work if the distribution of the population is not severely nonnormal [8]. In this section, we compare the results of 30 different sample sizes against smaller ones to see if there is a statistically significant difference inbetween. If there is no difference we can conclude that we do not have to take 30 different runs, so we can get the same results with smaller samples. The main random behaviours in the simulation of CvD are movements of the molecules and the bit sequence that the transmitter transmits to the receiver.

To compare the significance of different sample sizes, we use student t-test. Our null hypothesis is that the mean data rate of n samples ($n < 30$) is equal to mean data rate of 30 samples. For this purpose, we compare the mean data rates of all samples from 2 to 28 with 30 samples. Although we could not reject the null hypothesis in any of these tests, the p values of these tests differ from each other. The technical definition of the p-value is the smallest level of significance that would lead to rejec-

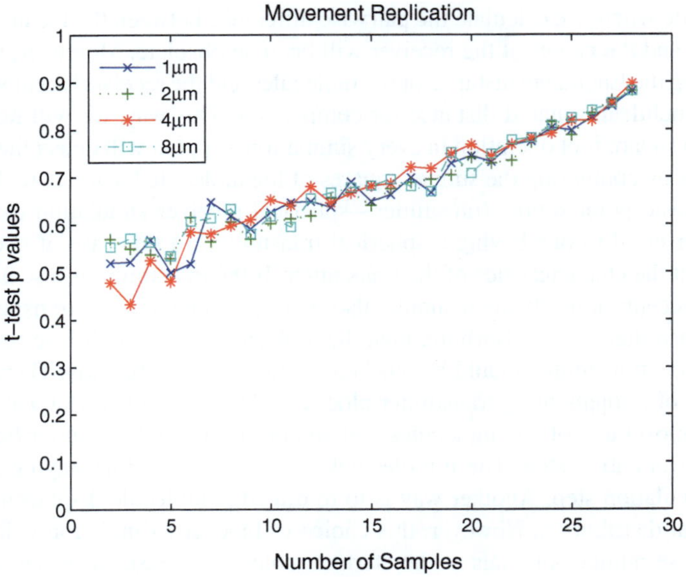

Fig. 2 Comparison of p-values for movement replication

tion of the null hypothesis [8]. Therefore, we can say the p-value is the power of the test in the sense that, as the p-value increases the test becomes more reliable.

Movement Replication

In Fig. 2, we present the comparison of the p-values for the replication of the movement of messenger molecules. We can conclude from the figure that the effect of the number of samples to the mean of the data rate is independent from the distance between the transmitter and the receiver. Moreover, as the number of samples increases (approaching 30), the p-value approaches to 1 as expected. As stated earlier, even for two samples the test does not reject the null hypothesis, i.e., the mean of the small sample is equal to the mean of the 30 samples. However, since the result is not as strong as larger number of samples, the researcher must decide how many samples would be sufficient regarding the power of the test with different sample sizes.

Bit Sequence Replication

In Fig. 3, similar to the movement replication, we observe that the effect of the number of samples is independent from the distance. Again, for bit sequence replication even two samples are sufficient for not-rejection of the null hypothesis, but the power of the test increases as the number of samples increases. The researcher must decide that how many sample are needed regarding the power of the test.

In conclusion, two different sample sizes must be selected for the implementation, one for replication of molecule movements and one for replication of bit sequences.

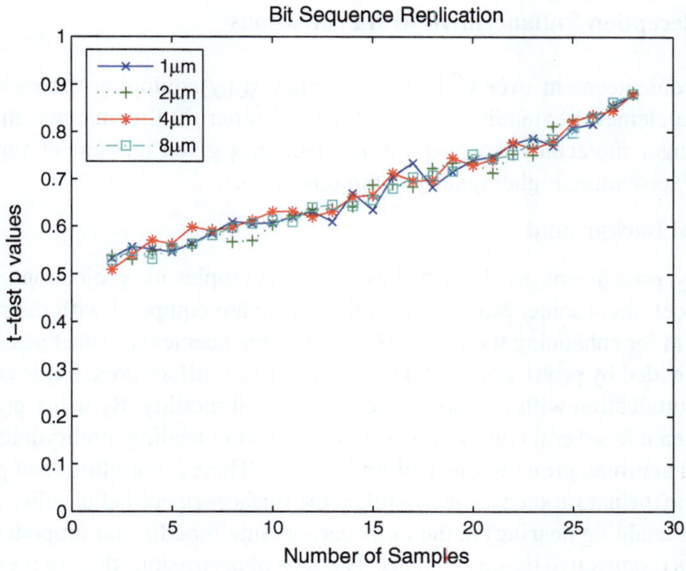

Fig. 3 Comparison of p-values for movement replication

Note that, if n is the selected number of samples for molecule movement replications and m is the number of bit sequence replications, in total $n \times m$ runs have to be run for all combinations of movement and bit sequence random seeds.

3 Simulation of Complex CvD Systems

3.1 Transmission and Reception Enhancements

Due to the Brownian motion characteristics in the diffusion environment, not all molecules reach the receiver body. Many of them scatter away, especially when the transmitter and receiver bodies are farther apart. As shown in [14], the distance between the transmitter and receiver bodies is a key factor that affects the propagation of molecules and increasing this separation degrades the reception performance substantially. Thus, enhancements at both the transmission and reception stages of 3-D diffusion are crucial for sustaining a successful communication. To this end, we present two manners of enhancements that can be implemented at either the transmission and/or the reception stage.

3.1.1 Reception Enhancement with Protrusions

The first enhancement over CvD can be achieved by equipping the receiver with protrusive elements extending towards the transmitter. In this manner, the number of messenger molecules reaching the receiver in a given amount of time can be increased, providing higher quality communication.

Biological Background

In biology, protrusions are described as arm-like cytoplasmic projections extending from the cell membrane. Many cells in the nature are equipped with the protrusion mechanism for enhancing their operations. The enhancements to the basic cell operations provided by protrusions include expanding the surface area, improving chemical communication with neighboring cells, and cell motility. By using protrusions, cells can enable several cellular processes like wound healing, embryonic development, and neuronal growth-cone pathfinding [16]. There is a multitude of protrusive elements in the nature such as microvilli on the surfaces of epithelial cells, stereocilia (hair cells enabling hearing) in the inner ear, or lamellopodia and filopodia found in fibroblasts (connective tissue cells) [5]. The type of protrusions that we mention here for enhancing CvD are based on the filopodia type protrusions.

The working principle of the protrusions is a continuous and ongoing process in the basic operations of a cell. A cell continuously monitors the environment it is in, always looking for signals that can arrive from the environment. These signals can be physical, chemical, or in some cases even luminous. A signal from the environment is detected by the receptors located on the cell membrane. Upon detection of a signal, another cellular process is triggered and the existence of a signal is transmitted to the cell interior. A cell equipped with the protrusion ability begins extending its protrusions once the internal signaling begins. Depending on the type of the cell and the type of the received signal from the environment, protrusions can be extended on the whole surface of the cell, or can also be directed towards the source of the external signal.

The filopodal protrusion mechanism in cells is driven by a branched network of actin filaments found in the cytoplasm. Once the decision of extending protrusions is made inside the cell, motor proteins called Myosin-X bundle polymerized actin filaments together and start pushing the cell membrane outwards [16]. To hold actin filaments together, Fascin protein is utilized and Dia2 proteins monitor the actin polymerization at the elongating tip. The base of the protrusion structure holds a cross-linked form for support, so that the actin filaments will not sink back into the cytoplasm. Steps of the protrusion extension mechanism can be observed in Fig. 4. We also note that while the protrusion is being extended, the external signal triggering this process is still continuously monitored with the receptors on the protruding cell membrane. In the nature, the filopodal protrusions may extend up to 40 μm in length and have a radius of several hundred nanometers. The versatility of protrusion size makes it a very useful tool in enhancing cellular operations.

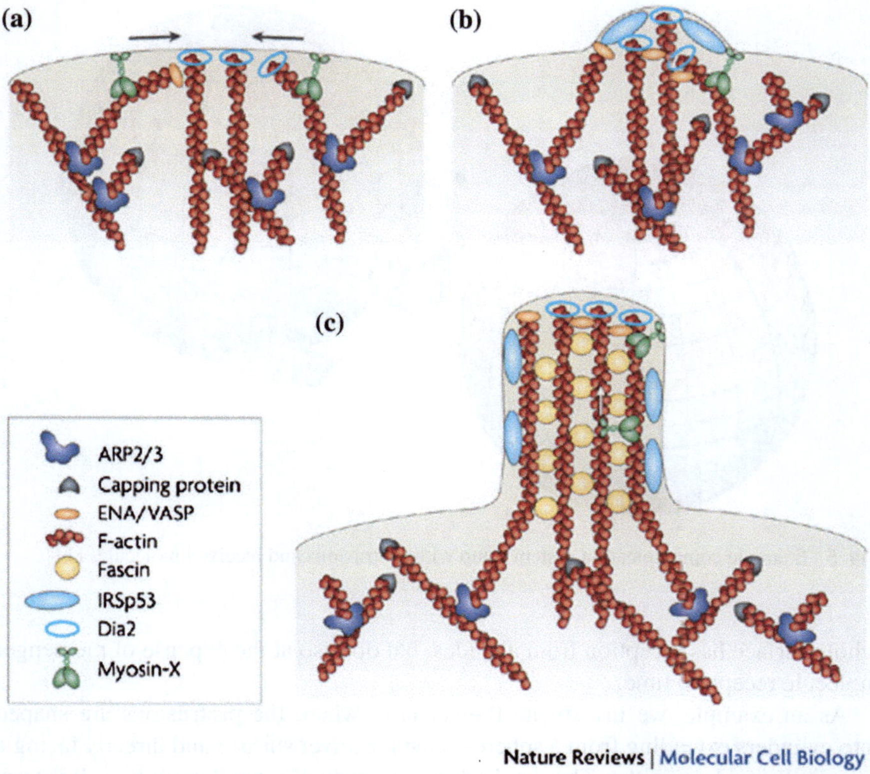

Fig. 4 The working model for filopodia formation [16]. **a** Myosin-X proteins pull actin filaments together towards the protrusion site. **b** Actin filaments push against the cell membrane. **c** Fascin proteins link the actin bundles to form a strong protrusion

CvD with Protrusion Enhancement

The protrusion mechanism in cells can be adapted to the CvD scenarios such that the receiver body is equipped with protrusions that are capable of messenger molecule reception. There are a multitude of ways for implementing protrusions on the receiver surface. The two most important points in protrusion enhancement are the deployment strategy on the surface and the geometric shape of these extensions. Protrusions can be deployed on the whole surface of the receiver either randomly or in a uniform manner. The deployment can also be made such that the part of the receiver facing the transmitter source is favored. Protrusions can be extended fully perpendicular to the receiver membrane, or they can face the transmitter source directly. Moreover, the geometric shape of the protrusions can be either conical or cylindrical. Each combination of deployment, direction, and geometric shape has various advantages and drawbacks. For example, protrusions favored on the transmitter-facing side of a receiver are very useful for capturing messenger molecules earlier, but the passerby molecules are ignored on the other sides of the receiver. A uniform distribution on the

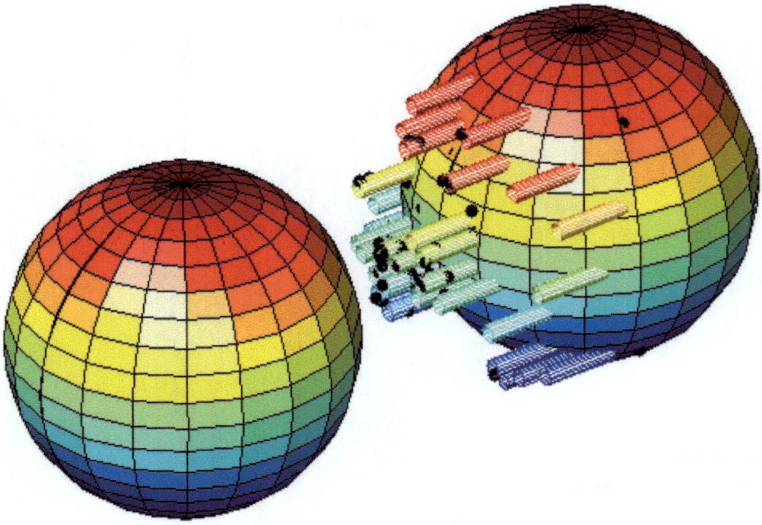

Fig. 5 Example communication system setup with protrusions and received molecules [11]

whole surface has reception from all sides, but does so at the expense of messenger molecule reception time.

As an example, we investigate the scenario where the protrusions are shaped into cylinders extending from a sphere-shaped receiver surface and directly facing a sphere-shaped transmitter. These cylinders have radii r_{prot} and length h_{prot}. It is very important to conserve the total volume of the receiver when extending protrusions for making sound comparisons between classical CvD and protrusion enhanced CvD approaches. Therefore, we conserve the total volume of the receiver NeN by extracting the volume occupied by the protrusions from the backside of the receiver.

Figure 5 shows an example communication system with a few tens of protrusions and some messenger molecules received either by the receiver surface or the protrusions. The protrusion sites are chosen at random, with the constraint that they face the transmitter NeN. The messenger molecules are removed from the environment once they come in contact with either the receiver surface or the protrusions. This way, the reception of messenger molecules is improved by decreasing the effective distance between the transmitter-receiver pair.

Performance Evaluation

We investigate the effect of protrusions on the CvD performance by simulations. In this setup, we have a spherical transmitter-receiver pair, where the receiver is equipped with protrusions facing the transmitter. The release of messenger molecules on the transmitter body is made from a single point located on the transmitter surface, directly facing the receiver. The performance metric we use is the probability of a messenger molecule hitting the receiver in a given amount of time t_s, called the symbol duration. The aim is to receive as many molecules as possible in a single

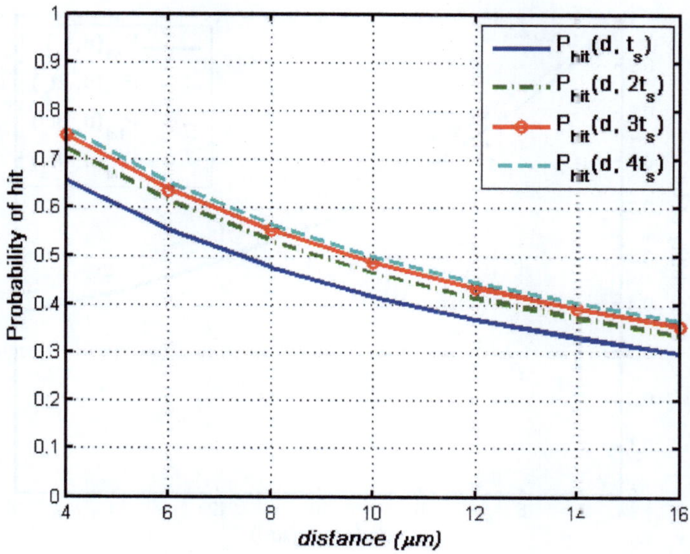

Fig. 6 Probability of hitting the receiver versus various transmitter-receiver pair distances without protrusion [11]

symbol duration, so that upcoming messages on consecutive symbol durations do not hamper each other's operations.

In Figs. 6 and 7, we observe the effect of protrusion on hitting probabilities for various amounts of distances between the transmitter and the receiver. The x-axis denotes the distance between the transmitter and receiver bodies, and the y-axis denotes the probability of a messenger molecule being received by either protrusions or the receiver surface. Each curve represents the probability of hit in a duration of 1–4 t_s. We observe that using protrusion enhances the probability of receiving a messenger molecule significantly. Moreover, the gaps between the curves are smaller, which means that the messenger molecules lagging behind and creating inter-symbol interference are reduced. Decreasing the inter-symbol interference and increasing the hitting probabilities play a crucial role in successful communication. We also observe that, keeping the protrusion length at 3.5 μm, the advantage obtained from extending protrusions decreases, and the gap symbolizing the inter-symbol interference increases.

3.1.2 Tunnel-Based Approach Using Destroyer Molecules

A second type of enchancement over CvD communication system can be achieved by shaping the molecular signal. This idea focuses on manipulating the molecular communication channel between the transmitter—receiver couple using a secondary type

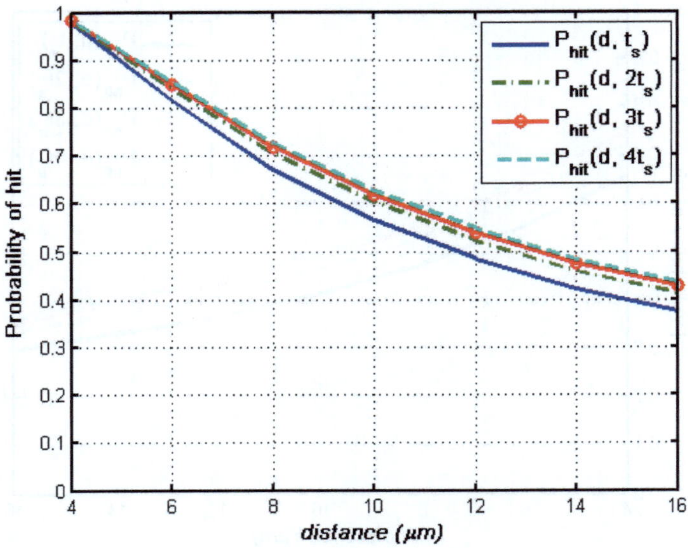

Fig. 7 Probability of hitting the receiver versus various transmitter-receiver pair distances with protrusion, $r_{prot} = 0.5\,\mu\text{m}, h_{prot} = 3.5\,\mu\text{m}$ [11]

of molecules, called Destroyer Molecules (DM), in an effort to shape the received signal according to the desired signal structure.

Biological Background

Neuromuscular junction (NMJ) is one of the occurrences in biological systems where two cells communicate with each other using an intermediary molecule that propagates in the extracellular environment following diffusion dynamics.

NMJ resides between a nerve and a muscle cell couple. When the muscle needs to be contracted, the nerve cell releases pre-synthesized special neurotransmitter molecules, called Acetylcholine (ACh), into the NMJ. These molecules propagate in this environment and when they get close to the cell membrane of the muscle cell, they bond with the transmembrane receptors, called the ACh receptors (AChR). The neurotransmitters stay in the bounded state for some time after which the bond degrades and the ACh molecules are again set free to the NMJ.

As seen in Fig. 8, the NMJ is a semi-closed environment and the molecules inside usually move between the two cells. Hence, after the degradation of the bonds between ACh and AChR, the neurotransmitter molecules are highly likely to re-bond with the receptors. Such re-bondings cause further unwanted muscle contractions, and after a few muscle contraction signals, the NMJ can be filled with ACh molecules. Thus all further contraction signals will be blocked after several contractions. To keep the communication between the nerve and muscle cell couple possible, the ACh molecules in the environment should be removed from the NMJ after the muscle cell is successfully contracted. This cleaning process is achieved through the use of a

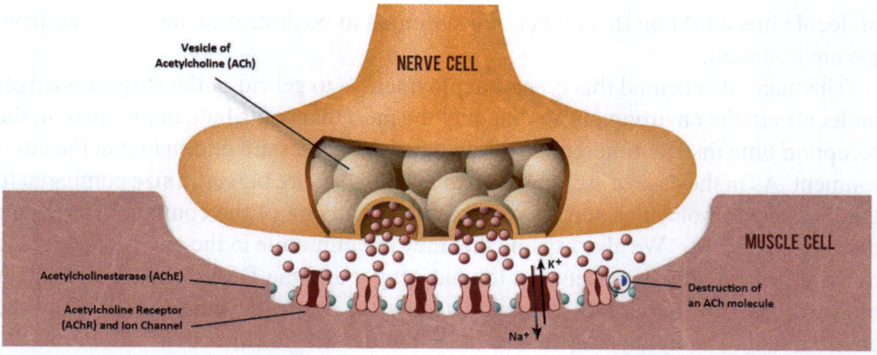

Fig. 8 Neuromuscular junction [13]

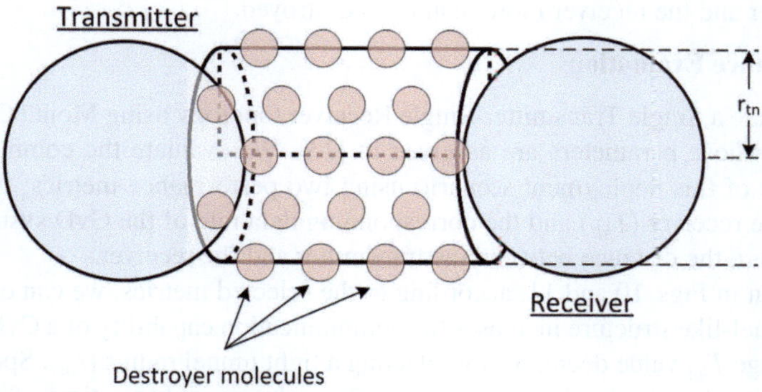

Fig. 9 Cylindrical tunnel environment for CvD [13]

secondary type of molecule, called Acetylcholinesterase (AChE). The AChE molecules interact with the ACh molecules and remove them from the NMJ by hydrolizing ACh into its basic components, acetate and choline. Thus, AChE molecules enable the muscle cell to be capable of responding to further contraction signals.

Tunnel Structure

If we approach the NMJ system from a communication perspective, the receiver can be considered utilizing specialized so-called Destroyer Molecules (DM) to control the shape of the signal and eliminate the undesired components of a signal. Also, the usage of these molecules reduces or eliminates the effects of the Intersymbol Interference (ISI) and allows the selection of shorter symbol durations. From a topological point of view, the destroyer molecules can be deployed in the environment in many different ways. As an example, we investigate the case where the DMs are deployed to form a cylindrical tunnel-like structure that forms a spatially restricted path between the transmitter and the receiver (Fig. 9). This structure follows biological extensions like filopodia and cytoneme. We assume that, when a messenger molecule hits a DM on this tunnel, it is assumed to be destroyed and removed from the environment.

The main idea behind this type of deployment is to get rid of the stray messenger molecules in the environment so that only the ones that contribute to the spike in the reception time distribution remain while other molecules are eliminated in the environment. As in the case of AchE, we assume that DMs are bigger in size compared to the messenger molecules and they are connected to one of the communicating pair through other DMs. We also assume that they are immobile in the environment. Due to the chemical attraction between the messenger and the DMs, when a messenger molecule gets close to a DM, it is attracted by the destroyer and removed from the environment. The cylindrical tunnel has a radius of r_{tn}, which is a variable in the simulations. In order to simplify the analysis, we assume that the whole tunnel is composed of DMs and any molecule that diverts from the shortest path between the transmitter and the receiver more than r_{tn} is destroyed.

Performance Evaluation

We simulate a Single Transmitter-Single Receiver topology using Monte Carlo simulations, whose parameters are as given in [13]. We evaluate the communication capability of this deployment scenario using two performance metrics: the hitting time at the receiver (T_{hit}) and the corresponding data rate of the CvD system based on r_{tn} and d, the distance between the transmitter and the receiver.

As seen in Figs. 10 and 11, according to the selected metrics, we can clearly see that a tunnel-like structure increases the communication capability of a CvD system. The average T_{hit} value decreases by selecting a tight tunnel radius (r_{tn}). Specifically, choosing r_{tn} close to d reduces the average T_{hit} value roughly ten times. This is due to the elimination of slow moving molecules from the environment by the DMs.

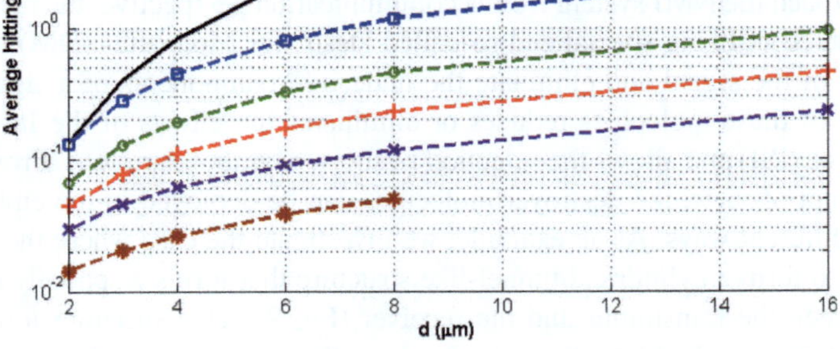

Fig. 10 Effect of destroyer molecules on average hitting time [13]

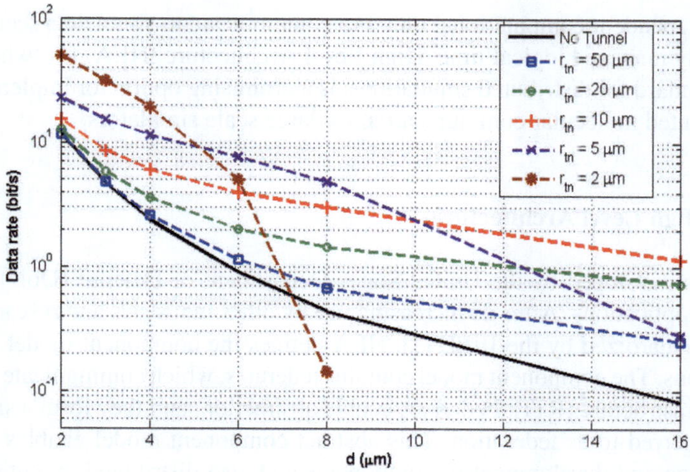

Fig. 11 Effect of cylindrical tunnel environment on data rate [13]

The last curve ($r_{tn} = 2$ m) does not yield any results since no molecules arrive at the receiver when the d/r_{tn} value is too high (i.e., all of them are destroyed due to the narrowness of the tunnel).

Furthermore, the data rate of the system considerably increases with the usage of the tunnel. This is also more prevalent with increasing d values. The d/r_{tn} limit is also apperant in this metric. Beyond $d = 4m$, the data rate of the tunnel environment with $r_{tn} = 2$ m drops very quickly. Similar to the average T_{hit}, this drop is due to the fact that in such environments the tunnel becomes too tight and the molecules hit the receiver with a very small probability.

Based on these results, it is apperant that this tunnel-like deployment of DMs greatly increases the performance of the CvD channel. However, in a realistic environment, the question of how a tunnel structure can be constructed between the communicating pair arises. Such a tunnel will also be costly in terms of energy when considering the energy cost of synthesizing the DMs. More relaxed and less restrictive DM deployment schemes must be investigated.

3.2 Distributed Simulations

Due to the need for simulating large number of nano-scale objects, the scalability of the simulation of molecular communication is an important concern. Flexibility, interoperability, and reusability are other design criteria for the simulation of molecular communication. Selecting the right architecture, which supports software component reuse and high level of scalability enables a growing library of simulation components. Distributed simulation can enable simulation of complex scenarios. A distributed simulation is a collaborative system in which each simulation unit runs on an independent computational unit and communicate to simulate a scenario in a commonly managed logical time. High Level Architecture (HLA) [1], which is the IEEE standard for distributed simulations, is a promising option for implementation of distributed molecular communication for large scale simulators.

3.2.1 High Level Architecture

HLA was developed by the United States Department of Defense (DoD) to cover defense applications. After its increasing use by other industries and research areas, it was standardized by the IEEE [1]. HLA defines the component model and their interactions. The component model contains federates, which communicate over Run Time Infrastructure (RTI). Federates enable abstraction, and they form a simulation model referred to as federation. This abstract component model enables independent design and development of components and also distributed execution of the simulation.

The RTI is the backbone of the federation, and provides synchronization, communication, and data exchange services to the federates [3]. Each federate can be an independent event or time driven simulations, real time simulation with human interaction, live system or equipment. The HLA does not restrict what is modeled in a federate; it defines the interaction among them. There are six classes of services that RTI provides [7]:

1. *Federation management:* Basic functionality required to create and execute a federation.
2. *Declaration management*: Management of data exchange between federates, using the information provided by federates.
3. *Object management*: Creation, deletion, identification, and other services at the object level.
4. *Ownership management:* The dynamic transfer of ownership of object/attributes during an execution.
5. *Time management:* Synchronization of runtime simulation data exchange.
6. *Data distribution management*: Routing of data among federates during federation execution.

The implementation of these services are not in the scope of HLA interface specification. The specification only focuses on the way these services are accessed.

3.2.2 A Distributed Simulation Design

Using HLA, it is possible to create a simulation design that focuses on interoperability, re-usability, and scalability. With such a design, it is possible for modules executing on different platforms to communicate. Common software libraries can be developed and used to create large scale simulations, and it is possible to run sim-

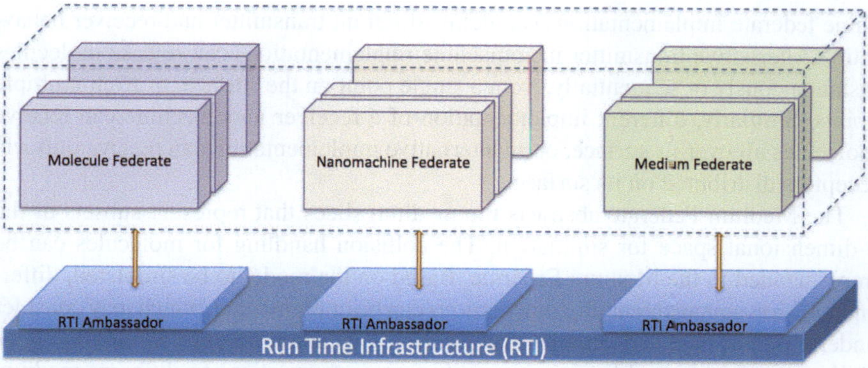

Fig. 12 A distributed simulation architecture

ulations on distributed computer systems. There are four primary principal benefits of executing a simulation program across multiple computers [10]:

1. *Reduced execution time.* It is possible to decrease execution time by dividing simulation computation into smaller sub-computations, and distributing these computational tasks to different computational units,
2. *Geographical distribution.* Running the simulation on geographically distributed computers enables interaction with users or live systems which have to be in different locations.
3. *Integrating simulators that execute on machines from different manufacturers.* A distributed approach enables interaction between different platforms. This approach enables the use of hetoregenous hardware and software system to execute a simulation.
4. *Fault tolerance.* If a unit fails while using multiple computational units, another unit can pick up the load of the failing unit.

A possible distributed architecture [4] is shown in Fig. 12. The simulation is defined as a Molecular Communication Federation. Separate federates are defined for molecules, nanomachines and the medium. The federates communicate with each other over RTI. Scalability is achieved by distributing molecule related tasks to Molecule Federates, nanomachine related tasks to Nanomachine Federates, and slicing the 3-D space and assigning a medium federate to manage each slice.

The Molecule Federate models the molecules in the environment and is responsible for the movements of the molecules. The molecules propagate in the three dimensional space. This movement can be modeled as three independent movements in each dimension as described in Sect. 2.1. The environmental parameters are defined in medium slices. The Molecule Federate subscribes to the Medium Federate attributes. Hence, when a molecule moves into another medium, the parameters are communicated to the Molecule Federate by RTI.

The Nanomachine Federate abstracts the nanomachines in the environment and is responsible for the transmission and the reception mechanisms. Different nanomachine federate implementations can define different transmitter and receiver behaviours. Alternative transmitter nanomachine implementations can release molecules instantaneously or sequentially, from a single point on the surface, or from multiple points. Similarly, different implementation of a receiver nanomachine can receive molecules all over its surface, or an alternative implementation can receive only via receptors distributed on its surface.

The Medium Federate abstracts the medium slices that represent subsets of the 3-dimensional space for simulation. The collision handling for molecules can be implemended in the Medium Federate. Based on the model to be simulated, different collision management implementations can be implemented, which may consider underlying physical and chemical laws during a collision process. Simulation scalability can be achieved by assigning different medium slices to different medium federates.

3.2.3 Performance Evaluation

For the performance evaluation of the distributed simulations, the speedup (S) can be defined as

$$S = \frac{T_s}{T_m} \tag{9}$$

where T_s is the execution time with a single node and T_m is the execution time with multiple nodes. Linear speedup is achieved when speedup is equal to the number of nodes used in the execution. Different algortihms or architectures for distributed execution of simulations for molecular communication can be compared using speedup metric, and optimal architecture can be selected.

4 Conclusions

This chapter approaches diffusion from a communication engineer's perspective and provides the researcher some useful hints and intricacies for designing molecular communication simulations in an effective manner. These hints and intricacies include the selection of the number of replications required in terms of movement and bit sequence for fast but meaningful results. We utilize this information further by discussing the design of more complex CvD systems such as tunnel-based approach that utilizes destroyer molecules and distributed simulator design based on HLA.

Acknowledgements This work has been partially supported by the State Planning Organization (DPT) of Republic of Turkey under the project TAM with the Project Number 2007K120610, Bogazici University Research Fund (BAP) under Grant Number 7436, and by the Scientific and Technical Research Council of Turkey (TUBITAK) under Grant Number 112E011. M. Şükrü Kuran partially carried out the work presented in this paper at LINCS (http://www.lincs.fr).

References

1. Modeling and simulation (m & s) high level architecture (hla) (IEEE 1516-2010 series). http://ieeexplore.ieee.org/xpl/mostRecentIssue.jsp?punumber=5553438. Accessed 29 Jan 2014
2. N3sim is a simulation framework for diffusion-based molecular communication in nanonetworks. http://www.n3cat.upc.edu/n3sim. Accessed 29 Jan 2014
3. AbouRizk S, Mohamed Y, Taghaddos H, Saba F, Hague S (2010) Developing complex distributed simulation for industrial plant. In: Proceedings of the 2010 winter simulation conference, pp 3177–3188
4. Akkaya A, Genc G, Tugcu T (2014) HLA based architecture for molecular communication simulation. Simul Modell Pract Theory. doi:10.1016/j.simpat.2013.12.012, http://www.sciencedirect.com/science/article/pii/S1569190X13001925
5. Ananthakrishnan R, Ehrlicher A (2007) The forces behind cell movement. Int J Biol Sci 3(5):303
6. Berg HC (1993) Random walks in biology. Princeton University Press
7. Dahmann JS (1997) High level architecture for simulation. In: First International workshop on distributed interactive simulation and real time applications, 1997, p 32
8. Douglas C, Montgomery GCR (2006) Applied statistics and probability for engineers. Wiley
9. Felicetti L, Femminella M, Reali G (2012) A simulation tool for nanoscale biological networks. Nano Commun Netw 3(1):2–18
10. Fujimoto RM (2000) Parallel and distributed simulation systems. Wiley
11. Genc G, Yilmaz HB, Tugcu T (2013) Reception enhancement with protrusions in communication via diffusion. In: 2013 First International Black Sea conference on communications and networking (BlackSeaCom), pp 89–93. IEEE
12. Gul E, Atakan B, Akan OB (2010) Nanons: a nanoscale network simulator framework for molecular communications. Nano Commun Netw 1(2):138–156
13. Kuran MŞ, Yilmaz HB, Tugcu T (2013) A tunnel-based approach for signal shaping in molecular communication. In: 2013 IEEE international conference on communications workshops (ICC), pp 776–781
14. Kuran MŞ, Yilmaz HB, Tugcu T, Özerman B (2010) Energy model for communication via diffusion in nanonetworks. Nano Commun Netw 1(2):86–95
15. Llatser I, Pascual I, Garralda N, Cabellos-Aparicio A, Alarcon E (2011) N3sim: a simulation framework for difusion-based molecular communication. In: IEEE Technical Committee on Simulation, vol 8
16. Mattila PK, Lappalainen P (2008) Filopodia: molecular architecture and cellular functions. Nat Rev Mol Cell Biol 9(6):446–454
17. Moore M, Enomoto A, Nakano T, Suda T, Kayasuga A, Kojima H, Sakakibara H, Oiwa K (2006) Simulation of a molecular motor based communication network. In: Bio-inspired models of network, information and computing systems
18. Redner S (2001) A guide to first-passage processes. Cambridge University Press
19. Saxton MJ (2007) Modeling 2d and 3d diffusion. In: Methods in membrane lipids. Springer, pp 295–321
20. Toth A, Banky D, Grolmusz V (2011) 3-d brownian motion simulator for high-sensitivity nanobiotechnological applications. IEEE Trans Nanobiosci 10(4):248–249
21. Tyrrell HJV, Harris K (1984) Diffusion in liquids. A theoretical and experimental study. Butterworth Publishers, Stoneham, MA

An Architecture of Calcium Signaling for Molecular Communication Based Nano Network

Amitava Mukherjee, Sushovan Das and Soumallya Chatterjee

1 Introduction

The Nobel laureate physicist Richard Feynman, in his famous speech in 1959 entitled "There's Plenty of Room at the Bottom", has pointed out the concepts in nanotechnology and described how the manipulation of individual atoms and molecules would give rise to more functional and powerful man-made devices. In his vision, he talked about having a billion tiny factories able to manufacture fully functional atomically precise nano-devices [1]. Several scaling issues would come up when reaching the nanoscale, which would require the engineering community to rethink totally the way in which nano- devices and nano-components are conceived. There is a need to rethink and redesign the way in which components and devices are created by taking into account the new properties of the nanoscale. A whole new range of applications can be enabled by the development of devices able to benefit from these nanoscale phenomena from the very beginning. These are the tasks at the core of the nanotechnology. Nanotechnology has been maturing since early 21st century.

A. Mukherjee (✉)
IBM India Private Limited, DLF IT Park I, Rajarhat Newtown, Kolkata 700156, India
e-mail: amitava.mukherjee@in.ibm.com

S. Das · S. Chatterjee
Department of Electronics & Communication Engineering, Jadavpur University, Kolkata 700032, India
e-mail: tapas1das2@gmail.com

S. Chatterjee
e-mail: sbchat13@gmail.com

J. Suzuki et al. (eds.), *Modeling, Methodologies and Tools for Molecular and Nano-scale Communications*, Modeling and Optimization in Science and Technologies 9, DOI 10.1007/978-3-319-50688-3_8

"Nanotechnology mainly consists of the processing of, separation, consolidation, and deformation of materials by one atom or by one molecule [1]." When the first simple structures on a molecular scale were obtained, the activities surrounding nanotechnology began to increase slowly and this term became more socially accepted in the early 2000s. The development of nanomachines, i.e., integrated functional devices consisting of nano-scale components which are able to perform simple tasks at the nano-level, is the aim of nanotechnology since 2000 onwards. Going one step ahead, the interconnection of nanomachines in a network or nanonetwork is proposed as the way to overcome the limitations of individual nano-devices.

2 Overview of Nanonetworks Comprising of Nanomachine (Node)

A nanonetwork is a system of interconnected or communicating nanomachines which may be conventional nanoelectronic devices, biological cells or biomimetic devices. From the term 'nano' it simply comes in our mind that one or more of the basic components of this network i.e., the transmitter, receiver, medium or message carriers are of nano-scale dimension (nano-scale refers to dimensions of 100 nm). Communication plays a very critical role in different aspects of nano-scale applications. Unlike conventional networks, nanonetworks use different physical principles than standard communication systems. These physical principles are suited to nanoscale systems, but have significantly different properties at the macroscale. In nanonetworks communication between two nano-machines can mainly takes place either using terahertz frequency electromagnetic wave or using molecular communication that are briefly discussed below.

2.1 Electromagnetic Communication

This is based on transmission and reception of classical communication principles need to undergo a change before being used in nanonetworks. Existing RF and optical transceivers suffer from several limitations that query the feasibility of EM communications among nano-devices. Few researchers have been working on the state of the art in molecular electronics of the terahertz band (0.1–10.0 THz) for EM communication among nanodevices [2]. A new propagation model for EM communications in the terahertz band is developed based on radioactive transfer theory and in light of molecular absorption [2]. This model accounts for the total path loss and the molecular electromagnetic radiations from components based on novel nano-materials. There are two ways for electromagnetic (EM) communication in the nanoscale: (i) Receive and demodulate an electromagnetic wave by means of a nano-radio, i.e., an electromechanically resonating carbon nanotube (CNT) that decodes an amplitude or frequency modulated wave and (ii) or by graphene-based nano-antennas used as potential electromagnetic radiators in the terahertz band. The tera-hertz band waves are very much prone to absorption noise. This is major limitation during propagation of Tera-hertz band waves over short distances [2].

2.2 Molecular Communication

Molecular communication is defined as the transmission and reception of information by means of molecules. Molecular communication is a new paradigm for communication between biological nanomachines over a nano and microscale range [3]. The molecular communication provides a mechanism for a nanomachine (i.e., a sender) to communicate information by propagating molecules (i.e., information molecules) that represent the information to a nanomachine (i.e., a receiver) [3].

Molecular communication involves four basic steps:

A transmitter encodes information in terms of different types of molecules or varying the concentration of molecules or ions, the transmitter nano-machine releases those stream of molecules known as molecular wave into an aqueous environment which acts as the propagation medium, this molecular wave propagates through this aqueous environment (propagation medium), a receiver nano-machine receives this molecular wave and finally the receiver nano-machine decodes the original information from the received molecular wave. Molecular transceivers are easy to incorporate in nano-machines due to their size are of nano-scale dimensions and molecular communication is also much more compatible compared to electromagnetic communication in nano-networks. Different molecular communication techniques are based on molecule propagation like walkway-based communication, flow-based communication and diffusion-based communication. Calcium (Ca^{2+}) signaling based communication occurring as biological cellular communication is a diffusion based communication (the propagation mechanism associated here is electro-diffusion).

A biological cell uses molecular communication that involves communication of information-carrying intracellular or intercellular molecular signals to accomplish sophisticated biological functions like respiration, nerve impulse conduction, hormone secretion, etc. However, understanding the role of cellular signaling in normal cell functioning and also under pathological conditions requires systematic modeling of the network (i.e., the interconnection) of cells and incorporating proper mathematical models for quantification of the associated electrochemical phenomena. One form of cellular signaling is calcium signaling in which the concentration of a stream of calcium ions (Ca^{2+}) is modulated spatio-temporally to bring about muscle contraction, cell differentiation, hormone secretion, etc. [4].

Besides the above two communication mechanisms mentioned there are some mechanisms in nano-networks. For example acoustic communication is also possible in communication between nano-machines but it is mainly based on transmission of ultrasonic waves. This mode of communication requires the nano-machines to be integrated with ultra-sonic transducers which should be capable to sense the rapid variations of pressure produced by ultrasonic waves and to emit acoustic signals accordingly. Again, the size of these transducers is a major problem during their fabrication and integration in the nano-machines.

In nanomechanical communication, the information is transmitted through hard junctions between linked nano-devices. The main drawback of this type of communication is that a physical contact is required between the transmitter and the receiver. Moreover, this physical connections should be precise enough to ensure that the desired mechanical transceivers are aligned correctly. This communication technique is not suitable in many application scenarios where nano-machines are deployed over large areas where physical-contacts between the interconnecting nano-machines are impossible. In Sect. 3, we briefly discuss the different types of communication scenarios in nano-network and highlight the basic features of those communications.

3 Communication Among Nano-Machines

A single nanomachine can perform simple tasks and in order to assemble more computational power for performing more complex tasks, it is imperative for nanomachines to be able to communicate with each other and work cooperatively. Commutation among nano-machines can incorporate two scenarios: (a) Communication between a nano-machine and a larger system, (b) Communication between two nano-machines [2]. As we have stated earlier that communications between nano-machines can be done in mainly in two ways. (a) Molecular communication, (b) Electromagnetic wave communication using terahertz frequency range.

3.1 Molecular Communication

In molecular communication, information is encoded in the type of molecules transmitted or in the concentration changes of ions. It is most suitable for communication in biological nanonetworks due to its bio-compatibility. This communication can take place over short range (from nm to mm) using molecular motors or calcium signaling, as well as over a long range (from mm to km) using pheromonal transport or molecular neuro-spike communication [2, 5]. The different types of molecular communication are discussed below.

3.1.1 Molecular Communication Using Molecular Motors

This is an analogue of wired communication for the nano-domain is a very useful way of molecular communication over short distance (nano-scale range). Most of intra-cellular communication are based on communication using molecular motors. In this form of molecular communication, different nanomachines are connected by microtubules which are like rails along which traffic of molecular motors can move. Molecular motors are protein based carriers of molecules that use chemical energy from ATP to walk along the microtubules carrying molecules from one point to another within the aqueous intracellular space as shown in Fig. 1. Different molecular motors have different step lengths and are capable of doing a certain amount of work. For instance, kinesin is a molecular motor that uses one ATP molecule to move steps of eight nm and in each step it generates six pN of force [6]. Molecular motors support bidirectional transport as they can move either from the centre of the cell to its periphery or the opposite and are said to be positive- ended or negative-ended. Specific molecular motors are able to carry different molecules using these intracellular molecular rails. Depending on the traffic, molecular motors move at a speed up to 400 mm/day [2]. By this ability of carrying molecules molecular motors act as carriers to transport information, i.e., molecules from the transmitter to the receiver nano-machine.

Basic communication features:

Molecular communication enabled by molecular motors takes place in aqueous medium [2]. The environment should include the necessary components and the biological and chemical conditions like temperature, humidity, medium viscosity and pH should be suitable for communication. As here the information is based on the chemical nature of the molecules, the nano network is highly sensitive to these conditions and the communication process can be adversely affected by sudden variations of these environmental conditions [2]. A proper network infrastructure should be developed before the beginning of the communication process. Depending

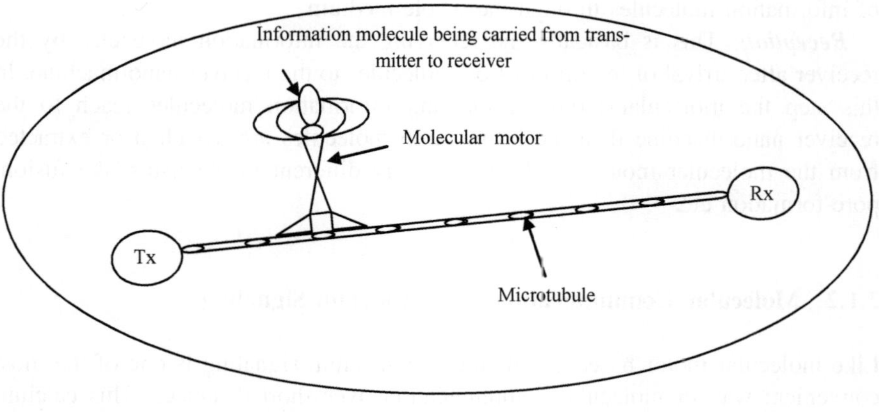

Fig. 1 Molecular communication over microtubules using molecular motors

on the nano-network infrastructure, a transmitter nano-machine will be able to support unicast or multicast communication [2]. To implement the unicast communication (i.e., communication from one single transmitter to one certain single receiver) the transmitter will have to be able to select a specific molecular rail to transmit the information molecule so that the information can reach to that particular destination receiver. To achieve a multicast communication(i.e. communication from one single transmitter to a certain group of receivers), the transmitter will have to release several molecules, each molecule containing the same data through different molecular rails so that each of these molecules reaches to different destination receivers. The propagation of molecular motors along a microtubule is unidirectional. But by indicating polarity we can also get bi-directional communication.

Communication process using molecular motors: Molecular motors carry the information molecules according to particular molecular rails from transmitter to receiver. To facilitate the reception, the transmitter uses protein tags that bind to specific receptors on receiver nano-machines [2]. The whole communication includes the following tasks:

Encoding: This is basically the generating of proper information molecule by selecting the right molecule according to the information (modulating signal). The entire process is related to the transmitter, when an external stimulus (information) is applied to the transmitter nano-machines the information molecules are generated by the transmitter. Thus it is possible to control the reaction of the reaction of the receiver according to the proper stimulus.

Transmission: This is basically the releasing of information molecules to the medium. We have stated earlier that molecular motors act as carrier for transmission of information from transmitter to receiver, so for successful transmission there will have to remain a high affinity between the information molecules and the molecular motors. Encapsulation techniques can be used to increase the affinity between the information molecules and molecular motors.

Propagation: This basically involves the movements of carriers (molecular motors) with information molecules. Here the micro-tubules or molecular rails controls the propagation direction restriction from diffusion of random movement of information molecules through the whole medium.

Reception: This is basically the receiving the information molecules by the receiver after arrival of the information molecules to the receiver nano-machine. In this step the molecular motors containing information molecules reach to the receiver nano-machine then the information molecules are detached or extracted from the molecular motors in the receiver by different mechanisms like fusion, pore-formation etc.

3.1.2 Molecular Communication Using Calcium Signaling

Like molecular motor based communication calcium signaling is one of the most convenient way of molecular communication over short distances. This calcium signaling is the basic mechanism of intercellular communication occurring in biological cells. In calcium signaling, a stimulus applied to the cell generates second messengers, inositol 1, 4, 5 triphosphate (IP_3), that bind to IP_3 receptors and trigger Ca^{2+} ion release from intracellular stores [4]. These Ca^{2+} diffuse through the aqueous medium in the cell. Their concentration, [Ca^{2+}], is modulated by cellular components (that act as a calcium signaling "toolkit") so as to produce different amplitudes or frequencies (spike rate) of the [Ca^{2+}]. This modulated [Ca^{2+}] is called the Ca^{2+} signal. According to the amplitude or frequency of the Ca^{2+} signal, cellular processes such as contraction, hormone secretion, differentiation, etc. are induced depending on the type of cell in which Ca^{2+} signaling is taking place [7]. This is short range communication [2] which can be both intracellular and intercellular. The entire framework of calcium signaling based network is discussed in Sect. 6. Figure 2 shows the steps in which calcium signaling takes place.

Communication features:

As stated earlier that calcium signaling is an approach of short distance molecular communication so naturally it implies that the transmitter and receiver nano-machines should be near each other. The propagation of information is also performed in two different ways:

Direct access: When the transmitter and receiver nano-machines are physically connected Ca^{2+} can flow from the transmitter to receivers through these connections (gates). These gates simply work as gap junctions allowing the flux of ions flowing from one nano-machine (transmitter) to another nano-machine (receiver) [2].

Indirect access: If there is not any physical connection between the transmitter and receiver then to transmitter will have to release the information based Ca^{2+} signal to the medium. Then the Calcium signal can flow through the medium by diffusion mechanism (more properly electro-diffusion) and ultimately received by the receiver.

From these two propagation schemes, it is clear that the main communication mechanism is diffusion and there are no pre-defined communication paths like molecular rails. These are discussed in Sect. 1.3.1.1. So this type of communication mainly supports braoadcast or multicast not unicast.

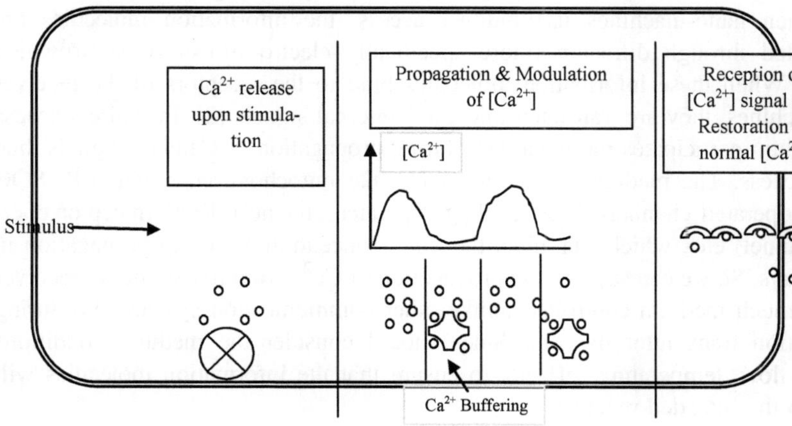

Fig. 2 Molecular communication using calcium signaling

Communication process using Calcium signaling: As discussed above that there two scenarios of information propagation in calcium signaling based nano-networks. In direct access the nano-machines are physically located one next to each other. The receiver can be any of these nano-machines members of these network. In indirect access method, the signals are directly released in the propagation medium (normally aqueous cytosol) by the transmitters and signals propagate through the medium and ultimately reaches to receiver. In both scenarios, the communication process contains following basic steps:

Encoding: This is the step involving generating information molecules. In calcium signaling based system when an external stimulus is applied the transmitter nano-machine encodes the information using Ca^{2+} concentration. The information is precisely encoded in amplitude and frequency of the function (modulating signal) describing the concentration of Ca^{2+} signal. An external stimulus is required to initiate the encoding process. It has been found that external stimulus applied to the cell cause the generation of IP_3 substance inside the cell. This presence of IP_3 unleashes the release of Ca^{2+} ions to the medium of Ca^{2+}.

Transmission: This involves the initiation of signalling. In direct access scheme, transmitter nano-machines stimulate neighbouring cells and consequently the signalling process starts. The signalling generates the initiation of propagation of Ca^{2+} waves [2]. Due to stimulus IP_3 is generated in the neighbouring cells. Now this presence of IP_3 cause more generation of Ca^{2+} ions by the neighbouring nano-machines. In the indirect access technique, transmitters may initiate signalling by releasing substances to the environment. Similar processes to cell fission or pore formation could be used by nano-machines to release the information molecules [2].

Propagation: In direct access when IP_3 is transmitted to neighbouring cells or nano-machines Ca^{2+} is released from the IP_3-sensitive Ca^{2+} stores. Then IP_3 is diffused to new neighbouring nano-machines and Ca^{2+} is released from these nano-machines. This chain reaction causes an increase of the Ca^{2+} concentration on the nano-machines which needs to be communicated and as a result, the Ca^{2+} wave propagates across the networked nodes (nano-machines) affected by IP_3. This propagation can be controlled varying the permeability of the gates or gap junctions [2]. When nano-machines use indirect access, the information molecules are propagated through diffusion (more specifically electro-diffusion) or Brownian motion. When these information molecules bind to the receptors of the receiver nanomachines, they are translated into Ca^{2+} internal signals [2]. In indirect access the medium participates more actively during propagation of Calcium signals than direct access. The medium has components like mitochondria, buffer, ER, SOC (source operated channel), VOC (voltage operated channel), ROC (receptor operated channel) etc., which can affect the Ca^{2+} concentration during propagation in many ways. So we can say that the propagation of Ca^{2+} from transmitter to receiver is very much medium controlled unlike other communication systems. So during propagation transmitter nano-machines should consider the medium conditions such as flow, temperature, pH etc. to ensure that the information molecules will arrive to the intended receiver.

Reception: In direct reception receiver nano-machine establishes gap-junctions with the neighbouring cells and perceives the Ca^{2+} concentration from inside of these cells [2]. After the message is received the receiver nano-machine closes the gates or gap junctions connecting with other nano-machines to stop further signal propagation. In the case of indirect reception receiver converts the information molecules to internal Ca^{2+} signals. A nano-machine can be equipped with different receptors to detect different information molecules [2]. This technique can be used to make parallel communication channels with different nano-machines.

Decoding: As we have stated earlier that an internal Ca^{2+} signal is generated according to the received information molecules. This Ca^{2+} signal can be encoded in amplitude and frequency to enable the activation of besides molecular motors and calcium signaling pheromones also act as a mode of molecular communication using nano-machines. Pheromones are nothing but chemicals released by organisms into the surrounding medium in order to convey messages to other members of the same species. The major difference of pheromone based communication over molecular motor based communication or Ca^{2+} signaling based communication is that as pheromones can diffuse through the surrounding medium over long distances so in this case the distance between two nano-machine are very much larger compared to the dimensions of the nano-machines whereas the distance between two nano-machines must be of nano-scale range for the other two cases. So we can say that pheromones can support communication from a transmitter nano-machine to a receiver nano-machine located at long distance apart from the transmitter [2].

3.1.3 Molecular Communication Using Pheromones

Different types of pheromones are capable of producing specific reactions in the receiving organism. Thus information is encoded in the type of pheromone used. The variety in the type of pheromones can be utilized for molecular division multiple access whereby, pheromones of different types can be released into the surrounding medium simultaneously without any interference [8]. Since the pheromones released by an organism can be sensed only by organisms of the same species, the information is secured from organisms of other species. The control and communication of nano-machines over long distances by this mechanism can be useful in many applications such as military field or environmental monitoring. Figure 3 shows pheromonal transport when there are two types of pheromones.

Communication features:
The communication is same as short distance communication implemented by molecular motors or calcium signaling but as the distance between the transmitter and receiver is very much so the channel is not deterministic here. The communication is also dependent on medium conditions like medium flow, temperature, humidity etc. Like all other molecular communication systems the message here is also encoded in terms of molecules. Since messages consist of molecules, there must be a huge number of possible combinations of molecules to encode messages.

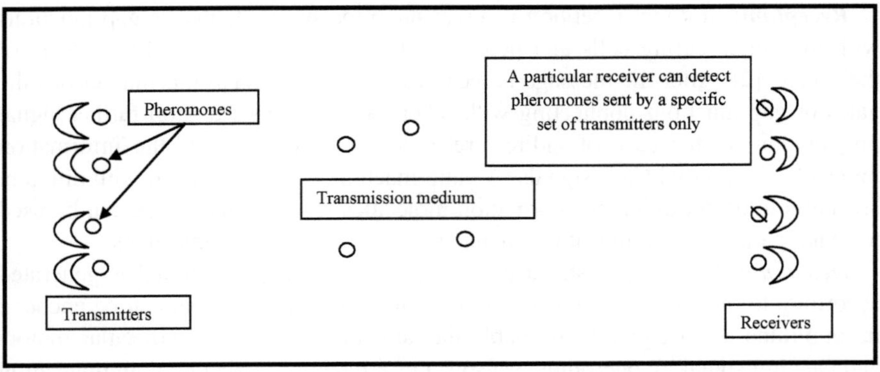

Fig. 3 Molecular communication using pheromones

Moreover, messages can be compounded by several different molecules allowing even more combinations to encode the information [2]. This reception of signals is basically ligand-receptor binding process same as in the case of calcium signalling. A ligand is a molecule that interacts and bind with a specific information molecule. In molecular communication using pheromones, the receptor proteins can be considered as the receiver nano-machine which converts the information contained in the message into a reaction at the receiver.

Communication process using pheromones: As other molecular communication processes the communication using pheromones can consists of five similar five taks which are discussed below briefly:

Encoding: Normally in a pheromone based communication a specific pheromone or appropriate molecule is used as a particular message. Encoding basically refers to the selection of proper molecules (pheromones) according to proper massages.

Transmission: This is basically the releasing of selected pheromones in the medium after the encoding process is over. The molecular messages are normally released in liquids or gases.

Propagation: The basic propagation mechanism of pheromones from transmitter to receiver is diffusion or Brownian motion. Environmental conditions like temperature, pressure, humidity etc., affect the propagation of pheromone molecules very much. Besides this, antagonist molecules present in the atmosphere may negatively the communication by modifying the information molecules before they reach the destination.

Reception: We have stated earlier that the reception of messages is a ligand-receptor binding process. Some receptor proteins do the work of this reception. When the pheromonal message reaches the receiver these receptor proteins detach the information molecules from the carrier. Generally receptor proteins are such molecules which have high affinity to the information molecules and help to detach the information molecules from the carrier.

Decoding: This is basically the extraction of information from the received message and this work is done is also done by the receptor. For example, the antennae and the maxillary palps of a fruit fly include 1300 olfactory receptor neurons (ORN) which can 40 different odours according to their information. The decoding system is embedded in the receptor organs and can be expressed as spatial-temporal activation pattern of the receiver.

3.1.4 Molecular Neuro-Spike Communication

This type of communication may be used when nanomachines are mobile. When an information-carrying nanomachine collides with another nanomachine, they adhere to each other. After adhesion, they establish a synapse which forms the communication medium between the nanomachines. The information is transferred in the form of an action potential which is an electrical pulse of about 80 mV. These action potentials are 'spikes' that are electrochemically transmitted from one neuron to other. The signal is transmitted by means of chemical messengers called neuro-transmitters [5]. This type of communication is inspired from neural communication in the nervous system. Thus, it is referred to as 'neuro-spike' communication [5].

Communication features and processes: Similar to other communication mechanisms, the communication involves four basic steps:

Encoding: Similar to other traditional digital communication systems, two bit levels are used: spike bit 1 and spike bit 0 corresponding to logic 0 and 1 respectively. Like calcium signaling, the information is encoded in terms of molecular concentration [5].

Transmission: This process involves initialization of electrochemical signalling. When a transmitter wants to send to a spike bit 1, it activates the release of vesicles containing neuro-transmitters [5].

Propagation: As stated above the released neuro-transmitters propagate through the synaptic channel formed between the adjacent nano-machines.

Reception: The neuro-transmitters propagating through the synaptic channels reach to the receiver and bind to the receptors on the receiver membrane. The binding of neuro-transmitters to receptors cause to the opening of the ligand gated channels. So ions flow into or out of the receiver. This flow of ions changes the membrane voltage.

Decoding: Receiver nano-machine monitors the plasma membrane voltage for certain time periods [5]. When a rapid change of the membrane voltage occurs the receiver nano-machine decides the received bit as spike bit 1, when no change occurs, it is decided as spike bit 0.

3.2 Nano-Electromagnetic Communication

This type of communication is suitable for nanomachines that are developed using nanoelectronic devices. This technology is very much similar to conventional networks using EM waves, only exception is that the device needs to be fabricated in nano-scale. A graphene based nano-antenna, carbon nanotubes (CNTs) or graphene nanoribbons (GNRs), have been shown to operate in the terahertz band (0.1–10.1 THz) [2, 9].

Communication features: Like the conventional electromagnetic communication systems here also a transmitter transmits a signal, the signal propagates through a medium (normally wire-less medium) and finally received by the receiver and the information is decoded. As the novel nano-antennas used in these cases can operate at Terahertz range frequencies the nano-transmitters and nano-receivers work in Terahertz frequency band unlike transmitters and receivers used in conventional communication systems which normally in Megahertz and Gigahertz frequency band. Though some electromechanical nano-trans-receivers are there which are able to operate at the upper range of Megahertz band but these are rarely used as their efficiency are very much low. Nano-machines may communicate using a binary stream where logic '1' is a femto-second long pulse as its frequency components are mainly in the THz band and for logic '0', non-transmission can be used. For nano-electromagnetic communication, it is also essential to develop nano-transceivers that are capable of receiving THz frequency.

Communication process:
Similar to other communication technologies this communication also includes five basic steps. The steps are discussed below:

Encoding: Encoding is the generating of proper signals corresponding to proper information. Different nano-sensors perform this work. Nano-sensors can be broadly classified in two classes: (i) Physical nano-sensors (Senses different physical quantities like mass, pressure, force etc. Working principal is based on the fact that the electronic properties of both CNTs and GNRs change with deformation). (ii) Chemical nano-sensors (Senses different chemical conditions like presence of certain molecule etc. Working principal is based on the change of number of electrons able to move through the carbon lattice of CNTs and GNRs when different molecules are adsorbed on their surface. When the information either in physical or in chemical form is sensed by the nano-sensors, they change their electronic property and generates signal in electromagnetic domain according to the information. Then the information is encoded.

Transmission: Several nano-trans-receivers (i.e., the work as both transmitters and receivers) do this job. For this job the EM trans-receiver should support baseband processing, frequency conversion, filtering, power amplification, modulation of the signals that have to be transmitted. Various graphene based FET nano-transistors are able to these works. After that the modulated signal is transmitted to the medium by nano-antennas. But all of the devices mentioned above should operate at Terahertz frequency range to incorporate them at nano-networks. For example, envisioned nano-antenna working at terahertz band frequency, RF FET transistors able to operate at these very high frequencies are necessary.

Propagation: Like other wireless communication systems, the signal propagation is the propagation of electromagnetic waves through the wireless media. We have stated that 0.1-10 THz frequency band can be used in nano-network applications. Some other factors come during propagation of these high frequency waves. For example:

(i) ***Path loss***: This is the addition of spreading loss and molecular absorption loss. First loss occurs due to spreading of EM waves during propagation and it depends on signal frequency and transmission distance and increases with both distance and frequency independently. Molecular absorption loss is the loss caused by absorption of energy by the molecules present in the medium and converting that energy into kinetic energy of those molecules. This is basically dependent on the molecular composition of the medium.
(ii) ***Noise***: In Terahertz frequency wave propagation the noise is basically the molecular noise generated by the molecules present in the medium. This noise is neither white nor Gaussian and the power spectral density of this noise is not flat but has peaks at certain frequencies.
(iii) ***Bandwidth and channel capacity***: Molecular absorption also determines the usable bandwidth. As molecular absorption is dependent on the molecular composition of the medium tho Bandwidth capacity is also medium sensitive.
(iv) ***Multi-path fading***: Depending on the scenario in which nano-sensor devices are deployed multiple replicas of the transmitted signal will be generated at the receiver. The combination of these replicas will cause oscillations of power detected in reception and cause degradation of the received signal.
(v) ***Particle scattering***: This occurs due to scattering of EM waves by the medium particles. As high frequency are more prone to scattering, so Terahertz frequency EM wave used here is also very much affected by scattering.

Reception: Like transmission this job is also done by nano-antennas and nano-trans-receivers. The work is basically the reverse of transmission. The electromagnetic wave propagated through the medium is received by a suitable nano-antenna and then demodulated to get the modulating signal.

Decoding: Finally the signal is decoded to get the original information by a suitable decoder. This information is used to do different operations or used for stored in nano-memories for future use.

Recent advancements in molecular and carbon electronics have opened a new age of generation of various nano-electronic devices like nano-batteries, nano-memories, nano-scale logical-circuitry including logic gates and even nano-antennas. From a communication point of view, the unique properties observed in novel nano-materials characterizes the specific bandwidths of electromagnetic radiation used, the time-lag of the emission of electromagnetic waves, time-delay produced due to propagation of these waves or the magnitude of the emitted power for a given input energy for a particular nano-network. All these advancements enhance a broad change in the present state of the art of analytical channel models, network architectures or communication protocols [10].

Table 1 Differences between electromagnetic communication and molecular communication

Communication	Electromagnetic communication	Molecular communication
Carrier	Electromagnetic waves	Molecules
Signal type	Electrical signal	Chemical signal
Propagation speed	As all are EM waves so propagation speed is very high	As the propagation is basically the diffusion of information molecules though the medium so speed is very low
Medium activity	Here medium is basically passive. It can only produce noise during propagation	Medium actively participates during propagation
Noise	External electromagnetic fields	Particles and molecules present in the medium

After discussing all the communication techniques we get some main differences between electromagnetic communication and molecular communication which are listed below in Table 1.

The next section, i.e., in Sect. 4, discusses a comparative study in between molecular and electromagnetic communication.

4 Advantages of Molecular Communication Over Electromagnetic Communication in Nano-Networks

We have discussed basic two techniques of communication in nano-networks. But molecular communication is more advantageous than electromagnetic communication during connection of nano-machines due to some reasons. The reasons are discussed below:

(i) ***Easy to integrate***: Molecular transmitters and receivers are much more easy to integrate than electromagnetic transmitters and receivers or trans-receivers in nano-machines.

(ii) ***Power consumption***: The molecular communications are entirely based on some chemical processes and reactions which consume very less amount of power compared to electromagnetic nano-devices. This is the main drawback of electromagnetic communication as more power consumption means high energetic power supply will be required which will more difficult to integrate in nano-machines.

(iii) ***Affection by noise***: As we have discussed above Terahertz frequency band used in electromagnetic communication in nano-networks is very sensitive to noise wheras molecular waves are very less sensitive to external molecular noise present in the medium.

(iv) ***Limitation of distance***: In electromagnetic communication the minimum distance between transmitter and receiver will be 1 m otherwise cross-talks and other negative effects will occur which will degrade the quality of the received signal. But in molecular communication as the information is encoded in terms of molecules so there is no limitation of distance between transmitter and receiver.

(v) ***Medium activity***: As the medium plays an active role in molecular signal propagation so signal propagation can be controlled with the help of the medium. But in electromagnetic communication the medium is passive so it cannot be used for controlling the signals.

(vi) ***Biocompatibility***: Most of the nano-networks used till nowadays are used for mainly biochemical and biological applications. Molecular communication is more compatible to biological processes, so most biological application based nano-networks are implemented by molecular communications.

Due to these reasons most of the nano-networks are implemented using molecular communication technologies. In the next section i.e. Sect. 4, we concentrate our discussion about the architecture of nano-networks using molecular communication with Calcium signaling.

5 Proposed Architecture for Nanonetwork Using Molecular Communication with Calcium Signaling

Presently, the standard architecture for nano-communication network is being studied for the development of a comprehensive framework which would address the unique features of communication in the nano-domain [11]. We propose an architecture for nanonetworks that uses molecular communication by means of calcium signals [12, 13, 14]. In nanonetworks, the channel plays a central role in signal processing unlike conventional networks where the channel is passive and ideally acts only as a transmission medium. The calcium signaling process involves: (i) generation of stimuli; (ii) modulation (encoding) of the amplitude or frequency of Ca^{2+} concentration $[Ca^{2+}]$ during propagation; (iii) transmission of the Ca^{2+} signal by diffusion through the intracellular space and over to the adjacent cells through the cell membrane or gap junctions; (iv) decoding of the Ca^{2+} signal [5]. In Ca^{2+} signaling, modulation of signals takes place along the channel rather than at the transmitter end. The components in the channel that take part in the modulation process may be regarded as active nodes within a passive diffusion channel. So, modelling of the overall nanonetwork as well as to model its different components is an important part. In this context, first we highlight the related works on different types of models for implementation of a nanonetwork and then we discuss our (authors of this chapter) proposed network architecture. And the protocol stack component model of the physical channel layer has been discussed in detail in the following sub sections.

Related works on different network models:

All over the world, group of researchers have been working with the different types of models for nano networks. Some of them are discussed below in brief.

(i) Based on the basic concepts of nano-networks a fast parallel multi-scale stochastic modelling based platform can be designed which will help to investigate the dynamics of large scale bacteria-based nano-networks. If the chemical signaling pathway inside each bacterium and the chemical gradients created by the receiver node can be accurately modelled the dynamics of a targeted drug-delivery system can be characterized [15].

(ii) A non-equilibrium statistical model can be developed for the dynamics of dense networks of bacteria. Secondly the chemotactic response of bacteria and their intercellular communication can be characterized for this purpose. A kinetic Monte Carlo method derived from standard Gillespie's algorithm can be used to generate realizations of the stochastic models [16].

(iii) The effect of Additive Inverse Gaussian Noise (AIGN) noise channel can be adopted in molecular communication based networks to know the corruption of information by molecular noise in the fluid medium. From this it can also be shown that use of multiple molecules leads to reduced error rate [17].

Our Proposed Architecture:

As we have pointed out, at the introduction of Sect. 5, Various processes such as Ca^{2+} release, Ca^{2+} buffering, spike generation, Ca^{2+} sequestration, sensing of Ca^{2+} signal, etc. are segregated and grouped into the four layers: Environmental Impact Control layer, Interface Control layer, Information Density Control layer and Physical Channel. This group of layers is the driver to propose this architecture of the network that is essentially channel-centric [18]. The environmental impact control layer has been proposed as the topmost layer of the architecture although it encloses the physical channel layer as it deals with all the secondary phenomena of the physical channel which affect calcium signaling but do not play a direct role in the communication process.

The different layers are discussed in detail as follows:

Physical channel

The physical channel for Ca^{2+} signaling is an aqueous medium through which Ca^{2+} ions diffuse from one region to another within a cell or among cells. Thus, it is possible to achieve intracellular as well as intercellular communication using Ca^{2+} signals. The cytosol acts as a passive, aqueous diffusion medium for Ca^{2+} signals. However, the channel, as a whole, is in no way passive since there are fixed and mobile active components within the cytosol that actively modulate the amplitude of Ca^{2+} signal during propagation. These active components are Ca^{2+} binding proteins present in the cytosolic channel, that act as Ca^{2+} buffers. Also, there are potential Ca^{2+} release sites like single membrane bound compartments like endosomes, Golgi vesicles, lysosomes, secretory granules and melanosomes which reside within the physical channel [18]. Thus, the physical channel is characterized by the distribution of active components in it. Cell organelles like mitochondria, endoplasmic reticulum (ER), inositol 1, 4, 5 triphosphate (IP_3) receptors and various Ca^{2+} pumps form feedback loops that regulate Ca^{2+} signal frequency locally. The resultant amplitude or frequency modulated Ca^{2+} signal propagates by electro-diffusion through the cytosol. The role of this layer is to model calcium dynamics in the channel and to mathematically model electro-diffusion.

Information density control layer

A Ca^{2+} signaling network uses a broadcast scheme of information transmission. Ca^{2+} signals from a stimulated cell is transmitted to adjacent cells through different channels such as connexin channels or ionic channels like NMDA, nicotinic, purinergic ionotropic channels. Ca^{2+} permeant channels like nicotinic receptors, NMDA receptors on adjacent cell membranes are gated by ATP, acetylcholine or small amino acids like glutamate [19]. Upon application of stimulus, Ca^{2+} release into the cytosol may be initiated through mainly three different types of channels: voltage-operated channels (VOCs), receptor-operated channels (ROCs), and store-operated channels (SOCs). These channels have different mechanism of activation [20]. The VOCs are activated when the membrane potential exceeds a threshold, i.e., by membrane depolarization [21]. The ROCs may be activated by Ca^{2+} itself or by messengers like IP_3 which bind to their respective receptors to initiate release of Ca^{2+} into the cytosol. The SOCs can release Ca^{2+} ions into the cytosol only until the store is depleted to a certain threshold, depletion below which causes Ca^{2+} uptake to replenish the store [Ca^{2+}].The endoplasmic reticulum (ER) is a SOC that leaks Ca^{2+} into the cytosol all the time while Sarco-endoplasmic Reticulum Ca^{2+} ATPase (SERCA) pumps continuously restore Ca^{2+} into the ER. Ca^{2+} release is also activated by Ca^{2+} itself [18]. This is known as calcium induced calcium release.

Thus, Ca^{2+} signals may propagate using the ROCs or take the VOC route that are activated when the plasma membrane of adjacent cells are depolarized [21]. The information density control layer deals with the broadcast range control by determining the effect of multiplexing Ca^{2+} signals from different channels on the spatio-temporal propagation of the Ca^{2+} signals.

When signaling is over, the cytosolic [Ca^{2+}] must be brought down to the resting level and global Ca^{2+} homeostasis is to be established. This is done by sequestering Ca^{2+} into the mitochondria or by the action of Plasma Membrane Ca^{2+} ATPase (PMCA) pumps, Na^+/Ca^{2+} exchanger, Na^+/Ca^{2+}-K^+ exchangers which remove Ca^{2+} from the cytosol and restore resting levels of [Ca^{2+}] to around 100 nM [18]. This may lead to activation of Ca^{2+} channels of the adjacent cells and thereby facilitate propagation of Ca^{2+} signals from the stimulated cell to the surrounding cells in a broadcast fashion.

In addition, Ca^{2+} sequestration by mitochondria plays an important role in maintaining global Ca^{2+} homeostasis [22]. The PMCA help to maintain low cytosolic [Ca^{2+}] in the long run. An estimation of the pump rate and the overall rate at which Ca^{2+} homeostasis can be achieved within the cell, determines the controllability of the information density. There are also parameters that increase the effectiveness of the information density control mechanisms. For instance, the presence of calmodulin increases the Ca^{2+} affinity of PMCA pumps as well as the pumping rate of ATPases [18].

Interface Control Layer

The amplitude or spike rate of the Ca^{2+} signal is sensed by Ca^{2+} sensors which initiate Ca^{2+} sensitive cellular processes based on the amplitude or spike frequency detected and the duration for which the signal persists in the cytosol [20]. The quantification of different types of stimuli and their relationship with the Ca^{2+} release pattern over time into the cytosol are to be determined in the interface control layer. The stimulus may be molecules of chemicals called ligands for activating the ROCs or voltage or a small triggering Ca^{2+} current. Each type of stimulus may activate Ca^{2+} release to different levels. At the receiving ends of the broadcast system, the role of this layer is to map various Ca^{2+} sensitive cellular processes with different amplitude and frequency of the Ca^{2+} signal.

The transduction of Ca^{2+} signals into cellular response in terms of stress (for contraction), volume of chemical release (for secretion), gene transcription, etc., requires modeling the associated cellular components. For instance, in cardiac cells Ca^{2+} signals bind to troponin-tropomyosin molecules followed by actin-myosin cross-bridge formation, thereby, resulting in contraction. For such a case, the role of this layer is to model the excitation-contraction (EC) coupling mechanism involving actin myosin proteins. Another task of this layer is to determine the sensitivity of different Ca^{2+} sensors (troponin C for cardiac cells) to the signal level, spike frequency and signal duration.

Environmental impact control layer

The investigation and analysis of the impact of different environmental parameters such as temperature, pH, etc. on the signaling rate or on the efficiency of information transfer is an important role of the environmental impact control layer. For instance, the transient receptor potential (TRP) channels get activated by environmental changes in temperature, pH, volatile chemicals, etc. [18]. The Ca^{2+} signals also interact with other signaling pathways such as mitogen-activated protein kinase, nitric oxide (NO), cyclic AMP, phosphatidylinositol-3-OH kinase, etc. [20]. The understanding of such interactions is crucial for determination of secondary effects that arise due to cross-talk between calcium and other channels.

We summarize the functionalities of four different layers for molecular communication based nanonetwork in Table 2.

5.1 Protocol Stack Components in Physical Channel Layer

The physical channel layer of the above mentioned four layered nano-network can be defined using the protocol stack components [11]. This stack is made up of the components (i) message carrier, (ii) perturbation, (iii) motion, (iv) field and (v) specificity. The physical channel for Ca^{2+} signaling is the cell cytosol through which Ca^{2+} ions diffuse from one region to another within the cell (intracellular) or

Table 2 Functionalities of four different layers for molecular communication

Nanonetwork reference model (Architecture)			
	Layer	Function	
5	Environmental impact control	Ability to control the channel's impact on the environment	Upper layer
4	Interface control	Ability to control the "injection" of information from the channel into the target node	
3	Information density control	Ability to control the channel signaling rate (corresponds to bandwidth)	
2	Direction control	Ability to control the direction of channel "flow" (corresponds to LINK/ROUTING)	Lower layer
1	Physical channel	The material comprising the nanonetwork channel that facilitates the flow of information	

from cell to cell (intercellular). There are also different active components like buffers, mitochondria, endoplasmic reticulum (ER), and receptors those help in the processing of the Ca^{2+} signal during propagation. The resultant modulated Ca^{2+} signal propagates by electro-diffusion through the cytosol. In other words, the physical channel layer deals with modeling of the physical processes involved in the communication of Ca^{2+} signals like binding of ligands with receptors, e.g., the binding of first messengers, IP_3, to IP_3 receptors, ensuring *specificity*; amplitude and frequency modulation of the $[Ca^{2+}]$ by different cell components e.g. mitochondria, ER, ER pumps, buffers etc. ensuring *perturbation;* propagation of Ca^{2+} through the cells by electro-diffusion (controlled by both drift and diffusion of ions) signifying *field* and *motion* respectively. And, most importantly this modulated Ca^{2+} is the *message carrier* in this communication process. So, to model the complete physical channel we have to model the components required as well as the dynamics of the Ca^{2+} propagation (electro-diffusion).

5.2 *Detail Discussion on Protocol Stack Components*

Specificity: The Ca^{2+} signaling toolkit includes receptors such as IP_3 receptors and ryanodine receptors; Ca^{2+} binding proteins such as parvalbumin, calbindin, calretinin, etc., that act as Ca^{2+} buffer; intracellular Ca^{2+} stores formed by sarco-endoplasmic reticulum, cell organelle like mitochondria and Ca^{2+} sensors such as troponin C, synaptotagmin, protein kinase C, Ca^{2+}/calmodulin dependent protein kinase II (CAMKII), etc. These components specifically interact with Ca^{2+}.

Perturbation: When a cell is stimulated, Ca^{2+} released into the cytosol through a single or through multiple channels, increases the cytosolic $[Ca^{2+}]$ from 50-100 nM to 500-1000 nM. A part of the cytosolic Ca^{2+} is bound by Ca^{2+} buffers, thereby, lowering the amplitude of the cytosolic Ca^{2+} signal. The amplitude of the Ca^{2+} signal depends on the concentration of buffers in the cell. Also, the duration of the Ca^{2+} signal varies with the concentration of fixed and mobile buffers [18].

Ca^{2+} spikes are generated by a feedback loop formed by the mitochondria, ER, IP_3 receptors and SERCA pumps. IP_3 generated upon externally stimulating the cell, binds to IP_3 receptors. This disables the SERCA pumps. Thus the ER releases Ca^{2+} into the cytosol due to its continuous leakage and is not replenished by the SERCA pump. This reduces the ER $[Ca^{2+}]$. The released ER Ca^{2+} increases the cytosolic $[Ca^{2+}]$ but it is taken up by mitochondria which again reduce the cytosolic $[Ca^{2+}]$ and cause further Ca^{2+} release from the ER [22]. However, when the ER $[Ca^{2+}]$ falls below a threshold the SERCA pumps are again activated, net Ca^{2+} release is reduced as the depleted store is replenished by the pumps. After replenishment of the store, if IP_3 is still present in the cytosol then the process is repeated. This gives rise to regenerative local Ca^{2+} spike generation. The algorithm for Ca^{2+} spike generation is explained with the help of a flowchart in Fig. 4 and the corresponding changes in $[Ca^{2+}]$ in the cytosol, ER and within the mitochondria are shown in Fig. 5 given below.

Discussion about the flow chart shown in Fig. 4:

When the external stimulus comes into the cell, IP_3 is started to be released and IP_3 is bounded with the IP_3 receptors. At that time, ER pumps are disabled and ER is started to release the Ca^{2+} into the cytosol. So, the $[Ca^{2+}]$ in the cytosol is increased. Now, while this process is going on, when the cytosolic $[Ca^{2+}]$ is crossed the threshold of Mitochondria, it is activated and starts to uptake Ca^{2+} from its surroundings. So gradually, the rate of increase of the cytosolic $[Ca^{2+}]$ is somehow gets lowered. After some time, the Ca^{2+} concentration in the ER will be lower than the its threshold, then ER pumps are become activated and the cytosolic Ca^{2+} gets pumped into the ER. On the other side Mitochondria is still up taking the Ca^{2+} from the cytosol. So, now, due to these cooperative effects, the cytosolic $[Ca^{2+}]$ is decreased. Thus the Ca^{2+} spike is generated and if there is more IP_3 still present in the cell, this process will be repeated again.

There are three phases of cytosolic Ca^{2+} concentration shown in Fig. 5:

(i) At first, only ER is activated and releasing Ca^{2+} into cytosol, ER pumps are disabled, so cytosolic $[Ca^{2+}]$ is increased at that phase.
(ii) At the next phase, the cytosolic $[Ca^{2+}]$ is crossed the threshold of Mitochondria, so Mitochondria is starting to uptake Ca^{2+} but till then ER is also releasing Ca^{2+} into cytosol, so effectively cytosolic $[Ca^{2+}]$ comes to the peak value.
(iii) At the 3rd phase, as $[Ca^{2+}]$ in ER is got lower than its corresponding threshold, ER pumps are enabled and ER stops leaking. So, now both mitochondria and ER pumps are taking Ca^{2+} from the cytosol. So, from the peak, the cytosolic $[Ca^{2+}]$ is become lowered.

Due to this overall process, the nature of the variation of cytosolic $[Ca^{2+}]$ w.r.t time will be like a sine wave and as the frequency of this wave is very high, so it will be looked like a SPIKE. This is called the spike generation process. This qualitative analysis will be cleared from the figure given below:

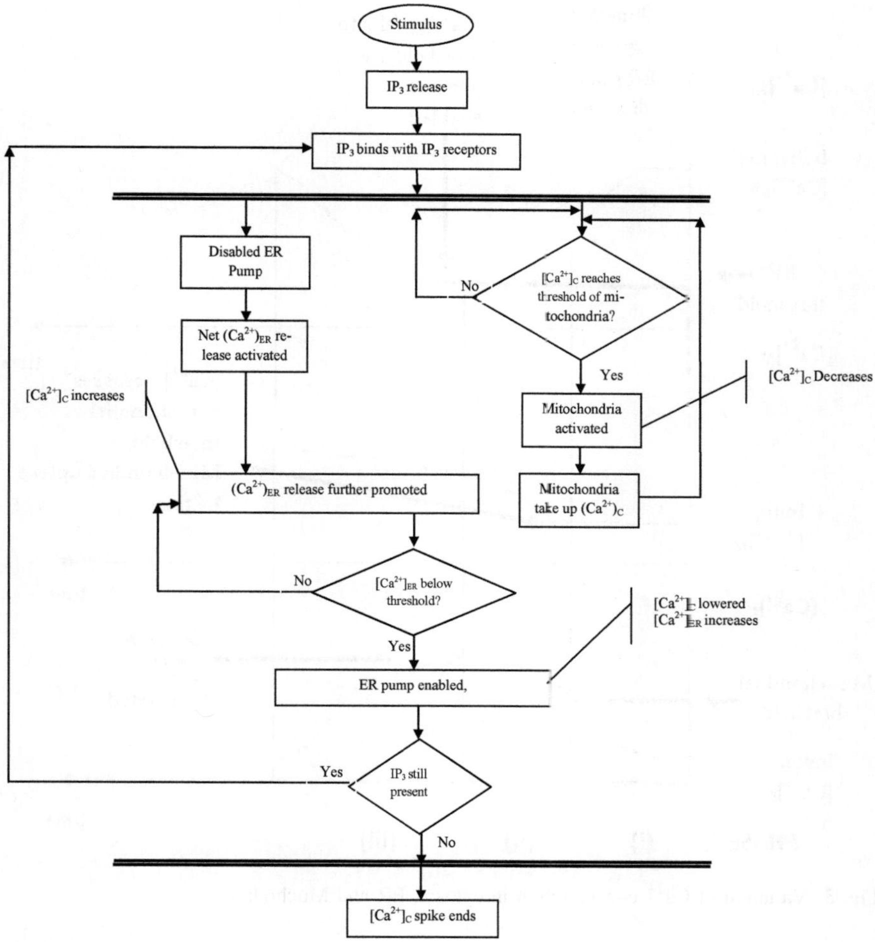

Fig. 4 Flow chart for the algorithm of Ca^{2+} spike generation

Figure 5 Calcium concentration in ER, $[Ca^{2+}]_{ER}$, mitochondria, $[Ca^{2+}]_M$, and spike generation in cytosolic concentration, $[Ca^{2+}]_C$.

Field and Motion: Ca^{2+} signals propagate within the intracellular space by electro-diffusion. Thus propagation is driven by field created by differences in potential in different regions of the cell. Also, the motion of Ca^{2+} takes place by diffusion which is governed by the concentration gradient of Ca^{2+}.

Message Carrier: The modulated concentration of Ca^{2+} constitutes the Ca^{2+} signal which carries the information regarding the cellular process to be triggered at the receiver. Hence, the Ca^{2+} ion is the message carrier propagating in the physical channel [11].

The functionalities of the protocol stack components is summarized in Table 3:

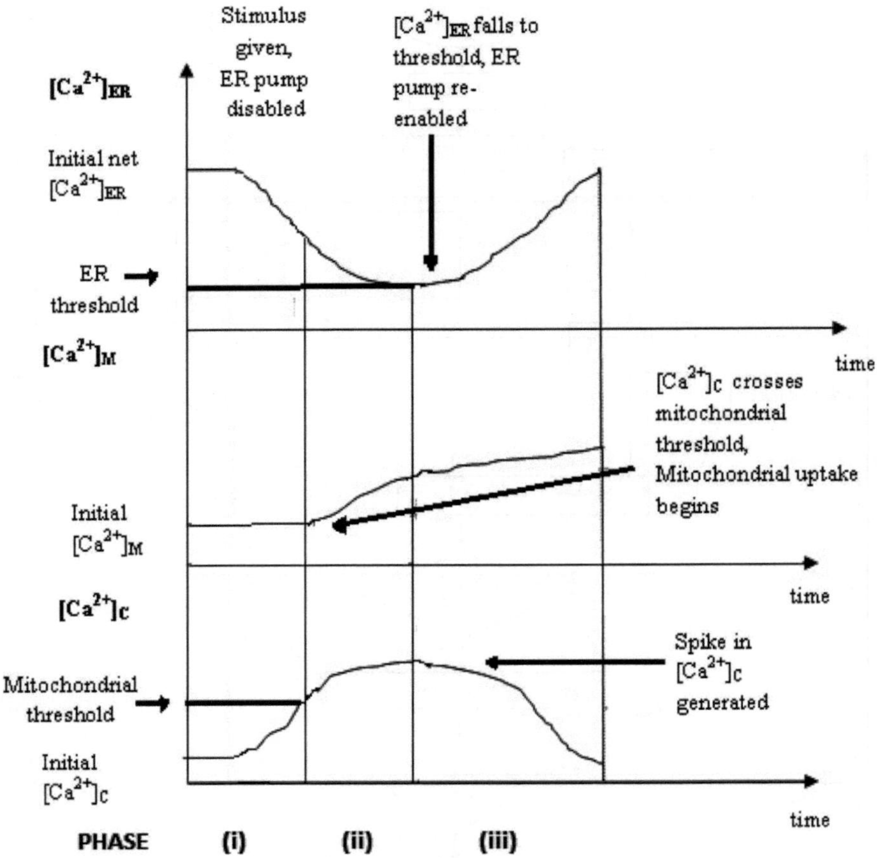

Fig. 5 Variation of Ca^{2+} concentration in cytosol, ER and Mitchodria

Figure 6 is the simplified outlook of the protocol stack components establishing the communication in the cell.

5.3 Channel Modeling and Solution Scheme

Physical channel modeling constitutes modeling components like receptors for specificity; signaling components like buffers, mitochondria, ER, ER pumps, etc. which modulate the amplitude or frequency of Ca^{2+} signals; and modeling of electro-diffusion which governs the motion of the message carrier, Ca^{2+} ions. We subdivide this task in two steps, namely, mathematical modeling of electro-diffusion and component modeling.

Table 3 Functionalities of the protocol stack components

Component level	Protocol stack component	Functionalities of protocol stack components
Component 0:	Mass or Energy [Message Carrier]	Fundamental Component: Message Carrier encodes the message. Molecular structure encodes information
Component 1:	Motion or Flow or Thrust (Force)	Builds upon Component 0. Flow rate (in any direction) caused by force/thrust of message carrier. This is the potential to form a channel Examples: Molecules diffusing through liquid, etc.
Component 2:	Field	Builds upon Component 1. This component provides organized flow. It may be thought of as routing or a virtual waveguide Examples: Fluid flow, applied EM field, etc.
Component 3:	Perturbation	Builds upon Component 2. Ability to vary message carriers as needed to represent a signal. This may be thought of as modulation (signal impression) Example: Controlled dense vs. sparse concentrations of molecules, etc.
Component 4:	Specificity	Builds upon Component 3. Ability to control sensing or attachment of message carrier to a target. This may be thought of as addressing Examples: The shape or affinity of a molecule to a particular target, etc.

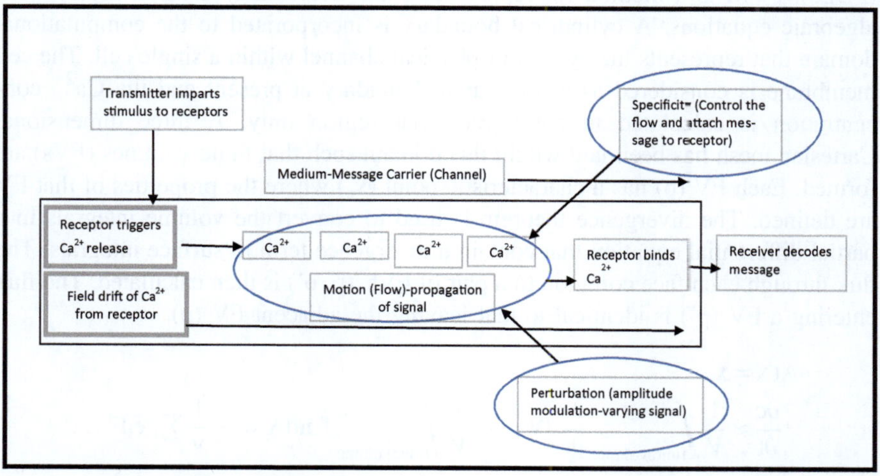

Fig. 6 Protocol stack components for calcium signaling

(i) ***Mathematical model for electro-diffusion through physical channel***

The intracellular space forms the physical channel for propagation of signals. We mathematically model the propagation of Ca^{2+} in the intracellular space by electro-diffusion using electro neutral model [23]. We model a cylindrical cell as a three-dimensional space with a cylindrical membrane. Upon reaching the membrane, these ions either add to the surface charge on the membrane or enter the extracellular space by flow of transmembrane current through the membrane ion channels. The membrane acts as a capacitor and maintains a membrane potential across it. In the electro neutral model, the ionic concentration follows ion conservation, drift-diffusion flux equation and electro neutrality condition given below:

$$\frac{\partial c}{\partial t} = -\nabla.\mathbf{f}\ldots \tag{1}$$

$$\mathbf{f} = -D\left(\nabla c + \frac{qzc}{k_B T}\nabla\boldsymbol{\Phi}\right)\ldots \tag{2}$$

$$0 = \rho_0 + qzc\ldots \tag{3}$$

Here, **f** denotes the flux, D is the diffusion coefficient, qz is the amount of charge of Ca^{2+}, where q is the elementary charge, i.e., the charge on a proton. qD (= k_BT) is the mobility of Ca^{2+} (Einstein relation) where k_B is the Boltzmann constant, and T the absolute temperature. ρ_0 is the background charge density.

Solution Scheme:
To solve the coupled P.D.Es the numerical scheme is adopted stated in [23], where a finite-volume method (FVM) is used to solve the partial differential equations (PDEs). FVM is a method for representing and evaluating PDEs in the form of algebraic equations. A cylindrical boundary is incorporated to the computational domain that represents the cytosol or physical channel within a single cell. The cell membrane is considered as a transparent boundary at present and the Ca^{2+} concentration is calculated in the intracellular region only. A three dimensional Cartesian mesh has been laid within this domain such that finite volumes (FVs) are formed. Each FV (p) has a characteristic point ($\mathbf{x}_c$) where the properties of that FV are defined. The divergence theorem is used to convert the volume integrals in a partial differential equation that contain a divergence term to surface integrals. The flux through each face common to a pair of FVs, (p, p') is then calculated. The flux entering a FV (p') is identical to that leaving the adjacent FV (p).

$$\text{At}\,x = \mathbf{x}_c,$$
$$\frac{\partial c}{\partial t} \approx \frac{1}{V}\int_{\text{finitevolume}} \frac{\partial c}{\partial t} dV = -\frac{1}{V}\int_{\text{finitevolume}} \mathbf{f}.\mathbf{n}d\mathbf{A} \approx -\frac{1}{V}\sum\nolimits_l e_l F^l \ldots \tag{4}$$

where $F^{(p,p')}$ is the flux density approximation from FV p to p′ as. The ionic concentration is conserved when

$$F^{(p,p')} = -F^{(p',p)}\ldots \tag{5}$$

$$\frac{\partial c^p}{\partial t} = -\frac{1}{V}\sum\nolimits_{p\neq p'}\left[hF^{(p,p')} + \gamma^{p,p'}G^{(p,p')}\right]\ldots \tag{6}$$

where h is the area of the face common to finite volumes p and p' and $G^{(p,p')}$ is the flux from a finite volume p to another finite volume p' that share a membrane of area $\gamma^{p,p'}$, so $G^{(p,p')}$ is termed as the membrane flux will make an effect only for the boundary FVs of the cell. For ordinary FVs in the intracellular space, $\gamma^{p,p'} = 0$, so the second term is zero. The ordinary flux $F^{(p,p')}$ is calculated using the equation

$$F = D\left[\frac{c^{p} - c^{p'}}{h} + \frac{qz(c^{p} + c^{p'})}{2K_{B}T}\frac{\varphi^{p} - \varphi^{p'}}{h}\right]\ldots \tag{7}$$

where D is the diffusion coefficient. $\varphi^{p} - \varphi^{p'}$ gives the potential difference between the representative points $\mathbf{x}_c$ for the finite volumes p and p'. $z = 2$ for Ca^{2+} as it is divalent. To calculate the concentration in the (n + 1)th instant from that in the nth instant of time we use the relation:

$$\frac{c^{p,n+1} - c^{p,n}}{\Delta t} = -\frac{1}{V}\sum\nolimits_{p \neq p'}\left[hF^{(p,p',n)}\right]\ldots \tag{8}$$

where Δt should have a value long enough for the ions to move over from one FV to an adjacent FV in this time period. However it should not be so long that ions can move over more than one FV.

(ii) ***Behavioral modeling for components of toolkit***

(a) ***Emitters***: The emitters are randomly oriented in the cellular space acting as basic transmitters. They emit Ca^{2+} to the medium, in order to modify the concentration in its environment. The emission patterns can be of different types like square wave, sine wave, random pattern etc. The emission pattern has been configured using different mathematical functions and in later section, the corresponding emission pattern has also been observed.

(b) ***Buffers***: The buffers are modelled as a storage component which has a predefined absorption capacity. It actually takes the Ca^{2+} ions from the background near to its surroundings until it reaches the absorption capacity and then it becomes inactive. So, due to the presence of buffers, the amplitude of Ca^{2+} is lowered and by increasing the number of buffers, naturally the amplitude is reduced further. The buffers may be positioned anywhere within the space provided it doesn't coincide with any other component. They may be modelled as static or mobile.

(c) ***Mitochondria***: The mitochondrion is modelled as a transceiver with a configurable absorption capacity and threshold. It accumulates Ca^{2+} ions from the background near to its surroundings when the background concentration exceeds the threshold and absorb until it reaches the absorption capacity and when the background concentration falls to the resting level it starts releasing the accumulated ions at a slow rate.

(d) ***Receptors***: The receptors are also modelled and placed randomly at different points in the cellular space. They accumulate and measure the concentration of particles within a predefined radius in their environment.
(e) ***ER and ER pumps***: The ER is modeled as a leaky store that releases Ca^{2+} continuously at a given leak rate until the store content falls to a threshold value, below which the leakage stops. During leakage, ER pumps restore the ER with Ca^{2+} at a given pumping rate.

5.4 Evaluation and Observations

Presently, we are trying to evaluate the above equations for modelling the physical channel layer of the architecture described in the previous sections and for testing the performance of the protocol stack components discussed before. A 3D cellular space is considered with a cylindrical transparent boundary in a cuboidal computational space as shown in Fig. 7i. A uniform initial background concentration has been introduced and Ca^{2+} ions have been modelled as divalent particles. The components are also being configured within the intracellular space with their respective parameters are given in Table 4 below and their different characteristics are implemented individually according to its nature and function in the signaling process. Each FV has been identified as inner, outer or boundary types on the basis of their position in the mesh. Each FV has been assigned a characteristic point, given by the geometric Centre of the FV, where the properties (potential, concentration, etc.) of the space within that FV are defined and used for calculations. The whole system should run for a certain time frame and all the components including the background concentration are updated to find the overall state of the cell in the next time frame. Figure 7ii shows the 2D cross-sectional view of the cell as a sum of finite volumes.

The different types of emission pattern observed after the evaluation are also given in Fig. 8. The different types of emission pattern observed after evaluation are shown in Fig. 8a, b, c.

Another important part is to model the membrane to implement the intercellular communication. To model the membrane we have to model the different type of ionic channels (e.g., ROC, SOC etc.) and then by spreading them on the membrane, the intercellular calcium signaling flow can be observed. There are another components like pumps (SERCA pump, PMCA pump etc.) and exchangers on the membrane controlling the intercellular Calcium signaling flow which are also to be modeled. So, our future work is to model the membrane by modeling the ion channels and the membrane components like pumps, exchangers etc.

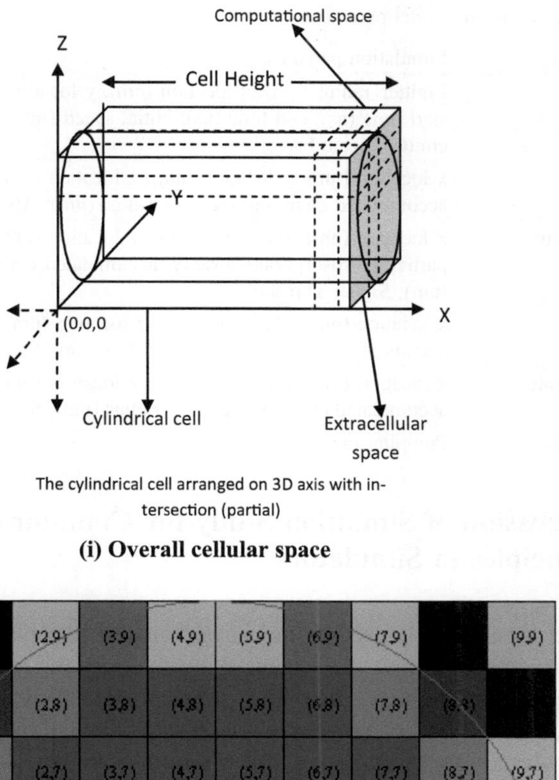

Fig. 7 **i** Overall cellular space. **ii** 2D cross-section of cellular space with mesh

Table 4 Component model parameters

Component	Simulation parameters
Emitter	Emitter_radius (nm); x location (nm); y location (nm); z location (nm); start time (ns); end time (ns); initial speed (m/s); punctual; concentration emitter; Scale Factor
Buffer	x location (nm); y location (nm); z location (nm); absorb particles; accumulated counting; receiver radius (nm); Absorption Capacity
Mitochondria	x location (nm); y location (nm); z location (nm); threshold; absorb particles; Absorption Capacity; accumulated counting; receiver radius (nm); Signal Threshold
Receptor	x location (nm);y location (nm); z location (nm);absorb particles; accumulated counting; receiver radius (nm)
Endoplasmic reticulum	x location (nm); y location (nm); z location (nm); absorb particles; accumulated counting; receiver radius (nm); Signal Threshold; Leak Rate
ER Pumps	Pumping rate

6 Discussion of Simultion Study for Communication Principles in Simulators

Computer simulation is used to model and analyze the physical systems. Applications of simulators into nano networking to study the behavior of its different components are relatively new. The principal idea is that if a system can be modeled, then features of the model can be modified and the corresponding results can be analyzed. As the process of model modification is relatively simpler than the complete real implementation, a wide variety of scenarios can be analyzed at the low cost relatively than to making similar changes to a real network.

Nano/molecular communication involves transmission of information at the nano-scale. This type of communication can be achieved by using terahertz frequency, transport of molecular motors, calcium signaling, pheromonal transport, etc. In order to explore the potential of nano/molecular communication for nano-networks, it is essential to study the mechanism of each of these modes of communication taking into account their unique features. A simulation framework has to be developed in order to extract the parameters that affect communication in the nano-network. However, the simulation models that take into account the unique features of nano/molecular communication are not easy to obtain. The simulation model may be based on mathematical formulation for terahertz frequency and laboratory based data for molecular communication. However, as the technology is still in the nascent stages obtaining real data may not always be possible for all forms of molecular communication. In such cases, the parameters affecting the communication are identified and the data input is made configurable by the user.

This section contains two sub-sections. In Sect. 6.1 we discuss about a number of simulators available in nano-communication research domain to use and in sub

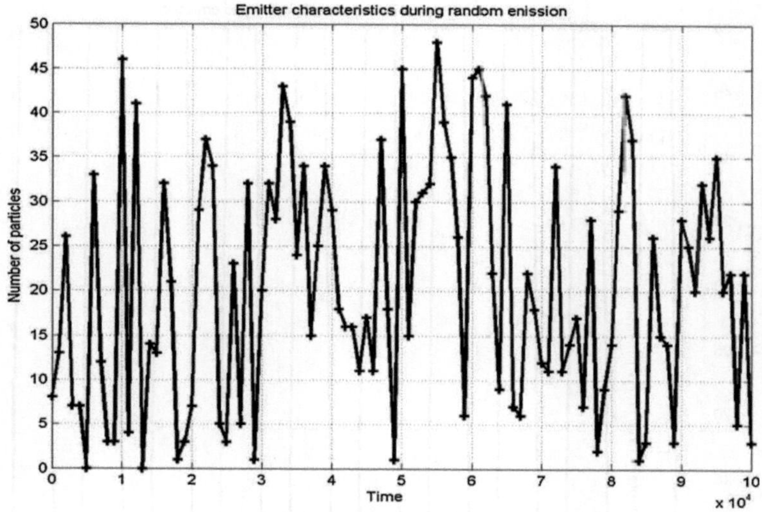

(a) Random emission

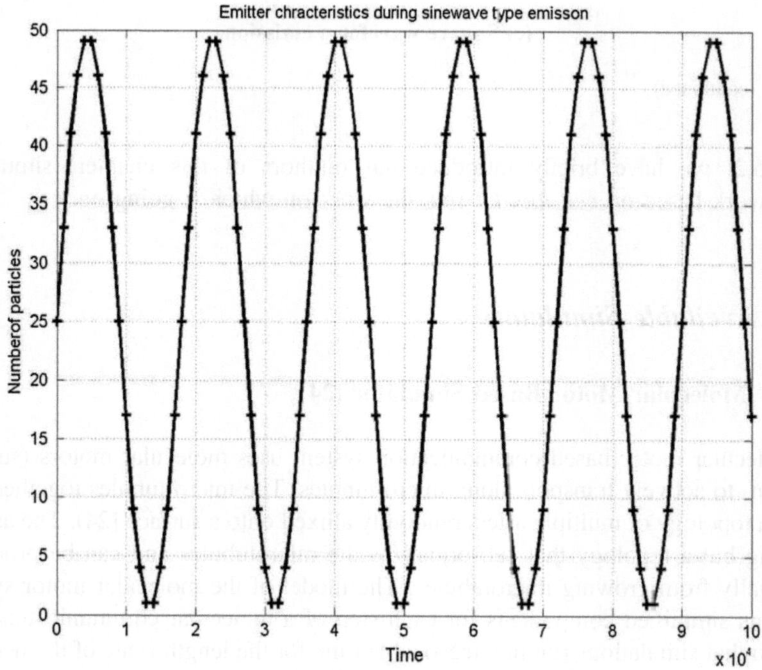

(b) Sinusoidal emission

Fig. 8 **a** Random emission. **b**. Sinusoidal emission. **c**. Square waveform emission

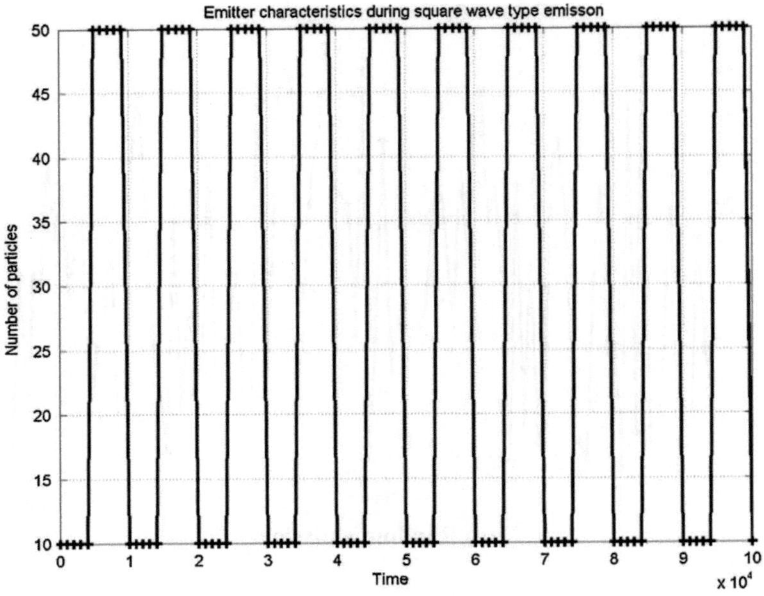

(c) Square waveform emission

Fig. 8 (continued)

Sect. 6.2, we have briefly introduce our (authors of this chapter) simulation framework based on *cell tool kit sim*, the work on which is going on.

6.1 Available Simulators

6.1.1 Molecular Motor Based Simulator [24]

A molecular motor based communication system uses molecular motors (such as kinesin) to actively transport along micro-tubules. The micro-tubules together may form a topology of multiple asters randomly affixed onto a surface [24]. The aster is nothing but a topology that self-organizes the microtubules and can be produced artificially from growing microtubules. The model of the molecular motor system contains simplified components for each step of a molecular communication process so that simulations run in a reasonable time for the length scale of the network (∼10–1000 μm) and the time scale of communication (∼1–100 s) [24]. The communication sender is abstracted as a component that generates a single kinesin molecule (representing the encoded information) on the molecular communication network. The communication receiver is assumed to receive and decode information molecule from nearby information molecules [24]. So one can easily design a simulator which can support the molecular motor based communication as discussed in [24].

6.1.2 3D Brownian Motion Simulator [25]

One important work in this aspect is the modelling of the three dimensional Brownian motion of the nanoparticles. The reliable modelling of this fast distribution is needed in the high sensitivity applications in molecular recognition [25]. In this model, the nanoparticles are placed uniformly random in the 3D container and they start to walk randomly until they find the target spot. The approach for this simulation model is that if the nano particle is far from the target point, it will be simulated in larger time step and when it reaches very near to the target point, the time step is taken smaller. This is called dual time-step approach. There is a possibility of collision for those particles if they hit the rough boundary of the container. This collision is modelled with equal probability to bind the particles.

6.1.3 N3 Sim [26]

For the analysis of diffusion based molecular communication, N3Sim is one of the most well-known simulation framework [26] available in nano-network research. In this framework, the nano machines communicating in the fluid media through molecular diffusion can be easily simulated. At the transmitter nanomachine, the information is modulated at first and then this modulated information propagates through the medium to the receiver. The receiver estimates the concentration of the particles and decodes the information [26].

To run this simulator, the user has to specify the different parameters like distribution of transmitters, receivers, size of emitted particles, diffusion coefficient of the medium etc. in a configuration file. Then the diffusion takes place by the help of diffusion simulator and the outputs are stored in the receiver files containing the concentration of nanoparticles measured by the receivers as a function of time. The variation can be visualized by graphically represent the results into a single plot. Figure 9 describes the N3 Sim architecture [26].

6.1.4 Simulator Based on Java [27]

Another simulator has been made in JAVA for simulating different communication types in nanoscale. It consists of a JAVA package able to provide a set of tools for simulating different nanoscale environments [27]. This framework is quite generic in the sense that it can be customized to analyze different scenarios with different modelling schemes for types of nanomachines, communication channel etc.

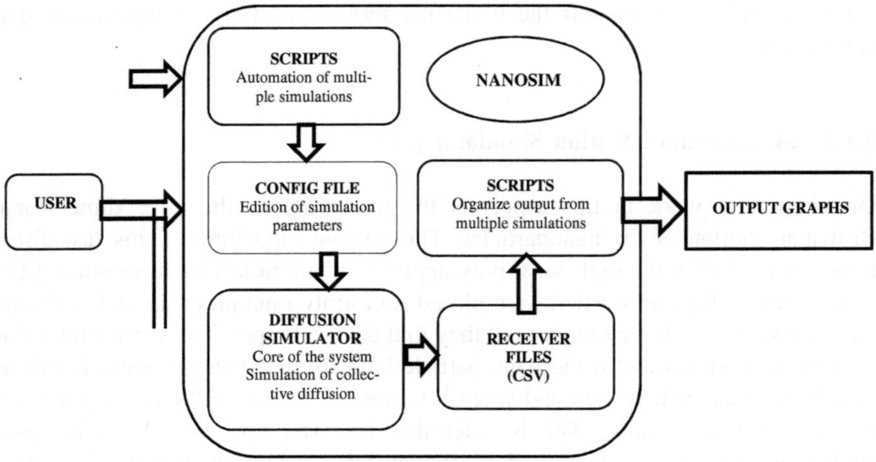

Fig. 9 N3 Sim architecture

(i) **Software architecture**:

The software package consists of different classes some of which are representing different network elements and the others are handling different operations. The software architecture has a multi-thread structure in order to optimize the performance of the computationally expensive operations, such as nano-object propagation and collision management [27]. The propagation of the nano particles are based on Brownian molecular diffusion process and there position and velocity are measured at each and every time step. Some of the classes are described as follows:

Manager class is the main class of the program. It learns from two XML files the configuration parameters that are necessary to setup the simulation environment. This class manages the three dimensional computational space (through the domain class)and controls the proper sequence of operation. The propagation phase and everything related to the simulated environment is managed by the **Mobility Model class** [27]. **Motion strategy class** determines the nature of movement of the nano-objects in the space. Carriers can be received by nodes only through specific receptors located on the outer surface of the node. The relevant code is implemented in the **Receptor abstract class** [27]. The specificity i.e. the compatibility in between the carrier and receptor is handled and taken care by **Carrier Observer Class.** Apart from them, there are **Living object observer class**, **Engine 3D class**, **Output strategy class**, **Nano object class** etc. Figure 10 represents the different classes implemented in this Java simulator.

(ii) **The software library**:

The software library of this simulator, which has been made in JAVA, provides a toolkit for simulating different types of nanonetworks and supports the possibility of simulating different types of nano-communications. The library provides also the

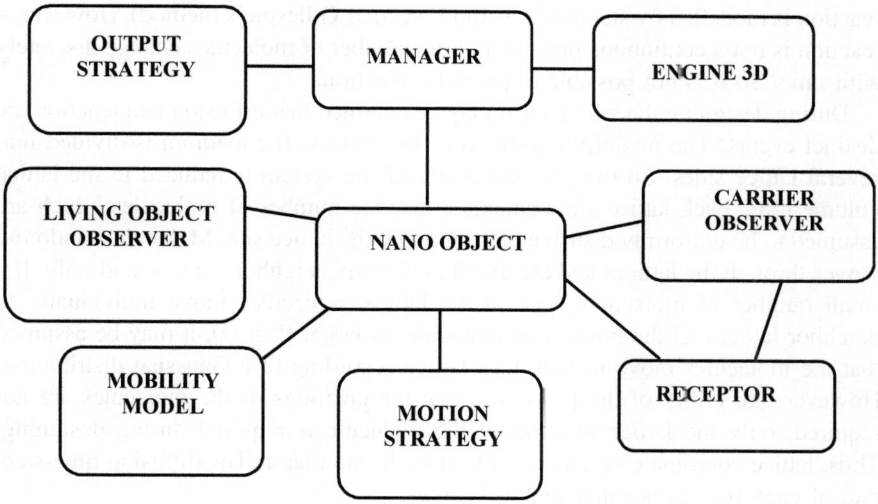

Fig. 10 A simplified picture of the class diagram

general rules for the specific interactions between nanomachines and carriers, carriers and carriers, nanomachines and nanomachines. These interactions realize a generalized behavior and do not limit the development of complex or custom scenarios [27].

Moreover, due to large number of simulated objects, the simulation can be done parallel by using multi-thread approach. Using this simulator, different case studies are also made e.g. in molecular communication, immune systems (Lymphocytes, Antibody response etc.).

6.1.5 NanoNS [28]

This simulator has been designed based on diffusion communication model. Diffusive molecular communication can be modelled according to ligand-receptor binding mechanism as it has been stated earlier that in diffusion based molecular communication system a transmitter releases ligand molecules to the medium. The released molecules diffuse in the environment and some of them bind to the receptor of a receiver. The binding event is a chemical reaction between ligand and receptor molecules and allows the receiver entity to capture the ligand molecules. After the receiver entity captures the molecules from the molecular channel, receiver decodes the information encoded in terms of molecules to fire an action potential [28]. Molecular diffusion is basically Brownian motion i.e., the random movement of molecules in a gas or liquid medium [28]. This diffusion process is formulated by using and solving the well-known Fick's law of diffusion. The reaction can be modelled by two approaches: deterministic approach (where the molecular reactions are continuous and predictable), stochastic approach (here the reaction is modelled by stochastic methods such as Gillespie's method). However, a reaction is not a continuous process and the number of molecules change discretely with time. So it is not possible to predict a reaction.

During designing the simulator it may be assumed that diffusion and reaction are distinct events. The model the basic diffusion process the medium is divided into several lattice sides. So the inhomogeneity of the system is reduced to the lattice volume [28]. Each lattice site contains a discrete number of molecules which are assumed to be uniformly distributed throughout the lattice site. Molecules randomly moves through the lattices and are distributed to six neighbor lattices randomly. If a small number of molecules exist in the lattice, molecules move individually to neighbor lattices. If the number of molecules is larger than 60, it may be assumed that the molecules move in bulk to a lattice according to a Gaussian distribution. However, the utility of the process is that the positions of the molecules are not required, only the lattice position of the molecule is required during designing. Thus, lattice coordinate system is utilized in the simulator. The diffusion time-step, τ_{D_S}, of each species is calculated as follows:

$$\tau_{D_S} = \frac{1}{2d}\frac{\lambda^2}{D_S}$$

where D_S is the diffusion coefficient of the species, λ is the length of each lattice, d is the dimension of the simulation medium [28].

To design a simulator with above characteristics any simulation environment can be chosen but *Network Simulator (ns-2)* is most advantageous because first of all ns-2 is an open source discrete event-driven network simulator providing the simulation of several networking layers like transport, routing, and multicast, protocols over wired and wireless networks. Secondly it supports development to model promising networks which are different from traditional communications. NS-2 is an object-oriented simulator which is written in C++ and an object-Tcl (OTcl) interpreter. To develop the simulator to support molecular communication one has to make some modifications on ns-2. First of all, a new library has to be developed. A new node structure supporting molecular communication requires to be designed which will be plumped in this library file, besides, new network components, parameters and methods for molecular communication are to be defined in this file. A separate class is required to model trans-receiver nano-nodes and another class to incorporate the features of carrier molecules such as proteins, ions or DNA. Subsequently a new class must be created to model the diffusion proposed above. Besides these a separate class also needs to be created to relocate every object position to lattices. The trans-receiver nano-nodes may be created in the lattice space by volume occupation facility. Nano-nodes are normally assumed to be static. For design purpose the shapes of the nano-nodes may be considered as spheres.

In the simulator each nano-node object contains a position object that points to the center of itself, a ligand and receptor molecule pointers [28]. The radius of the nano-node and the number of receptor molecules of the nano-node should remain constant. Likewise, there are some pointers which point to the network components in nano-node. The diffusion and molecular reactions can be modelled by adopting proper and suitable numerical methods. The ligand molecules in the simulator are considered as spheres. Thus Diffusion coefficient for the ligand molecules can be derived from the Stokes-Einstein formula

$$D = \frac{kT}{6\pi r\eta}$$

where k is Boltzmann constant, T is the absolute of the medium, r is the radius of the sphere of the ligand molecules, η is the viscosity of the aqueous medium.

6.2 Cell Tool Kit Sim

To understand the behavior of cell mechanism in molecular communication, laboratory based techniques offer a good insight but the development of required infrastructures and the availability of equipment are difficult in many cases. Although these methods are accurate, but they are expensive, time consuming, labor intensive and less feasible in real time scenario. The ***Cell Toolkit Sim*** is being designed as a simulator for researchers where one can simulate the proposed architecture model discussed in Sect. 5 so that the behavior of calcium signaling mechanism performed in cell can be simulated and analyzed. In the 3D cellular space described in Sect. 5.4, all the cell toolkit components (e.g., Emitter, Buffer, Mitochondrion, ER, ER pumps etc.) described in Sect. 5.3 are being programmed individually in accordance with their corresponding modelling parameters and the dynamics of Ca^{2+} flow through the channel can be programmed using the electro-diffusion equation described in that section.

It is being written in Java to take the advantages of object oriented programming (OOP) and multi-threading. Most of the characteristics of all the components can be configured individually to analyze the performance of the cell architecture mechanism. The way to configure the mechanism is by providing the values from the real experiments.

Software architecture:
This software architecture has three basic components:

(a) input configuration file
(b) architecture space and
(c) trace file (see Fig. 11).

These three components are briefly describes below and Fig. 11 represents the simplified block diagram of this software architecture.

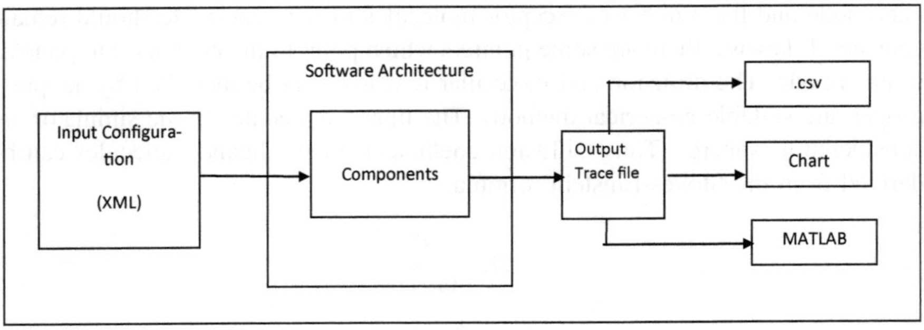

Fig. 11 Block diagram of CELL TOOL KIT SIM

(a) ***Input configuration file***:
In input configuration files, whole configuration of the cell (number of components along with their attributes) is given. In this file, mitochondria, buffer, ER, ER pumps, background, cell dimensions and their individual characteristics can be represented in XML format. Thus, this file describes the environment of the cell to the simulator.

(b) ***Architecture space***:
The second component is the core part of the simulator called as architecture space. This space is to coordinate and control the behavior of each component. In the component part of the space, all the components can be defined with their own activities. Their activities will be affected and completely managed by the architecture space.

(c) ***Trace file***:
Finally, the architecture space generates trace files where behavior of each component at every instant of time can be stored. They may be in CSV format or they may be presented in graphical chart or in appropriate format for MATLAB. If they are saved as CSV (**comma separated value**) file, each column is created by each comma and each row is created by each new line. Therefore, if the file is opened in a spreadsheet program such as Microsoft Excel, it would create a table that would help to make the graphs.

7 Conclusion

In this chapter, we have discussed in Sect. 3 about the various communication techniques those take place among nano machines. Among these techniques, molecular communication is considered as the practically suitable for its bio compatibility and other related reasons discussed in Sect. 4. Thereafter we have focused on molecular communication through calcium signaling and proposed the four architectural layers for nano networks in Sect. 5. Our goal is to model all these layers and virtually realize the overall architecture. Initially we have dealt with the realization of the physical layer by implementing the different cell components according to their modelling discussed in the later subsections of Sect. 5. The further work is to implement the dynamics of Ca^{2+} flow through the physical channel using electro-diffusion model equations as described before and then to observe the intracellular concentration at different points in the cell at each and every timing instances. The observation of the background concentration by varying the parameters of different cell components e.g. for different emission pattern of emitters; for different absorption capacity, position, size and density of buffers, ER, for different threshold, release rate etc., of Mitochondrion may also be an important study. The measurement of received signal power by varying the characteristics of receptors e.g. their sensing capacity, distance from the emitter etc. can give some important information about the quality of the communication and also validation of the discussed model. Furthermore, by using the real laboratory data as the input to the simulator, the obtained result can be compared with the practical one which may help the different branches of medical science. In near future, the readers may also try to implement the whole process by modelling all the necessary components in the proper simulation environment. They can make their own simulator for this and can view the overall process virtually by taking the help of the simulation frameworks discussed above and that would be beneficial for the mankind.

References

1. Akyildiz IF, Jornet JM, Pierobon M (2011) Nanonetworks: a new frontier in communications. CACM, 54(11):84–87
2. Akyildiz Ian F, Brunetti F, Blázquez C (2008) "Nanonetworks: A new communication paradigm". Comput Netw 52:2260–2279
3. Moore MJ, Suda T, Oiwa K (2009) Molecular communication: modeling noise effects on information rate. IEEE Trans Nanobioscience 8(2):169–180
4. Berridge MJ, Lipp P, Bootman MD (2000) The versatility and universality of calcium signalling. Nat Rev Mol Cell Biol 1:11–21
5. Guney A, Atakan B, Akan OB (2012) Mobile ad hoc nanonetworks with collision-based molecular communication. IEEE Trans Mob Comput 11(3)
6. Visscher K, Schnitzer MJ, Block SM (1999) Single kinesin molecules studied with a molecular force clamp. Nature 400:184–189
7. Berridge MJ (1997) The AM and FM of calcium signaling. Nature 386:756–780
8. Parcerisa Ll (2009) Molecular communication options for long range nanonetworks. Master's Thesis, School of Electrical and Computer Engineering, Georgia Institute of Technology
9. Jornet JM, Akyildiz IF (2011) Channel modeling and capacity analysis for electromagnetic wireless nanonetworks in the terahertz band. IEEE Trans Wirel Commun 10(10):3211–3221
10. Akyildiz IF, Jornet JM (2010) Electromagnetic wirewless nanosensor networks. Nano Commun Netw 1:3–19
11. IEEE Group 1906.1: IEEE communications society standards board wiki. (https://standardswiki.comsoc.org/wiki/Framework_proposal)
12. Chakraborty D (2013) Modeling the architecture of calcium signaling from the perspective of nanonetworks. Calcutta University, M. Tech. Thesis

13. Chakraborty D, Mukherjee A, Sadhu S, Ray SS, Das S, Chatterjee S, Naskar MK (2013) Modeling the architecture of calcium signaling from the perspective of nanonetworks. ICC MONACOM. (accepted)
14. Chakraborty D, Mukherjee A, Sadhu S, Chakraborty A, Das S, Chatterjee S, Naskar MK (2014) Physical channel study for calcium signaling based communication in nanonetworks. In: Globecom 2014—symposium on selected areas in communications: GC14 SAC nanotechnology (submitted)
15. Wei G, Bogdan P, Marculescu R (2013) Efficient modeling and simulation of bacteria-based nanonetworks with BNSim. IEEE J Sel Areas Commun 31(12):868–878
16. Wei G, Bogdan, P, Marculescu R (2013) Bumpy rides: modeling the dynamics of chemotactic interacting bacteria. IEEE J Sel Areas Commun 31(12):879–890
17. Srinivas KV, Eckford AW, Adve RS (2012) Molecular communication in fluid media: the additive inverse gaussian noise channel. IEEE Trans Inf Theory 58(7):4678–4692
18. Clapham DE (2007) Calcium signaling, cell 131:1047–1058
19. Bush F (Unpublished) Nanoscale communication network definition & framework. IEEE group P1906.1
20. Berridge MJ, Lipp P, Bootman MD (2000) The versatility and universality of calcium signaling. Nat Rev Mol Cell Biol I:11–21
21. Bidaud I, Mezghrani A, Swayne LA, Monteil A, Lory P (2006) Voltage-gated calcium channel in genetic diseases. Biochem et Biophys Acta 1763:1169–1174
22. Vay L, SanMiguel EH, Santo-Domingo J, Lobaton D, Moreno A, Montero M, Alvarez J (2007) Modulation of Ca^{2+} release in HeLa cells and fibroblasts by mitochondrial Ca^{2+} uniporter stimulation. J Physiol 580(1):39–49
23. Mori Y, Peskin CS (2009) A numerical method for cellular electrophysiology based on the electrodiffusion equations with internal boundary conditions at membranes. Commun Appl Math Comput Sci 4(1):85–134
24. Moore M, Enomoto A, Nakano T, Suda T, Kayasuga A, Kojima H, Sakakibara H, Oiwa K (2006) Simulation of a molecular motor based communication network
25. Totha Á, Bánky D, Grolmusz V (2011) 3D Brownian motion simulator for high sensitivity nano-biotechnical applications. IEEE Trans Nano Biosci
26. Llatser I, Pascual I, Garralda N, Cabellos-Aparicio A, Alarcon E (2011) N3 Sim: a simulation framework for diffusion-based molecular communication. IEEE Trans Nano Biosci
27. Felicetti L, Femminella M, Reali G (2012) A simulation tool for nanoscale biological networks. J Nano Commun Netw 3:2–18. (Elsevier)
28. Gul E, Atakan B, Akan OB (2010) NanoNS: a nanoscale network simulator framework for molecular communications. Nano Commun Netw I:138–156
29. Garralda N, Llatser I, Cabellos-Aparicio A, Pierobony M (2011) Simulation-based evaluation of the diffusion-based physical channel in molecular nanonetworks. In: Proceedings of INFOCOM
30. Chakraborty D, Mukherjee A, Sadhu S, Ray SS, Das S, Chatterjee S, Naskar MK (2013) Modeling the physical channel in the architecture of nanonetworks using molecular communication with Ca^{2+} signaling. IEEE Trans NanoBioscience. ((submitted), August 2013)
31. Jornet JM, Pujol JC, Pareta JS (2012) PHLAME: a physical layer aware MAC protocol for electromagnetic nanonetworks in the terahertz band. Nano Commun Netw 3:74–81
32. Nakano T, Suda T, Moore M, Egashira R, Enomoto A, Arima K (2005) Molecular communication for nanomachines using intercellular calcium signaling. In: Proceedings of 2005 5th IEEE Conference on Nanotechnology, July 2005
33. Clapham DE (2007) Calcium signaling. Cell 131, 1047–1058
34. Chakraborty D, Mukherjee A, Sadhu S, Ray SS, Das S, Chatterjee S, Naskar MK (2013) Physical channel study for calcium signaling based communication in nanonetworks. IEEE, ICC'14. ((submitted), November, 2013)
35. Leeson MS, Higgins MD (2012) Error correction coding for molecular communications. In: International workshop on molecular and nanoscale communications

36. Moore MJ, Suda T, Oiwa K (2009) Molecular communication: modeling noise effects on information rate. IEEE Trans Nanobiosci 8(2):169–180
37. Leeson MS (2000) Performance analysis of direct detection spectrally sliced receivers using fabry-perot filters. J Lightwave Technol 18(1):13–25
38. Kuran MS, Birkan Yilmaz H, Tugcu T, Özerman B (2010) Energy model for communication via diffusion in nano networks. Nano Commun Netw 1(2):86–95
39. Costello DJ, Jr Hagenauer J, Imai JH, Wicker SB (1998) Applications of error-control coding. IEEE Trans Inf Theory 44(6):2531–2560
40. Wang X, Song J, Liu J, Wang ZL (2007) Direct-current nanogenerator driven by ultrasonic waves. Science 316(5821):102–105
41. Moon TK (2005) Error correction coding: mathematical methods and algorithms. Wiley, New York, NY. (Chapter 14)
42. *Ibid*, Chapter 3
43. Nakano T, Suda T, Koujin T, Haraguchi T, Hiraoka Y (2007) Molecular communication through gap junction channels: system design, experiments and modeling. In: Bionetics'07. Budapest, Hungary. (December 10–13, 2007)
44. Akyildiz IF, Jornet JM (2010) Graphene-based nano-antennas for electromagnetic nanocommunicationsin the terahertz band. In: Proceedings of 4th european conference on antennas and propagation, EUCAP, pp. 1–5. (April 2010)
45. Rosenau da Costa M, Kibis OV, Portnoi ME (2009) Carbon nanotubes as a basis for terahertz emitters and detectors. Microelectron J 40(4–5):776–778
46. Zhou G, Yang M, Xiao X, Li Y (2003) Electronic transport in a quantum wire under external terahertz electromagnetic irradiation. Phys Rev B 68(15):155309
47. Woolard D, Zhao P, Rutherglen C, Yu Z, Burke P, Brueck S, Stintz A (2008) Nanoscale imaging technology for terahertz-frequency transmission microscopy. Int J High Speed Electron Syst 18(1):205–222
48. Akyildiz IF, Jornet JM (2010) The Internet of Nano-Things. IEEE Wirel Commun Mag 17 (6):58–63
49. Jornet JM, Akyildiz IF (2010) Channel capacity of electromagnetic nanonetworks in the terahertz band. In: Proceedings of IEEE international conference on communications, ICC, pp. 1–6. (May 2010)
50. Garrald N, Llatser I, Cabellos-Aparicio A, Pierobony M (2011) Low-weight channel coding for interference mitigation in electromagnetic nanonetworks in the terahertz band. In: Proceedings of IEEE INFOCOM 2011
51. Dressler F, Kargl F (2012) Towards security in nano-communication: challenges and opportunities. Nano Commun Netw 3(3):151–160
52. Fall K (2009) The ns manual (formerly ns Notes and Documentation), in: The VINT Project. (January 2009)
53. https://standardswiki.comsoc.org/wiki/Framework
54. https://standardswiki.comsoc.org/wiki/Nano_Sim_Simulator
55. https://standardswiki.comsoc.org/wiki/NS-3_simulation_framework
56. http://www.n3cat.upc.edu/n3sim
57. http://web.njit.edu/~matveev/calc.html
58. Gaurav G, Shivendra T, Pardasani KR (2009) Calora: a software to simulate calcium diffusion. J Comput Inf Sci 2(2):20–30
59. Berridge MJ, Lipp P, Bootman MD (2000) The versatility and universality of calcium signalling. Nat Rev Mol Cell Biol 1:11–21

Part II
Molecular Communication in Biology

On Regulation of Neuro-spike Communication for Healthy Brain

Mladen Veletić, Pål Anders Floor, Rié Komuro and Ilangko Balasingham

1 Introduction

In this chapter strategies for controlling neuronal communication and behavior from a theoretical neuroscience perspective are discussed. The main motivation for controlling neuronal networks is to slow down, halt or reverse mental diseases like senile dementia and Alzheimer's disease (AD (a list of acronyms most used in this chapter is provided in Table 1)), as well as other mental diseases that reduce quality of life. The concept of neuronal networks used in this chapter denotes a group of interconnected biological neurons and must be distinguished from the concepts of neural networks (group of interconnected nerves) and artificial neural networks (group of interconnected "neurons" used in computer science).

Neuronal communication can be affected by drugs, usually affecting synaptic transmission between neurons. Direct stimulation by current injection is another possibility. It is also of interest to develop non-invasive methods that can control neuronal communication from outside the body, possibly by electromagnetic—or alternating magnetic fields. Recent studies performed on gene modified AD mice showed that electromagnetic radiation (EMR) exposure at 918 MHz with

M. Veletić (✉) · P. Anders Floor · R. Komuro · I. Balasingham
Department of Electronics and Telecommunications, Norwegian University of Science and Technology, Trondheim, Norway
e-mail: mladen.veletic@iet.ntnu.no

P. Anders Floor
e-mail: andflo@rr-research.no

R. Komuro
e-mail: rkom004@aucklanduni.ac.nz; rie.komuro@econ.kyushu-u.ac.jp

I. Balasingham
e-mail: ilangko.balasingham@medisin.uio.no

P. Anders Floor · I. Balasingham
Intervention Center, Oslo University Hospital, and Institute of Clinical Medicine, University of Oslo, Oslo, Norway

R. Komuro
Department of Economic Engineering, Kyushu University, Fukuoka, Japan

J. Suzuki et al. (eds.), *Modeling, Methodologies and Tools for Molecular and Nano-scale Communications*, Modeling and Optimization in Science and Technologies 9, DOI 10.1007/978-3-319-50688-3_9

Table 1 List of acronyms

Acronym	Description
AD	Alzheimer's disease
EMR	Electromagnetic radiation
SAR	Specific absorption rate
MF	magnetic field
VCCG	Voltage controlled calcium ion-gate
$A\beta$	Amyloid-beta
AP	Action potential
EMF	Electromagnetic field
GIF	Generalized integrate-and-fire
ARP	Absolute refractory period
RRP	Relative refractory period
AMPA	α-Amino-3-hydroxy-5-methyl-4-isoxazolepropionic
NMDA	N-Methyl-D-aspartate
STP	Short term potentiation
STD	Short term depression
LTP	Long term potentiation
LTD	Long term depression
AM	Adjacency matrix
FF-	Feed forward-
REC-	Recurrent-
BCM	Bienestock, cooper and munro

217 Hz modulation frequency with specific absorption rate (SAR) of 0.25 W/kg ±2 dB improved the cognitive abilities of mice compared with control mice without EMR exposure [1–4]. Furthermore, it has been reported that the brain temperature decreases during exposure, eliminating the possibility that the positive effects are caused by increased temperature. Another non-invasive strategy, that has been shown to reduce symptoms in chronic pain patients, is stimulation with alternating or pulsed magnetic fields (MF) [5]. Possible explanations on why EMR or MF affect neuronal communication are that they affect ion gates or receptors embedded in the neuronal membrane and thereby the membrane conductance (see [6] Chaps. 2 and 6 with references therein). If the membrane conductance in certain synapses in the brain can be controlled, it may be possible to affect plasticity (dynamic wiring of neuron connections) in certain parts of the brain, and thereby rewire it.

An ideal invasive futuristic approach is to develop nanomachines that can deliver drugs, mimic neurotransmitters, or expose neurons to EMR/MF at very specific targets, minimizing risk of side effects. Due to all the possibilities of controlling neurons with such devices one becomes less dependent on specific physical relationships (like the EMR/MF interaction with cells).

In this chapter we summarize some of the existing theories behind EMR and MF effects, and how they are thought to interfere with neurons. We also discuss control of neuronal behavior by current injection and possible future strategies for applying nanomachines. Furthermore, existing theoretical models are summarized for neuronal signal propagation and synaptic transmission that can indicate what changes one can expect both in single neurons and networks of neurons by controlling ion gates. Any ion gate or receptor could potentially be controlled, but we emphasize on calcium here for reasons described in the following.

Calcium plays a crucial role in neuronal signal transmission and memory formation. The intracellular and extracellular calcium concentration is, among many other things, affecting how strongly two neurons are wired together through chemical synapses. It is therefore important that calcium concentration is well regulated for neuronal communication to function properly. It is believed that any disruption in the processes that regulates calcium concentration levels (referred as *calcium dysregulation*), will lead to dramatic changes in neuronal functioning. Studies have revealed that there is an alteration of calcium influx into neuronal cells in aging brains [7]. Recent studies have also shown that drugs blocking voltage controlled calcium ion gates (VCCG), which makes the neuron membrane permeable to calcium ions, lead to a reduction in the progression of AD [8]. Their results support the so-called *calcium hypothesis*, stating that transient or sustained increase in intracellular free calcium in aging brains leads to impaired functioning and eventually cell death [7–9]. This effect is most aggressive in brains with AD. It is also thought that amyloid-beta ($A\beta$) deposits, one of the most essential hallmarks of AD, lead to *calcium dysregulation* [10]. There is also increasing evidence that *calcium dysregulation* in fact leads to an increased production of $A\beta$ [11]. The *calcium hypothesis* shows the importance of calcium regulation in the brain.

Some theoretical models hypothesize that calcium can be controlled by external EMR or MF, by regulating ion gates or ligand binding, although the exact mechanism is unknown. Some experiments have also shown a regulation of calcium influx and efflux under EMR exposure (see [6, 12] and references therein). Some of these results are somewhat controversial, however [6, Chap. 6].

A possible treatment in brains with calcium dysregulation is to develop strategies that re-establish proper calcium regulation, whether it is through drugs, external fields or nanomachines. The objective of this chapter is to design and develop a possible strategy of controlling calcium by deliberately affecting the neuronal electrical properties and VCCGs. We deploy the communication theory aspects integrated into neuronal biological system and re-evaluate existing mathematical models in the literature expressing important neuronal processes as function of intra—and extracellular calcium concentrations in order to link between the time-varying stimuli at targeted neuron, its response patterns, and calcium concentration. Observing the neuronal response when an alternating electric current enters a cell, we reveal the neuronal communication as frequency-dependent. Obtained results provide a strategy of manipulating the spiking frequency and thereby intracellular calcium concentrations, $[Ca^{2+}]_i$, at targeted neuron via controlled current injection. Moreover, a framework presented for synaptic transmission and modification of memory formation and storage, indicates targeted neuron being able to control the behavior of its postsynaptic neurons, as long as current is held long enough to make relevant processes reach the steady state.

The hippocampus, located in the temporal lobe of the brain, is central in the process of memory formation and consolidation from short term to long term memory. It is also one of the the areas of the brain most severely struck by senile dementia and AD. Accordingly, we emphasize on hippocampal neurons and neuronal networks.

2 Neuronal Communication at Cellular Level

Neurons, whose anatomical structure is shown in Fig. 1, are specialized cells that generate electrical pulses called action potentials (APs), or spikes, via its membrane. An AP is a response to chemical inputs (usually) collected by compartments called dendrites (serve as receiver front-end), and is transmitted to presynaptic terminal down though the nerve fiber called axon (serves as communication channel connecting the neurons).

The difference in electrical potential between the interior of a neuron and the surrounding extracellular medium forms the AP pulse shape. Two processes are of special relevance here: the hyperpolarization, that makes the neuron's membrane potential more negative due to the positively charged ions (Na^+, Ca^{2+}, and K^+) flowing out or negatively charged ions (Cl^-) flowing in, and depolarization, that makes the membrane less negative. Fluctuation of hyper—and depolarized electrical potential

Fig. 1 Neuronal structure [13]

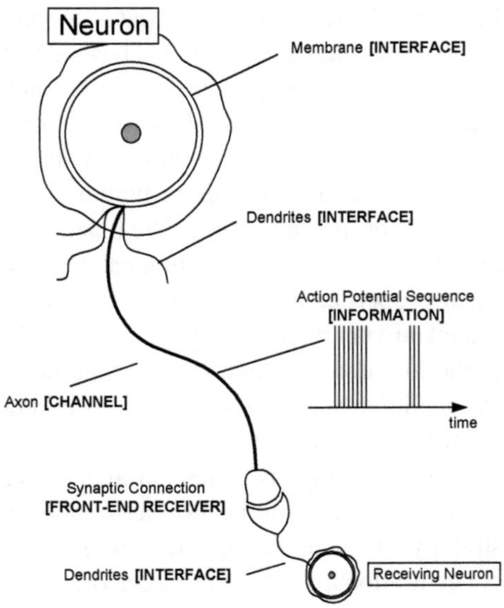

defines an AP. An electric potential difference between the cell and its surroundings additionally produce one more effect, called ephaptic coupling [14]. Since neurons are surrounded by a conducting medium, they can sense electric gradients generated during a neuronal processing and accordingly change their electrical properties. Such a way of communicating, distinct from direct point-to-point communication, can alter the functioning of neurons strongly entraining synchronization and timing of spikes [15].

A structure that further allows a neuron to pass a message generated, i.e. an electrical pulse, to another cell is called synapse (see Sect. 3.2). Synaptic transmission is based on exchange of molecular particles called neurotransmitters and corresponds to the concept of *molecular communication*. Regular synaptic transmission greatly depends on positively and negatively charged ions, predominately calcium ions, Ca^{2+}. Synaptic transmission is closely linked to molecular biology.

The communication process between neurons can be defined as follows: (1) The postsynaptic membrane, which is a part of the neurons dendrites, can be seen as the *receiver front end* receiving information from another neuron (through the synaptic connection). (2) The dendrites and the soma's (cell body) membrane can be seen as the *information processing unit* and *transmitter*. (3) The Axon as well as the synapse can be seen as the *communication channel* between neurons.

2.1 Calcium in Neuronal Communications

The neuron's external and internal calcium concentration, $[Ca^{2+}]_o$ and $[Ca^{2+}]_i$, affects the neurons capability to communicate. For instance, $[Ca^{2+}]_i$ affects the probability of neurotransmitter release at the presynaptic terminal and helps modify the conductivity of the neuron membrane at the postsynaptic terminal. Both determine the synaptic strength, which again determine whether or not two neurons will be connected, and how reliably two neurons will be able to communicate.

Calcium gates are located on dendrites, soma, and axon terminal, but few (if any) are located along the axon. The magnitude of $[Ca^{2+}]_i$ is determined by calcium coming from the outside by an Ca^{2+} ion current flowing into the neuron whenever an AP arrives and opens the VCCGs, the binding of Ca^{2+} to intracellular buffers, and releases from intracellular storage facilities. We mainly consider changes in $[Ca^{2+}]_i$ due to calcium currents in this chapter.

The calcium current, I_{Ca}, is mainly flowing into the neuron since $[Ca^{2+}]_o \gg [Ca^{2+}]_i$. Typical resting values are $[Ca^{2+}]_o \sim 1$ mM and $[Ca^{2+}]_i \sim 100$ nM. Since positive current direction is defined outwards from cells, the calcium current is usually negative in sign. Since dendrites are tiny with small volume, rather few ions are needed in order to raise $[Ca^{2+}]_i$ significantly. A concentration of around 10 μM require about 300–400 Ca^{2+}.

The $[Ca^{2+}]_i$ can be expressed by the following differential equation [16]

$$\frac{d[Ca^{2+}]_i(t)}{dt} = -\gamma I_{Ca}(t) - \frac{[Ca^{2+}]_i(t)}{\tau_{Ca}}, \tag{1}$$

where I_{Ca} is the total calcium current through the membrane, γ is a factor that converts from [coulombs/second] to [mol/second] and τ_{Ca} is a time constant reflecting how fast the intracellular Ca^{2+} is removed (e.g., binding to internal buffers). The solution to (1) is

$$[Ca^{2+}]_i(t) = e^{-\frac{t}{\tau_{Ca}}} \int_0^t e^{\frac{t'}{\tau_{Ca}}} \gamma I_{Ca}(t')dt'. \tag{2}$$

The Goldman, Hodgkin and Katz equation [16] relates I_{Ca} to $[Ca^{2+}]_o$ and $[Ca^{2+}]_i$

$$I_{Ca} = \mathscr{P}_{Ca} 2\nu F \frac{[Ca^{2+}]_i - [Ca^{2+}]_o e^{-\nu}}{1 - e^{-\nu}}, \tag{3}$$

where $\nu = 2V_m F/(RT)$, V_m is the membrane potential, R is the gas constant and T the absolute temperature. $\mathscr{P}_{Ca}$ is the cell membranes permeability to Ca^{2+}.

If the $[Ca^{2+}]_o$ is reduced, it will cause lower efficiency of the relevant synapse. An excess of $[Ca^{2+}]_o$, on the other hand, can lead to cell death due to a too high influx of Ca^{2+}. In AD one observe excess $[Ca^{2+}]_o$, and thereby $[Ca^{2+}]_i$, due to loss of calcium ion-gate inhibition. This often leads to a "runaway effect" of neurotransmitter release where excess neurotransmitters will seep into extra-neuronal space making surrounding neurons overexcited, leading to more calcium influx etc. This is known as exitotoxity and leads to neuron death.

2.2 Electromagnetic Exposure on Neuronal Systems

Electrical properties of neurons, and, consequently ions' flows, directly depends on electromagnetic fields (EMF) generated by electronic devices; therefore, the effects of EMF on human bodies are of great interest. An induced EMF has thermal and non-thermal components. The non-thermal effects are due to the extremely low frequencies, that is, modulation frequencies of the information signal. Although no definitive evidence has been found, it is generally considered that exposure to low energy EMF could be a risk to human health [17]. The non-thermal effects are still not well-known; how and why neurons and networks can function as electromagnetic receivers and demodulators for given electromagnetic transmitter characteristics have not yet been scientifically determined. It is considered that the EMF generated near the head penetrate the skull and reach neurons in the brain and that the current induced by the EMF could stimulate neurons and generate an AP. Thus, if the mechanism of the EMF effects on the neuronal network is elucidated, we could make use of the induced current to fire APs, and this could enable scientists to invent treatments of new types in the future.

There are quite a few different mathematical descriptions for neuronal behavior; e.g., models based on the integrate-and-fire model and the Hodgkin-Huxley (HH) type models. Which model to use depends on the purpose [18]. The simulation can be executed using an open-source simulator such as NEURON (www.neuron.yale.edu) and GENESIS (www.genesis-sim.org). The HH model describes the physiological process, but since it was originally developed for the axon of a squid [19], it is not necessarily the best model for simulations on humans. There are some models developed for humans [20]. In order to simulate more detailed EMF effects on a neuron or a neuronal network, it may be required to consider a mathematical model coupled with biophysical phenomena on the ribonucleic acid (RNA) level.

2.3 Possible Mechanisms Behind Neuronal Effects in Response to EMR and MF

The exact mechanism behind the effects of EMR and MF observed in neurons is unknown. There are also controversies surrounding some of the effects reported, since certain experiments have not been able to reproduce similar effects, or observe any effect at all (see [6, Chap. 6]). There are, however, lots of evidence supporting non-ionizing low intensity radiation effects on neurons, specifically related to therapeutic effects (see [6] and references therein), that does not have to do with temperature increase.

It is likely that the mechanisms explaining every aspect of EMR/MF interactions with neurons is complicated. However, there exists some simplified mathematical models in the literature able to reproduce some of the effects observed in experiments. Although there are controversies around these models, they will anyway serve as a starting point for further investigation. It is natural that models trying to describe unknown mechanisms are up for debate, and in time some of these models may turn ut to have something on them (and can be extended further), while others will have to be rejected since they are based on grounds that do not correspond with reality. This emphasizes the importance of conducting experiments in parallel with the development of theoretical models, since it will make the chances of ending up in a blind alley much smaller.

We will briefly summarize the main ideas behind some of the existing approaches of theoretical modeling here, and also mention some of the controversies surrounding these models. All models assume that the place of EMF interaction is the cell membrane where ion binding and transport is affected. It is also possible that EMF affects distribution of protein and lipids in the membrane itself by altering the kinetics of binding.

One model by Panagopoulos et al. [6, Chap. 2.2] hypothesize that ions used in the neuronal communication process (K^+, Na^+, Ca^{+2} etc.) experience forced vibrations in and around ion gates due to constant frequency harmonic—or pulsed EMF. Forced vibrations will affect the opening and closing of voltage controlled gates since they disrupt the electrochemical balance of the neuronal membrane. Forced vibrations are governed by the wave equation, and show that the necessary stimuli intensity needed to achieve forced vibrations, like harmonic oscillators, increases linearly with frequency. Solutions to the same equation also show that pulsed fields double the impact of the stimuli (compared to harmonic oscillators) because of transient effects. This corresponds with many experiments conducted. One of the main criticisms of this model is that fluctuations due to thermal effects are larger than the amplitudes of the forced vibrations. Panagopoulos et al. further claimed that thermal effects will average to zero over time, since they present random forces in all directions. Then, since forced vibration is a coherent motion (superimposed on the thermal motion), it will result in a net effect on average.

Another model by Liboff et al. [21], [6, Chap. 2.4] hypothesizes that MF interaction with neuronal tissue seen in experiments can be explained by ion cyclotron resonance described by the Lorentz force equation $\mathbf{F} = q(\mathbf{v} \times \mathbf{B})$. $\mathbf{B}$ is the magnetic field, $\mathbf{v}$ is the particles velocity vector and q is charge (see [21] and [6, Chap. 2.4]). The Lorentz force is resulting from the acceleration a charged particle experience in a MF, where the acceleration is proportional to the charge-to-mass ratio. With an electric field, $\mathbf{E}$, present the Lorentz force becomes

$$\mathbf{F} = q(\mathbf{E} + \mathbf{v} \times \mathbf{B}). \tag{4}$$

This force will in principle interfere with the ion transport through and around the membrane which further alter biological responses. It is believed that the α-helical configuration of the proteins making up the ion channels places regular constraints on the ionic motion. Liboff et al. further hypothesize that if the protein channel structure forces ions to move in helical paths, then with the right direction on the magnetic (and electric) field, the Lorentz force may result in a set of gyro-frequencies (harmonics) for the channel, where each frequency corresponds to different cyclotron resonance conditions (eigenfrequencies). I.e., the channels frequency spectrum becomes quantized. The main criticism against this model is that collisions with other particles and thermal effects will eliminate the hypothesized effects [22].

Yet another model, named Electrochemical Information Transfer Model (EIT), proposed by Pilla [23], [6, Chap. 2.3] hypothesizes that low intensity EMF modulate the rate of binding of ions to receptor sites. The EIT formulates a membrane current which depends on change in membrane surface concentration of the relevant binding ion over time, the surface concentration itself, and frequency. The change in concentration is again dependent on membrane voltage and the following biochemical reactions. This model can be illustrated by an equivalent circuit diagram as that in Fig. 2. C_d is the dielectric membrane capacitance and R_M the leaky resistance (due to ions passing through ion gates under normal circumstances). The ion-binding pathway is represented by R_A and C_A, and C_B and R_B represents follow-up biochemical reactions. This leads to two time constant R_AC_A and R_BC_B in addition to the (normal) leaky pathway and membrane capacitance. This model has been applied to successfully construct magnetic and electromagnetic pulses for therapeutical

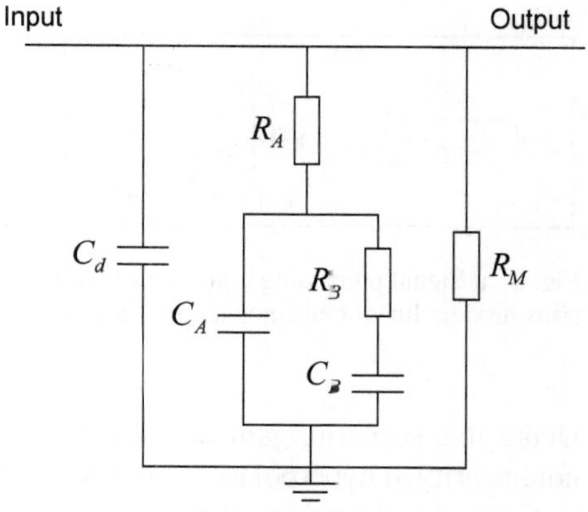

Fig. 2 Equivalent circuit diagram for (single compartment model) cell membrane exposed to EMR. The figure is based on Fig. 4 in [23]

applications, and the resulting pulses has shown to be efficient in treating bone diseases. The model may also be applied for other cells and tissues.

Another relevant model that evaluate EMR effects at the cellular membrane level is given in [24]. This model is also validated with experimental data.

3 Neuron's Message Transmission

3.1 Fundamental Neuronal Operating Principles

According to (3), I_{Ca} directly depends on the membrane potential. One possible way of controlling $[Ca^{2+}]_i$ is through change in membrane voltage, since it also drives the $[Ca^{2+}]_i$ according to (2). To change the neuronal membrane potential, it is sufficient to inject or induce a convenient current.

Referring to previous work in neuroscience, biology and chemistry, it appears to be hardly possible to determine an analytical expression for the neuron response when it is affected by specified stimulus. The reason is confined to the large trial variability directed to the biological systems even when the same stimulus is applied repeatedly [25]. In this section, we endeavour to inspect the oscillatory behavior of a biophysically realistic sample neuron, utilizing the engineering concepts integrated in a neuronal structure and keeping the analysis as simple as possible.

Soma is the cell body part of a neuron where summation of all the dendrite inputs takes place. It can be represented via a cascade (linear filter and static gain function), similarly to that introduced by Hunter and Korenberg in 1986 intended to general biological systems [26]. In describing neuronal processing (see Fig. 3a), we can introduce the quadratic form of gain function confined to habituation of (almost) all biological micro- and macro-systems. The nearly straight-line of preprocessing gain has been shown to correspond to the low-variance input [27]. The function, however, allows nonlinearities in the amplitude response when high-variance input is applied.

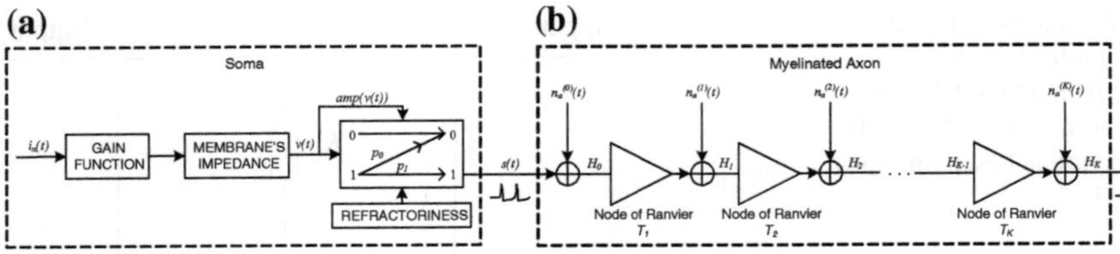

Fig. 3 **a** Signal processing unit: cascade of gain function and membrane's impedance; **b** Axonal transmission: linear multi-hop amplify-and-forward channel

Under this scenario, gain function curve exhibits reduced slope and compressive nonlinearity at hyperpolarized and depolarized potentials.

Generalized versions of deterministic IF model, GIF, formulates the membrane's impedance (transfer function) to characterize the neuron subthreshold dynamic [28]. Since the model is capable of introducing frequency-dependent filtering of input signal, GIF model is convenient from the communication theory viewpoint. Cognizance of type-specific parameter values, such as variables characterizing the membrane dynamics, conductances, and parameters proportional to conductances, analytically define the impedance magnitude an phase as

$$|Z(\omega)| = \left(\frac{\tau}{C}\right)\sqrt{\frac{(1+\tau^2\omega^2)}{(\alpha+\beta-\tau^2\omega^2)^2+\tau^2\omega^2(1+\alpha)^2}},$$
$$\phi(\omega) = \arctan\left(\tau\omega\frac{\beta-(1+\tau^2\omega^2)}{\beta+\alpha(1+\tau^2\omega^2)}\right). \tag{5}$$

Parameter τ stands for time scale of the variable characterizing the membrane dynamics, C represents membrane capacitance, parameter α is proportional to effective leak conductance, whereas β is proportional to conductance measuring the membrane potential variation [28]. Note that both the α and β can be experimentally measured for hippocampal neurons. Since the findings presented in [27] indicate the neurons properties remains the same during stimulation, it appears that the membrane's impedance does not differ between the low- and high-variance input. Moreover, it implies the membrane's impedance is uniquely defined with neuron's type and properties, which makes the analysis presented so far fully deterministic.

As stated previously, due to the many potential sources of response variability, including variable levels of arousal and attention, randomness associated with various biophysical processes that affect neuronal firing, and the effects of other cognitive processes taking place during a stimulation [25], a randomness in neuronal behavior must be accounted for in order to produce a realistic model. The neuron will typically fire, i.e., generate an action potential only when its somatic potential, $v(t)$, reaches a threshold value of about −55 to −50 mV. However, this is not the only condition; neuron firing is controlled by a refractory period (RP), i.e., a time period during which a cell is incapable of or inhibited from repeating a spike generation.

Two RPs are defined: the absolute refractory period (ARP), as the interval during which a new spike cannot be initiated, regardless of the intensity of stimulation, and relative refractory period (RRP), as the interval during which a new spike generation is inhibited but not impossible. The RRP follows immediately after the ARP.

One way of introducing the randomness is through the state machine mechanism that seems to be adequate for spike generation process (see Fig. 3a). A binary 1 at the input symbolizes the somatic voltage $v(t)$ with an amplitude above the threshold, whereas 0 at the input symbolizes the voltage is beyond the threshold value. Corresponding alphabets at the output symbolize the events when spike is generated, and spike is not generated, respectively. Spontaneous generation of spikes at the time of subthreshold voltages at the input is ignored. Probabilities p_1 and p_0 account for randomness. The spike generation probability, denoted as p_1, is set to 1 when both the ARP and RRP have expired and when somatic voltage is above the threshold. During the ARP, the probability p_1 is 0, as well as when the voltage $v(t)$ is beyond the threshold no matter what period is valid. During the RRP, the threshold is higher immediately after a spike and decays (usually exponentially) then back to its resting value. Equivalently, probability p_1 increases proportionally to the time and suprathreshold amplitude of $v(t)$ and takes the value from 0 to 1, i.e.,

$$p_1(t) = 1 - \mathrm{e}^{-\frac{t}{\tau}}, \tag{6}$$

where the time constant τ follows the rule $1/\tau \propto$ stimuli intensity. Transient probability p_0, in contrast, exhibits the opposite behavior relative to that of the p_1, i.e., $p_0 = 1 - p_1$. Eventually, note that not only does the suprathreshold voltage drive the probabilities p_1 and p_0 directly, but also indirectly, through the impact on duration of the ARP. It has been shown that the larger the suprathreshold amplitude, the smaller the ARP, or equivalently, the higher the firing rate [29]. Note, however, that the ARP has a lower bound limit because the neuron cannot fire with the rate higher than that physiologically determined.

At this point we are able to build up strategies of affecting the membrane conductances and consequently $[Ca^{2+}]_i$ via VCCGs. The analysis performed helps to eliminate signals that lead to disruption in communication, or equivalently, to select those adequate for VCCGs control. Elaboration presented so far helps not only to grasp the neuronal encoding and defines strategies of controlling the $[Ca^{2+}]_i$, but also to decently mimic the neuronal behavior while designing the information that might be followed during fabrication and assembly of nanotechnology-based neuron. Performance of introduced model, however, demand further analysis and comparison with measurements obtained in vivo.

3.1.1 Axonal Transmission

An axon is a nerve one-way channel that conducts the spikes typically in a direction outward from the soma. This propagation is called orthodromic propagation [25].

Antidromic propagation in the reverse direction is not excluded. For proper communication and functioning of the nervous system, many axons in vertebrates are wrapped with a myelin sheath that increases the speed at which impulses propagate along the fiber by hops or saltation. In order to decrease the impact of channel attenuation due to the possible significant axon length, and let the signal be amplified, electrically insulating myelin sheath is interrupted with nodes of Ranvier. There is a high density of fast voltage-dependent Na^+ at these points making the axonal membrane uninsulated and capable of regenerating spike impulses. Axonal channel is highly reliable and adjusted to its frequency band-limited input. Owing to its properties, we refer an axonal channel as the one-way linear multi-hop amplify-and-forward channel, as shown in Fig. 3b. All relay terminals, i.e., nodes of Ranvier T_k, $k = 1, 2, \dots, K$, are located on a straight line along the axon from the source terminal T_1 to the destination terminal T_K. Relaying gain of terminal T_k is denoted as G_k, the channel coefficient from T_k to T_{k+1} terminal as H_k, and communication noise components as ${n_a}^{(k)}$. With ${n_a}^{(k)}$ we refer to axonal noise caused by disturbances affecting the Na^+ gates. Further, it is assumed that the kth terminal can only hear and amplify the signal transmitted by the $(k-1)$th terminal. Hence, the signal received by the presynaptic terminal can analytically be expressed as

$$\hat{s}(t) = \left(\prod_{k=1}^{K} G_k\right)\left(\prod_{k=0}^{K} H_k\right) s(t) + \sum_{j=1}^{K+1}\left(\prod_{k=j}^{K+1} G_k\right)\left(\prod_{k=j-1}^{K} H_k\right) {n_a}^{(j-1)}(t), \tag{7}$$

where $s(t)$ denotes the spike sequence entering the axon, and $G_{K+1} = 1$.

Prospective usage of nano-scale devices that interconnect with neurons and re-establish an information propagation through the network would set up a ground-breaking ICT-inspired approach in treatment of neurodegenerative diseases (see Sect. 5). Moreover, design of such nano-scale devices, i.e., *synthetic neurons*, can directly implement the previously defined theoretical concepts confined to soma and axon. Since myelination is thought to inhibit not only the ephaptic interactions but also other sources of disturbances making the axonal transmission very reliable, human implementation of *synthetic neuron* with linear multi-hop amplify-and-forward axonal channel should find out an efficient approach in reducing the noise components ${n_a}^{(k)}$.

3.2 Synaptic Transmission

Chemical synapses transfer signals from one neuron to another through neurotransmitter molecules released into the synaptic cleft. Figure 4a shows a simplified synaptic model explained in the following.

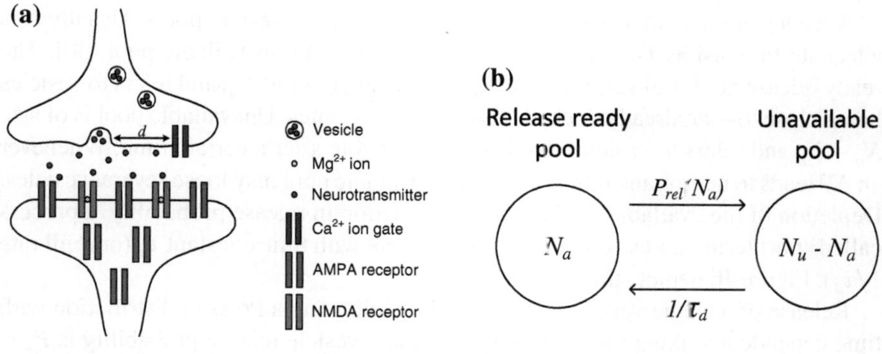

Fig. 4 **a** Simplified sketch of synapse. **b** State-diagram for vesicle pools

When an AP reaches the presynaptic membrane, it will depolarize, leading to opening of VCCGs and an influx of Ca^{2+} into the presynaptic terminal. This increases the probability for neurotransmitter release. There exits a myriad of neurotransmitter molecules and many postsynaptic receptors that neurotransmitters activate, making synaptic communication difficult to fully comprehend. To simplify the discussion we focus on one neurotransmitter; glutamate, one of the most important neurotransmitter for modifiable synapses. We consider only two receptors AMPA (α-amino-3-hydroxy-5-methyl-4-isoxazole propionate) and NMDA (N-methyl-D-aspartate), that are crucial for memory formation and learning.

Neurotransmitters are stored in synaptic vesicles in the presynaptic terminal. We refer to the set of neurotransmitters released by one vesicle as *quanta*. The vesicles are transported to the synaptic terminal and attached to the presynaptic membrane at release sites. The number of *quanta* released per AP, N_q, varies throughout the central nervous system, typically ranging from one to a few *quanta* per AP. This results in variations in reliability of synapses, with $N_q = 1$, the least reliable.

Synaptic strength is determined by N_q, the postsynaptic conductance, and the open ion channel probability P_o. P_o is the probability of finding an arbitrary gate among a large number of gates in open state. For synapses $P_o = P_{\text{rel}} P_{\text{post}}$, where P_{rel} is the probability for neurotransmitters release when an AP is fired, and P_{post} is the open probability for postsynaptic ion gates given that neurotransmitters were released. Typically the synaptic strength/weight is proportional to the mean postsynaptic action $N_q P_o G$, with G some postsynaptic effect, like peak conductance (or current) [16, p. 91].

3.2.1 Presynaptic Terminal and Neurotransmitter Release

A rise in $[Ca^{2+}]_i$ from its resting concentration of about $0.1\,\mu$M increases the probability that vesicles releases neurotransmitters. Since neurotransmitters are released probabilistically, synaptic transmission is stochastic in nature.

Vesicles are usually divided into several sets named vesicle pools. Usually it is adequate to consider two pools, release ready pool and unavailable pool [30]. The ready release pool is of variable size N_a, with an upper limit N_u, and refers to vesicles being close to—or already docked to vesicle release sites. Unavailable pool is of size $N_u - N_a$ and refers to vesicles that become available after a certain time. Whenever an AP leads to neurotransmitter release, the available pool may loose several vesicles. Depletion of the available pool leads to a reduction in release probability, a process called short term depression (STD), and recovers with time constant τ_d (or refill rate $1/\tau_d$). Figure 4b depicts this process.

Release of a single vesicle during an AP is following a Poisson distribution with time dependent spiking rate $r(t)$ [30]. The single vesicle release probability is $P_v = 1 - e^{-\alpha_v}$, where $e^{-\alpha_v}$ is the single vesicle failure probability, and

$$\alpha_v = \int_0^{\Delta_t} r(t)\mathrm{d}t \tag{8}$$

is named *vesicle fusion rate*, where Δ_t is the duration of the AP. With N_a vesicles in the release ready pool, and with $N_q = 1$ release sites, we get the release probability

$$P_{\text{rel}}(N_a) = 1 - e^{-\alpha_v N_a}. \tag{9}$$

In the hippocampus, for instance in the synapses connecting CA3 (*Cornu Ammonis*) pyramidal cells with CA1 neurons, $N_q = 1$, and two neurons usually form only one synapse with each other. This implies a high failure probability $P_e = 1 - P_{\text{rel}} = e^{-\alpha_v N_a}$. P_{rel} can, however, increase when a high frequency spike train or a burst of spikes reach the presynaptic terminal, since $[Ca^{2+}]_i$ builds up when many AP's are fired over a short time interval, a process named short term potentiation (STP).

Short Term Synaptic Plasticity

STP: The duration of STP increases with the frequency of AP's, f_{AP}, as well as the duration of the stimuli. Let P_0 denote the baseline release probability. After a burst of high frequency APs, the release probability will be modified to $P_p > P_0$. After this $P_p \rightarrow P_0$ with time constant τ_p ranging from 100ms to minutes depending on f_{AP} and stimuli duration. Since the time constant for onset $\ll \tau_p$, we get

$$P_{\text{rel}_p}(t) = P_0 + (P_p - P_0)e^{-\frac{t}{\tau_p}}. \tag{10}$$

STD: With $1/\tau_d$ the vacancy refill rate (τ_d in the order of seconds), and $P_d < P_0$, the minimal release probability after depression, then

$$P_{\text{rel}_d}(t) = P_0 - (P_d + P_0)e^{-\frac{t}{\tau_d}}. \tag{11}$$

Relation Between Release Probability and Presynaptic Calcium

The expression in (9) does not explicitly take the $[Ca^{2+}]_i$ into account, although it will indirectly affect both N_a and α_v. The relation between P_{rel} and $[Ca^{2+}]_i$ is complicated, and for this reason, no exact mathematical formula has been found at present (at least to our knowledge). Experiments have revealed a highly nonlinear relationship between the presynaptic Ca^{2+} current (influx) and the postsynaptic potential [31]. As mentioned by Wu and Saggau in [31], the Ca^{2+} influx caused by an AP will activate certain Ca^{2+} sensors, which further trigger the processes that cause vesicle fusion and neurotransmitter release. P_{rel} is therefore a product of two terms $P_{\text{rel}} = P(Ca^{2+}) \cdot P(\text{Ves})$, where $P(Ca^{2+})$ is the probability reflecting the Ca^{2+} binding process (to internal buffers and sensors) and $P(\text{Ves})$ reflects the vesicle release process (dependent on vesicle pool dynamics).

One reason that complicates the relation between P_{rel} and $[Ca^{2+}]_i$ is that the effective concentration around the vesicle release sites depends on the distance d to the VCCG [32] (see Fig. 4a). $[Ca^{2+}]_i$ needs to reach levels of about 100 μM in the vicinity of a vesicle release site for neurotransmitters to be released. If Ca^{2+} enters at about $d \approx 10$ nm, from the vesicle release site, a release of neurotransmitters usually follow, since then $[Ca^{2+}]_i > 100$ μM. If d is larger than about 100 − −200 nm, fast intracellular buffers will bind most of the calcium before $[Ca^{2+}]_i$ gets even closer to the 100 μM level.

A mathematical relation between P_{rel} and $[Ca^{2+}]_i$ must take the following interrelated features into account: (1) $[Ca^{2+}]_i$ that depends on d. (2) Ca^{2+} binding through internal buffers. (3) Ca^{2+} release from endoplasmic reticulum. (4) Interaction between vesicle pools. (5) STP and STD.

3.2.2 Postsynaptic Terminal

For simplicity, it is assumed that the conductance of the postsynaptic terminal is determined by AMPA's and that NMDA determines how conductance of AMPA gates change. This is motivated by the fact that NMDA is about 10x slower than AMPA, and so the AMPA's are nearly closed when the NMDA's start to conduct. In a more exact model, the NMDA's contribution to membrane conductance should also be evaluated.

AMPA Receptors

AMPA is a fast ion conducting receptor permeable to sodium (Na^+), potassium (K^+) (and in some cases also Ca^{2+}), whenever glutamate binds to it.

The current through the AMPA's is typically given by

$$I_{\text{AMPA}}(t) = G_{\text{AMPA}} \cdot P_{\text{AMPA}}(t) \cdot (V_{\text{post}} - E_x), \tag{12}$$

where G_{AMPA} is a modifiable conductance factor, $P_{\text{AMPA}}(t)$ is the open probability, V_{post} is the postsynaptic potential and E_x is the reversal potential of the relevant ion (Na^+ or K^+). The rise-time in open probability is almost instantaneous compared to the decay time, τ_{AMPA}, and so

$$P_{\text{AMPA}}(t) \approx P_{\max} e^{-\frac{t}{\tau_{\text{AMPA}}}}. \tag{13}$$

G_{AMPA} can be altered in two ways: (1) Modification of the conductance of AMPA receptors embedded into the postsynaptic membrane. (2) Insertion of several new AMPA receptors into the postsynaptic membrane (see Fig. 4a). In reality, both of these processes take place, referred to as AMPA receptor trafficking (a graphical explanation is given in [9, pp. 783–784]). Mathematically, these processes are modeled differently, but the end result is almost the same [33].

(1) Modification of AMPA receptor conductance: The following model was presented by Castellani et al. in [34]. The AMPA are composed of 4 protein compartments GluR1-GluR4. Different configurations of these lead to different classes of AMPA. In the hippocampus, AMPA receptors consisting of GluR1 and GluR2 are abundant. AMPAs consisting of GluR2 are impermeable to Ca^{2+}. GluR1 is essential to plasticity since it can be modified depending on $[Ca^{2+}]_i$ levels. Let GluR1 be denoted by A:

(i) When $[Ca^{2+}]_i > \theta_p$ the process of protein kinase is initiated, which means that phosphate groups are added to Serine-831 and Serine-845 on the GluR1. This process is named phosphorylation. There are two kinase processes: Kinase A, denoted $\mathscr{K}_1$, which phosphorylate Serine-845 and Kinase C, denoted, $\mathscr{K}_2$, which phosphorylate Serine-831. Both increases the AMPA conductance. Denote the phosphate groups due to these kinase-processes p_1 and p_2. There are 4 possible phosphorylation processes: $A \xrightarrow{\mathscr{K}_1} A_{p_1}$, $A \xrightarrow{\mathscr{K}_2} A^{p_2}$, $A^{p_2} \xrightarrow{\mathscr{K}_1} A_{p_1}^{p_2}$, $A_{p_1} \xrightarrow{\mathscr{K}_2} A_{p_1}^{p_2}$. $A_{p_1}^{p_2}$ is called double phosphorylation, and results in the highest conductance.

(ii) When $\theta_d < [Ca^{2+}]_i < \theta_p$ the process of protein phosphatase is initiated, in which phosphate groups are removed from Serine-831 and Serine-845 on the GluR1 protein, denoted $\mathscr{P}_1$ and $\mathscr{P}_2$ respectively. This process is named dephosphorylation and decreases the AMPA conductance. There are 4 possibilities $A \xleftarrow{\mathscr{P}_1} A_{p_1}$, $A \xleftarrow{\mathscr{P}_2} A^{p_2}$, $A^{p_2} \xleftarrow{\mathscr{P}_1} A_{p_1}^{p_2}$, $A_{p_1} \xleftarrow{\mathscr{P}_2} A_{p_1}^{p_2}$.

(iii) When $[Ca^{2+}]_i < \theta_d$, the AMPA stays unchanged.

The AMPA conductance is given by [34]

$$G_{AMPA} = A + 2(A_{p_1} + A^{p_2}) + 4A_{p_1}^{p_2}, \tag{14}$$

which was shown to be quite consistent with experiment. A, A_{p_1}, A^{p_2}, and $A_{p_1}^{p_2}$ are all functions of the kinase and phosphatase processes $\mathscr{K}_1$, $\mathscr{K}_2$, $\mathscr{P}_1$ and $\mathscr{P}_2$ given as a solution to a system of 4 differential equations [35]. $\mathscr{K}_1, \mathscr{K}_2, \mathscr{P}_1, \mathscr{P}_2$ are all functions of $[Ca^{2+}]_i$, determined by experiment. More exact models, built on *Michaelis-Menten kinetics*, are derived by Castellani et al. in [35].

(2) Insertion and removal of AMPA receptors: The following model was presented by Shouval et al. in [33]. When $\theta_d < [Ca^{2+}]_i < \theta_p$, some of the AMPAs already attached to the membrane will become detached (see Fig. 4a).

Let B_M denote the number of AMPA receptors attached to the postsynaptic membrane, and B_I denote the AMPAs available for insertion. It is assumed that $B_M + B_I = B_T$, where B_T is a constant, corresponding to conservation of protein. Further, let $K_I = K_I([Ca^{2+}]_i)$ and $K_R = K_R([Ca^{2+}]_i)$ denote the kinetic constants for AMPA insertion and removal respectively. It was shown in [33] that

$$B_M(t) = \left(B_M(0) - B_M^{f_p}\right) e^{-\frac{t}{\tau_{[Ca]_i}}} + B_M^{f_p}, \tag{15}$$

where $B_M^{f_p} = B_T K_I/(K_I + K_R)$ and $\tau_{[Ca]_i} = 1/(K_I + K_R)$ i.e., the steady state response and the time it takes to reach steady state. The kinetic constants K_I, K_R as functions of $[Ca^{2+}]_i$ are found by experiment [33].

NMDA Receptors

NMDA receptors differ from AMPA in several ways (in addition to being slower):

(I) When glutamate locks onto NMDA, it opens, but is still blocked by Mg^{2+} present in extracellular space (see Fig. 4a). To remove the Mg^{2+}, the postsynaptic neuron must depolarize in order to "kick" the Mg^{2+} out. A current through the NMDA receptor therefore requires that glutamate is released from the presynaptic terminal and that the postsynaptic neuron subsequently depolarizes.

(II) In addition to Na^+ and K^+, the NMDA is permeable to Ca^{2+} (about 7% of the NMDA current). Since AMPAs that contain GluR2 (which are the only ones considered in this chapter) are impermeable to Ca^{2+}, the amount of Ca^{2+} influx through NMDA receptors signals the level of pre- and postsynaptic co-activation.

The NMDA calcium current is typically given by

$$I_{Ca(NMDA)} = G_{NMDA} \cdot P_{NMDA} \cdot \mathscr{H}(V_{post}). \tag{16}$$

G_{NMDA} is the maximal conductance and P_{NMDA} is the open probability,

$$P_{NMDA}(t) = P_{max} \cdot \left(e^{-\frac{t}{\tau_s}} - e^{-\frac{t}{\tau_f}}\right). \tag{17}$$

The rise time, τ_f, is fast compared to the decay time τ_s, but considerably slower than the AMPA rise time. $\mathscr{H}(V_{post}) = H_{NMDA}(V_{post})(V_{post} - E_{Ca})$, where E_{Ca} is the reversal potential for calcium and $\mathscr{H}(V_{post})$ is a "gain" term that takes the Mg^{2+} blockage into account [36]

$$H_{NMDA}(V_{post}) = \left(1 + \eta \left[Mg^{2+}\right]_o e^{-\gamma V_{post}}\right)^{-1}. \tag{18}$$

This implies that the higher $[Mg^{2+}]_o$ is, the higher V_{post} must be before the NMDA gate becomes conductive. Under normal conditions, typical values at temperature 37 °C are $\gamma = 0.06/\text{mV}$, $\eta = 0.33/\text{mM}$ and $[Mg^{2+}]_e \approx 1$ mM.

A relation between the postsynaptic $[Ca^{2+}]_i$ due to the NMDA current and the frequency of a presynaptic spike train with constant frequency f_{AP} held long enough for calcium concentration to reach steady state was derived by Castellani et al. in [34]:

$$\overline{[Ca^{2+}]_i} = \mathscr{H}(V_m) \cdot \tau_{Ca}\left(\tau_f G_f + \tau_s G_s\right) \cdot f_{AP}, \tag{19}$$

where G_f, τ_f and G_s, τ_s are magnitudes and time constants for fast rising- and slowly decaying conductance component respectively. This formula was found by extending the one-spike NMDA calcium current in (16) to a Poisson distributed spike train. When the current is found one can apply (2) to find the concentration. One must also determine relation between spike time and f_{AP} (see [34] for details). (19) shows that there is a linear relationship between steady state postsynaptic calcium concentration and the presynaptic spiking frequency.

Like AMPA, NMDA can be modified. A model was presented in [34], in which G_s and G_f depend on the "history" of cortical/neuronal activity.

Long Term Synaptic Modification

The more active AMPA receptors present in the postsynaptic terminal, the higher the synaptic strength becomes, and the probability that the postsynaptic neuron will fire in response to glutamate release increases. This is the process underlying Long Term Potentiation (LTP). In the opposite case, when the number of active AMPAs are reduced, a long lasting weakening of the synapse takes place, the process behind Long Term Depression (LTD). This change in conductance can last for hours, days and even weeks to years. LTP and LTD wire neurons together in a dynamical way which is essential for the brains ability to learn. It is the patterns of connections that the LTP and LTD processes create in the brains network that represent memories.

Donald Hebb's hypothesis [37] for synaptic long term modification is: (I) synaptic connections between two neurons are strengthened when firing of the presynaptic neuron leads to immediate or slightly delayed firing of the postsynaptic neuron, i.e., when pre- and post synaptic activity are highly correlated. (II) synaptic strength is weakened when the postsynaptic neuron fires before the presynaptic neuron, or long enough after to only create moderate increase in postsynaptic potential. I.e., when the correlation between pre- and postsynaptic activity is low.

Since NMDA receptors require subsequent depolarization of pre- and postsynaptic neuron to become fully conductive, they will function as *Hebbian detectors* for correlated pre- and postsynaptic activity: Large influx of Ca^{2+} initiate the biochemical mechanisms behind LTP and requires strong NMDA activity. For LTD, on the other hand, the probability for finding an arbitrary NMDA gate in open state is

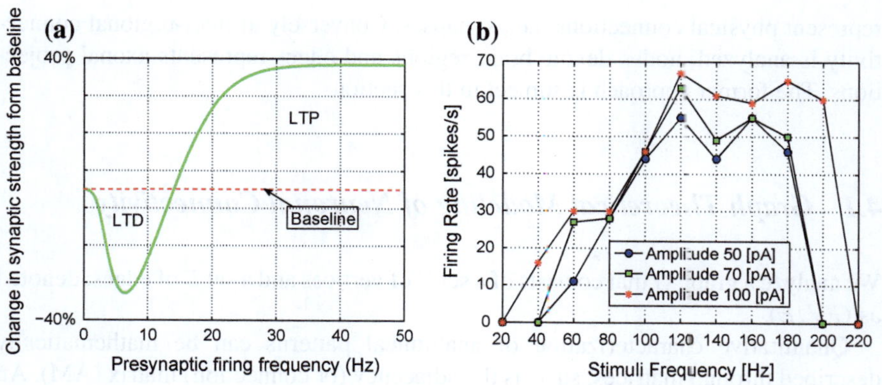

Fig. 5 **a** The process of LTP and LTD. Steady state synaptic strength (AMPA conductance) is plotted as a function of presynaptic spike train frequency f_{AP}. **b** Relation between spiking frequency and current stimuli frequency injected at soma

smaller than under LTP. That is, some of the Mg^{2+} blockages are removed, but less so than under LTP. With no influx of Ca^{2+} through the NMDA, the synapse stays unchanged.

Figure 5a shows the steady state change in synaptic strength, i.e. the change in G_{AMPA}, as a function of presynaptic spiking frequency, f_{AP}. The plot was generated by (19) and (14) using results and data from Castellani et al. [34]. Observe that LTP happen at higher f_{AP}. This is natural, since when consecutive spikes are close in time, the excitatory postsynaptic potential will accumulate more easily, and thereby it is more likely that presynaptic firing leads to postsynaptic firing, in line with Hebb's hypothesis. Note that if the NMDA gates are modified (by changes in G_s and G_f in (19)), the frequency at which the synapse goes from LTD to LTP will be shifted up or down [34]. As stated in [34], the curve shown in Fig. 5a is also quite consistent with the experimental data in [38].

To get an idea on what stimuli frequencies lead to LTP or LTD we have plotted the spiking frequency as a function of the soma current stimuli frequency in Fig. 5b for several current amplitudes. By comparing Fig. 5a and b one can observe that for stimuli current amplitude of 100 pA, the change from LTD to LTP will happen at about 40 Hz. For stimuli current amplitude of 50 pA, this transition takes place at around 65 Hz.

4 Neuro-Spike Communications at Network Level

The anatomical configuration of brain networks can quantitatively be characterized by a graph representing either inter-neuronal- or inter-regional connectivity. If inter-neuronal connectivity is analyzed, the graph's nodes denote neurons and the edges represent physical connections, i.e., synapses. Conversely, if inter-regional connectivity is analyzed, nodes denote brain regions and edges represents axonal projections. The former approach is subject in this section.

4.1 Graph Theoretical Modeling of Neuronal Connectivity

We analyze a graph G that consists of a set V of vertices and a set E of edges, denoted as $G(V, E)$.

Quantitative characterization of anatomical patterns can be mathematically described through matrices, such as the adjacency (or connection) matrix (AM). An AM is with non-zero entries W_{ij} if a connection is present between neurons i and j; otherwise, W_{ij} is zero. Although we lack a formula describing P_{rel} as a function of $[\mathrm{Ca}^{2+}]_i$, and thereby W_{ij}, a formula for W_{ij} as a function of extracellular $[\mathrm{Ca}^{2+}]_o$ has been found by Dodge and Rahminoff [39]

$$w_{ij} = R \sim \left(\frac{[\mathrm{Ca}^{2+}]_o / K_c}{[\mathrm{Ca}^{2+}]_o / K_c + [\mathrm{Mg}^{2+}]_o / K_m + 1} \right)^a . \tag{20}$$

K_c and K_m are equilibrium constants of calcium and magnesium respectively, and a is the number of independent sites Ca^{2+} must bind to in order to release neurotransmitters. Note that synaptic connections are either excitatory or inhibitory, which corresponds to positive and negative W_{ij} values, respectively.

Anatomically, many compartments (e.g., many synapses from one axon) of single presynaptic neuron i can be connected to postsynaptic neuron j making a numerous synaptic connections. In our analysis, we assume that all such connections are superimposed, resulting in one weight value W_{ij}. Furthermore, each two neurons can both be the pre- and postsynaptic ones to each other, in which case they mutually share two connections represented by usually different values of W_{ij} and W_{ji} in AM. Additionally, the graph $G(V, E)$ representing neuronal network is directed, due to one-way axonal transmission. This ultimately makes the $G(V, E)$ directed, weighted, and signed. Excluding the self-connections, all diagonal elements of AM are zero.

With the neuronal network mapped (usually by means of functional magnetic resonance imaging (fMRI) or electrophysiological techniques such as electroencephalography (EEG) and magnetoencephalography (MEG)), one can reveal the connectivity patterns and structure formed of densely connected clusters linked together in a small-world network [40]. Small-worldness combines high levels of local clustering among neurons, associated with high efficiency of information transfer and robustness, and short average path lengths, that link all neurons of the regional network. Small-world organization is intermediate between that of random networks, characterized with short overall path length and low level of local clustering, and that of regular networks, characterized with long path length accompanied by high level of clustering. Such a structure is essential not only for regions of specialized neurons but also for the brain in general, since it combines two fundamental functioning aspects: *segregation*, where the similar specialized neuronal units are organized in densely connected groups that are interconnected, and the functional *integration*, which allows the collaboration of a large number of neurons to build cognitive states [40].

A set of graph theory concepts of special relevance to the computational analysis of connectivity patterns and strategies of diagnosis and ICT-oriented treatment of neurodegenerative diseases will be presented later in Sect. 5.2.

4.2 *Memory Networks*

In memory networks so-called *feed-forward* (FF)- or *recurrent* (REC) networks are usually considered. The graph representation of a combined FF- and REC network is depicted in Fig. 6. In a FF network information coming from one stage along an *information processing path* (e.g., from one part of the brain to another) that has been processed by several neurons are forwarded to another set of neurons at a later stage along the processing path. A REC network refers to connections between several neurons on the same "stage" along a processing path.

Memory networks are usually modeled as a combinations of FF and REC networks, where memory is stored as connection configuration in the REC network, quantified through entries in the AM denoted M. Changes in connection patterns are governed by LTP and LTD.

A deterministic input output relation will be presented here for the network in Fig. 6, following along the lines of Dayan and Abbott in [25]. To analyze very large networks, one usually have to resort to stochastic network models (e.g., Boltzmann machines [25, pp. 273–276]).

4.2.1 Mathematical Modeling of Memory Networks

Consider Fig. 6. **u** and **v** denote the input and output of the network respectively. **u** is an N_u dimensional vector with components u_j denoting presynaptic firing rate at

Fig. 6 Network consisting of a feed forward and recurrent stage

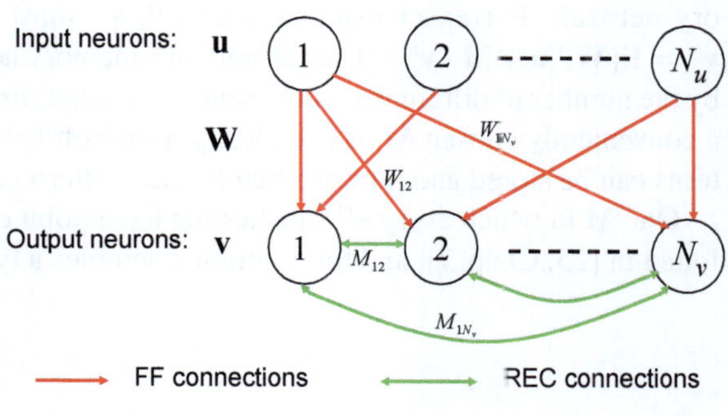

synapse j. $\mathbf{v}$ is an N_v dimensional vector with components v_j denoting postsynaptic firing rate at synapse j. $\mathbf{W}$ is the connection matrix of the feed-forward stage of the network with entries W_{ij}, denoting synaptic strength from input unit j to output unit i. $\mathbf{M}$ is the connection matrix of the recurrent stage of the network with entries M_{ij}, denoting the synaptic strength between output unit j and output unit i.

To model the input-output relationship, the total soma-current for neuron k, $I_{S(k)}$, resulting from all inputs from other neurons must be found. This current will determine the output firing rate, $v = F(I_{S(k)})$, for some nonlinear function F. Let $\rho_j(\tau) = \sum_i \delta(\tau - t_i)$ denote the neuronal response function (a simplified expression for the spike sequence, where each AP is approximated as a Dirac-delta function) for input neuron j. For neuron k, a simplified expression for the soma current is given by [25, p. 233]

$$I_{S(k)} = \sum_{j=1}^{N_u} W_{kj} \int_{-\infty}^{t} K_s(t-\tau)\rho_j(\tau)\mathrm{d}\tau. \tag{21}$$

One simplification made in (21) is that the time response of each synaptic connection, $K_s = (1/\tau_s)e^{-t/\tau_s}$, is approximately the same for all synapses. τ_s denotes the time it takes for the soma current to reach steady state. For dendrites with AMPA receptors τ_s is the time constant for the decay of synaptic conductance. The dynamics of the soma current is typically given by [25, p. 235] $\tau_s \dot{I}_{S(k)} = -I_{S(k)} + \mathbf{w} \cdot \mathbf{u}$. Then the steady state firing rate at the output of neuron k becomes $v_\infty = F(\mathbf{w} \cdot \mathbf{u})$, since $I_{S(k)}(t) \approx \mathbf{w} \cdot \mathbf{u}$ for large t.

Under the assumption that the soma current reaches steady state long before the firing rate, $\tau_s \ll \tau_r$, one can derive that the input output relation for the network in Fig. 6 is governed by [25, p. 260]

$$\tau_r \frac{\mathrm{d}\mathbf{v}}{\mathrm{d}t} = -\mathbf{v} + \mathbf{F}\big(\mathbf{W} \cdot \mathbf{u} + \mathbf{M} \cdot \mathbf{v}\big). \tag{22}$$

F makes (22) a nonlinear differential equation. One should choose F so that it corresponds well to reality. It is also important that conditions for network stability are satisfied, so that the network converges to fixed points. For instance, an F that does not allow negative firing rates and saturates at a certain (positive) firing rate leads to a stable network [25, pp. 263–264].

Let $\mathbf{v}^m$, $m = 1, 2, \ldots, N_{\text{mem}}$ denote the set of possible memory patterns in a memory network. For exact memory retrieval, $\mathbf{v}^m$ must be fixed points of (22), i.e., $\mathbf{v}^m = \mathbf{F}\big(\mathbf{W} \cdot \mathbf{u} + \mathbf{M} \cdot \mathbf{v}^m\big)$. The capacity of a memory network is typically determined by the number of different $\mathbf{v}^m$ simultaneously satisfying this fixed point condition for a conveniently chosen $\mathbf{M}$. It is typically a tradeoff between how many memory patterns can be stored and how accurately each pattern can be represented.

One $\mathbf{M}$ in which every $\mathbf{v}^m$ satisfies the fixed point condition (approximately) was found in [25, Chap. 7], and the solution resembles a typical learning rule.

4.2.2 Learning Rules

A learning rule typically describes the change in synaptic strength over time as a function of pre- and postsynaptic activity. We consider a single neuron here (for simplicity) with output v receiving N_u inputs from other neurons (see [25, pp. 301–313] for an extension to several output neurons). The synaptic weight from neuron j is denoted W_j, and all weights are gathered in the vector $\mathbf{w}$ (w stands for "weight" here, not necessarily a FF connection. The rules described works just as well for REC connections). We also assume that stimuli is held long enough for the network to reach steady state, and so $v = \mathbf{w} \cdot \mathbf{u}$.

Bienenstock, Cooper and Munro proposed a learning rule, named *BCM rule* [41]: $\tau_w \dot{\mathbf{w}} = v\mathbf{u}(v - \theta_v)$, where θ_v is a threshold on output firing rate above which LTD changes to LTP and τ_w is controlling the rate at which synapses change. If θ_v varies with synaptic strength and grow more rapidly than v as the output firing rate grows large, the BCM rule gives stable solutions for $\mathbf{w}$. This is achieved if θ_v is a low-pass filtered version of v^2, $\tau_\theta \dot{\theta}_v = v^2 - \theta_v$, where τ_θ is the time scale for threshold modification. The BCM rule incorporates competition between synapses since strengthening of some synapses leads to an increase in postsynaptic firing rate, which raises θ_v, and makes it more difficult for other synapses to be strengthened.

Shouval et. al derived a calcium based BCM-type learning rule under the assumptions that calcium entry is through NMDA receptors only, and that the presynaptic spikes are generated from a non-stationary Poisson distribution with rate, $r(t)$, varying slowly enough to consider calcium levels at steady state [33]. Based on the AMPA insertion/removal process of Sect. 3.2.2, assuming that $W_j = \beta \cdot B_M$ for some proportionality constant β, the following rule results

$$\tau_{[\mathrm{Ca}]_i} \frac{\mathrm{d}W_j}{\mathrm{d}t} = \left(\Omega_{[\mathrm{Ca}]_i} - W_j\right), \tag{23}$$

where $\Omega_{[\mathrm{Ca}]_i}$ is the steady state response of W_j, $\Omega_{[\mathrm{Ca}]_i} = \beta B_M^{f_p}$, with $B_M^{f_p}$ as in (15). $\tau_{[\mathrm{Ca}]_i}$ is the convergence time to Ω as in (15).

Since calcium concentrations vary from synapse to synapse, a learning rule for several synapses must incorporate such variations. In a similar way as (19), Shouval et al. showed that the steady state calcium concentration as a function of the presynaptic firing rate of neuron j, r_j, is given by [33]

$$\overline{[\mathrm{Ca}^{2+}]_{i(j)}} = \mathscr{H}(V_m) \cdot \tau_{\mathrm{Ca}} \cdot G_{\mathrm{NMDA}} \cdot S_j(r_j), \tag{24}$$

where $S_j(r_j) = P_{\mathrm{inc}}\tau_N r_j/(1 + P_{\mathrm{inc}}\tau_N r_j)$. τ_N is the NMDA open probability decay time and P_{inc} is the increase in open probability for presynaptic AP's leading to neurotransmitter release.

Shouval et al. [33] further claimed that an explicit expression of Ω and $\tau_{[\mathrm{Ca}]_i}$, quantitatively similar to experimental observations, is the quadratic form $\Omega_{\mathrm{Ca}(j)} = [\mathrm{Ca}^{2+}]_{i(j)}\left(([\mathrm{Ca}^{2+}]_{i(j)} + \theta_0\right) + \Omega_0$ with Ω_0 is some baseline synaptic strength, θ_0 some baseline threshold (where LTD changes to LTP), and the linear relation $\tau_{[\mathrm{Ca}]_i} = \tau_0/[\mathrm{Ca}^{2+}]_{i(j)}$. By inserting this and (24) into (23), a calcium based learning rule for several input synapses is achieved (see [33] for details).

To make this into a BCM rule with varying threshold, Shouval et al. [33] further defined $\theta_M = \theta_0/(G_{\mathrm{NMDA}}\tau_{\mathrm{Ca}})$, following the differential equation $\tau_M\dot{\theta}_M = \left(\mathscr{H}(V_m)^{\mu} - \theta_M\right)$, where τ_M must be small enough so that θ_M grows much faster than v as v gets large. This rule works since the NMDA conductance gain increases as the firing rate increases. Note that is was assumed that $\theta_M \sim \langle \mathscr{H}(V_m)^{\mu}\rangle_{\tau_M}$, which is natural since $\mathscr{H}(V_m)$ increases as the firing rate increases.

5 ICT-Inspired Treatment of Neurodegenerative Diseases

5.1 Complementary Techniques of Calcium Controlling

We summarize possible communication-wise contributors in the following without going into detailed biophysics (see Fig. 7).

5.1.1 Non-invasive Techniques

Human neuronal communication network is (willingly or aversely) exposed to external EMR that can be generated by several devices for many purposes. The electricity power supply and all appliances using electricity are the main sources of extremely low frequency fields that range up to 300 Hz. Other technologies are sources of

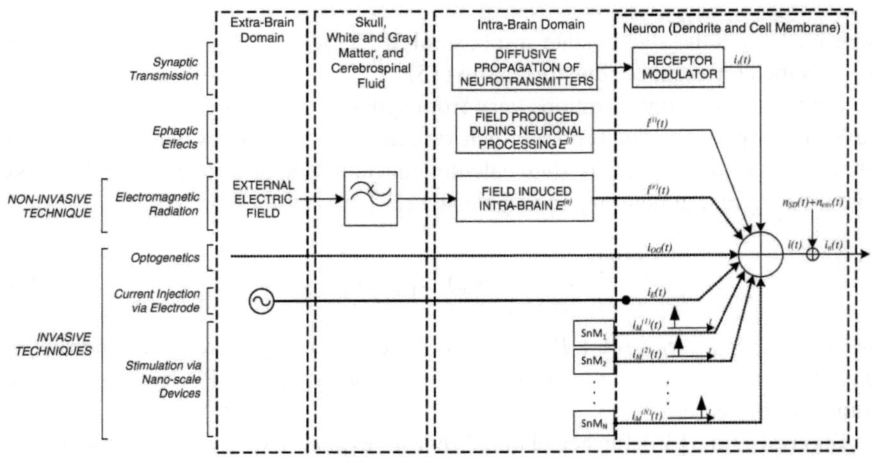

Fig. 7 Possible means of controlling $[\mathrm{Ca}^{2+}]_i$

intermediate frequency (IF) fields that range from 300 Hz up to 10 MHz, whereas the simple mobile phones and Wi-Fi devices produce the radio-frequency (RF) fields with frequencies from 10 MHz to 300 GHz. Depending on an electric field level, frequency and electrical properties of the human body, fields interact with the body in different ways. As a consequence, electric currents, sufficient to produce a range of effects, are induced. Apart from thermal effects, which are quantified by SAR coefficient that concern the possible alterations on cells, recent findings implies there are specific but still uncovered non-thermal interactions between RF exposure and neurons [6].

A considerable amount of research work is performed to see whether EMR can be of any damage to brain, metabolism, neuron's membrane, DNA alterations, which further causes cancer and reduces fertility, etc. As mentioned in the introductory part, it has been even shown that neurocognitive systems affected by neurodegenerative disease can benefit from RF exposure. A framework we present considers the non-thermal effects and scenario with an EMR able to affect the neuronal communication network, i.e., to alter or even generate a certain amount of current in the brain.

To quantify the current produced by a field generated externally but induced intra-brain, natural parameters regarding the filtering features of biological tissues must be considered. The resistivity of the skull is an important physical parameter since the skeleton has a great impact on studies of bioelectricity, the biological effects of the electromagnetic field, and modeling of the electrical properties of the human head [42]. The inhomogeneity of the skull due to the variation of the structure at different positions complicates analysis. Fortunately, this issue was subject to several studies in the past [42–44].

The equivalent circuit models are often applied to describe the macroscopic electrical properties of biological tissues. Akhtari et al. established the three-element RC model for one- and three-layer human skulls in the 10–90 Hz frequency range [43]. Recently, Tang et al. used Cole-Cole equations to establish the three-element CPE model. The CPE is a non-intuitive circuit element that compensates for inhomogeneity in the system. The impedance and resistivity spectroscopy data of the skull samples with six types of different structures (standard tri-layer, quasi-tri-layer, standard compact, quasi-compact, dentate suture and squamous suture) are obtained from 30 Hz to 3 MHz [44]. According to the findings, real part of all impedance loci act as a low-pass filter which filters out the high frequency components of the external stimulus. Since the resistivity of the skull is much higher than that of other tissues in a human head (white and gray matter, cerebrospinal fluids, and scalp), we approximate the effects of all tissues on the pathway from outside environment to the sample hippocampal neuron as the low-pass filter, as shown in Fig. 7. Findings presented in [45] additionally support the concept since authors identified 250–300 MHz frequency band to induce the largest electric field intensity while exposing a human head to sagittal- and coronal-plane polarized waves within range of 100–1000 MHz.

Once the frequency-dependent resistivity/conductivity of the tissue is defined and the spatio-temporal electric field is evaluated inside the human head, $E^{(e)}(x, y, z, t)$, quantification of the spatio-temporal electric current density is given [45]

$$J^{(e)}(x, y, z, t) = \sigma_t E^{(e)}(x, y, z, t). \tag{25}$$

Superscripts (e) tag the components resulting from externally generated field, while σ_t denotes the equivalent conductivity. To assess a single contribution of induced component, the total current injecting a sample neuron is derived from the current density as

$$i^{(e)}(t) = J^{(e)}(t)A_C, \tag{26}$$

where A_C denotes the total membrane area of the neuron (usually several tens of μm^2). If particular hippocampal neuron is analyzed, spatial dependency might be excluded to keep the reasoning as simple as possible.

The internally induced field may also be used in some of the models described in Sect. 2.3 in order to estimate possible change in Neuron behavior. These effects would then have to be validated by experiment in order for the model to be rendered useful.

5.1.2 Invasive Techniques

Optogenetics: Controlled pulses of light are used in neuroscience as spike triggers. Unlike inevitable ephaptic coupling and nearly inevitable EMR effects, this type of neuronal stimulation requires devoted human intervention. Among the light-oriented techniques, an optogenetics is recent neuromodulation technique that combines optics and genetics to control and monitor the activities of individual neurons [46]. Even a single cell can be targeted via genetic photo-sensitization and remote optical activation (see Fig. 7) [47]. The photo-sensitization of cells is achieved using the optogenetic actuators like protein channelrhopsin-2 (ChR2) and light-gated chloride pump halorhodopsin (NpHR), whereas the spatial selectivity can be achieved by focusing the light beam onto targeted area, e.g., soma, dendrite or axon.

Evaluation of optogenetic illumination effects can be given through definition of the ionic current that further superimposed to those invoked synaptically and ephaptically. Expression is given in [48]

$$i_{OG}(t) = g_{OG}\psi(\phi, t)f(v)A_C. \tag{27}$$

Conductance of optogenetic ion channels is denoted as g_{OG}. Functions ψ and f describe how the current depends on the photon flux $\phi(t)$ and membrane voltage of targeted neuron $v(t)$.

Direct Current Injection We present and show in Fig. 7 two approaches of direct injection of controlled current into neurons, one currently used in neuroscience, and one futuristic.

Electrodes: The "Current clamp" technique is commonly used in physiological studies deployed in neuroscience. This technique is used to study how a cell responds when electric current enters a cell by means of recording glass micro-pipettes (electrodes). Using findings presented in Sect. 3.1, an identical approach can be used in order to control $[Ca^{2+}]_i$ values as desired.

Nano-scale devices: Architecture development of nano-scale devices, which can be used intra-body to interact at cellular level, is intriguing scientific topics. Hybrid structures consisting of arrays of nano-wire field-effect transistors (FET) integrated with the individual mammalian neurons, used for stimulation, and/or inhibition of neuronal signal propagation, are tested by Patolsky and his team [49]. In terms of future-oriented works, preliminary modeling of communication between nano-scale device and sample neuron [50] encouraged the definition of conceptual nano-machine-to-neuron communication interface based on nano-scale stimulator device called synaptic nano-machine (SnM) [51]. Although there is still much to be done in this field, ranging from isolation of nano-materials to biocompatibility issues, a nano-scale device intended to either directly apply or indirectly induce a controlled amount of current will presumably be developed in this decade. According to information built into such devices, controlled alternating currents can be induced directly into neurons.

One issue to be accounted for when using any stimulation technique is the signal dependant encoding noise [52] that reflects to the input as $i_n(t) = i(t) + n_e(t) = i(t) + n_{SD}(t) + n_{env}(t)$. $i(t)$ denotes a noise-free current, $n_{SD}(t)$ is the signal-dependent noise component derived from an auto regressive (AR) random process $w(t)$ as $n_{SD}(t) = i(t) * w(t)$, and $n_{env}(t)$ is the environmental noise. Since our study accounted for ephaptic stimulation, it is reasonable to assume $n_{env}(t)$ as zero mean Gaussian noise. Another non-trivial issue is the synchronization between stimulus sources since high levels of desynchronization can delay the generation of spikes in the targeted neuron and lead to unrestrained control.

5.2 *Graph-Based Strategies of Diagnosis and Treatment of Neuronal Disorders*

As previously shown in Sect. 4.1, graph interpretation of neuronal connectivity provides researchers with connectional relationships between individual neurons. Although our research is mostly confined to hippocampal neurons, without any loss of generality we use two-dimensional spatial representations of a local subnetwork of 131 neurons with 764 unidirectional connections within *Caenorhabditis elegans* rostral ganglia from [53] (and references therein) to show how graph theory methods can be utilized in formation of adequate strategy assisting in diagnosis and treatment of neuronal diseases. Spatial two-dimensional positions represent the position of the soma of individual neurons. It is not established which graph theory measures are most appropriate for the analysis of the brain. Nevertheless, we explore a set of those that might have relevance for neuroscience applications.

Depending on the nature of the ion flow, the synapses representing connections can have either an excitatory, depolarizing, or an inhibitory, hyperpolarizing, effect on the postsynaptic neuron. Thereby, to keep the reasoning as realistic as possible, we modify 20% of the number of synapses in AM giving them negative weights according to (20). Note that distribution of inhibitory and excitatory cells is not temporal, unlike the time-variable synaptic weight values (temporal synaptic weights form dynamic wiring of neuron connections, i.e., the plasticity).

Spatial and temporal visualization can be used in neuroscience as one possible way of diagnosis. For instance, significant reduction of clustering and loss of small-worldness might provide a clinically useful diagnostic marker indicating AD [54]. Moreover, graph theory can be used to create a certain visualization tool that might assist in neuronal treatment. As an example, we elaborate on treatment based on regulation of calcium.

Calcium signaling is crucial in neurotransmitters transmission and memory formation. The intracellular and extracellular calcium concentration level is affecting how strongly two neurons are wired together through chemical synapses. Disruption in the processes that regulates calcium concentration levels (calcium dysregulation), lead to dramatic changes in neuronal functioning and potentially weak connections between cells or clustering regions. This further lead to impaired functioning and cell death, which is the effect in brains with AD. Potential remedy of AD lies in direct regulation of calcium concentration levels or in control of amyloid-beta (Aβ) deposits, that lead to calcium dysregulation [10]. Therefore, an efficient and effective human treatment can directly explore particular graph theory measures that are able to mark critical areas and cells in network.

5.2.1 Centrality Criteria

As historically first, conceptually simplest, and very useful measure of centrality, degree is first applied here to find the neurons with most synaptic connections. Additionally, betweenness centrality is also deployed in what follows since introduced as a "measure for quantifying the control of a human on the communication between other humans in a social network" [55]. Conversely, closeness and eigenvector centrality do not qualify themselves as a usable criteria in this study due to the unmatched physical interpretations. For instance, closeness can be regarded as a measure of how long it will take to spread information from analyzed vertex to all other vertices sequentially, thereby being completely irrelevant in $[Ca^{2+}]_i$ regulation strategy.

The degree of a neuron is the sum of its indegree and outdegree values defined as the number of incoming (afferent) or outgoing (efferent) edges, respectively. We are interested in outdegrees, $od(v)$. They are subject to constraints due to growth, tissue volume or metabolic limitations [56], but have obvious functional interpretations: high outdegree indicates a large number of potential functional targets. Outdegree criterion further qualifies an adequate neuron which $[Ca^{2+}]_i$ is to be regulated. This criterion visualizes the $G(V, E)$ in a way shown in Fig. 8a.

Betweenness centrality measures a vertex's or edge's importance in network. Node betweenness quantifies the number of times a vertex participate in the shortest path between two other vertices. Analogously, the betweenness centrality of edges is calculated as the number of shortest paths among all possible vertex couples that

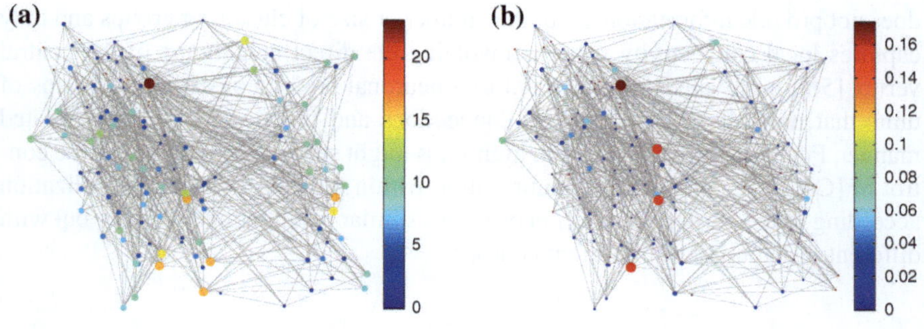

Fig. 8 **a** Representation of neurons according to the outdegree criterion. **b** Representation of neurons according to the node betweenness centrality criterion

pass through the given edge. Vertices and edges with the high centrality values are assumed to be crucial for the graph connectivity and thereby shall be monitored in potentially malfunctioning brains. Neurodegenerative disease removing important neurons and synaptic connections lead to both anatomically and functionally disconnected clusters. The most important edges, called bridges, are particularly important for decreasing the average path length among neurons in a network, and for speeding up the information transmission. Region of *Caenorhabditis elegans* network is visualized in Fig. 8b applying the node centrality criteria. For analyzed network consisting of 131 frontal neurons, 4 cells are identified as important for proper communication. In general, the higher the number of important neurons, the lower the network robustness.

Average path length is one of the most important measures, and can be obtained from distance matrix that is result of the Floyd-Warshall algorithm (particularly applied here to determine node betweenness). Average path length is what makes the brain efficient in information transfer from one part to the other. It is also a function of degree (e.g., for a random network $l_{avg} \approx \log(N)/\log(<k>)$) ($<k>$ is the average degree). In AD brains, one can typically observe an increase on average path length, as described in [54]. For network analyzed here $l_{avg} = 3.1277$.

5.2.2 Clustering Coefficient Criterion

The cluster coefficient of a vertex, $\gamma(v)$, is defined in [57]. As a measure of the degree to which vertices in a $G(V, E)$ tend to cluster together, it indicates how many connections are maintained between vertex neighbors. In particular scenario with directed graph on board, neighbours are all those neurons that are connected through an incoming or an outgoing synaptic connections to the central neuron v. The ratio of actually existing connections between the neighboring neurons and the maximal number of such connections possible defines the neuron's cluster coefficient. If neuron does not have any neighbors, then $\gamma(v) = 0$ by convention. The average of the cluster coefficients determines the cluster coefficient of the $G(V, E)$. Although $\gamma(v)$

does not provide information about the number or size of clustering groups and only captures local connectivity patterns involving the direct neighbours of the central vertex [56], high values of $\gamma(v)$ point to a neuronal region consisting of groups of units that mutually share structural connections and function in a closely related manner. Furthermore, neurons with high $\gamma(v)$s might serve as candidates whose control of $[Ca^{2+}]_i$ regulates the communication within the entire cluster. Visualization according to the clustering criterion produces similar result to that in Fig. 8, but with different distribution of important neurons.

6 Chapter Summary

Calcium is important in signal transmission in and between neurons. The neuron's intracellular calcium concentration, $[Ca^{2+}]_i$, determines synaptic strength. High concentration levels usually lead to synaptic strengthening, whereas moderate levels lead to a weakening. The $[Ca^{2+}]_i$ depends on both the spike train frequency as well as stimulus frequencies. In general, higher spiking and stimulus frequencies lead to significantly increased $[Ca^{2+}]_i$ and therefore to a synapse strengthening. Lower frequencies, on the other hand, lead to a synapse weakening. If one can find means of controlling voltage controlled calcium ion-gates (VCCGs), it will be possible to regulate $[Ca^{2+}]_i$ and therefore affect the neurons ability to communicate with other neurons.

In this chapter we identified that changes in neuron's membrane potential drive the magnitude of $[Ca^{2+}]_i$. The findings show the neuronal communication phenomena being frequency-dependent not only on the stimulus frequencies, but also on the spiking frequency. Therefore, problems in aging brains, strongly noticeable in brains with neurodegenerative diseases such as Alzheimer's disease (AD), may be overcome via devoted human regulation of $[Ca^{2+}]_i$. From what we have presented in this chapter, it is likely that the frequency of the external stimuli is crucial in processes that may take place within the brain. The findings thus can pave the way to the definition of novel treatment tools for medical applications. Furthermore, synchronized or non-synchronized neuronal activity that underlies long term potentiation (LPT) and long term depression (LTD) processes, respectively, can be controlled at postsynaptic neurons through convenient stimulation of presynaptic neuron. Since LTP and LTD patterns are confined to memory formation and learning processes, one can exploit mapping between spiking- and stimulus frequencies presented here to drive changes in those processes. Moreover, study presented here helps in development of a graph-based strategy of neurons selection whose electrical properties should be altered in order to further evoke certain changes in $[Ca^{2+}]_i$. Obviously, it is likely that neuronal network can be described as graph, and quantitative characterization of anatomical patterns through the adjacency matrix (AM). Moreover, specified graph theory concepts have been utilized in formation of adequate strategy assisting in diagnosis and treatment of neuronal diseases. A set of graph theory measures that might have certain relevance for neuroscience applications have been stated and deployed in this study to create a visualization tool that helps in selection of targeted neurons whose $[Ca^{2+}]_i$ is to be controlled. Possible non-invasive and invasive means of controlling neuronal communications are also envisioned.

To keep the theory as realistic as possible, noise effects referred to disturbances on the stimulus currents and synchronization issues are not discussed in this chapter and are to be accounted for and thoroughly elucidated. To complete the calcium based synaptic model, it is also necessary to determine a mathematical relation between neurotransmitter release probability and the level of $[Ca^{2+}]_i$. Computationally demanding handling of huge AM describing large-scale networks, that can be recorded with functional magnetic resonance imaging, electroencephalography and magnetoencephalography, is another issue. These issues are not trivial and demand more research.

In terms of the applicability of concepts and results presented in this chapter, it is possible to make their use in communication controlling of either artificial nano-network composed of nano-devices fabricated following a design presented in Sect. 3, or neuronal biological network. Findings can also be used for implementation of novel communication technique between nano-scale devices.

An important problem to pursue through future research is to evaluate the performance of neural information processing seen from a communication perspective.

References

1. Arendash GW, Sanchez-Ramos J, Mori T, Mamcarz M, Lin X, Runfeldt M, Wang L, Zhang G, Sava V, Tan J, Cao C (2010) Electromagnetic field treatment protects against and reverses cognitive impairment in alzheimer's disease mice. J Alzheimer's Dis 19:191–210
2. Arendash GW (2012) Transcranial electromagnetic treatment against alzheimer's disease: why it has the potential to trump alzheimer's disease drug development. J Alzheimer's Dis 32: 243–266
3. Dragicevic N, Bradshaw PC, Mamcarz M, Lin X, Wang L, Cao C, Arendash GW (2011) Long-term electromagnetic field treatment enhances brain mitochondrial function of both alzheimer's transgenic mice and normal mice: a mechanism for electromagnetic field-induced cognitive benefit. Neuroscience 185:135–149
4. Juutilainen J, Hoyto A, Kumlin T, Naarala J (2011) Review of possible modulation-dependent biological effects of radiofrequency fields. Bioelectromagnetics 32:511–534
5. Rosen AC, Ramkumar M, Nguyen T, Hoeft F (2009) Noninvasive transcranial brain stimulation and pain. Curr Brain Headache Rep 13(1):12–17
6. Stavroulakis PE (2003) Biological effects of electromagnetic fields. Springer
7. Disterhoft JF, Moyer JR, Thompson LT (1994) The calcium rationale in aging and Alzheimer's disease. Annals of the New York academy of sciences—calcium hypothesis of aging and dementia, vol 747(2) pp 382–406
8. Goodison WV, Frisardi V, Kehoe PG (2012) Calcium channel blockers and Alzheimer's disease: potential relevance in treatment strategies of metabolic syndrome. Annals of the New York academy of sciences—calcium hypothesis of aging and dementia 30(2):269–282
9. Bear MF, Connors BW, Paradiso MA (2006) Neuroscience: exploring the Brain 3rd edn. Lippincott
10. Green KN, LaFerla FM (2008) Linking calcium to A-beta and alzheimer's disease. Neuron 59:190–194

11. Yu J, Chang RCC, Tan L (2009) Calcium dysregulation in Alzheimer's disease: from mechanisms to therapeutic opportunities. Prog Neurobiol 89:240–255
12. Adey WR (1981) Tissue interactions with nonionizing electromagnetic fields. Physiol Rev 61(2):435–514
13. Mesiti F, Balasingham I (2011) Novel treatment strategies for neurodegenerative diseases based on RF exposure. In Proceedings of the 4th international symposium on applied sciences in biomedical and communication technologies ser, ISABEL 11, pp 100:1–100:5. http://doi.acm.org/10.1145/2093698.2093798
14. Buzsaki G, Anastassiou CA, Koch C (2004) The origin of extracellular fields and currents. Nat Rev Neurosci 13:407–420
15. Anastassiou CA, Perin R, Markram H, Koch C (2011) Ephaptic coupling of cortical neurons. Nat Neurosci 14:217–223
16. Koch C (1999) Biophysics of computation: information processing in single neurons. Oxford University Press, Inc
17. Repacholi MH (1998) Low-level exposure to radiofrequency electromagnetic fields: health effects and research needs. Bioelectromagnetics 19(1):1–19
18. Izhikevich EM (2004) Which model to use for cortical spiking neurons? IEEE Trans Neural Networks 15(5):1063–1070
19. Hodgkin AL, Huxley AF (1952) A quantitative description of membrane current and its application to conduction and excitation in nerve. J Physiol 117(4):500–544
20. Warman EN, Durand DM, Yuen GL (1994) Reconstruction of hippocampal CA1 pyramidal cell electrophysiology by computer simulation. J Neurophysiol 71(6):2033–2045
21. Liboff AR, McLeod BR (1988) Kinetics of channelized membrane ions in magnetic fields. Bioelectromagnetics 9:39–51
22. Halle B (1988) On the cyclotron resonance mechanism for magnetic field effects on transmembrane ion conductivity. Bioelectromagnetics 9:381–385
23. Pilla AA (2006) Mechanisms and therapeutic applications of time-varying and static magnetic fields. In: Handbook of biological effects of electromagnetic fields. CRT Press
24. Appollonio F, Liberti M, d'Inzeo G, Tarricone L (2000) Integrated models for the analysis of biological effects of em fields used for mobile communications. IEEE Trans Microwave Theory Tech 48:2082–2092
25. Dayan P, Abbott LF (2001) Theoretical neuroscience: computational and mathematical modeling of neural systems. The MIT Press, Cambridge, Massachusets, London, England
26. Hunter IW, Korenberg MJ (1986) The identification of nonlinear biological systems: wiener and hammerstein cascade models. Biol Cybern 55:135–144
27. Cook EP, Guest JA, Liang Y, Masse NY, Colbert CM (2007) Dendrite-to-soma input/output function of continuous time-varying signals in hippocampal CA1 pyramidal neurons. J Neurophysiol 98:2943–2955
28. Richardson M, Brunel N, Hakim V (2003) From subthreshold to firing-rate resonance. J Neurophysiol 89:2538–2554
29. Mesiti F, Floor PA, Kim AN, Balasingham I (2012) On the modeling and analysis of the RF exposure on biological systems: a potential treatment strategy for neurodegenerative diseases. Elsevier Nano Commun Networks 3:103–115
30. Matveev V, Wang X-J (2000) Implications of all-or-none synaptic transmission and short-term depression beyond vesicle depletion: a computational study. J Neurosci 20(4):1575–1588
31. Wu LG, Saggau P (1997) Presynaptic inhibition of elicited neurotransmitter release. Trends Neurosci 20(5):204–212
32. Neher E (1998) Vesicle pools and Ca^{2+} microdomains: new tools for understanding their roles in neurotransmitter release. Neuron 20:389–399
33. Shouval HZ, Castellani GC, Blais BS, Yeung LC, Cooper LN (2002) Converging evidence for a simplified biophysical model of synaptic plasticity. Biol Cybern 87:383–391
34. Castellani GC, Quinlan EM, Cooper LN, Shouval HZ (2001) A biophysical model of bidirectional synaptic plasticity: Dependence on AMPA and NMDA receptors. PNAS 8(22):1272–12777

35. Castellani GC, Quinlan EM, Bersani F, Cooper LN, Shouval HZ (2005) A model of bidirectional synaptic plasticity: from signalling network to channel conductance. Learn Mem 12:423–432
36. Jahr CE, Stevens CF (1990) Voltage dependence of NMDA-activated macroscopic conductances predicted by single-channel kinetics. J Neurosci 10(9):3178–3182
37. Hebb DO (1949) The organization of behavior: a neuropsychological theory. Wiley, New York
38. Dudek SM, Bear MF (1992) Homosynaptic long-term depression in area CA1 of hippocampus and effects on N-methyl- D-aspartate receptor blockade. In: Proceedings of the National Academy of Sciences. USA, 89:4363–4367
39. Dodge FA, Rahamimoff R (1967) Cooperative action of calcium ions in transmitter release at the neuromuscular junction. J Physiol 193:419–432
40. Bullmore E, Sporns O (2009) Complex brain networks: graph theoretical analysis of structural and functional systems. Nat Rev Neurosci 10:186–198
41. Bienenstock EL, Cooper LN, Munro PW (1982) Theory for the development of neuron selectivity: orientation specificity and binocular interaction in visual cortex. J Neurosci 2:32–48
42. Tang C, You F, Cheng G, Gao D, Fu F, Dong X (2009) Modeling the frequency dependence of the electrical properties of the live human skull. Physiol Meas 30:1293–1301
43. Akhtari M, Bryant HC, Emin D, Merrifield W, Mamelak AN, Flynn ER, Shih JJ, Mandelkern M, Matlachov A, Ranken DM, Best ED, DiMauro MA, Lee RR, Sutherling WW (2003) A model for frequency dependence of conductivities of the live human skull. Brain Topogr 16(1):39–55
44. Tang C, You F, Cheng G, Gao D, Fu F, Yang G, Dong X (2008) Correlation between structure and resistivity variations of the live human skull. IEEE Trans Biomed Eng 55:2286–2292
45. Khaleghi A, Eslampanah MS, Sendi, Chavez-Santiago R, Mesiti F, Balasingham I (2012) Exposure of the human brain to an electromagnetic plane wave in the 100–1000 MHz frequency range for potential treatment of neurodegenerative diseases. IET Microwaves Antennas Propag 6:1565–1572
46. Liu X, Ramirez S, Pang PT, Puryear CB, Govindarajan A, Deisseroth K, Tonegawa S (2012) Optogenetic stimulation of a hippocampal engram activates fear memory recall. Nature 484:381–385
47. Grossman N, Poher P, Grubb MS, Kennedy GT, Nikolic K, McGovern B, Palmini RB, Gong Z, Drakakis EM, Neil MA, Dawson MD, Burrone J, Degenaar P (2010) Multi-site optical excitation using Ch R2 and micro-led array. J Neural Eng 7(1):381–385
48. Nikolic K, Jarvis S, Grossman N, Schultz S (2013) Computational models of optogenetic tools for controlling neural circuits with light. In: Annual international conference of the IEEE, EMBS
49. Patolsky F, Timko BP, Yu G, Fang Y, Greytak AB, Zheng G, Lieber CM (2006) Detection, stimulation, and inhibition of neuronal signals with high-density nanowire transistor arrays. Science 313:1100–1104
50. Galluccio L, Palazzo S, Santagati GE (2011) Modeling signal propagation in nanomachine-to-neuron communications. Nano Commun Network 2:213–222
51. Mesiti F, Balasingham I. Nanomachine-to-neuron communication interfaces for neuronal stimulation at nanoscale. J Sel Areas in Commun—Special Issue on Emerging Technologies in Communications In Press
52. Jabbari A, Balasingham I (2013) Noise characterization in a stochastic neural communication network. Nano Commun Network 4(2):65–72
53. Kaiser M, Hilgetag CC (2006) Nonoptimal component placement, but short processing paths, due to long-distance projections in neural systems. PLoS Comput Biol 2(7):e95. doi:10.1371/journal.pcbi.0020095
54. Stam C, Jones BF, Nolte G, Breakspear M, Scheltens P (2007) Small-world networks and functional connectivity in alzheimer's disease. Cereb Cortex 17(1):92–99
55. Freeman LC (1977) A set of measures of centrality based upon betweenness. Sociometry 40:35–41

56. Sporns O (2003) Graph Theory methods for the analysis of neural connectivity patterns. neuroscience database. Springer US, Ch
57. Watts D, Strogatz S (1998) Collective dynamics of 'small-world' networks. Nature (London) 393:440–442

Molecular Dynamics Simulations of Biocorona Formation

Rongzhong Li, Cody A. Stevens and Samuel S. Cho

Abstract The development and advancement of nanomedicine has opened up many exciting, new applications of nanoparticles such as sensing, imaging, delivery, and therapy. However, their ability to readily enter cells and organelles that allow these nanomedical applications also opens up the possibility of unintended adverse nanotoxicity. The interaction between nanoparticles and biomolecules results in biocorona formation on the nanoparticle surface that is very different from adsorption of biomolecules on a flat surface. It remains a great challenge to understand the applications and risks associated with nanoparticles being in contact with biological systems beyond experimental methods that have limited resolution of the interactions and conformational changes involved. Recently, biomolecule-nanoparticle molecular dynamics (MD) simulations are becoming a viable approach for a detailed view of biocorona formation. In this review, we present the advantages and challenges of several MD simulation approaches for the study of biomolecule-nanoparticle interactions. In particular, we argue for the development of GPU-optimized MD simulations as a critical step in the study of biocorona formation. We discuss recent successes on how integrated computational and experimental studies are important to establish how the structure and functions of biomolecules are affected by nanoparticle interactions with the biomolecules.

1 Biomolecule-Nanoparticle Interactions

The development of nanomaterials and the characterizations of their properties have become a highly attractive subject that spans physics, chemistry, biology, and engineering [1–3]. Nanoparticle properties have opened up industrial products

R. Li · S.S. Cho (✉)
Department of Physics, Wake Forest University, Winston-Salem, NC 27106 USA
e-mail: choss@wfu.edu

C.A. Stevens · S.S. Cho
Department of Computer Science, Wake Forest University, Winston-Salem, NC 27106, USA

J. Suzuki et al. (eds.), *Modeling, Methodologies and Tools for Molecular and Nano-scale Communications*, Modeling and Optimization in Science and Technologies 9, DOI 10.1007/978-3-319-50688-3_10

involving nanomaterials, and there is particular interest in the development of biomedical applications involving nanoparticles due to their small size that allows them to interact with cellular machinery that was previously thought to be inaccessible [4]. On the other hand, as nanoparticles become widely used in many different industries, they also have the potential for unintended exposure to the environment and living matter that could result in deleterious toxicological effects. Their increased applications in daily life inevitably will lead to their accumulation in the environment [5], and their subsequent entry into biological systems, raising important questions about their nanotoxicity [6]. As such, there is a strong motivation to develop safe nanomedical applications that limit their negative effects. To understand the biological impact of nanoparticles, we must develop a molecular level understanding for their interactions with biological media.

Nanoparticles have very different properties from bulk materials. Due to their high surface area to volume ratio, nanoparticles have very active surface chemistry interactions with biological media that reduces their surface energy. As a result, their surfaces end up being rapidly covered by a complex layer of biomolecules [5]. The absorption of biomolecules to surfaces confers a new "biological identity" that results in different cellular responses. Of particular interest in recent studies is that formation of a "biological corona" on the nanoparticle surface that is very different from when biomolecules adsorb to a flat surface [7]. The biocorona formation is a dynamic process by which proteins and other biomolecules compete to interact with the nanoparticle surface. Recently, functionalized molecules conjugated on nanoparticles were shown to lose their function when placed in a biological medium presumably because protein corona formation screened the molecules from interacting with their intended target [8]. Therefore, the safe development of nanomedicine requires the study of nanoparticle interactions in the biological medium.

When biomolecules interact with nanoparticles, their structure, dynamics, and function are expected to change, which could further impact recognition of the biomolecules by their intended receptors. It is now well established that protein and RNA fold into specific structures in the cell to perform their biological functions [9, 10]. The main forces that stabilize the protein structure, such as the hydrophobic, electrostatic, van der Waals, and hydrogen bond interactions can also compete with the nanoparticle surface. The competition results in a perturbation of the protein structure such that it may not longer be able to perform its function.

For example, enzymes catalyze reaction in a well-defined, specific active site, but their structure may be destabilized such that the active site is significantly changed or completely lost upon interaction with the surface of a nanoparticle (Fig. 1a). Even if the active site were to be preserved, the substrate entrance or exit may be obstructed or destabilized. The end result is a loss of catalytic activity due to the presence of nanoparticles. The adsorption of lysozyme on a negatively charged silica nanoparticles resulted in 70% loss of α-helical structure, which accompanied 40% loss of catalytic activity [11].

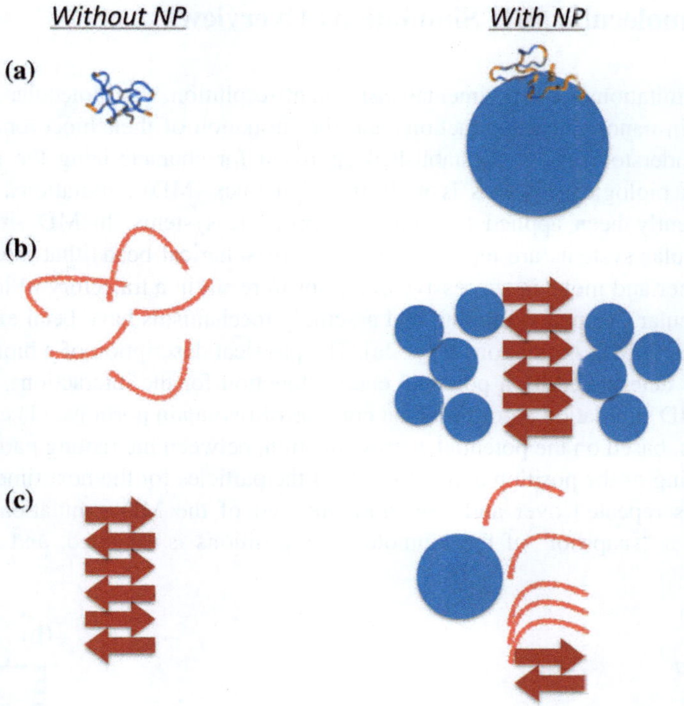

Fig. 1 Modes of biomolecular structural changes upon interaction with nanoparticles (*solid blue circles*). **a** A biomolecule can undergo structural changes upon binding to the nanoparticle surface, which can diminish or abolish their original biological function or introduce an unintended cellular response. **b** Biomolecular interaction with nanoparticles can induce fibril formation for peptides that would otherwise be unstructured. **c** Nanoparticle interaction can disrupt the formation of amyloid fibrils

A well known issue with protein folding is that some amyloidogenic proteins midfold and aggregate to form fibrils, which are thought to play important roles in neurodegenerative diseases such as Alzheimer's Parkinson's, and Huntington's diseases [12]. The introduction of nanoparticles may induce amyloidogenic peptides to form β-sheet structures [13] (Fig. 1b), which are the hallmark of fibril formation that lead to diseases.

The competition of native protein interactions with nanoparticles can also have a beneficial result too. Ikeda et al. showed that cholesterol-bearing pullulan nanogels inhibit protein aggregation in the case of Aβ peptides. The natively coiled Aβ peptides associated into protofibrils with a β-sheet structure that can act as nucleation centers for fibril formation. However, in the presence of hydrophilic nanogel particles, the β-sheet formation and therefore fibril nucleation was inhibited, presumably by competing with the hydrogen bonds that stabilize the β-sheet structure [14] (Fig. 1c).

2 Biomolecular MD Simulations Overview

Due to limitations of experimental instrument resolution, the molecular details of the protein-nanoparticle interactions and the formation of their biocorona remains poorly understood. A well-established approach for characterizing the molecular details of biological systems is molecular dynamics (MD) simulations, and they have recently been applied to protein-nanoparticle systems. In MD simulations, biomolecular systems are represented as sets of spherical beads that interact with one another and move in successive timesteps to result in a trajectory of its motion. Biomolecular dynamics, folding, and assembly mechanisms have been extensively studied using MD simulations (Fig. 2a). The physical description of a biomolecular system is determined by a potential energy function for the interactions.

The MD simulation algorithm then consists of two main portions: (1) computing the forces, based on the potential energy function, between interacting particles, and (2) updating of the position and velocities of the particles for the next timestep. The process is repeated over and over until the end of the MD simulation. In each timestep, a "snapshot" of the biomolecule's positions is obtained, and all of the

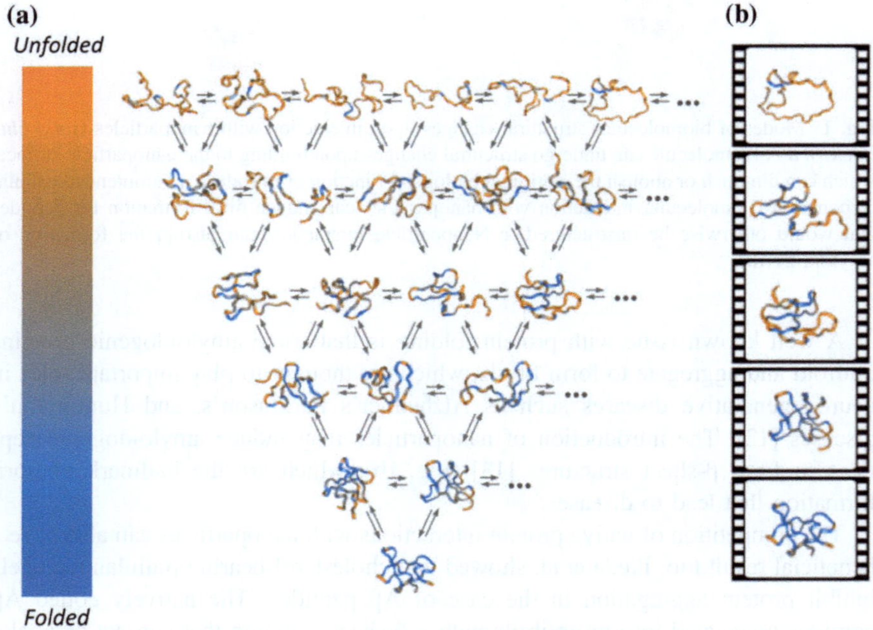

Fig. 2 Biomolecular folding mechanisms and MD simulations of their folding. **a** A schematic of a funneled energy landscape for folding where many possible unfolded states are directed towards a single, well-defined folded structure that corresponds to a function. For each protein structure from a coarse-grained Go-type MD simulation shown, the degree of local native structure formed is colored according to whether it is partially folded (*blue*) or unfolded (*orange*). **b** Representative structures from an MD simulation trajectory are shown

snapshots can be stitched together into a "movie" to result in a trajectory of the biomolecule's motion that is based on the potential energy function (Fig. 2b).

MD simulations are commonly implemented with an empirical force field. In this approach, a biomolecular system is described at atomistic resolution by an energy function that consists of bonds, angles, dihedral, electrostatic, and van der Waals interactions. The parameters for the energy function are derived from quantum mechanical calculations of model compounds or empirical data when reliable measurements are available. Popular atomistic empirical force field based MD simulation programs include CHARMM [15], AMBER [16], and NAMD [17], and the force field parameter sets are available for proteins [18], nucleic acids [19], and other biological systems.

In general, there are two main computational challenges for MD simulations of biomolecular systems to observe biologically relevant functional events (Fig. 3). The first is the system size because a large number of proteins interact with the nanoparticle. The second is the timescale required for the proteins to encounter the nanoparticle and interact with it. Large structural rearrangements or unfolding events are considered very challenging, even for relatively small systems (~100 residues) or small timescales (~μs-ms). Advanced sampling methods such as replica exchange can increase computational sampling by minimizing the time spent in trapped states, but the kinetic information is lost [20]. Others use advanced computing infrastructures such as the distributed computing Folding@Home approach [21] or the Anton supercomputer [22].

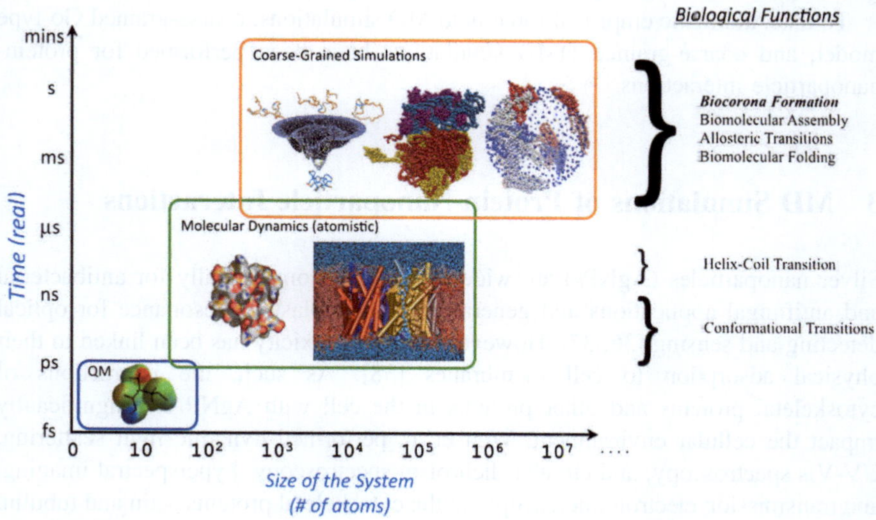

Fig. 3 Length- and time-scales of MD simulations and biological functions. **a** A graph of the approximate size of the system versus the length of time one can simulate using MD simulations using current computing standards using quantum mechanical, atomistic empirical force field, and coarse-grained Go-type MD simulations. **b** Some characteristic biological functions and the approximate corresponding timescales that can be observed in MD simulations

To increase the size and the timescales of the system one can simulate, some researchers use coarse-grained MD simulations where the residues or nucleic acids are represented as a single bead. A coarse-grained MD simulation approach that has been developed and rigorously tested is the Go-type model, which is defined by an energy function such that the long-range interactions are globally attracted to the native basin [23]. Based on the Funneled Energy Landscape Theory of biomolecular folding [10], the Go-type model MD simulation potential energy function consists of bonds, angles, dihedral, and Lennard-Jones terms whose global minimum corresponds to the folded state. For proteins, Go-type model MD simulations typically represent protein or RNA as a one bead per residue or nucleotide representation [24–26], but different degrees of coarse-graining such as a two bead per residue (backbone and sidechain) [27], three bead per nucleotide (base, sugar, and phosphate) [28–30], or even atomistic representations have been developed [31–33]. Although the size of the system and timescale one can simulate varies with the complexity of the Go-type model, researchers can readily perform MD simulations of 100–1,000s of residues, and the timescales are sufficiently large that one can observe multiple folding and unfolding events in a single trajectory.

Another simplification of MD simulations is to use discrete square-well potentials instead of a continuous potential for interactions between particles. In this discrete molecular dynamics (DMD) approach [34, 35], collision events determine the movement of the particles. A local search algorithm identifies particles that could possibly participate in a collision and only those particles are required to update their positions if a collision occurs. The fewer calculations required allows greater sampling at the expense of some accuracy.

To date, atomistic empirical force field MD simulations, coarse-grained Go-type model, and coarse-grained DMD simulations have been performed for protein-nanoparticle interactions.

3 MD Simulations of Protein-Nanoparticle Interactions

Silver nanoparticles (AgNPs) are widely produced commercially for antibacterial and antifungal applications and generating surface plasmon resonance for optical detecting and sensing [36, 37]. However, their cytotoxicity has been linked to their physical adsorption to cell membranes [38]. As such, the interactions of cytoskeletal proteins and other proteins in the cell with AgNP can significantly impact the cellular environment. Wen et al. performed dynamic light scattering, UV-Vis spectroscopy, and circular dichroism spectroscopy, hyperspectral imaging, and transmission electron microscopy on the cytoskeletal proteins actin and tubulin. In both cases, the cytoskeletal proteins readily interacted with citrate coated AgNPs, and they observed conformational changes of a decrease in α-helical content and an increase in β-sheet content upon binding to AgNPs. They suggested that the electrostatic, van der Waals, and hydrophobic interactions between the two species were the dominant interactions [39].

DMD simulations have been performed for a number of proteins interacting with a AgNP to characterize the consequence of biocorona formation. Ding et al. recently showed that for ubiquitin, interaction with AgNP results in a decrease in α-helical content and an increase in β-sheet too, and they observed a stretched exponential binding kinetics [40]. Interestingly, the secondary structure was only marginally affected in both experiments and simulation by interacting with AgNP for the firefly luciferase protein [41].

Auer et al. performed DMD simulations of peptides interacting with a nanoparticle [42]. They observe highly ordered β-sheet structure that eventually forms fibrils. Therefore, the presence of the nanoparticle catalyzes fibril formation by increasing the rate of peptide aggregate formation, and they attribute fibril formation enhancement to the increase in local concentration of the peptides upon interacting with the nanoparticle.

4 GPU-Optimized MD Simulations

Graphics Processing Units (GPUs) are becoming a widely accessible and mature computational architecture for performing MD simulations, and many studies have recently shown that MD simulations on GPUs can have a significant performance increase over the traditional CPU-only approach. Note there are other review articles on the development of GPU and can be found elsewhere [43–45], and we will briefly review the background here. Since the GPU architecture is significantly different from the CPU, implementing programs on the GPU often requires substantial changes in the implementation and the development of new algorithms that are optimized for the GPU architecture.

GPUs were originally designed for rendering computer graphics to offload and alleviate the computational burden of CPUs. The GPU is essentially an accelerator that is optimized for performing parallel floating-point calculations across individual geometric primitives due to the independent nature of rendering graphical images. As such, a standard GPU contains a few hundreds to thousands of independent cores that can work in parallel. This is in contrast to CPUs, which typically contain around 1–16 processors. The GPU processors are significantly slower than on a CPU, but there are so many of them so the overall floating point operation rate is higher. Therefore, the GPU architecture is ideally suited for parallel algorithms.

In order to utilize the resources found on the GPU to implemented general purpose parallel algorithms, NIVIDIA has developed their own programming language known as the Compute Unified Device Architecture (CUDA) so that software can be developed, compiled, and executed on NVIDIA GPUs. CUDA as a programming language is an extension of the C language with features that allow the program to execute code on both the CPU and GPU. The software instructions on the GPU architecture follow the Single Instruction Multiple Data (SIMD) paradigm where a kernel instruction performs the same instruction on different data (Fig. 4a). The GPU is able to execute code through the use of kernel calls that are

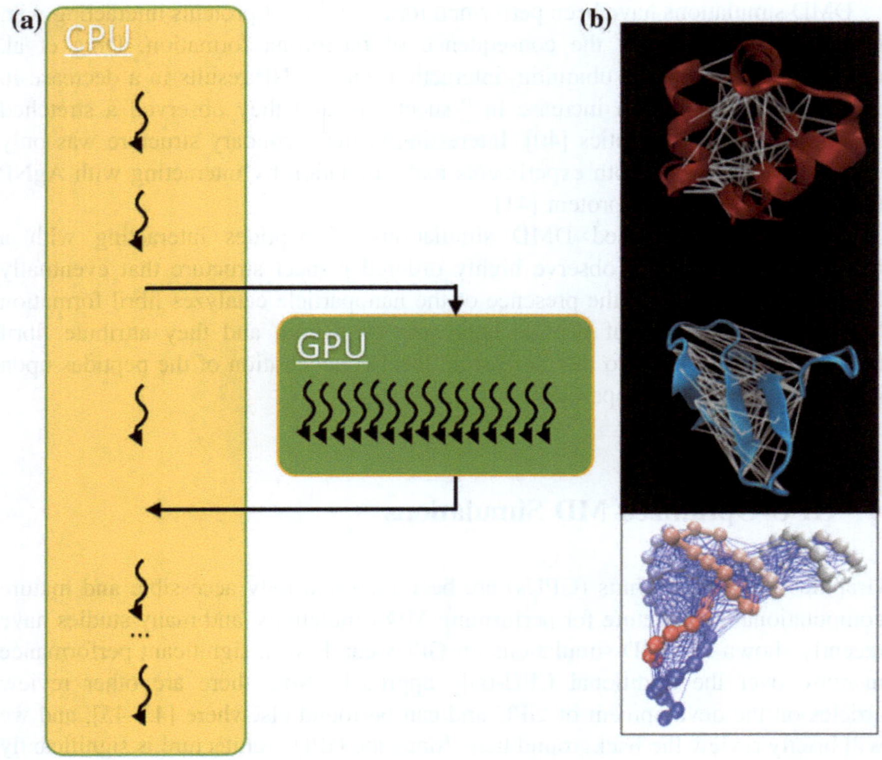

Fig. 4 Schematic of the GPU programming paradigm for MD simulations. **a** In a heterogeneous CPU-GPU architecture, a CUDA program executes on a CPU as a serial code. For portions of the code that is parallelizable, the program will transfer information to the GPU device and perform kernel operations (single instruction) in parallel on the (multiple) data using many independent threads. The data is then transferred back to the CPU. The MD simulation algorithm is parallelizable because the interactions in a single timestep are independent. **b** Representative α-helical (*top*) and β-sheet (*middle*) proteins and an RNA with their interactions shown with *lines*. These interactions are independent so the forces due to these interactions can be computed independently in a parallel algorithm

equivelant to functions for the CPU. When a kernel call is made on the CPU, the GPU allocates a certain number of threads that will each execute the kernel function in parallel. These threads are arranged in a grid system containing blocks of threads with each thread containing its own unique thread id. Threads are executed concurrently in warps where each thread executes all code within the kernel call. Before kernel calls are executed, all data that is used during parallel computation must be transferred to the GPU from the CPU. Similarly, all data that was generated by the GPU must be transferred back to the CPU once a kernel call is finished.

The MD simulation algorithm is well-suited for implementing on the GPU architecture. The calculation of the forces between interacting particles in a biomolecule is the most computationally demanding portion of the algorithm because it considers all possible pairs of interacting beads, a $O(N^2)$ computation. The main advantage of performing MD simulations on the GPU is that the calculation of the forces between interacting particles is a parallelizable algorithm because those interactions are independent within a single timestep such that those calculations can be distributed among the different processors (p) of the GPU: $O(N^2)/p \sim O(N)$. Furthermore, cutoff methods that first evaluate whether an interaction is significant before actually computing the force, such as the well known Verlet neighbor list algorithm [46] or the cell list algorithm [47], can also improve performance without sacrificing accuracy if they can also be implemented on the GPU, as we will discuss below.

While there may be many benefits of performing simulations using CUDA on GPUs, these approaches also have their issues. The data transfer between the CPU and GPU use the PCI Express Bus, which is one of the main performance bottlenecks of any GPU program. Ideally, programs written for the GPU should minimize data transfer to avoid the performance cost. Another significant issue of GPU-optimized programs is the limited amount of memory available on the GPUs. The latest in GPUs offered by NVIDIA, the Titan X, is able to support 12 gigabytes of memory while most NVIDIA GPU models can only support 1-6 gigabytes of memory. If the problem size grows past the hardware memory limit, it is necessary to partition the data and (1) perform the calculation sequentially for every partition, (2) perform the calculation while other data is being transferred so that the GPU is performing calculations while the data transfer is occurring to hide the latency, or (3) use multiple GPU with OpenMP or MPI to perform the calculations, although a multi-GPU approach has its own issues because the information transfer across GPUs also is a performance bottleneck. There are now several atomistic empirical force field and coarse-grained Go-type or general MD simulation packages that are now available including ACEMD, AMBER, HOOMD-Blue, LAAMPS, NAMD, and SOP-GPU. Here, we will briefly review our own in-house coarse-grained Go-type (SOP model) MD simulation software for the GPU [48–50], but our discussion is general and applicable to all MD simulation codes for the GPU architecture (Fig. 5).

Software	Type	URL
ACEMD	Atomistic	http://multiscalelab.org/acemd
AMBER	Atomistic	http://ambermd.org
HOOMD-Blue	General	http://codeblue.umich.edu/hoomd-blue/
LAMMPS	General	http://lammps.sandia.gov
NAMD	Atomistic	http://www.ks.uiuc.edu/Research/gpu/
SOP-GPU	Coarse-grained	http://faculty.uml.edu/vbarsegov/gpu/sop/sop.html

Fig. 5 A table of MD simulation software available for the GPU architecture

We recently implemented our own version of a Go-type (SOP model) MD simulation software, and we introduced two novel GPU-optimized MD simulation algorithms, while inspired by CPU versions, is specifically optimized for the GPU architecture: (1) Parallel Verlet Neighbor List [48] and (2) Parallel Hybrid Neighbor/Cell List [50] Algorithms. In the original Verlet neighbor list algorithm, distance cutoffs were used to determine whether two particles were sufficiently close enough to have a significant interaction. Two lists were maintained, an outer "skin" layer that was updated every n timesteps and an inner "cutoff" layer such that the forces of only pairs of particles within the inner distance cutoff would be calculated. The distance cutoffs were chosen such that pairs of particles beyond the distance cutoff had forces that were essentially zero and no particle could leave the cutoff layer and move beyond the skin layer within the n timesteps.

When we compared our performance to a CPU version of the same MD simulation code, we observed an N-dependent speedup with about 30× speedup for the largest system we simulated (~10,000 beads). For the smallest system we simulated, we actually observed a speed-down such that it took longer to perform the MD simulation on the GPU than on the CPU only. The performance gain from running the MD simulation on the GPU does not outweigh the cost of transferring the information to the GPU. We also developed a related Parallel Hybrid Neighbor/Cell List algorithm, and we observed about 10% speedup over the Parallel Verlet Neighbor List algorithm [50].

The Verlet neighbor list algorithm has been a staple of MD simulations for a very long time, but it is inherently serial because generating a subset list requires copying those pairs of interactions from the list of all pairs to an iteration dependent location in the subset list of pairs. We instead performed a parallel key-value sort of all pairs of interactions and placed all pairs of interactions to be copied to the subset list at the top. We then copied those pairs of interactions at the top in parallel over to the subset list. While the parallel version of the algorithm requires more steps, they can be performed entirely on the GPU in parallel.

Another significant issue of implementing software on the GPU involves accuracy of the floating point operations. Since the original purpose of the GPU is the rendering of graphical images, the floating point operations on older model GPUs did not need to be IEEE compliant like CPUs. In some implementations of MD simulations using early NVIDIA Tesla model GPUs, researchers observed an "energy drift" indicating that detailed balance was not being preserved due to errors in the calculations [51]. More recent models of GPUs are now IEEE compliant so this is no longer an issue. A related issue also exists where a large performance difference results between single and double precision calculations. MD simulations on CPUs have traditionally used double precision calculations because the performance difference was very minor. As such, there can be a performance gain if one chooses to implement MD simulations using single precision operations only.

We must note that the choice of double precision calculations over single precision calculations for MD simulations is arbitrary, and we are unaware of a significant reason for the choice. We therefore implemented our MD simulation code using single precision calculations only and double precision calculations only by starting off with the same exact positions and velocities. Even though the initial conditions were identical, we do not expect the same results even if we ran two trajectories with double precision calculations because the order of the floating point operations would be different in two executions. We therefore used that situation as a control and compared differences to single versus double precision only MD simulations. To quantify our MD simulations, we computed order parameters that could be directly compared to experiments, specifically the radius of gyration that is measured in SAXS experiments and end-to-end distances that is measured in single molecule FRET experiments. We observed minimal differences between the single and double precision only MD simulations for our coarse-grained model [48]. Walker and coworkers introduced a mix single-double precision method for a GPU-optimized version of AMBER. They observed minimal energy drift and minimal differences with respect to RMSD and RMSF while preserving a performance that is similar to single precision only calculations [52].

5 GPU-Optimized MD Simulations of Biocorona Formation

We developed a novel GPU-optimized MD simulation approach for studying biocorona formation. Specifically, we simulated the interaction of a silver nanoparticle with apolipoprotein A-1, which is the main component of a high-density lipoprotein. In a CD spectroscopy experiment, with AgNP/apolipoprotein molar ratios of 1:300 and 1:600, the spectra reveals that the presence of AgNP reduces the ratio of α-helical content as compared to when there is no AgNP in the solution [53].

We developed a new computational model of nanoparticle interactions with proteins for the purpose of performing MD simulations of biocorona formation [53]. The model consists of 1 Ag nanoparticle (AgNP) and 15 apolipoproteins. We model the citrate-coated AgNP using 500 individual negatively charged spherical beads randomly distributed 10 nm from the center. And for the 243-residue apolipproteins, we developed a coarse-grained MD simulation based on well-established Go-type models of protein folding [23, 54].

In addition, we modeled the interactions between the AgNP and proteins by introducing two sets of terms: [1] excluded volume and [2] electrostatic interactions.

The excluded volume forbids the proteins getting too close and overlapping with the AgNP surface, and we modeled it using a hard sphere potential:

$$H_{\mathrm{EV}} = \sum_{i,j}^{\text{protein}-\text{NP}} \left(\frac{\sigma}{r_{\mathrm{ij}}}\right)^{12}$$

where r_{ij} is the distance between two interacting beads and σ is set to 3.8 Å. We also modeled the electrostatic interaction between the negatively charged AgNP and positive amino-acid residues with the Debye-Huckel potential:

$$H_{elec} = \sum_{i,j}^{protein-NP} \frac{z_i z_j e^2}{4\pi\epsilon_0\epsilon_r r} e^{-r/l_D}$$

where z_i and z_j are the charges of the interacting protein and nanoparticle beads and r is the distance between them. l_D is the Debye length, which can be tuned by the ion-concentration of 0.02 M [Na^+]. In our simulation, we used a relatively low ion-concentration for stronger attraction. We performed Langevin MD simulations in the underdamped limit (i.e., low friction coefficient) for effective sampling at 300 K (Fig. 6).

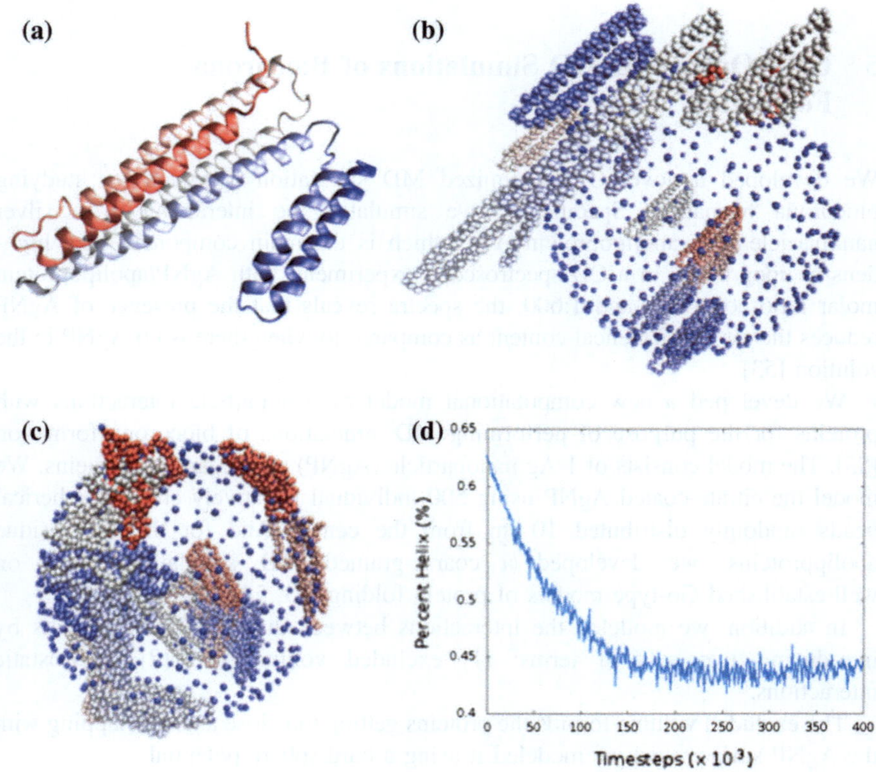

Fig. 6 MD simulations of biocorona formation on a GPU. **a** A ribbon diagram of apolipoprotein A-1. **b** The starting coordinates for coarse-grained MD simulations of 15 apolipoproteins around a negatively charged spherical model of AgNP composed of 500 beads. **c** A snapshot from GPU-optimized coarse-grained MD simulations of the AgNP-apolipoprotein biocorona. **d** A graph of the percent helicity of the apolipoprotein over a representative MD simulation trajectory. We observe a decrease in the helical content upon binding to the AgNP

The resulting system is comprised of 500 beads for the AgNP, and 3,645 beads (=243 residues/protein × 15 proteins) for the 15 apolipoproteins, which is computationally demanding even with a coarse-grained MD simulation approach. We therefore implemented our MD simulation model on a GPU-optimized parallel GPU code to improve the performance.

To quantify the secondary structure change during the biocorona formation, we monitored the torsional angles formed among the four successive beads in helical region, and we define an α-helical angle to be between 45°–60°. In our MD simulations, we observe a decrease in the percentage of the α-alpha helical angles from 65 to 45%, which is similar to experimentally observed values from CD spectra [53].

6 Conclusions

The development of nanomedicine and the widespread use of nanoparticles for industrial use have opened up new advances but it is also pressing to evaluate their biological exposure to determine how unintentional accumulation and exposure to nanoparticles could lead to toxicological effects due to their interaction with biological systems. Experiments have shown that nanoparticles readily interact with the biological medium to form a biocorona that assumes a new biological identity. To understand the consequences, we must evaluate nanoparticle interactions with biomolecules using an interdisciplinary approach that spans physics, chemistry, biology, engineering, and computer science.

The relatively large size of the biocorona requires new and advanced approaches for the study of their formation from MD simulations. Coarse-grained MD simulations are powerful tools that can access the timescales necessary to study these events while still reproducing experimental observables. More detailed description of the phenomenon requires advanced computing power, and the GPU architecture has already demonstrated to be a valuable tool for the study of biocorona formation that can reproduce experimental observables and provide new insight at a molecular resolution. These computational advances can give much needed biophysical insight into our understanding of biological and environment consequences of nanomaterial interactions.

Acknowledgements The National Science Foundation (CBET-1232724) financially supported this work. RL and SSC acknowledge financial support from the Wake Forest University Center for Molecular Communication and Signaling (CMCS). CAS was supported by a CMCS graduate research fellowship. SSC appreciates fruitful conversations with Pu-Chun Ke.

References

1. Alivisatos AP (1996) Perspectives on the physical chemistry of semiconductor nanocrystals. J Phys Chem 100:13226–13239
2. Dujardin E, Mann S (2002) Bio-inspired materials chemistry. Adv Mater 14:775–788
3. Nirmal M, Brus L (1999) Luminescence photophysics in semiconductor nanocrystals. Acc Chem Res 32:407–414
4. Ye D et al (2013) Nanoparticle accumulation and transcytosis in brain endothelial cell layers. Nanoscale 5:11153–11165
5. Cedervall T et al (2007) Understanding the nanoparticle–protein corona using methods to quantify exchange rates and affinities of proteins for nanoparticles. Proc Natl Acad Sci 104:2050–2055
6. Schrurs F, Lison D (2012) Focusing the research efforts. Nat Nanotechnol 7:546–548
7. Morriss-Andrews A, Bellesia G, Shea J-E (2011) Effects of surface interactions on peptide aggregate morphology. J Chem Phys 135:085102–085109
8. Salvati A et al (2013) Transferrin-functionalized nanoparticles lose their targeting capabilities when a biomolecule corona adsorbs on the surface. Nat Nanotechnol 8:137–143
9. Anfinsen CB (1973) Principles that govern the folding of protein chains. Science 181:223–230
10. Wolynes PG, Onuchic JN, Thirumalai D (1995) Navigating the folding routes. Science 267:1619–1620
11. Vertegel AA, Siegel RW, Dordick JS (2004) Silica nanoparticle size influences the structure and enzymatic activity of adsorbed lysozyme. Langmuir 20:6800–6807
12. Dobson CM (1999) Protein misfolding, evolution and disease. Trends Biochem Sci 24: 329–332
13. Linse S et al (2007) Nucleation of protein fibrillation by nanoparticles. Proc Natl Acad Sci 104:8691–8696
14. Ikeda K, Okada T, Sawada S, Akiyoshi K, Matsuzaki K (2006) Inhibition of the formation of amyloid β-protein fibrils using biocompatible nanogels as artificial chaperones. FEBS Lett 580:6587–6595
15. MacKerell AD et al (1998) All-atom empirical potential for molecular modeling and dynamics studies of proteins. J Phys Chem B 102:3586–3616
16. Weiner PK, Kollman PA (1981) AMBER: assisted model building with energy refinement. A general program for modeling molecules and their interactions. J Comput Chem 2:287–303
17. Phillips JC et al (2005) Scalable molecular dynamics with NAMD. J Comput Chem 26: 1781–1802
18. MacKerell AD, Feig M, Brooks CL (2004) Improved treatment of the protein backbone in empirical force fields. J Am Chem Soc 126:698–699
19. Denning EJ, Priyakumar UD, Nilsson L, Mackerell AD (2011) Impact of 2'-hydroxyl sampling on the conformational properties of RNA: update of the CHARMM all-atom additive force field for RNA. J Comput Chem 32:1929–1943
20. Sugita Y, Okamoto Y (1999) Replica-exchange molecular dynamics method for protein folding. Chem Phys Lett 314:141–151
21. Voelz VA, Bowman GR, Beauchamp K, Pande VS (2010) Molecular simulation of ab initio protein folding for a millisecond folder NTL9(1-39). J Am Chem Soc 132:1526–1528
22. Dror RO, Dirks RM, Grossman JP, Xu H, Shaw DE (2012) Biomolecular simulation: a computational microscope for molecular biology. Annu Rev Biophys 41:429–452
23. Clementi C, Nymeyer H, Onuchic JN (2000) Topological and energetic factors: what determines the structural details of the transition state ensemble and "en-route" intermediates for protein folding? An investigation for small globular proteins. J Mol Biol 298:937–953
24. Chavez LL, Onuchic JN, Clementi C (2004) Quantifying the roughness on the free energy landscape: entropic bottlenecks and protein folding rates. J Am Chem Soc 126:8426–8432

25. Karanicolas J, Brooks CL (2002) The origins of asymmetry in the folding transition states of protein L and protein G. Protein Sci 11:2351–2361
26. Hyeon C, Dima RI, Thirumalai D (2006) Pathways and kinetic barriers in mechanical unfolding and refolding of RNA and proteins. Structure 14:1633–1645
27. Cheung MS, Garcia AE, Onuchic JN (2002) Protein folding mediated by solvation: Water expulsion and formation of the hydrophobic core occur after the structural collapse. Proc Natl Acad Sci USA 99:685–690
28. Hyeon C, Thirumalai D (2005) Mechanical unfolding of RNA hairpins. Proc Natl Acad Sci 102:6789–6794
29. Cho SS, Pincus DL, Thirumalai D (2009) Assembly mechanisms of RNA pseudoknots are determined by the stabilities of constituent secondary structures. PNAS 106:17349–17354
30. Li R, Ge HW, Cho SS (2013) Sequence-dependent base-stacking stabilities guide tRNA folding energy landscapes. J Phys Chem B 117:12943–12952
31. Li L, Shakhnovich EI (2001) Constructing, verifying, and dissecting the folding transition state of chymotrypsin inhibitor 2 with all-atom simulations. PNAS 98:13014–13018
32. Whitford PC et al (2009) An all-atom structure-based potential for proteins: bridging minimal models with all-atom empirical forcefields. Proteins Struct Funct Bioinforma 75:430–441
33. Feng J, Walter NG, Brooks CL (2011) Cooperative and directional folding of the preQ 1 riboswitch aptamer domain. J Am Chem Soc 133:4196–4199
34. Dokholyan NV, Buldyrev SV, Stanley HE, Shakhnovich EI (1998) Discrete molecular dynamics studies of the folding of a protein-like model. Fold Des 3:577–587
35. Proctor EA, Ding F, Dokholyan NV (2011) Discrete molecular dynamics. Wiley Interdiscip Rev Comput Mol Sci 1:80–92
36. Choi O, Hu Z (2008) Size dependent and reactive oxygen species related nanosilver toxicity to nitrifying bacteria. Environ Sci Technol 42:4583–4588
37. Jin X et al (2010) High-throughput screening of silver nanoparticle stability and bacterial inactivation in aquatic media: influence of specific ions. Environ Sci Technol 44:7321–7328
38. Zhang W, Yao Y, Sullivan N, Chen Y (2011) Modeling the primary size effects of citrate-coated silver nanoparticles on their ion release kinetics. Environ Sci Technol 45: 4422–4428
39. Wen Y et al (2013) Binding of cytoskeletal proteins with silver nanoparticles. RSC Adv 3:22002–22007
40. Ding F et al (2013) Direct observation of a single nanoparticle–ubiquitin corona formation. Nanoscale 5:9162–9169
41. Käkinen A et al (2013) Interaction of firefly luciferase and silver nanoparticles and its impact on enzyme activity. Nanotechnology 24:345101
42. Auer S, Trovato A, Vendruscolo M (2009) A condensation-ordering mechanism in nanoparticle-catalyzed peptide aggregation. PLoS Comput Biol 5:e1000458
43. Anderson JA, Lorenz CD, Travesset A (2008) General purpose molecular dynamics simulations fully implemented on graphics processing units. J Comput Phys 227:5342–5359
44. Liu W, Schmidt B, Voss G, Müller-Wittig W (2008) Accelerating molecular dynamics simulations using graphics processing units with CUDA. Comput Phys Commun 179: 634–641
45. Xu D, Williamson MJ, Walker RC (2010) In: Wheeler RA (ed) Annual reports in computational chemistry, Elsevier, pp 2–19
46. Verlet L (1967) Computer "experiments" on classical fluids. I. Thermodynamical properties of Lennard-Jones molecules. Phys Rev 159:98
47. Allen MP, Tildesley DJ (1989) Computer simulation of liquids. Oxford University Press, USA
48. Lipscomb TJ, Zou A, Cho SS (2012) Proceedings of the ACM conference on bioinformatics, computational biology and biomedicine, BCB'12. ACM, New York, NY, USA, pp 321–328
49. Proctor AJ, Lipscomb TJ, Zou A, Anderson JA, Cho SS (2012) 2012 ASE/IEEE international conference on biomedical computing (BioMedCom), pp 14–19

50. Proctor AJ, Stevens CA, Cho SS (2013) Proceedings of the international conference on bioinformatics, computational biology and biomedical informatics, BCB'13. ACM, New York, NY, USA, pp 633:633–633:640
51. Bauer BA, Davis JE, Taufer M, Patel S (2011) Molecular dynamics simulations of aqueous ions at the liquid–vapor interface accelerated using graphics processors. J Comput Chem 32:375–385
52. Le Grand S, Götz AW, Walker RC (2013) SPFP: speed without compromise—a mixed precision model for GPU accelerated molecular dynamics simulations. Comput Phys Commun 184:374–380
53. Li R et al (2013) Computational and experimental characterizations of silver nanoparticle-apolipoprotein biocorona. J Phys Chem B 117:13451–13456
54. Cho SS, Levy Y, Wolynes PG (2006) P versus Q: Structural reaction coordinates capture protein folding on smooth landscapes. Proc Natl Acad Sci 103:586–591

Modeling Cell Communication by Communication Engineering

Jian-Qin Liu and Wuyi Yue

Abstract In molecular biology of the cell, cell communication is defined as the process carried out by chemical signals within and among cells. The informatics issue of cell communication in this book chapter is to uncover the principles of the bioinformatics of cell communication by means of communication engineering, e.g., the statistical tool for performance analysis of communication processes. As we know well by now, the state of the art of molecular science has been reshaped by advanced technologies since the genome sequencing became a reality. In accordance with nowadays available nanotechnology for molecular signal detection, we apply communication engineering technology in the theoretical analysis of cell communication whose goal is to discover the mechanism of cell communication that determines the cellular functions connected with applications in medicine. Though intensive research has been devoted to the biochemistry of signaling pathways, laying a strong scientific foundation for the informatics study of communication processes of the cell in the form of signaling pathways, the study of the communication mechanism of signaling pathway networks in the cell—cell communication—by means of communication engineering is still a relatively new field, where supporting technologies from multiple disciplines are needed. In this book chapter, the formulation of the cell communication mechanism of signaling pathway networks using martingale measures for random processes is proposed and the performance of the cell communication system constructed by the signaling pathways in simulation studies is evaluated from the viewpoint of communication engineering. From the computational analysis result of the above cell communication process, it is concluded that the modeling method in this study not only is

J.-Q. Liu (✉)
Center for Information and Neural Networks, National Institute of Information and Communications Technology, Kobe 651-2492, Japan
e-mail: liu@nict.go.jp

W. Yue
Department of Intelligence and Informatics, Konan University, Kobe 658-8501, Japan
e-mail: yue@konan-u.ac.jp

J. Suzuki et al. (eds.), *Modeling, Methodologies and Tools for Molecular and Nano-scale Communications*, Modeling and Optimization in Science and Technologies 9, DOI 10.1007/978-3-319-50688-3_11

efficient for bioinformatics analysis of biological cell communication processes but also provides a reference framework for brain communication towards its application in molecular biomedical engineering.

1 Introduction

It is well known that cell communication [1] plays an important role in the specificity of cellular functions. One of the most promising technologies in the research of cell communication is the super high-resolution imaging of molecules in living cells [2], especially the GPS-like molecular sensing technology for the neurons in the brain, which offers a new approach to exactly locate and trace the neuronal signals leading to the development of neurodegenerative diseases [3]. This approach of molecular signal measurement, which provides a fine-grained image of molecular signal distribution within the cells, inspires us to study cell communication. The physical basis supporting this study consists of the knowledge of diffusion process and active transportation process of molecular motors [4]. Though diffusion process theory is well-established, the mathematical description of the transportation processes within the cell is still a challenging theme in the frontier of structural molecular biology [5], for example, the structural-biological explanation of the cargo transportation mechanism of molecular motors in the intracellular environment where the signaling dynamics of ER and Golgi apparatus cannot be neglected. In this book chapter, we will apply mathematical theory of random processes [6] to the biomolecular signal analysis of cell communication. Nowadays, with the advances of nanotechnology, the knowledge of protein folding [7] is applicable for understanding the malfunction of protein folding on the neuro-degenerative disease because of the conservativeness at the genome level among different species.

Systematically understanding cell communication through communication engineering contributes to apply the biomolecular signal analysis to nano-medicine. Owing to the existence of the information fusion of the neural signals at different levels by the genomes of brain cells [8], it is natural to extend molecular signal analysis to neural signal analysis by molecular neuroscience to "unlock the brain" [9] to explore the adaptation of the neural memory circuits for learning. As the intrinsic spontaneous network of the brain, the so-called default mode network [10] acts as the bridge between molecular signaling and neural signaling. Since knowledge in multiple disciplines [11–22] (imaging [11], measurement [12], molecular biology for the model organisms such as yeast [13, 14], mathematics [15], bioinformatics [16], signal processing [17], single-molecule FRET [18], advanced biotechnology [19], advanced computing technology [20], computer science [21], cell biology for biological nanomachines [22], molecular communication [23, 24]) is required, the integration of informatics and biology is the basis of the study on cell communication which is helpful to understanding the principle of the brain at the level of molecular neuroscience.

2 Informatics Principles of Cell Communication: Complex Network of Cellular Signaling Pathways

The nanobio-world provides a unique scenario for the biological nano-imaging [11, 12] by which the spatial information of molecular concentration in the cell enhances our ability to analyze the molecular signal transmission processes called the "intra-cell communication" and "inter-cell communication". The feasibility of the super high resolution of molecular imaging makes it possible to study the biomolecular signaling pathways of the cell through the structural biological aspects of cellular signaling interactions. The signaling cascade within the cell is a major instance of signaling pathways. Figure 1 shows the pathway network of kinases and Table 1 gives its related energy values calculated by using MOE®.

It is noticeable that the necessary knowledge of the cell communication mechanism by using the cells as the potential implementation materials will help to explore the possible design of a controllable cell communication system. Strictly speaking, the identical information processing mechanism between biological cell communication and expected protocol of controllable cell communication in engineering is a necessary condition of the practice on any "wet" form of "engineered" controllable cell communication systems in the near future, which can be expressed in two aspects—mathematical modeling and computational analysis of

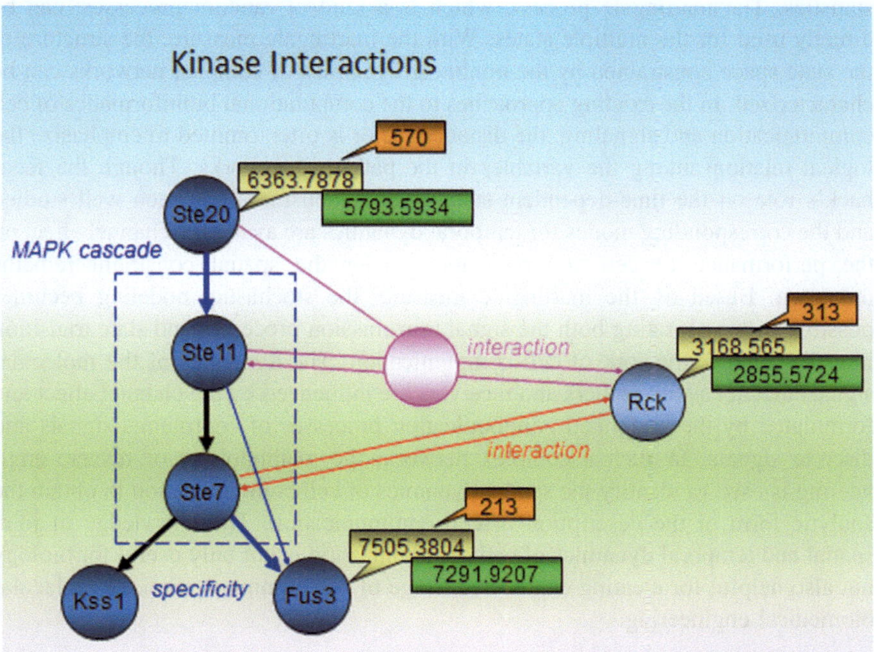

Fig. 1 Kinase interaction

Table 1 E^P (potential energy) of kinases

Kinase	E^P		
	E_l^P	E_h^P	$\Delta E_h^P = (E_h^P - E_l^P)$
Ste20	5793.5934	6363.7878	570.1946
Fus3	7291.9207	7505.3804	213.4593
Rck	2855.5724	3168.5650	312.9926

cellular signaling pathway networks. The converged form of these two aspects is how the molecules realize the cell communication at the molecular levels from the information represented and processed in the signaling pathways.

In systems biology, the nonlinear dynamics of signaling pathways has been studied based on biochemical measures such as the Michaelis-Menten equation and the Hill equation. Before the theory of linear systems is used to handle the case of the stochastic modeling of cellular signaling pathways under the stationary condition by the asymptotical approximation approach, the dynamics of molecular signals in cellular signaling pathways has to be modelled by random processes. Markov process is one of the methods to describe the underlying cell communication process which is applicable for the discrete form of the biological networks. In our study, we assume a complex system that varies among different steady states from which the sustainable robustness exists. We will introduce the martingale formulation for the signaling molecular transportation process of cell communication systems, which is the third major type of random processes widely applied in statistics. The martingale process, which is a kind of random processes, can be directly used for the multiple states. With the martingale measure, the structure of the state space constrained by the nonlinear dynamics of complex networks can be characterized. In the existing approaches to the computational bioinformatics of cell communication and signaling, the distance factor is often omitted to emphasize the logical relation among the variables in the pathway networks. Though the feedback's role on the time-dependent stability and robustness has been well studied and the corresponding modes for temporal dynamics are available, channel effect on the performance of cell communication under the spatial constraint remains unknown. Based on the martingale measure, the stochastic modeling becomes possible with integrating both the signal transmission processes and state transition into the same framework of cell communication. The dynamics of the molecular signals that act as the senders and receivers are influenced by the channel effect and formulated by the integrated communication processes of continuous signals and discrete signals. In such a complex network, the methodology of reverse engineering is used to identify the spatial dynamics of cell communication to obtain the analytic form of the description of cell communication. The knowledge of joint spatial and temporal dynamics of cell communication is not only useful for biology but also helpful for creating engineered usage of cell communication for molecular biomedical engineering.

2.1 Discrete State Transition of Cell Communication in Terms of Complex Systems

Among the differences between the robust cellular signaling networks and the robust industrial control networks, the most obvious one is that the state transition in biological systems is discrete, i.e., no continuous function exists for the state transition under the unknown objective function. Here the "configuration" of the network dynamics is adapted in response to the uncertain environment. To analyze the signaling mechanism of the adaptation process corresponding to the state transition requires the nonsmooth analysis in mathematics. From the viewpoint of mathematics, the nonsmooth analysis is a kind of special differential analysis, which uses the so called generalized directional derivative and other measures derived from it. Being different from the existing approaches to model complex networks, the nonsmooth analysis approach to model the signaling pathway networks captures the essence of the discrete and discontinuous characteristics of their signal evolution processes. Here the concept of evolution refers to the changes of the signals that correspond to the steady states (SS) of the underlying complex networks. The conventional notations for discrete event systems are integrated with the ones for discontinuous and continuous systems.

In systems biology, the analysis of the signaling mechanism of the complex biomolecular networks is often carried out by the most observable measure—stability, from which robustness can be easily quantified. Robustness here is defined as a mechanism that guarantees the stability under the condition of variation of the parameters and structure of the network dynamics of the underlying system. Let $X(t)$ be the state vector that refers to the signaling molecules in the cell. Based on the state representation for a nonlinear system, the network dynamics of $X(t)$ is modeled as follows

$$\frac{d}{dt}X(t) = f(X(t), v(t)), \tag{1}$$

$$\text{a.e. } 0 \leq t \leq T. \tag{2}$$

Though the stability of state $X(t)$ provides the indication of the variation of the steady states subject to the variation of the parameter set $v(t)$, the existing robustness analysis methods in systems biology cannot be directly applied to the case of the complex networks with discontinuous states and discrete state transitions in our study, mainly owing to the fact that the stepwise approximation method cannot exactly explain the discrete state transition processes between different steady states. Thus, it is necessary to select the nonsmooth analysis representation for the formulation of the complex networks.

According to the notation in nonsmooth analysis [15], (1) subject to the constraint $v(t)$ which characterizes the configuration of the network dynamics. The notation "a.e." refers to "almost everywhere". The variable $X(t)$ is not necessarily smooth; i.e., it can be nonsmooth. "Nonsmooth" means that the function is not differentiable. In the nonsmooth analysis of the robust network, the problem of robustness becomes a problem of connectiveness of multiple steady states.

In the state space, the Lyapunov method is feasible for the stability analysis for genetic regulatory networks as reported in [25], which implies that the increase of the strength of stability can be achieved by the analytic design of inequality that directs the approximation process of the searching procedure for the solution to the stability problem. But, any integrated analytic approach for joint continuous signaling dynamics and discrete state transition is still unavailable for the Lyapunov-method-oriented robustness analysis. The kernel method for robustness analysis of complex networks we proposed is the integration of the formulation of the signal dynamics by random processes and the corresponding channel constructed by information theory. In sequence, the simulation of complex networks with discrete state transition is the procedural synthesis based on partial analytic results of the complex networks. With the computational analysis of the signaling dynamics, it becomes possible to quantitatively describe the molecular signals going through the different places within the cells where the martingale measure is used to describe the state transition processes.

In order to describe the variation of the molecular signals we observed, we need an indicator, that is, a function of the random variable. Let different steady states be Y_0, Y_1, …, Y_n, then the bounded Z_n is defined as a martingale process that satisfies

$$Z_n = E(Z_n + 1 | Y_0, Y_1, \ldots, Y_n), \tag{3}$$

where Z_n refers to the measure of robustness that converges in different steady states. One of the configuration of Z_n is the indicator of the stability described in (1), i.e.,

$$E(Z_n) = \frac{d^2}{dt^2} X(t) < 0, \tag{4}$$

$$\text{s.t. } \frac{d}{dt} X(t) = 0, \tag{5}$$

where $X(t)$ is the state within the steady states.

The increase of the strength of robustness indicated by the above-mentioned measure can be achieved by increasing the range of the parameters for steady states. The maximum criterion of mutual information for the channel is consistent with the minimum of the martingale given above in (4) depending on the constraint of information theory on the channel from which the cell communication processes coincide with the stable signals going through the channel. Thus, the channel effect enhances the robustness of cell communication under the condition of the convergence of the martingale measure for the robustness. An example for this is the interference channel in information theory, in which the convergence of the martingale process can be characterized by the multiplication of multiple Poisson processes which correspond to the multiple states of the cell communication processes.

In accordance with theoretically obtained convergence of formulated cell communication processes using the martingale measure mentioned above, we need to give a mathematical explanation on the active molecular transportation which is autonomously self-organized. From an informatics viewpoint, the convergence indicates that the existence of the stability of the cell communication. This argument equals to the argument that the stable cell communication can be achieved under stochastic channel dynamics. This argument needs to be tested by the following two aspects: algebraic system description for communication protocol and observable signals for formulated communication processes.

(1) Algebraic system description for communication protocol

An algebraic system $<S, Q>$ is adopted to describe the cell communication process where S is a set of letters and Q is a set of operators. Being different from the automata in discrete mathematics, algebraic system for cell communication is assigned with a nonlinear continuous function which is used to connect the continuous signals in signaling pathways and the discrete state transition of the cell communication process. We define set S as the set of $\{a, b, c, d, e\}$ whose elements are given as follows:

a steady state of signaling pathway,
b transition state among different steady states,
c unsteady state,
d molecular transportation,
e signal transduction in signaling pathways

The operators of Q include the follows:

q_1 $a \rightarrow a$, s.t. $b/d/e$,
q_2 $c \rightarrow a$, s.t. $b/d/n$,
q_3 $c \rightarrow c$, s.t. d/n,
q_4 $a \rightarrow c$, s.t. $b/d/n$,
q_5 $a \rightarrow d$,
q_6 $d \rightarrow a$,
q_7 $c \rightarrow d$,
q_8 $d \rightarrow c$,
q_9 $e \rightarrow a$,
q_{10} $e \rightarrow c$,
q_{11} $a \rightarrow e$,
q_{12} $c \rightarrow e$,

where n: (internal and/or external) noise.

The representation for the operations of the model of algebraic system we designed is given in Fig. 2.

The above-mentioned algebraic system representation is the primitive form of the formulated cell communication process. By inferring the partial relation *R'*

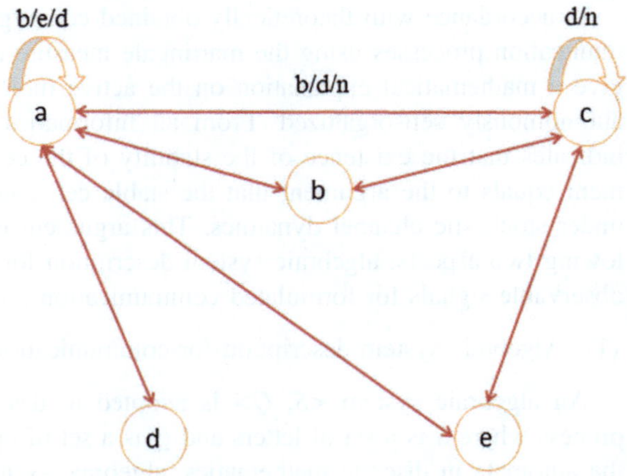

Fig. 2 Representation for the operations of the model of algebraic system

explicitly described by Fig. 2, the three orders of relationship *R'* compared with the basic two orders *R* defined by <*A*, *Q*> can be written by the form of

w_1 $a\ d\ a \rightarrow a\ d\ a,$
w_2 (repeat) $a\ d\ a \rightarrow a\ d\ a,$
w_3 $a\ d\ a \rightarrow a\ c\ a,$
w_4 $a\ c\ a \rightarrow a\ c\ a,$
w_5 (repeat) $a\ c\ a \rightarrow a\ c\ a,$
w_6 $a\ d\ a \rightarrow a\ c\ a \rightarrow a\ d\ a,$
w_7 $a\ d\ c \rightarrow a\ e\ c.$

The rules labeled coincide with the rules of MFL (McNaughton Families of Languages).

(2) Observable signals for formulated communication processes

The signal transmission can be described by the Smoluchowski equation

$$\frac{\partial}{\partial t}p = \frac{\partial}{\partial x}(p\frac{\partial}{\partial x}\phi) + \frac{\partial^2}{\partial x^2}p, \tag{6}$$

where

ϕ—an external field,
$p(x, t)$—defined by

$$p(x,t) = c(x,t) / \int_0^L c(x,t)dx, \tag{7}$$

$c(x, t)$—"concentration" of signaling molecules,
x—position of the signaling molecule, $0 \le x \le L$,
t—time.

(2.1) Biochemical reaction for the global signal transmission

Let p in (6) be σ^{32} in Fig. 4 and field ϕ be the influence of *DnaK* which is in the form of biochemical reaction of the channel effect. Because the gene regulation of σ^{32} and *DnaK* is stable during the period of heat shock response, the second order derivative of σ^{32} on the distance of the channel is zero (that corresponds to the condition of Z (i.e., $\{Z(n)\}$ in (3)) mentioned in previous subsection), i.e., the speed of the molecular movement in the channel is constant. Then, we have that

$$\frac{\partial}{\partial t}\sigma^{32} = \frac{\partial}{\partial x}(\sigma^{32}\frac{\partial}{\partial x}\phi), \tag{8}$$

and

$$\frac{1}{\sum_x \sigma^{32}}\int_x (\frac{\partial}{\partial t}\sigma^{32})\partial x = Cp\int_x \frac{\partial}{\partial x}(\sigma^{32}\frac{\partial}{\partial x}\phi)\partial x, \tag{9}$$

where $x = 0$ and $x = L$ refer to the sender and receiver, respectively. C_p is a constant.

Let σ^{32*} be the estimated value, i.e., the expectation of σ^{32}

$$\sigma^{32*} = \frac{1}{\sum_x \sigma^{32}}\int_x \sigma^{32}\partial x. \tag{10}$$

Then we can write that

$$\frac{\partial}{\partial t}\sigma^{32*} = C_p(\sigma^{32}\frac{\partial}{\partial x}\phi). \tag{11}$$

(2.2) Variation of the local signal σ^{32} caused by the channel effect

The local effect of the gene regulation on channel is described by

$$\frac{\partial}{\partial x}\phi = -C_x[DnaK], \tag{12}$$

where C_x is a constant. Here the channel is defined as the transmission between the sender side of individual σ^{32} without any binding and the receiver side of σ^{32} bounded with *Dnak*. At $x = L$, σ^{32*} is the estimated value. At $x = 0$, *DnaK* is 0 because *Dnak* hasn't been sent out from the sender side of σ^{32}. Thus, the estimated value of the variation of σ^{32} caused by the channel effect is given as follows

$$\frac{\partial}{\partial t}\sigma^{32*}(channel) = -C^*[DnaK](G(\sigma^{32})), \tag{13}$$

where C^* is a constant and $G(.)$ is the mean of σ^{32} under certain distribution of the variable in the channel.

Thus, at the receiver we can get that

$$\frac{\partial}{\partial t}\sigma^{32*} = -C^*[DnaK](G(\sigma^{32})) + C_p\beta\sigma^{32}, \tag{14}$$

provided that the condition $\frac{\partial}{\partial x}\phi = \beta$ exists for the receiver whose signals are not constrained by the channel.

When $G(.)$ is converged as a constant, the observed value given by the left side of (14), which is the expectation flow of σ^{32} (i.e., $E(\sigma^{32})$), will proportionally converge to the ideal value of σ^{32}. This result indicates the existence of the observability of the cell communication process formulated above, which is the theoretical basis for our simulation presented in Sect. 2.3.

2.2 An Example of the Complex Mechanism of Cell Signaling

Here we consider an example for our study on the cell communication by using the cellular signaling process of the HSR (heat shock response) pathway. The stress signal of the cell is a signal from the environment going irregularly beyond the normal threshold. As a kind of response to stress, heat shock response (HSR) is a cellular function in which the cell can sustain the protein folding when the temperature is higher than the normal range (as illustrated in Fig. 3). We select the HSR (heat shock response) pathway network of *E. coli* to study the robustness. In heat shock response of *E. coli*, we use the signaling molecule σ^{32} as the indicator of the robustness.

According to the methodology of systems biology, the HSR signaling pathway is modeled as a controller in which one of the molecular signal is *DnaK* and the channel is embedded in it. The dynamical variation of *DnaK* caused by the intrinsic thermodynamics corresponds to the phase transition of the dynamics of the HSR pathway. In order to model the cell communication process of the HSR pathway, the signals of sender (denoted as *Tx*) and receiver (denoted as *Rx*) are represented by the concentration of σ^{32}, which is spatially distributed in the cell and the channel between them located in the cell is controlled by the HSR network to transport the signaling molecules σ^{32} and *DnaK*. Then will consider the channel effect. Because the noise is caused by thermodynamics, the capacity of the channel is dependent on the active/inactive state of the molecules. The signal loss in the channel constrained by the channel capacity is mainly caused by non-specific interactions of signaling molecules with respect to the specificity of signaling pathways. The feedback is a factor that determines the stability of the underlying network. With the parametric variation of the network, the robustness is described by different states in which the set of steady states correspond to the feature patterns of different SS. In the field of

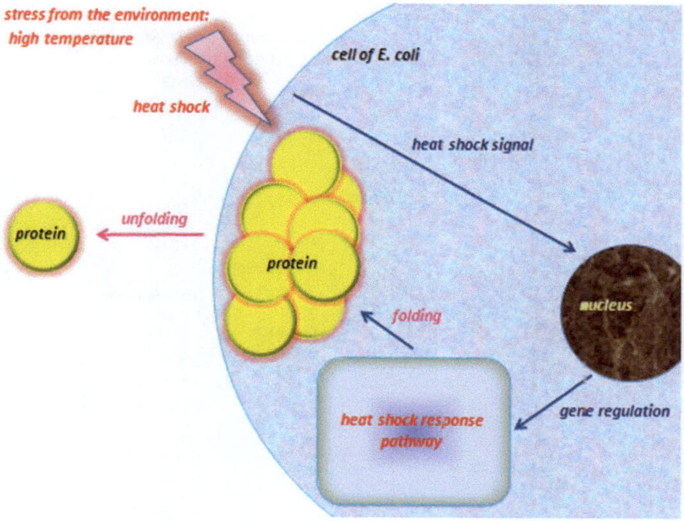

Fig. 3 Heat shock response of *E coli*

systems biology, the univariate statistical analysis and multivariate statistical analysis have been applied for the computational study on the behavior of the complex networks under same steady state (SS) and different steady states (SSs), respectively. The process of state transition is formulated by random processes, whose selection is made according to the stopping time estimated by the martingale measure. The complex network in this study is a kind of nonlinear systems constrained by stochastic components. As emphasized in the methodology of post-complexity systems, the stochastic modeling by random processes is used to estimate the probability of state transition of the complex networks. In the living cell, the signaling pathways can be modeled as a network of networks, i.e., an equivalent extracted meta-network constructed by functional modular networks. Hence the problem of robustness analysis can be transformed into a problem of the estimation of the parameters in the phase space structurally constrained by the "evolution" process of nonlinear dynamics.

The stress signal of the cell is a signal from the environment going irregularly beyond the normal threshold. As a kind of response to stress, heat shock response (HSR) is a cellular function in which the cell can sustain the protein folding when the temperature is higher than the normal range (as illustrated in Fig. 3). We select the heat shock response pathway network of *E. coli* to study the robustness. In heat shock response of *E. coli*, we use the signaling molecule σ^{32} as the indicator of the robustness.

The intrinsic fluctuation is the reason to explain why *DnaK* degenerates to a less stable state while it is still stationary. The robustness of the heat shock response pathway quantitatively described by the dynamical variation of *DnaK* is caused by the intrinsic thermodynamics, which corresponds to the phase transition of the

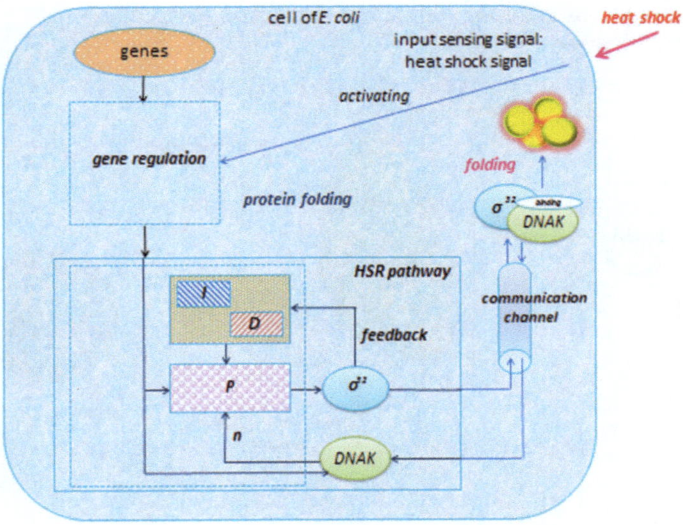

Fig. 4 Heat shock response pathway of *E. coli*

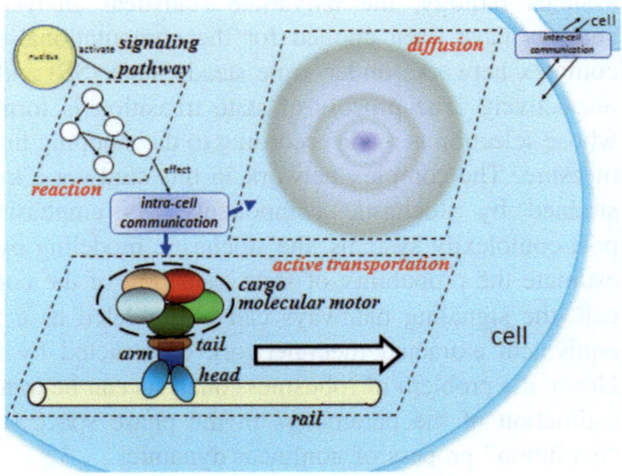

Fig. 5 Molecular signaling mechanism of cell communication

dynamics of the heat shock response pathway (Cf. Figs. 3 and 4, in Fig. 4 the notations of *P*, *I*, and *D* refers to the units of a controller, *n* means that DnaK (i.e., DNAK in Fig. 4) can be assumed to be a kind of noise when univariate statistics is used to estimate the value of σ^{32}). From such a framework of stochastic modeling, both spatial and temporal information about the signaling dynamics of the cellular signaling networks in Figs. 5 and 6 can be measured.

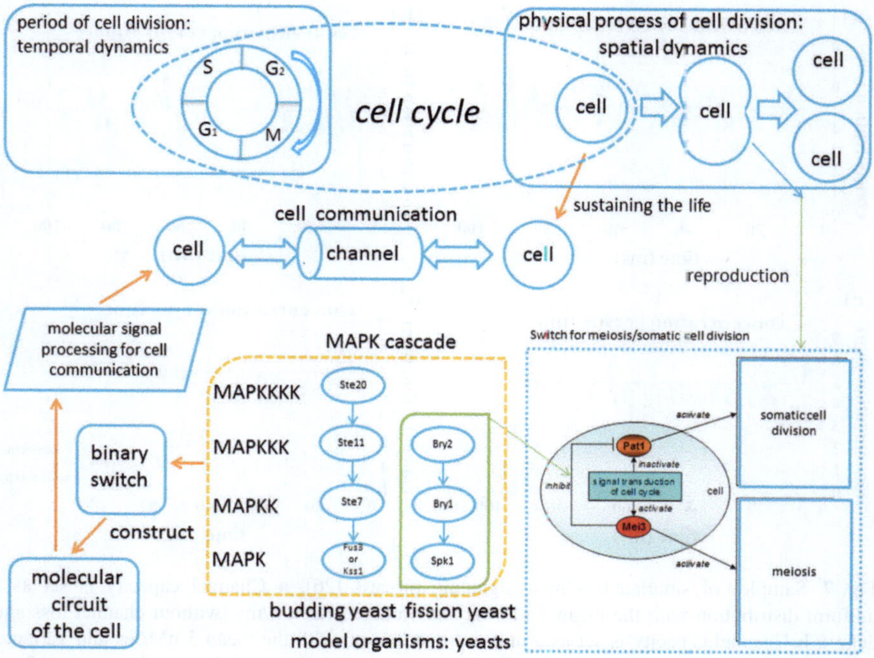

Fig. 6 Temporal-and-spatial dynamics determining the fate of the cell. (For detail on fission yeast, reference [14] is suggested. For detail on budding yeast, reference [27] is suggested.)

2.3 *Computational Analysis of Signaling Pathway Networks in Terms of Communication Networks*

Our computational study on signaling pathway networks using the method of communication engineering consists of three aspects: network architecture, simulation and performance evaluation (Cf. the simulation results given in Fig. 7a–c) in which the crucial factor for understanding the behavior of the signaling mechanism in communication processes is channel capacity [26]. In the simulation, the data structure of a queue is adopted within the nodes to simulate the temporal dynamics where the channel effect is set by probability distribution. The uniform and Gaussian distributions for channel capacity are used in the simulation shown in Fig. 7d. The relative stable signals during certain periods are observed. The effects of channel delay and loss on the signaling processes imply that the channel delay and loss influence the signaling dynamics of signaling pathway networks, although they are not general phenomena. With the temporal and spatial dynamics of signaling pathway networks, a channel unit plays an important role in cellular communication processes.

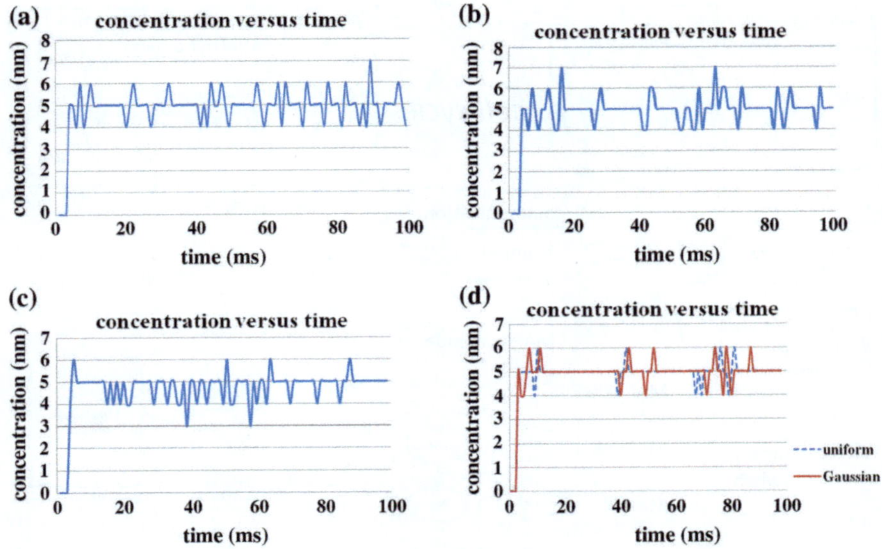

Fig. 7 Samples of simulated cellular signaling process [26]. **a** Channel capacity is set as a uniform distribution with the mean 5 nM/ms and variance 1.2 nM/ms (without channel loss and delay). **b** Channel capacity is set as a uniform distribution with the mean 5 nM/ms and variance 1.2 nM/ms; channel delay is set as a uniform distribution with the mean 1 ms and variance 0.2 ms (without channel loss). **c** Channel capacity is set as a uniform distribution with the mean 5nM/ms and variance 1.2 nM/ms; channel delay is set as a uniform distribution with the mean 1 ms and variance 0.2 ms; channel loss is set as a uniform distribution with the mean 0. 2nM/ms and variance 0.001 nM/ms. **d** Comparison of two cases where channel capacity is set as a uniform distri-bution with the mean 5 nM/ms and variance 1.05 nM/ms and a Gaussian distribution with the mean 5 nM/ms and variance 0.6 nM/ms (without channel loss and delay)

3 Conclusion

In this book chapter, a framework for the study of the quantitative analysis of signaling pathway networks is presented in term of communication networks, in which a communication-network-based method for cellular communication is proposed and discussed from the point of view of simulation and performance evaluation. The simulation results show the proposed method is efficient for the bioinformatics analysis of signaling pathway networks by modelling cell communication process in terms of communication engineering. It is helpful for establishing a mathematics-based "systematical" bioinformatics theory for cell communication and extend it to the whole cell simulation for understanding the principle of communication between brain's cells.

References

1. Bruce A, et al (2008) Molecular biology of the cell, 5th edn. Garland Science, New York
2. The Nobel Prize in Chemistry (2014). http://www.nobelprize.org/nobel_prizes/chemistry/laureates/2014/
3. The Nobel Prize in Physiology or Medicine (2014). http://www.nobelprize.org/nobel_prizes/medicine/laureates/2014/
4. Tyreman MJA, Molloy JE (2003) Molecular motors: nature's nanomachines. IEE Proc Nanobiotechnol 150(3):95–102
5. Zhang QC, Petrey D, Deng L (2012) Structure-based prediction of protein–protein interactions on a genome-wide scale. Nature 490:556–560
6. Grimmett GR, Stirzaker DR (2001) Probability and random processes, 3rd edn. Oxford University Press, Oxford
7. Prahlad V, Morimoto RI (2009) Integrating the stress response: lessons for neurodegenerative diseases from *C. elegans*. Trends Cell Biol 19(2):52–61
8. Cover Story (2012) Nature, vol 489, no 7416
9. Bardin J (2012) Unlocking the brain. Nature 487:24–26
10. Buckner RL, Andrews-Hanna JR, Schacter DL (2008) The brain's default network: anatomy, function, and relevance to disease. Ann NY Acad Sci 1124:1–38
11. Mukamel EA, Babcock H, Zhuang X (2012) Statistical deconvolution for superresolution fluorescence microscopy. Biophys J 102(10):2391–2400
12. http://www.spring8.or.jp/en
13. Breitkreutz A, Choi H, Sharom JR, Boucher L, Neduva V et al (2010) A global protein kinase and phosphatase interaction network in yeast. Science 328(5981):1043–1046
14. Yamamoto M (1996) Regulation of meiosis in fission yeast. Cell Strcut Funct 21(5):431–436
15. Clarke F (2008) Nonsmooth analysis in systems and control theory. In: January 2008, The encyclopedia of complexity and system science. Springer. http://math.univ-lyon1.fr/~clarke/Clarke_Control.pdf
16. Liu J-Q (2010) Molecular informatics of nano-communication based on cells: a brief survey. J Nano Commun Netw 1(2):118–125
17. Oweiss KG (ed) (2010) Statistical signal processing for neuroscience and neurotechnology, burlington. Academic Press, MA
18. Roy R, Hohng S, Ha T (2008) A practical guide to single-molecule FRET. Nat Meth 5:507–516
19. Isalan M (2012) A cell in a computer. Nature 488:40–41
20. http://www2.nict.go.jp/isd/ISDS-contents/english/
21. Liu J-Q, Shimohara K (2007) Biomolecular Computation *for Bionanotechnology*. Artech House, Boston and London
22. Wang Z, Edwards JG, Riley N, Provance DW, Karcher R et al (2008) Myosin Vb mobilizes recycling endosomes and AMPA receptors for postsynaptic plasticity. Cell 135(3):535–548
23. Nakano T, Moore MJ, Wei F, Vasilakos AV (2012) Molecular communication and networking: opportunities and challenges. IEEE Trans Nanobiosci 11(2):135–148
24. Nakano T, Eckford AW, Haraguchi T (2013) Molecular communication. Cambridge University Press, Cambridge
25. Zhang X, Wu L, Cui S. An improved integral inequality to stability analysis of genetic regulatory networks with interval time-varying delays. IEEE/ACM Trans Comput Biol Bioinform Early access at IEEE Xplore®
26. Liu J-Q (2011–07) On communication mechanism of signaling pathway networks. IEICE Technical Report, MBE2011-37, pp 91–94
27. Elion EA, Qi M, Chen W (2005) Signaling specificity in yeast. Science 307(5710):687–688

Quantifying Robustness in Biological Networks Using NS-2

Bhanu K. Kamapantula, Ahmed F. Abdelzaher, Michael Mayo, Edward J. Perkins, Sajal K. Das and Preetam Ghosh

Abstract Biological networks are known to be robust despite signal disruptions such as gene failures and perturbations. Extensive research is currently under way to explore biological networks and identify the underlying principles of their robustness. Structural properties such as power-law degree distribution and motif abundance have been attributed for robust performance of biological networks. Yet, little has been done so far to quantify such biological robustness. We propose a platform to quantify biological robustness using network simulator (NS-2) by careful mapping of biological properties at the gene level to that of wireless sensor networks derived using the topology of gene regulatory networks found in different organisms. A Support Vector Machine (SVM) learning model is used to measure the correlation of packet transmission rates in such sensor networks. These sensor networks contain important topological features of the underlying biological network, such as motif abundance, node/gene coverage, and transcription-factor network density, which we use to map the SVM features. Finally, a case study is presented to evaluate the NS-2 performance of two gene regulatory networks, obtained from the bacterium *Escherichia coli* and the baker's yeast *Sacharomyces cerevisiae*.

B.K. Kamapantula (✉) · A.F. Abdelzaher · P. Ghosh
Virginia Commonwealth University, Richmond, VA, USA
e-mail: kamapantulbk@mymail.vcu.edu

A.F. Abdelzaher
e-mail: abdelzaheraf@mymail.vcu.edu

P. Ghosh
e-mail: pghosh@vcu.edu

M. Mayo · E.J. Perkins
Environmental Laboratory, US Army Engineer Research and Development Center, Vicksburg, MS, USA
e-mail: Michael.L.Mayo@usace.army.mil

E.J. Perkins
e-mail: Edward.J.Perkins@usace.army.mil

S.K. Das
Missouri University of Science & Technology, Rolla, MO, USA
e-mail: sdas@mst.edu

J. Suzuki et al. (eds.), *Modeling, Methodologies and Tools for Molecular and Nano-scale Communications*, Modeling and Optimization in Science and Technologies 9, DOI 10.1007/978-3-319-50688-3_12

Studies have shown that the current *Homo Sapiens* have an estimated 250,000 years of evolution. However, the exact functioning of a Human body eludes us till date. Human body is an intricate system of complex mechanisms that continues to interest scientists including biologists, computational and medical researchers. To address this, an ambitious project called *The Human Genome Project* (HGP) was announced by the US Department of Energy in 1985. Its mission was to identify all the genes in the human genome. HGP was completed in the year 2003 [5]. However, the genome of every human being is unique and the data is still being refined to date [16]. Intensive research in structural genomics and their functional significance followed the completion of HGP creating the new field of Systems Biology, wherein the goal is to study the behavior and dynamics of complex biological systems.

Human body is made up of trillions of cells [23]. Each cell is comprised of genes in which information is encoded. The function of a cell varies depending on the level of gene expression that is regulated by a set of transcription factors. Such interacting genes and transcription factors can be represented as a Gene Regulatory Network (GRN). GRNs have been extensively explored by researchers as it is believed that they hold the key to unravel the mystery behind the working of a human body. Although a major portion of gene-gene interactions is still unknown for higher order organisms, the scientific community has recently focused on simulating the dynamics of GRNs from lower order organisms. In such simulations, it is essential to consider the topological characteristics of GRNs that contribute to their robustness in information transport.

Our contribution lies at this realm of GRNs and in-silico experiments. We propose a framework to quantify biological robustness using NS-2, a network simulator. NS-2 has been primarily used to simulate different computer networks including Wireless Sensor Networks (WSNs). Information in this chapter is categorized as follows. Section 1 presents a discussion on the state of computational modelling of biological systems. Section 2 presents similarities and differences between GRNs and WSNs thereby enabling a way to map a GRN to a WSN for simulation. Section 3 details the simulation setup including the parameters used and the assumptions. Section 3 also explains the network generation procedure and the sink selection strategies necessary for a network to be simulated. A case study is presented in Sect. 4 to identify the suitable model organism, between the bacterium *Escherichia coli* (abbreviated *E. coli*) and the baker's yeast *Saccharomyces cerevisiae*, for mapping purposes. Finally, future research directions are presented in Sect. 5.

1 Computational Modelling

Ordinary differential equations (ODE) based computational models of biological systems, termed reaction rate equations or mass action kinetics, has received much attention [26]. Here, a homogeneous biological system is represented as a group of biochemical reactions and its dynamics are explored in the continuous-deterministic realm. However, ODE-based models are limited to study the underlying stochastic present in many biological processes such as gene expression and protein synthesis [6]. The limitations of ODE-based models for biological systems are detailed in [26].

Ghosh et al. [7] describes the advantages of using discrete event simulators for modeling biological systems. A fundamental challenge in computational systems biology [12] is the simplification of the biological system complexity without losing the ensemble dynamic behavior. In the system engineering view of complex processes [29], the key notion is to abstract the complexity of the system as a set of discrete time and space variables (random variables), which capture the behavior of the system in time. The entire system is a collection of functional blocks or modules, which are driven by a set of events, where an event defines a large number of micro level state transitions between a set of state variables accomplished within the event execution time. The underlying assumption driving this abstraction is the segregation of the complete state space into such disjoint sets of independent events which can be executed simultaneously without any interaction. The application of this technique in large complex communication networks has demonstrated the accuracy of the approach for the first and higher order dynamics of the system within the limits of input data and state partitioning algorithms [30]. For example, discrete event based system modeling has been effectively applied for designing routers, the key components responsible for routing traffic through the Internet. Discrete event based simulation techniques have also been used in a wide variety of manufacturing processes and studying the system dynamics of complex industrial processes.

Researchers have also tried to adapt existing simulation platforms to model molecular communication. NanoNS is one such [8] simulation framework to model molecular communications. The framework is built over NS-2 software and uses a diffusive molecular communication channel. Researchers in [8] present an extensive review of communication models in nanoscale networks and out of three possible molecular communications, namely diffusive, motor-based and gap junction-based, their work is focused on diffusive-based molecular communication. As an extension of this work, researchers presented a case to build models for a variety of molecular communication channels, intra-body molecular nanonetworks and the network of such intra-body nanonetworks in [17]. This work comprehensively showcases the significance of modelling nanonetworks. Efforts are currently underway to simulate wireless nano sensor networks using NS-3 software (*next version of NS-2*) [24]. In this work, wireless nano sensor networks are modelled using electromagnetic communication instead of molecular communication as mentioned above. As it is evident by now, the challenges in achieving a simulation framework for communications in molecular networks are multifold [21]. Our core goal here is to identify ground rules for GRN-based robustness—the ability of a biological state to persist despite component errors—by setting up a generic NS-2 simulation platform, rather than developing more detailed molecular communication channels.

Network simulator, NS-2 (*NS-2*), is a discrete event simulator widely used for studying wireless networks. NS-2 has been used by researchers to model communication in wireless networks and embedded devices. This simulator continues to evolve with the active support of the research community. Taking a step forward, we have used NS-2 as an in-silico platform for quantifying the robustness of biological networks. Specifically, since the primary objective of a wireless sensor network is information transport to specific sink nodes, and because they operate under similar noisy and error prone conditions as biological networks, we define robustness of biological networks as the ability for each node in the network to deliver packets with minimal packet loss. Before envisioning a model for any time-varying functional biological system, it is important to illustrate the preliminary model for the biological system in NS-2. While exclusive simulators to model a molecular network are not present currently, existing simulators can be adjusted to model the desired network. It should be noted that this might not be the perfect approach, but the opportunity to explore the qualitative and quantitative dynamics of molecular networks is not lost. Scenarios are presented below whenever applicable to demonstrate the use of NS-2 to quantify biological robustness.

2 Mapping GRNs to WSNs

Transmission inconsistencies frequently plague WSNs where they suffer from signal disruptions due to sensor failure or from the absence of routing protocols that are sufficiently insensitive to local as well as global network conditions. In a WSN, nodes sense, process and communicate information with each other. Structurally, a GRN can be related to a WSN where every gene or transcription factor is a sensor. Signal transmission within a GRN can be considered as packet transmission in a WSN. The fundamental assumptions in modelling a bio-inspired WSN are [10]:

1. GRN node structure is preserved in WSN.
2. Interactions among nodes in WSN are based on the existing connections in the GRN.

The physical signaling structure of sensors within the WSN must be adapted to reflect the communication between genes in the GRN. If gene $G1$ up-regulates $G2$, then the equivalent interaction in the WSN is that sensor $S1$ sends a packet to $S2$ according to specific probability distribution defined by gene-gene interactions. For homogeneous sensor nodes, each up-regulation edge in a GRN is replaced by a bi-directional edge; if we allow sensor $S1$ to send a packet to $S2$, then $S2$ should also be able to send a packet to $S1$. For heterogeneous sensor nodes, however, it is not necessary that both $S1$ and $S2$ possess the same transmission radii, giving a directed edge from $S1$ to $S2$ and not vice versa.

We recognize that WSNs conceptually operate under noisy and/or adverse conditions similar to the stochastic cellular environment encountered by GRNs. We hypothesize that if it is possible to exploit the simulation platform used for WSNs, namely NS-2, to assess the signal transmission robustness in GRNs, then any observed robust qualities can be explained by fundamental biological processes, such as transcription. The process where signals from nearby neighbors in the form of transcription factors stimulate/inhibit other genes by generating mRNA molecules is transcription. Thus, GRN nodes communicate with one another by sending signals (transcription factors), which are in return processed into output signals (mRNAs). This process is similar to WSNs where sensors receive packets from its neighbors with packet forwarding instructions to other destination nodes. As a result, any node in a network (GRN or WSN) can affect the decision of other nodes and hence the overall network performance.

Here, we considered the transcriptional regulatory network (TRN) of the bacterium *E. coli* to generate the sample GRN graphs. Such TRNs bear the actual topology of the GRNs with any gene-gene and gene to transcription factor edges deleted. Thus, in such TRNs, a single transcription factor can regulate other transcription factors and genes, while genes do not directly regulate other nodes. Note that our earlier work on WSNs derived from GRN topologies actually considered the TRNs from *E. coli* which were shown to achieve high packet transmission efficiency [10]; hence such TRNs exhibit the desired biological robustness measures that we seek to model here. The transcription factor molecules having half-lives $T_{1/2} = \ln 2/k$ [2], where k represents the decay rate constant, are subject to degradation if held at the transcriptional regulation queue. Similarly in the case of WSNs, packets are forwarded from source nodes to sink nodes using multiple hops and can be dropped at intermediate nodes if they exceed the queue length. Hence, genes can be considered as sink nodes and transcription factors as the source nodes. On that account, we describe our measure of robustness in WSNs that adopt the GRN topologies as the ability for each node in the network to deliver information to their local sinks with minimal packet loss.

3 NS-2 Simulation Setup

Consider a biological network topology derived from a well studied organism, *E. coli*. Sub-networks that are extracted from *E. coli* comprise of interactions among genes. Let us call this extracted network a Gene Regulatory Network (GRN). Such GRNs comprise two classes of nodes: transcription factors and genes. A transcription factor either up-regulates or down-regulates one (or more) gene. The packet transmission rates are assumed to be identical in NS-2, for all the non-sink nodes; however, in a real biological setting, such rates are directly proportional to the rate constants associated with every edge in the network along with the concentration of the molecules associated with a node. This however creates a roadblock for existing biological network simulators as each of these rate constants need to be experimentally validated which is not currently feasible for the different sample networks generated in this work. The simulation also assumes all packets transmitted to be identical in type and size which correspond to similar signaling molecules affecting the different nodes in the GRN in the context of biological robustness.

Queue limit in NS-2 is useful to limit the number of packets that can be queued at a node. Queue limit in the corresponding GRN represents the half-life of each signal sent from one node to another node. Although this is another approximation in the simulation set-up, it is impossible to characterize all such signaling molecules accurately in the different extracted GRNs. In summary, our proposed NS-2 set-up makes broad assumptions for the pertinent details of biological network signaling but we feel that this is indeed necessary for studying the qualitative dynamics of many sample GRNs wherein such details are not known at length.

Traditionally, robustness of biological networks has been measured by its static graph theoretic characteristics such as network diameter, average shortest path [22], network efficiency [15] amongst others. A network with negligible change in its diameter is considered to be robust when it loses node(s) after an attack. Similarly, negligible change in average shortest path and network efficiency under network perturbations related to temporal fluctuations in the node and/or link availability is attributed to robust networks. Packet receival rate is the ratio of the number of packets received in the network to the number of packets sent. Higher the packet receival rate of a GRN, higher its robustness. Randomly generated WSNs and GRN-derived networks are compared with respect to the packet receival rate. This section discusses the methods used for random network generation to be used as a wireless sensor network. In addition, approaches used to identify the sink node in the network are detailed. A new algorithm for biological network generation is presented in Sect. 3.2.1.

3.1 Network Generation

A script written in the Python programming language [25] is used to generate networks modelled as WSNs. Here, two different nodes within the network are chosen at random, and a link is established between them with probability p.[1] Networks with 100, 150, 200, 250 and 300 nodes were generated for demonstration purposes as representing "medium" sized sensor networks. 25 networks of each size (100, 150, 200, 250 and 300 nodes) are considered to illustrate the sink node selection approach. Networks of a certain size are spread over an area with specific node transmission range. For example, 25 different networks of size 150 nodes are spread over $3{\cdot}6 \times 10^5 \text{m}^2$ (with $x = 600$ m and $y = 600$ m) with a node transmission range of 85 m. Node range for a network has been assigned based on the work by [9]. Similarly, networks of size 200 are spread over area of 4.9×10^5 m^2 (with $x = 700$ m and $y = 700$ m) with a node transmission range of 90 m and networks of size 250 are spread over 8.1×10^5 m^2 (with $x = 900$ m and $y = 900$ m) with a node transmission range of 90 m. Networks of size 300 are spread over an area of 10^6 m^2 (with $x = 1000$ m and $y = 1000$ m) with a node transmission range of 110 m. Few assumptions are made for simplicity. The directionality of the links between the nodes is ignored. Self-edges,[2] edges with same source and destination nodes, are removed from the network. Nodes in model

[1] $p(K)$ is the probability to find a node of degree K in a network that follows the power law distribution $p(K) \sim K^{-\gamma}$.

[2] In a biological context, self-edges for a gene refers to auto-regulation of expression.

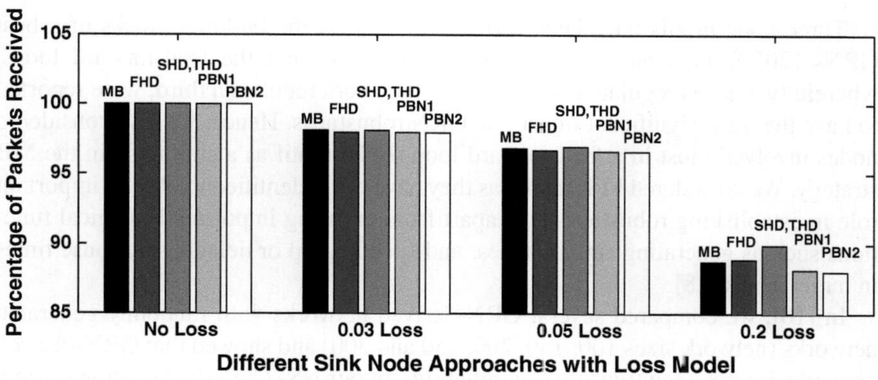

Fig. 1 Sink node selection and respective packet receival rates for different loss models—GRN of 20 nodes [10]

organisms such as *E. coli* and S. cerevisiae self, up- or down-, regulate themselves. However, we ignore self-edges in this case of WSN simulation. In order to compare similar entities, only networks with same number of nodes and edges are considered for comparison. All 25 networks of the same size have exact number of edges. Each network generated using this approach is considered to be a Random Wireless Sensor Network (RWSN).

3.2 Sink Selection Strategy

Sink node selection strategy is critical for optimal GRN performance. In [11], we listed three sink selection strategies: (a) Highest Degree (HD), (b) Highest Coverage (HC) and (c) Motif-based (MB) and identified HD strategy as the best approach to provide higher robustness for NS-2 based simulation of GRNs. Nodes with highest degree are selected as a sink node in the HD strategy. Node involved in any three-node motif is selected as a sink node in the HC strategy. Figure 1 shows the comparison of sink selection strategies for a GRN-derived of twenty nodes [10]. In this figure, FHD stands for the node with First Highest Degree, SHD stands for the node with Second Highest Degree, THD stands for the node with Third Highest Degree and PBN stands for node identified with Probabilistic Boolean Network. A PBN is a formalism where set of functions define the expression value of genes in the network. The node with the highest expression is selected as sink node. For detailed information on sink selection strategies including PBN, refer to [10]. Intuitively, using FHD node as a sink makes sense since the node is regulated, in a biological context, by several other regulators and are critical for important biological functionalities. Such nodes also act as hubs in a network.

Three-node motifs have been earlier identified as the building blocks of robust GRNs [20] from a purely topological perspective, and the feed-forward loops, wherein two genes regulate each other and they both regulate a third, were reported to have the most significant impact on GRN robustness. Hence, we also considered nodes involved most in a feed-forward loop (FFL) motif as a sink node in the MB strategy. We considered FFL motifs as they have been identified to play an important role in establishing robustness [13] apart from ensuring important biological functions such as generating signal pulses, and speeding up or delaying response times in target genes [18].

In [10], we compared several GRN-derived networks with randomly generated networks (network sizes 100, 150, 200, 250 and 300) and showed that GRN-derived networks improve the transmission reliability in our NS-2 based simulation setting. The procedure for generating random networks is described in Sect. 3.1. Figures 2 and 3 present a comparison for best, mean and worst performing RWSNs and GRN

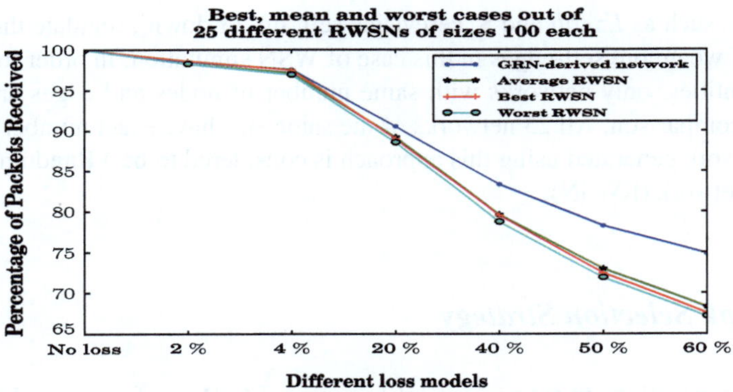

Fig. 2 Comparison of best, mean and worst (out of 25 networks) performing RWSNs to GRN—network size 100 [10]

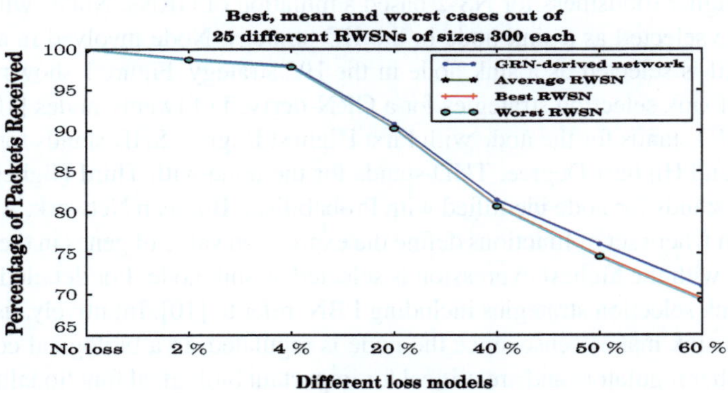

Fig. 3 Comparison of best, mean and worst (out of 25 networks) performing RWSNs to GRN—network size 300 [10]

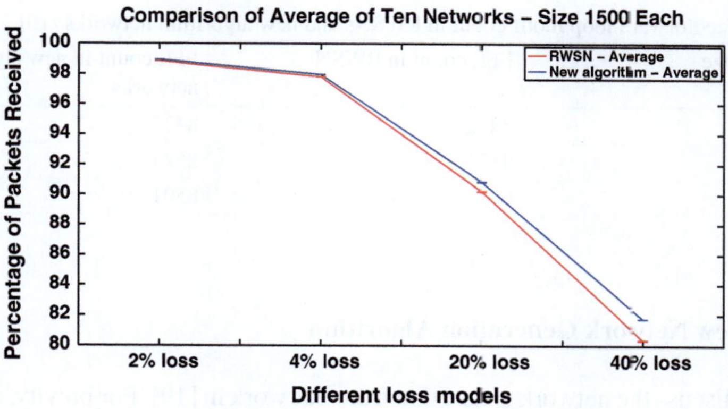

Fig. 4 Comparison of average of ten networks between new algorithm and RWSN—network size 1500

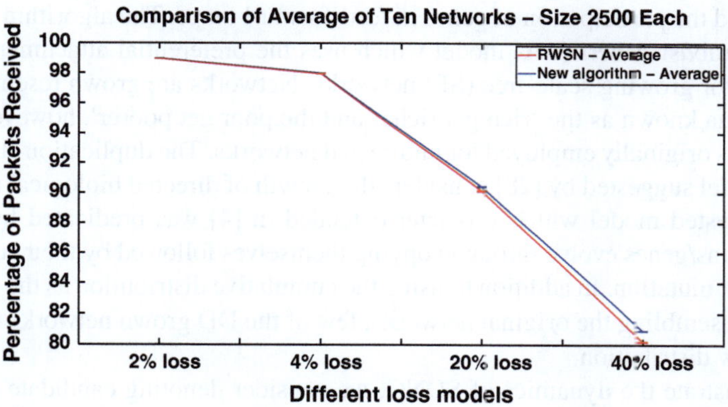

Fig. 5 Comparison of average of ten networks between new algorithm and RWSN—network size 2500

of network sizes 100 and 300 respectively. For this experiment, a total of 25 RWSNs are considered and three cases are presented. Comparisons are also made for large-scale predicted GRNs (network size 1500, 1750, 2000, 2250 and 2500). The performance of GRN versus RWSNs in large scale networks (network size 1500 and 2500) is presented in Figs. 4 and 5. The graphs for network sizes 1750, 2000 and 2250 are not reported since they follow similar trend as networks of size 1500 and 2500. This might be possible due to the presence of higher number of FFLs in GRN-derived networks as compared to randomly generated networks. The abundance of FFL motifs in random networks and networks derived from new algorithm[3] is presented in Table 1. The counts reported in the table are averaged, and approximated to nearest decimal, across ten different networks of a particular type.

[3]Algorithm proposed by [19] is explained in Sect. 3.2.1.

Table 1 Feedforward loop motif count in RWSNs and new algorithm networks [10]

Network size	FFL count in RWSN	FFL count in new algorithm networks
1500	3972	8429
2000	4125	8524
2500	6742	8591

3.2.1 New Network Generation Algorithm

Here we discuss the network generation from our work in [19]. For brevity, the Scale-free Directed Network Generator is referred to as SDNG. The algorithm can be utilized to expand existing networks as well as generating directed networks emulating the different distributions of *E. coli*, namely in-degree, out-degree, cumulative degree and the participation of genes in feed-forward loops. The algorithm is similar to the Barabắsi-Albert (BA) model which uses the preferential attachment mechanism [1] for growing scale-free (SF) networks. Networks are grown resembling the phenomena known as the 'rich get richer and the poor get poorer', however the BA model was originally employed for undirected networks. The duplication-divergence (DD) model suggested by [28] considers the growth of directed biological networks. The suggested model which was later extended in [4] was predicated by the fact that proteins/genes evolve through copying themselves followed by their subsequent infrequent mutation. In addition to using the cumulative distribution as the sole measure for resembling the original networks, few of the DD grown networks retained a power-law distribution.

To illustrate the dynamics of SDNG, we consider denoting candidate nodes for preferential attachments in an existing network of size n with subscript i, wherein K_i and R_i label the out- and in-degrees respectively. The probability for a candidate node to be connected to a node foreign to the existing network with an edge directed from the candidate node to the foreign node is given by $A(K_i, R_i)$. The probability that a link is drawn from the foreign node to the candidate node is given by $B(K_i, R_i)$. Each probability is normalized against all nodes of the existing network to form attachment kernels [14], and their formulas are listed in Table 2.

For this particular work, we considered the power-law attachment kernel for calculating the edge probabilities. Starting with a fully connected eight node network, a candidate node is picked at random with equi-probability. Next, a random number d is selected with equi-probablity from the interval $d \in (0, 1)$. An edge is drawn from the candidate node to the foreign node if $d \leq A(K_i, R_i)$. This process is then repeat for an edge drawn out of the foreign node to the candidate node, provided the probability satisfies $d \leq B(K_i, R_i)$. The above steps are then reiterated $m_i - 1$ times, wherein m_i is an another number selected at random from an exponential probability distribution $\rho(m_i) = (f^{\frac{1}{1-m_0}} - 1)f^{-m_i/(1-m_0)}$. The decay of this distribution resembles the degree distribution of *E. coli*. Here, we considered values of $f = \frac{1}{4}$ and $m_0 = 2$.

Table 2 Attachment kernels

Functional type	Attachment kernels	
	$A(K_i, R_i)$	$B(K_i, R_i)$
Linear	$\frac{K_i}{\sum_{i=1}^{n} K_i}$	$\frac{R_i}{\sum_{i=1}^{n} R_i}$
Power-law	$\frac{K_i^{0.8}}{\sum_{i=1}^{n} K_i^{0.8}}$	$\frac{R_i^{0.8}}{\sum_{i=1}^{n} R_i^{0.8}}$
Sigmoid	$\frac{K_i}{\sum_{i=1}^{n} (K_i+R_i)}$	$\frac{R_i}{\sum_{i=1}^{n} (K_i+R_i)}$

3.3 SVM Validation

While the network evaluations presented in Figs. 2 and 3 establish the significance of GRN-derived networks, only one sink operates in those networks which is not the case in functional GRNs. To address this, we used multiple sink nodes to model GRN communication. An Support Vector Machine (SVM) model, built using LibSVM [3], is then used to investigate the relative efficiency of packet receival rates based on topological metrics such as network density, genes coverage, transcription factor network density, motif abundance and genes percentage, defined below.

For this, GRNs of varying sizes, $100 < n < 500$ were used, where n is the number of nodes in the GRN. Transmission is considered from source nodes (similar to transcription factors) to sink nodes (similar to gene nodes). 410 out of the 490 networks are used to train the learning model and remaining networks are used to test the model. The directionality of the links between the nodes is considered. The topological metrics used in the learning model are briefly described below.

3.3.1 Network Density (ND)

ND is a ratio of the number of edges present in the network to the total number of edges possible in the network.

3.3.2 Genes Coverage (GC)

GC is the summation of the ratios of in-degree of each sink node to the ratio of source nodes having a path to that particular sink node.

3.3.3 Transcription Factor Network Density (TND)

TND is the ratio of the number of edges that transcription factor nodes participate to the total number of edges in the network.

3.3.4 Motif Abundance

Motif abundance is the ratio of abundances of FFL (R^{FFL}) and bifan (R^{BF}) motifs that relate to the number of nodes.

3.3.5 Genes Percentage (GP)

GP is the ratio of number of gene nodes to the total number of nodes in the network.

3.4 Contributions of Topological Metrics to GRN Robustness

These topological metrics are then used to construct the SVM learning model. Cross validation is used in the training stage; test data is then used to predict the robustness of the networks. The relative importance of the features used in the model in the decreasing order is as follows: ND, R^{BF}, GP, TND, GC and R^{FFL}. Figure 6a shows the weight w_i of features divided by the maximum weight (w_{ND}): $|wi/w_{ND}|$. Figure 6b shows same ratio but the directions of the weights are considered. It should be noted that a GRN is more communicative when it is sparse implying low ND and high R^{BF} as shown in Fig. 6b.

4 Case Study: Comparison of Derived Networks from E. Coli and Yeast

We have demonstrated the performance of NS-2 as a platform to quantify robustness in biological networks. In order to exploit the principles of a biological network, it is crucial to evaluate the model organisms. For this purpose, we compare net-

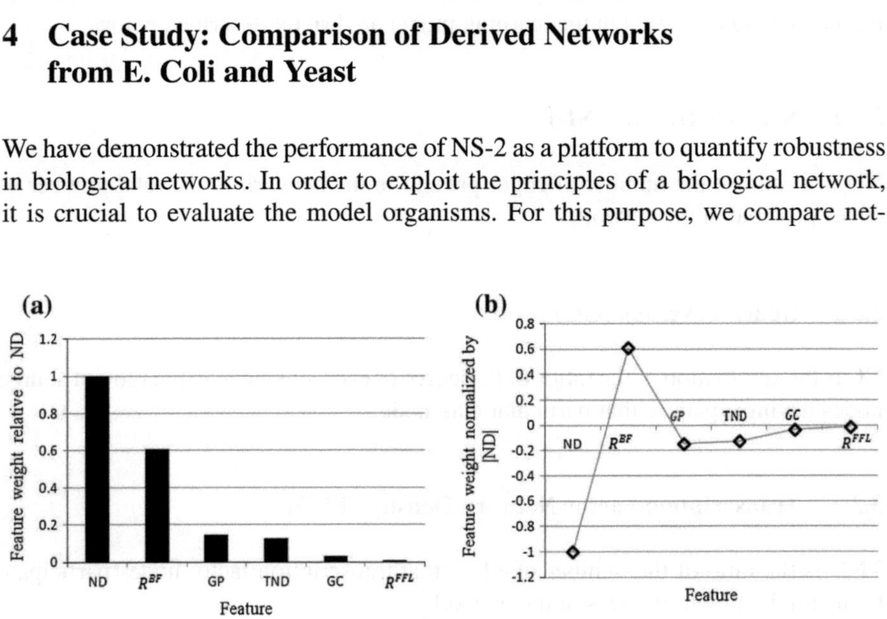

Fig. 6 **a** Relative importance of the feature weights, **b** Relative importance of feature directions

works derived from two well studied model organisms, *E. coli* and S. cerevisiae, of sizes consisting 100, 200, 300, 400 and 500 nodes using GeneNetWeaver software [27]. One hundred networks of each size are generated and NS-2 simulations are performed on each of these networks. As comparing the average performance of all networks may not distinguish the performance of the derived networks properly, we compared the best perfoming, average performing and least performing networks. The directionality of the links between the nodes is ignored. The simulation parameters are as follows:

1. Bandwidth = 1Mb
2. Delay = 1.0 ms
3. Queue limit = 5
4. Packet size = 900 bytes

Figure 7 shows the best performing derived networks from *E. coli* and S. cerevisiae for network sizes: 100, 200, 300, 400, and 500 (nodes) w.r.t. 20, 35 and 50% loss. While the performance of S. cerevisiae derived networks is consistently higher for 500 node network under 20 and 35 and 50% loss, *E. coli* derived networks perform better, in almost all cases except for 200 network size at 20% loss, for networks of size 100, 200, 300 and 400.

Figure 8 shows the mean performing derived networks from *E coli* and S. cerevisiae for network sizes: 100, 200, 300, 400, and 500 (nodes) w.r.t 20, 35 and 50% loss. It can be clearly observed from the figure that the performance of *E. coli* derived network is better at 20 and 35% loss and S. cerevisiae derived network performs better for higher loss percentage (50%). The difference in performance is ~ 0.51 at 20 %

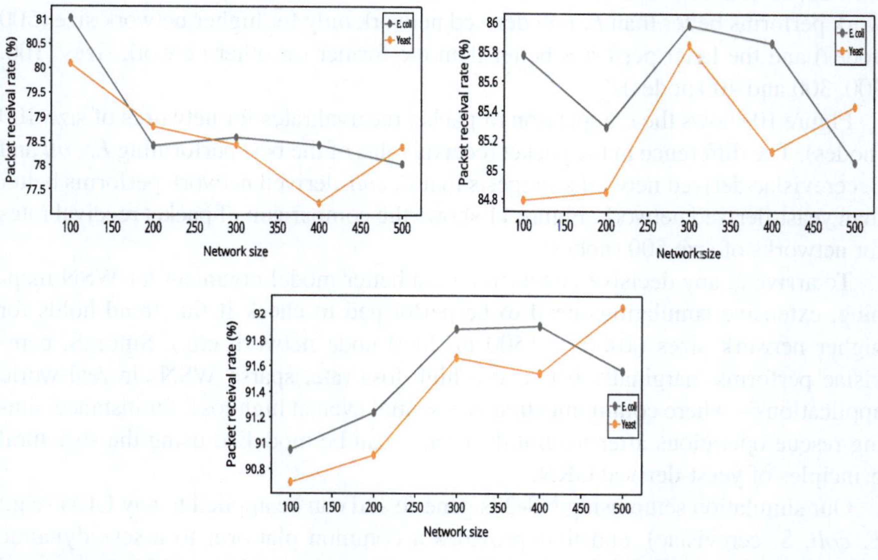

Fig. 7 Comparison of best performing networks derived from E. coli and Yeast—20, 35 and 50% loss

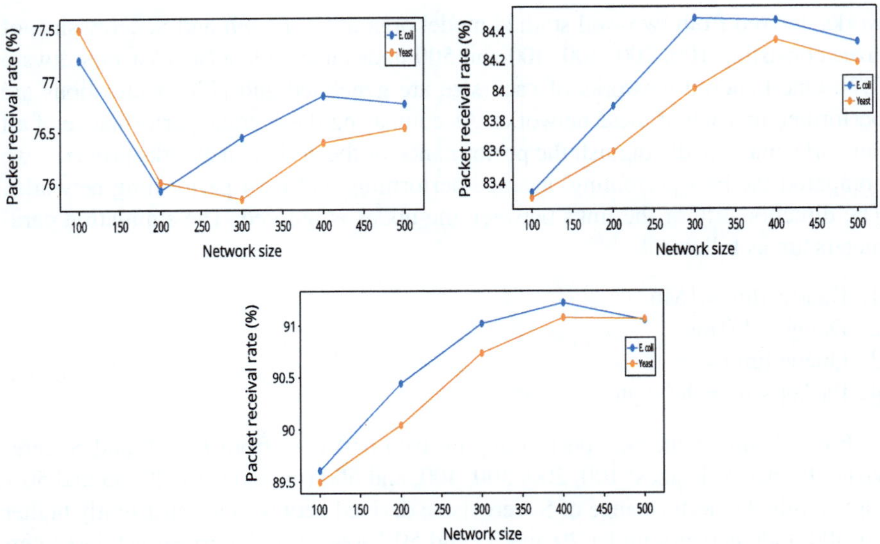

Fig. 8 Comparison of mean performing networks derived from E. coli and Yeast–20, 35 and 50% loss

loss (*E, coli*), ∼ 0.38 at 35 % loss (*E. coli*), ∼ 0.359 for 50% loss (*Yeast*). It appears that S. cerevisiae derived network performs better than *E. coli* derived network at higher loss percentage. Similarly, Fig. 9 shows the worst performing *E. coli* and S. cerevisiae derived networks for network sizes: 100, 200, 300, 400, and 500 (nodes) w.r.t 20, 35 and 50% loss. It can also be noticed here that S. cerevisiae derived network performs better than *E. coli* derived network only for higher network size (500 nodes) and the latter performs better than the former for other network sizes (100, 200, 300 and 400 nodes).

Figure 10 shows the comparison of packet receival rates for networks of size 100 (nodes). The difference in the packet receival rates of the best performing *E. coli* and S. cerevisiae derived networks suggests that *E. coli*-derived network performs better than yeast-derived network. Figure 11 shows the comparison of packet receival rates for networks of size 500 (nodes).

To arrive at any decisive conclusion on a better model organism for WSN mapping, extensive simulations need to be performed to check if this trend holds for higher network sizes (1000 or 1500 or 2000 node network etc.). Since S. cerevisiae performs marginally better at a high loss rate, sparse WSNs in real-world applications—where communication is essential even at high loss, for instance, during rescue operations after natural disasters—can be modelled using the structural principles of yeast-derived GRN.

Our simulation setup using NS-2 is generic and can be applied to any GRN (e.g.: *E. coli*, S. cerevisiae), and thus provides a common platform to assess dynamic robustness of biological networks. This also allows to sample several extracted and

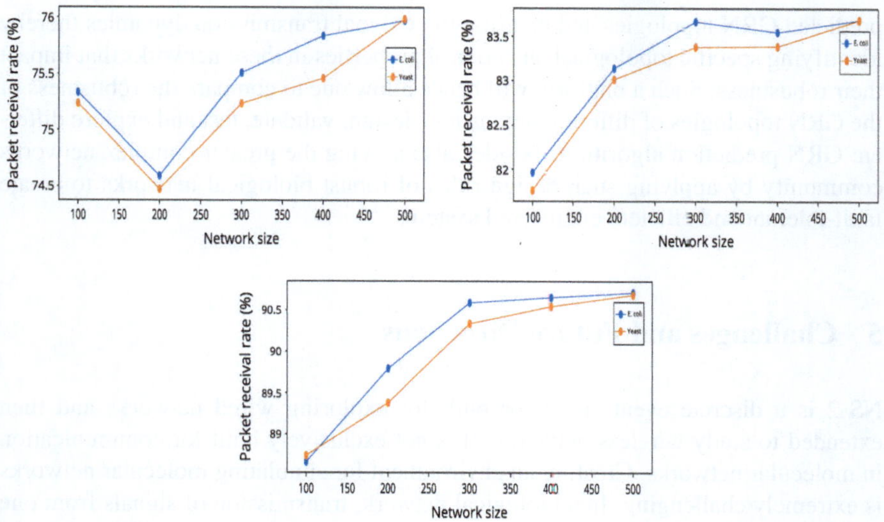

Fig. 9 Comparison of worst performing networks derived from E. coli and Yeast—20, 35 and 50% loss

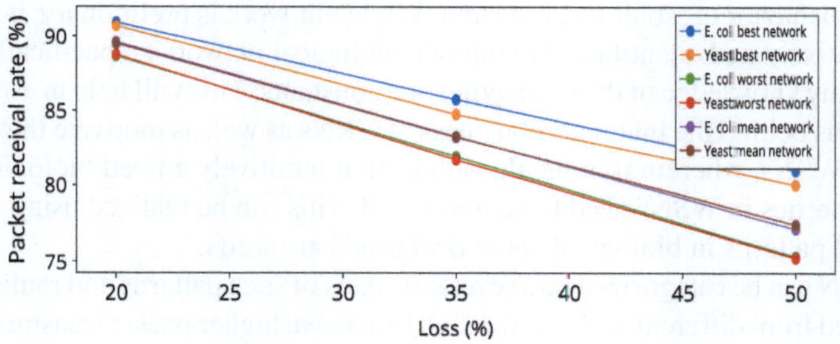

Fig. 10 Comparison of 100 node networks derived from E.coli and Yeast respectively—20%, 35%, 50% loss

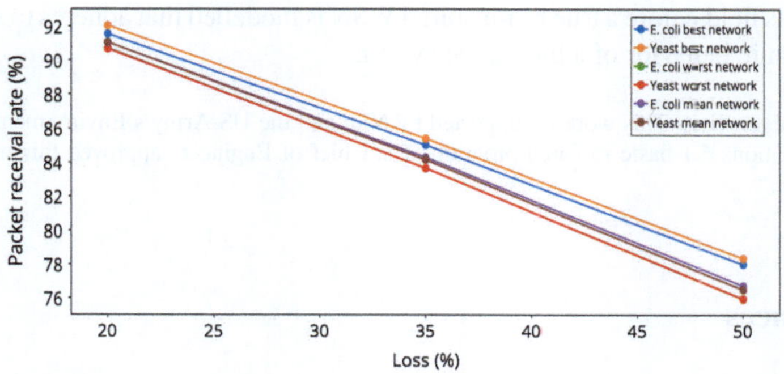

Fig. 11 Comparison of 500 node networks derived from E.coli and Yeast respectively—20%, 35%, 50% loss

predicted GRN topologies and measure their signal transmission dynamics thereby identifying specific topological and control properties in these networks that impact their robustness. Such a platform will hence allow one to compare the robustness of the GRN topologies of different organisms, design, validate, test and explore different GRN prediction algorithms besides also serving the greater complex networks community by applying such design rules of robust biological networks to create fault-tolerant and efficient engineered systems.

5 Challenges and Future Directions

NS-2 is a discrete event simulator built for exploring wired networks and then extended to study wireless networks. It is not exclusively built for communication in molecular networks. Creating an environment for simulating molecular networks is extremely challenging. In a biological network, transmission of signals from one node (transcription factor) to another (gene/transcription factor) occurs at a rate that has not been determined yet. Active effort by researchers is focused on estimating such rate constants. Determining the rate constants is critical for modelling the dynamic behavior of a biological system. While our work is preliminary, it allows us to qualitatively and quantitatively simulate biological networks (specifically GRNs) without any knowledge of the underlying rate constants. This will help in establishing the reasons behind the inherent robustness of GRNs as well as motivate the design of efficient WSNs, wherein routing algorithms that intuitively embed biological structural properties in WSNs need to be developed. This can be realized using repeating structural patterns in biological networks termed as motifs.

A WSN can be categorized into several pockets of such patterns and routing can be introduced from different nodes to the sink to achieve higher packet transmission efficiency. Adaptive routing mechanisms can be imagined to improve WSN efficiency. Bandwidth limitations on edges and nodes in a regulatory network need to be studied before bandwidth based studies can be carried out in WSNs. Much needs to be realized in this field before a true bio-inspired WSN is modelled that adheres to structural and dynamic behavior of a biological system.

Acknowledgements This work is supported by NSF and the US Army's Environmental Quality and Installations 6.1 basic research program. The Chief of Engineers approved this material for publication.

References

1. Barabáasi, A-L, Albert (1999) Emergence of scaling in random networks. In: Science 286.5439, pp 509–512
2. Belle A, Tanay A, Bitincka L, Shamir R, OShea EK (2006) Quantification of protein half-lives in the budding yeast proteome. In: Proceedings of the National Academy of Sciences 103.35,

pp 13004–13009. doi:10.1073/pnas.0605420103. eprint: http://www.pnas.org/content/103/35/13004.full.pdf+html. url: http://www.pnas.org/content/103/35/13004.abstract
3. Chang Chih-Chung, Lin Chih-Jen (2011) LIBSVM: a library for support vector machines. ACM Trans Intell Syst Technol (TIST) 2(3):27
4. Chung, F, Lu L, Dewey TG, Galas DJ (2003) Duplication models for biological networks. J Comput Biol 10(5):677–687
5. Collins, FS, Morgan M, Patrinos A (2003) The human genome project: lessons from large-scale biology. Science 300(5617):286–290
6. Ghosh Preetam, Ghosh Samik, Basu Kalyan, Das Sajal K, Zhang Chaoyang (2010) Discrete di usion models to study the e ects of Mg2+ con- centration on the PhoPQ signal transduction system. BMC Genom 11(Suppl 3):S3
7. Ghosh S, Ghosh P, Basu K, Das SK, S Daefler S (2011) A discrete event based stochastic simulation platform for in silico study of molecular-level cellular dynamics. J Biotechnol Biomater 6:2
8. Gul E, Atakan B, Akan OB (2010) NanoNS: a nanoscale network simulator framework for molecular communications. Nano Commun Netw 1(2):138–156
9. Han B, Leblet J, Simon G (2009) Query range problem in wireless sensor networks. Commun Lett IEEE 13(1):55–57. doi:10.1109/LCOMM.2009.081546. Institute, Information-Sciences. NS-2. http://isi.edu.nsnam/ns
10. Kamapantula BK, Abdelzaher A, Ghosh P, Mayo M, Perkins EJ, Das SK (2012a) Leveraging the robustness of genetic networks: a case study on bio-inspired wireless sensor network topologies. J Amb Intell Hum Comput 1–17
11. Kamapantula BK, Abdelzaher A, Ghosh P, Mayo M, Perkins E, Das SK (2012b) Performance of wireless sensor topologies inspired by E. coli genetic networks. In: 2012 IEEE International conference on Pervasive Computing and Communications Workshops (PERCOM Workshops). IEEE, pp 302–307
12. Kitano H (2002) Computational systems biology. Nature 420(6912):206–210
13. Kitano H (2007) Towards a theory of biological robustness. Mol Syst Biol 3(1)
14. Krapivsky Paul L, Redner Sidney, Leyvraz Francois (2000) Connectivity of growing random networks. Phys Rev Lett 85(21):4629
15. Latora V, Marchiori M (2004) The architecture of complex systems. Oxford UP
16. Lunshof Jeantine E, Bobe Jason, Aach John, Angrist Misha, Thakuria Joseph V, Vorhaus Daniel B, Hoehe Margret R, Church George M (2010) Personal genomes in progress: from the human genome project to the personal genome project. Dialog Clin Neurosci 12(1):47
17. Malak D, Ozgur BA (2012) Molecular communication nanonetworks inside human body. Nano Commun Netw 3(1):19–35
18. Mangan S, Uri A (2003) Structure and function of the feed-forward loop network motif. In: Proceedings of the National Academy of Sciences, vol 100, no. 21, pp 11980–11985
19. Mayo M, Abdelzaher A, Perkins EJ, Ghosh P (2012) Motif participation by genes in E. coli transcriptional networks. Front Physiol 3(357). ISSN: 1664-042X. doi:10.3389/fphys.2012.00357. http://www.frontiersin.org/fractal_physiology/10.3389/fphys.2012.00357/abstract
20. Milo R, Shen-Orr S, Itzkovitz S, Kashtan N, Chklovskii D, Alon U (2002) Network motifs: simple building blocks of complex networks. Science 298(5594):824–827
21. Nakano T, Moore MJ, Wei F, Vasilakos AV, Shuai J (2012) Molecular communication and networking: Opportunities and challenges. NanoBiosci IEEE Trans 11(2):135–148
22. Ng Alex KS, Efstathiou Janet (2006) Structural robustness of complex networks. Phys Rev 3:175–188
23. NIH (2013) Cells and DNA—Genetics Home Reference. http://ghr.nlm.nih.gov/handbook/basics?show=all
24. Piro G, Grieco LA, Boggia G, Camarda P, DEE-Dip di Elettrotecnica (2013) Simulating wireless nano sensor networks in the NS-3 platform. In: Proceedings of Workshop on Performance Analysis and Enhancement of Wireless Networks, PAEWN, Barcelona, Spain
25. Python, Software Foundation (1991) Core Python Programming. http://www.python.org

26. Samoilov MS, Arkin AP (2006) Deviant effects in molecular reaction pathways. Nat Biotech 24(10):1235–1240
27. Schaffter T, Marbach D, Floreano D (2011) GeneNetWeaver: in silico benchmark generation and performance profiling of network inference methods. Bioinformatics 27(16):2263–2270
28. Vázquez A, Flammini A, Maritan A, Vespignani A (2002) Modeling of protein interaction networks. Complexus 1(1):38–44
29. Zeigler BP, Praehofer H, Kim TG et al (1976) Theory of modeling and simulation, vol 19. John Wiley, New York
30. Zeng X, Bagrodia R, Gerla M (1998) GloMoSim: a library for parallel simulation of large-scale wireless networks. In: Twelfth Workshop on Parallel and Distributed Simulation, 1998. PADS 98. Proceedings. IEEE, pp. 154–161

Part III
Electromagnetic-Based Nano-scale Communication

Fundamentals of Graphene-Enabled Wireless On-Chip Networking

Sergi Abadal, Ignacio Llatser, Albert Mestres, Josep Solé-Pareta, Eduard Alarcón and Albert Cabellos-Aparicio

Abstract In the broad sense of the term, *nanonetworks* may refer not just to networks composed of nanosized devices, but also to communication networks enabled by nanotechnology. Nanoscale communication techniques can be suitable to interconnect elements far larger than a few square micrometers in applications subject to strong size constraints or bandwidth requirements. Here, the concept *Graphene-enabled Wireless Network-on-Chip* (GWNoC) is introduced as a clear example of this category. In GWNoC, graphene plasmonic antennas are used to wirelessly communicate the components of a multicore processor, which are located in the same chip. This shared medium approach is opposed to current chip communication trends and aims to reduce many of the issues that hamper the development of scalable multiprocessor architectures. In this chapter, we describe the scenario and the communication requirements that justify the employment of nanonetworking techniques, as well as the main challenges that still need to be overcome in this new research avenue.

1 Introduction

Parallelization has been the natural trend in microprocessor architecture design for the last decades and it is expected to continue in the near future. Parallelism can be found and exploited at different granularities, being the instruction-level parallelism the traditional approach which takes advantage of the potential overlap among simple instructions. However, fundamental limits at this level rapidly caused diminishing

---

Ignacio Llatser is not with N3Cat anymore. He was with N3Cat at the time the book chapter was prepared. He is now in the industry.

S. Abadal (✉) · I. Llatser · A. Mestres · J. Solé-Pareta · E. Alarcón · A. Cabellos-Aparicio
NaNoNetworking Center in Catalunya (N3Cat), Universitat Politècnica de Catalunya, Barcelona, Spain
e-mail: abadal@ac.upc.edu

J. Suzuki et al. (eds.), *Modeling, Methodologies and Tools for Molecular and Nano-scale Communications*, Modeling and Optimization in Science and Technologies 9, DOI 10.1007/978-3-319-50688-3_13

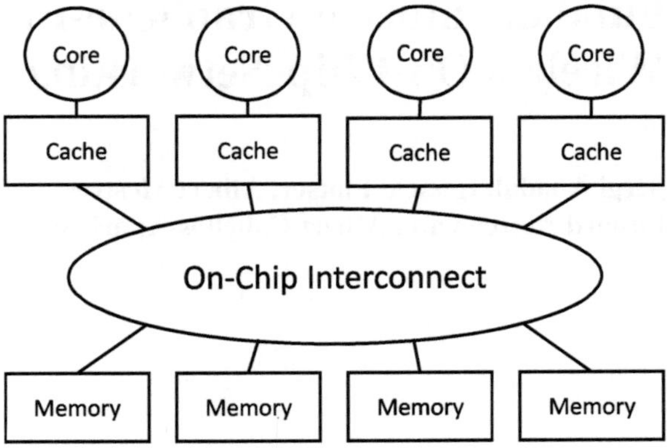

Fig. 1 Schematic diagram of a shared-memory multiprocessor

returns in its exploitation and finally caused power consumption in uniprocessors to grow way faster than its performance [27].

Alternatively, single-chip multiprocessor architectures have emerged aiming to keep pace with the performance trends predicted by Moore's Law, while maintaining an acceptable energy footprint. Parallelism is exploited either by simultaneously executing instances of independent applications or by dividing the application in a set of tasks and processing them in a collaborative manner. To do so, multiprocessors or multicore processors consist of the interconnection of a given number of independent processor *cores* and a memory system within a single die. Figure 1 shows a generic scheme of a shared-memory multicore processor, wherein the memory system is generally hierarchical with some levels of the hierarchy being shared by all the processors. The shared memory paradigm is widely used in current multicore processors and will be the architecture assumed throughout the chapter.

The on-chip interconnect is a central element of a multicore processor since it implements the communication between cores and memory and has a large impact on performance. In shared memory schemes, communication between cores actually occurs implicitly as a result of conventional memory access instructions. Cooperation and coordination among threads is accomplished by reading and writing shared variables [17]. The presence of caches within the memory system decreases the average latency of such memory accesses but, at the same time, it also introduces the problem of cache coherence. Multiple copies of the same shared data may be distributed in a plurality of caches, so that different cores may be seeing different values if this data is modified. Cache coherence protocols are designed to enforce that a read to a shared variable returns the last written value at the expense of generating extra communication. Other issues such as data consistency or synchronization among threads are equally critical for the operation of a multicore processor, as well as additional sources of traffic that the on-chip interconnect must deal with [17, 27].

Given the direct relation between memory architecture, communication and overall performance, the research focus in multiprocessors has gradually shifted from how cores compute to how cores communicate. Buses were first widely considered

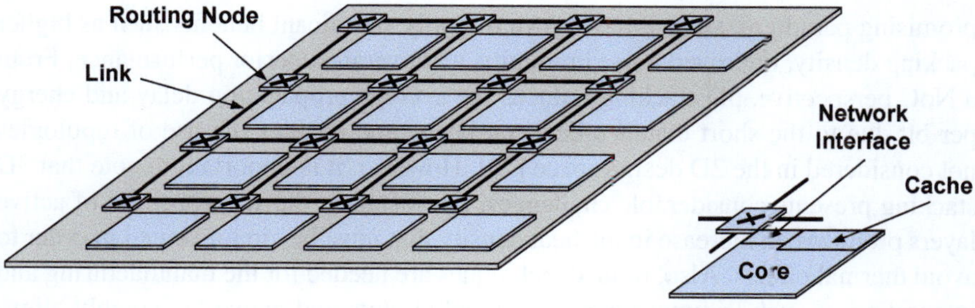

Fig. 2 Schematic diagram of a conventional network-on-chip (NoC)

for the implementation of the on-chip interconnect, but their use is restricted to small-scale architectures given their limited scalability beyond a few cores [9]. Instead, Network-on-Chip (NoC) has been widely adopted as the paradigm of choice for on-chip interconnection networks. NoC can be defined as the application of networking theory and methods to on-chip communication and it generally consists in the employment of point-to-point packet-switched schemes. Figure 2 represents a simple example of NoC, where a given number of on-chip Resistive-Capacitive (RC) wires interconnect the cores (and caches) by means of their respective network interfaces and passing through a network of routing nodes. The interconnection to main memory and the I/O system is omitted for simplicity. Such designs offer improvements in fault tolerance, in modularity and, most importantly, in the overall scalability of the interconnection network; still, it remains unclear whether NoCs based on RC wires will be able to meet the increasingly stringent requirements of next-generation multiprocessors. There are numerous reasons, the most important being the expected increase in delay and power consumption of the wires [55].

As we approach the manycore era, where chips will integrate thousands of cores, several challenges need to be addressed in order to prevent communication to become the performance bottleneck of multicore processors. On-chip interconnects must provide higher throughput levels while maintaining a low latency on a chip-wide basis, taking into consideration that the area and power of the solution must remain bounded (see Sect. 2 for more details). Given that on-chip wires will not be able to cope with such combination of demands, considerable research efforts are devoted to extending the original NoC paradigm with interconnect technologies yielding improved performance. Four emerging alternatives, namely, 3D stacking, Radio Frequency (RF) interconnects, wireless on-chip communication and nanophotonic communication, appear as serious contenders for this regard and are briefly presented next. These will provide both important improvements at the physical layer due to their higher bandwidth per area densities or energy efficiencies, as well as additional degrees of freedom for the design of scalable network architectures.

First, three-dimensional stacking consists in the superposition of different layers of active devices. These layers are separated by just a few tens of micrometers and are vertically interconnected by means through-silicon vias [12] or near-field coupling schemes [15]. The creation of 3D integrated circuits has proved to be a promising paradigm, since it has shown to imply significant benefits such as higher packing density, improved noise immunity, and overall superior performance. From a NoC perspective, 3D stacking reduces the average propagation delay and energy per bit due to the short distance between layers and enables the use of topologies not considered in the 2D design space [21]. However, it is important to note that 3D stacking presents considerable challenges. For instance, the superposition of active layers produces an increase in the heat density that must be circumvented in order to avoid thermal effects. Also, refined techniques are needed for the manufacturing and integration of such tridimensional integrated circuits and networks, notably alignment methodologies for accurate placement of the vertical vias.

Second, the RF interconnection paradigm is presented as an alternative to traditional voltage and current signaling through metallic wires. Baseband signals are modulated using gigahertz carriers and then sent at the speed of light through transmission lines printed in the chip surface [55]. In long-range links, the improvement in terms of propagation time is very large with respect to the delay introduced by the modulation process and, therefore, the communication latency can be effectively reduced. RF interconnects also enable multiple access schemes in shared transmission lines, e.g. by means of frequency-multiplexing or code-multiplexing schemes. Each core is assigned a set of channels, enabling the possibility of interconnecting several cores using the same transmission line and therefore reducing the number of wires. Further, the bandwidth for each core could be dynamically assigned according to real-time demands. Due to these advantages, complementing a baseline NoC with an overlaid global RF interconnect has been proposed [16]. The main downturn is that the physical topology must be carefully designed as impedance mismatch reflections at the transmission line ends may generate interferences, limiting the number of practical network architectures and their scalability.

A possible solution to the RF-interconnect issues is to transmit the signals wirelessly instead of through transmission lines. The resulting Wireless Network-on-Chip (WNoC) approach not only inherits the advantages of RF-interconnects, but also adds natural adaptability and broadcast capabilities to the equation as no path infrastructure is needed. WNoC is feasible due to the availability of both on-chip antennas [46] and high-speed transceivers [23], and has recently given rise to a plethora of proposals (see [18] and references therein). However, as we will see in the following sections, the size of current and future metallic on-chip antennas largely limits the potential of this approach and motivates the employment of nanoscale communication techniques.

Last but not least, nanophotonics is enabling the creation of CMOS-compatible optical building blocks for, among others, on-chip communications [8]. Chip-scale transmissions at speeds of 50 Gbps have been accomplished thus far [45], whereas potential for energy figures several orders of magnitude lower than conventional interconnects is envisaged [13]. In light of the promise that this technology shows for low energy per bit communications, intense research efforts have been directed towards creating *photonic NoCs* by means of the integration of nanophotonic devices. Apart from yielding an outstanding potential for low power consumption, such networks also maintain the main advantages of RF interconnects as signals can be wavelength-multiplexed. Such feature provides both potential for extremely high bandwidth per area, as well as a wide range of possibilities from the network architecture perspective. Extensive works in the design and development of photonic NoCs, including a wide variety of topologies and network architectures, serve as proof of this trend (see [3, 36, 54, 62] and references therein). It must be noted that these works, in most cases, aim to overcome the main limitations of the nanophotonic approach. Existing on-chip laser sources are excessively large in terms of area or involve costly integration processes; whereas the implementation of all-optical packet routing schemes remains as a grand challenge at the chip level.

Even though considerable advances have been accomplished in the field of on-chip networking, efficiently delivering multicast and broadcast traffic remains an open challenge at the time of writing this book chapter. The case is particularly concerning within manycore settings, where one-to-many communications will play a crucial role (see Sect. 2 for more details), and even considering the new interconnect technologies mentioned above. While one may think that the advent of WNoC would solve this issue given the inherent broadcast capabilities of this technology, the reality is that the size of metallic antennas prevents the integration of one antenna per core to fully take advantage of such competitive advantage.

In this chapter, we present the concept *Graphene-enabled Wireless Network-on-Chip* (GWNoC), which aims to address this grand challenge by providing each core with wireless communication capabilities and sharing the medium [2]. The approach is enabled by graphene antennas, whose plasmonic effects allow them to radiate electromagnetic waves in the terahertz band (0.1–10 THz) while occupying an area up to two orders of magnitude lower than metallic antennas for the same radiation frequency [40, 59]. This way, the stringent requirements of the scenario in terms of area and bandwidth, which are detailed in Sect. 2, can be met. Section 3 presents a description of GWNoC and its advantages over emerging alternatives, as well as an outline of the main communications and networking design considerations. Section 4 concludes the chapter.

2 Open Issues in Communication Within Manycore Chip Multiprocessors

Taking into consideration several physical constraints such as the thermal design power, technology improvements are foreseen to steadily provide a scaling of at least 1.4X, per technology generation, of the number of cores that can be integrated within a chip [30]. However, the entire system must scale before this trend translates into effective parallel performance improvement. This implies solving the open issues that are found when scaling aspects such as parallel programming models, the memory system or the on-chip interconnect fabric. In this chapter, we focus upon the on-chip interconnect while being aware of the memory system, since the performance of a multicore processor is largely dictated by how fast both memory accesses and the traffic generated by these accesses are served.

From an architectural perspective, a balance must be struck between the effectiveness of the memory system and the communication requirements cast upon the on-chip interconnect. However, this task becomes especially challenging as the number of cores per chip grows, since the communication demands of existing architectures exponentially increase when upscaled. In this regard, unconventional and less communication-intensive architectures need to be explored. From a communications perspective, the main objective is to match the performance of the on-chip interconnect with the potential communication demands placed by the architecture, while complying with some design constraints. For instance, it has been widely proved that chip communication mainly occurs among neighboring cores due to the spatial locality of code [27, 56] and initial NoC designs were better suited for this type of traffic [49]. However, the complexity of this matching process grows with the number of cores, as the architectural aspects that impact upon the characteristics of the on-chip traffic may change significantly. New interconnect solutions will be therefore required.

In the following, we detail the physical constraints and driving requirements that challenge the design of on-chip interconnects for manycore settings, putting special attention upon the case of multicast and broadcast on-chip communication.

Power Consumption

Thermal effects are a primary concern when designing a processor. In order to hold down the costs of thermal cooling, manufacturers generally impose constant power limits across generations. The energy efficiency of the on-chip interconnect will need to be improved, since the communication demands are foreseen to sharply increase with the number of cores. Projections derived from the International Technology Roadmap for Semiconductors (ITRS) calculate that transmitter energies of between 10 to 100 fJ/bit must be targeted in the near future [44].

Efficiencies around 200 fJ/bit have been demonstrated using conventional interconnects [52]. Even though these figures can be still improved, it remains unclear whether it is possible to meet future energy requirements without largely affecting other metrics such as the data rate. This situation has been the main driving force behind proposing nanophotonics for on-chip communication, as it promises to push figures down to around 1 fJ/bit [13].

Power consumption is also a concern at the network level as the core density and the complexity of NoCs increase. The average number of hops of widely-used mesh topologies increases with the number of cores, incurring in a proportional increase in power as routers consume a significant fraction of energy. One of the first implementations of NoCs for manycore chip multiprocessors is described in [29], where the NoC consumes approximately 40% of the total 100 W chip power. This suggests that alternative topologies (perhaps enabled by 3D stacking) may be needed in manycore settings to reduce the number of hops and, therefore, the average power consumption [21, 49]. However, the energy savings are generally traded off with area as these topologies require additional wires and more complex routers. In the case of photonic NoCs, all-optical alternatives at the network level are reduced and do not scale due to current laser and router complexity limitations [3]. Designs combining electrical and optical planes offer higher degrees of freedom and have recently been considered instead [36, 54]. Tools are available for the evaluation of their energy efficiency [14, 58].

Area

In order to ensure a growing yield, manufacturers aim to keep the die size as small as possible. Processor dies are currently on the order of a few hundreds of square millimeters and grow slower than the area occupied by cores for each technology generation [30]. Therefore, the area overhead of the on-chip interconnect is a critical evaluation factor as chip real estate becomes an extremely scarce resource in many-core environments. For instance, the high bandwidth per area figures of nanophotonic interconnects is one of the reasons for considering them among the plethora of emerging contenders. Technological models are broadly used for the early-stage evaluation of the area in conventional and photonic NoCs [33, 58].

Closely related to the area constraint is the wiring complexity. NoCs based on wires or waveguides may need to include a large number of links to implement a complex topology suitable to the demands of manycore chip multiprocessors. Regardless of whether the area limitations are respected or not, finding an appropriate layout strategy may be unfeasible due to the increasing wire routing congestion. A potential solution would be to replace part of the wiring with wireless RF interconnects [18]. However, the size of on-chip antennas limits the usefulness of the WNoC approach and motivates the use of graphene-enabled wireless communications.

Performance

The multicore scenario imposes a set of general requirements to the on-chip interconnects. Cores generally send messages after a given computation and stop their execution until a response is received. A slow or lossy delivery of these messages must be avoided, as it will cause the cores to reduce their speed and therefore to reduce general performance. Hence, any on-chip interconnect must guarantee a given performance in terms of latency, throughput, and losses. The main challenge here is to provide solutions that will allow us to maintain these conditions as the number of cores grows.

Latency is arguably the most important constraint in on-chip networks despite the strong requirements in area, power, and bandwidth. The communication delay in operations that are in the critical path of the processor will directly impact upon its performance. Therefore, latency must be kept within certain bounds (ideally constant) when scaling the number of nodes. This is not possible in conventional mesh designs due to the increase of the average hop count. Again, alternative topologies or the use of RF/nanophotonic long-range links has been proposed to improve the overall latency [36, 38, 55, 62].

Secondly, multicore processors are extremely data intensive scenarios and, as a result, the bandwidth of the interconnect is also crucial. A rule of thumb is that its throughput should scale at least proportionally with the number of cores. Conventional NoCs meet such scalability demands, but optimizations are still needed as chip resources become more scarce. Overprovisioning is generally employed in order to avoid the network to saturate in high contention phases, which are typical in parallel programs and generate large bursts of communication. Fine-grained reconfigurable links have been proposed in order to save area and power wasted in such process [28]. Nanophotonics are also taken into consideration as they yield much improved bandwidth per area figures.

Finally, all packets need to be delivered free of errors in order to guarantee a correct operation of the processor. At the link level, on-chip interconnects are designed to operate with a bit error rate (BER) around 10^{-15} [38] and generally apply Forward Error Correction (FEC) schemes to correct infrequent errors. At the network level, congestion may cause packets to be discarded due to network buffers being overrun, motivating the need for flow control mechanisms and retransmission policies.

Multicast/Broadcast

Area, power, and performance are general requirements that apply to all the traffic generated by the memory system, regardless of its characteristics. As the core density grows, the general tendency is to scale current multicore architectures and then to address the resulting increase in communication by means of the improved on-chip networks. However, it occurs that the traffic may not necessarily scale in the same direction than the interconnect performance does. One clear example is the communication between topologically distant cores: whereas the number of these transmissions increases with the core density, the performance of conventional NoCs worsens in this situation. Although a possible strategy is to design a memory system or a programming model aiming to reduce long-range communication, this implies facing additional challenges that are out of the scope of this chapter.

Within this context, a particularly concerning case is that of multicast and broadcast communication. From a computer architecture standpoint, broadcast communications have been traditionally regarded as expensive and its use is avoided whenever possible. However, operations such as thread communication or data synchronization generate a significant amount of multicast messages even in moderately size multiprocessors [20]. As the core density grows, one-to-many traffic will increase not only in number of messages but also in number of destinations of each message. Figure 3 exemplifies this trend by plotting both metrics as a function of the number of cores for a set of applications from the SPLASH-2 and PARSEC benchmarks [10, 64]. The results are obtained by means of full-system simulation using gem5 [11] and assuming two different types of coherence. The simulated architecture consists of N cores, each of which accounts for two private 32-kB 2-way associative L1 caches (one for instructions and another one for data), as well as a bank of shared 8-way associative L2 cache of size 512 kB. We modified the network interfaces in order to register the characteristics of the multicast traffic generated by the cache [6].

Whereas the importance of multicast and broadcast increases with the core density, the performance of NoCs is likely to decrease. Conventional designs are based upon point-to-point links and messages with M destinations are generally treated as M unicast messages. At low core counts, the impact of such type of traffic can be neglected. Nevertheless, the interconnect fabric will saturate as the number of multicast packets and the average number of destinations grow for high core counts.

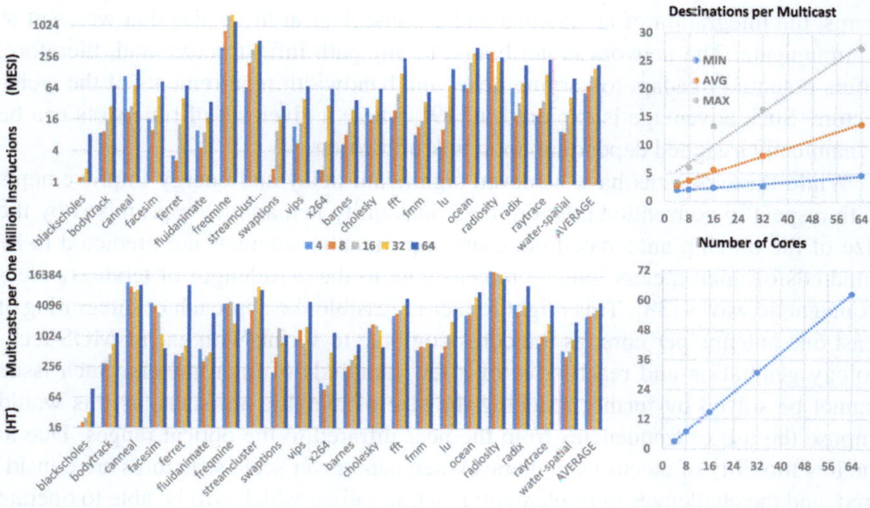

Fig. 3 Number of multicast messages per instruction (*left*) and average number of multicast destinations (*right*) as a function of the number of cores for different benchmark applications assuming MESI and HyperTransport coherence [6]

The size and wired nature of current NoCs render broadcast and multicast communications excessively costly and motivate the need for an alternative and efficient broadcast platform. The transition of broadcast from a constraint to an opportunity will not only provide means for the scaling of current architectures, but will also open a vast design space for the design and development of new architectures [20].

Although proposals to improve the performance of multicast and broadcast have been formulated for conventional NoCs [35, 50], RF NoCs [16] and photonic NoCs [36, 62], their scalability in terms of performance and cost remain largely unexplored. In light of the growing importance of one-to-many communications in the manycore scenario, a cost-effective solution is required. In the next section, we aim to address this issue by proposing the application of nanoscale communication techniques for chip-scale communications.

3 Graphene-Enabled Wireless Network-on-Chip

Among the plethora of emerging alternatives for on-chip communication, WNoC stands as a promising approach to complement existing wired interconnects. Wireless long-range links can be used to considerably decrease the multihop latency of a conventional NoC and even to provide one-hop communication for delay-critical traffic. Existing proposals adopt such approach by either placing antennas in a regular layout [38, 43] or following the principles of small-world networks [22], whereas multiple access is achieved by means of frequency or time channelization. Another advantage of WNoC is that implementing wireless links only requires, in physical terms, the integration of an antenna and a transceiver at the nodes that we want to communicate. The network is not bound to any path infrastructure and, therefore, offers potential to adapt to varying delay and bandwidth requirements of the architecture. Such advantage is explored in [19], where a given set of time slots can be dynamically assigned depending upon link utilization.

While these designs have achieved significant delay and energy improvements with respect to conventional NoCs, their scalability is mainly compromised by the size of the on-chip antennas. Future on-chip metallic antennas are predicted to be hundreds of micrometers long, commensurate to the wavelength of terahertz electromagnetic waves [38]. This might render unfeasible the approach of integrating at least one antenna per core, as the cores continue to shrink with each CMOS technology generation and reach sizes of a few hundreds of micrometers. Such issue cannot be solved by further reducing the size of metallic antennas, as this would impose the use of frequencies from the near infrared to the optical ranges. Due to the low mobility of electrons in metals when nanometer scale structures are considered, and the challenges in implementing a transceiver which will be able to operate at this extremely high frequency, the feasibility of wireless communications at the core level would be compromised if this approach would be followed. Given these constraints, the current approach when integrating hundreds or thousands of cores is to use wireless links among sets of cores and then internally communicate these sets using on-chip wires [22, 43]. A packet may therefore propagate through the wired plane, then traverse a wireless link and finally return to the wired plane; whereas broadcast packets are distributed from the sender to the rest of sets and then internally within each set. In all cases, the performance improvements are ultimately limited by the performance of the wired network.

Instead, we propose to apply novel nanoscale communication techniques seeking to enable the integration of one or more antennas per core. This approach, to which we already referred to as GWNoC, consists in delivering core-level broadcast capabilities by means of the employment of graphene planar antennas. Antennas based upon a graphene patch just a few micrometers in size, i.e. two orders of magnitude below the dimensions of future metallic on-chip antennas, are expected to radiate in the terahertz (0.1–10 THz) band. These unique characteristics will both enable size compatibility with each processor core and offer enough bandwidth in massively parallel settings [31]. With a proper protocol stack, the latter will lead to low-latency and high-throughput schemes while complying with the severe area and power constraints of the manycore scenario.

Figure 4 shows the schematic representation of GWNoC within a manycore processor. We assume a hybrid approach, where the GWNoC is used to transport control flows and significant part of the broadcast-based data, and is deployed over a state-of-the-art NoC which serves heavy flows of data (not represented for simplicity). Each core is equipped with a network interface, a transceiver and at least one graphene antenna. Upon the release of a packet from a core, its network interface decides whether it must be transmitted through the wired or the wireless plane; in the second case, the transceiver modulates the information to be sent through the graphene antenna. At the receiver side, the graphene antenna picks up the wireless

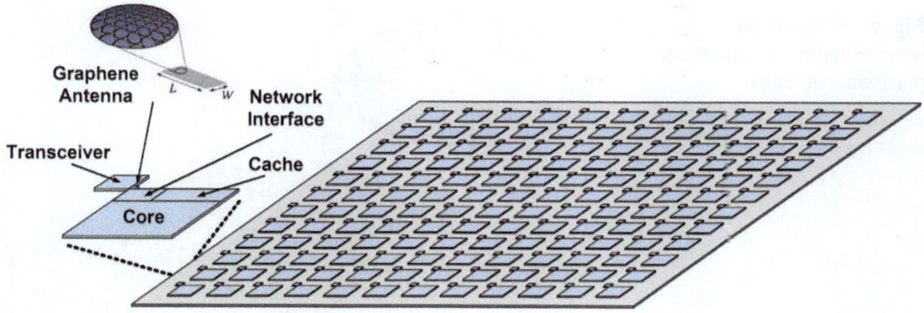

Fig. 4 Schematic Diagram of a 144-core graphene-enabled wireless network-on-chip

signal and passes it to the transceiver, which demodulates the data. The network interface then checks the address of the packet and decides whether it must be delivered to the core or discarded.

Since the information is radiated and can be received by any receiver within the chip, GWNoC not only provides native broadcasting capabilities, but also makes data transmission transparent with respect to the location of data. This heavily alleviates the constraints of parallel architecture design, therefore reducing the complexity of parallel programming and impacting upon the performance of virtually any future application. Further, the integration of a wireless communication unit on a per-core basis confers replicability and modularity to the on-chip design by means of the concept of *wireless core*. A library of general-purpose or specific wireless cores could be created, allowing the building of custom multicore processors by the integration and configuration of a set of such pre-designed cores.

3.1 Modeling GWNoC Communications

Communications in the GWNoC scenario are unique since they are enabled by novel antennas and occur within a unique environment and in the terahertz band. Understanding these aspects is an important step before the actual implementation of such communications can be addressed. Conventional models, methods and tools cannot be used and need to be profoundly revised to this end. In the following, we detail the characteristics, requirements, and potential impact of each element involved in the communication upon the system performance. Models and design methodologies are briefly summarized whenever available.

3.1.1 Graphene Plasmonic Antennas

As conceptually represented in Fig. 5, a graphene antenna is composed of a finite-size graphene layer, mounted over a metallic flat surface (the ground plane) with a dielectric material layer in between, and an ohmic contact. These antennas are the

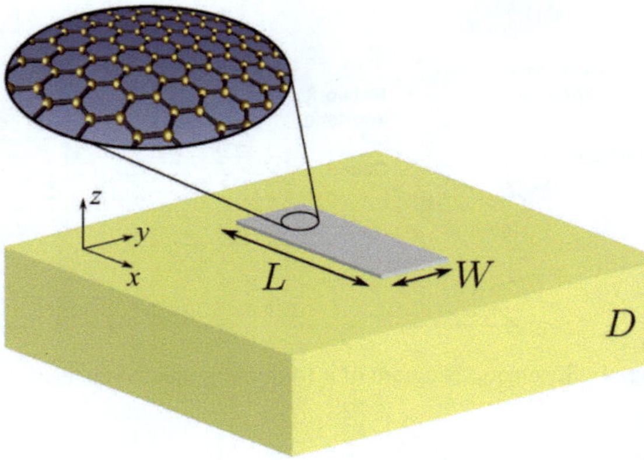

Fig. 5 Conceptual representation of a graphene plasmonic antenna

main enabler of the GWNoC approach due to their unique relation between size and radiation frequency. On the one hand, being up to two orders of magnitude smaller than metallic antennas for the same resonant frequency allows the integration of one or more antennas per processing core. On the other hand, the potential to radiate in the terahertz band provides a huge transmission bandwidth, allowing not only the transmission of information at extremely high speeds, but also the design of ultra-low-power and low-complexity schemes.

The reason behind such subwavelength behavior is that graphene antennas support the propagation of tightly confined Surface Plasmon Polariton (SPP) waves. Such phenomenon occurs at the interface between any metallic and dielectric material pair when an electromagnetic wave impacts upon the metal (graphene in our case). The wavelength of the SPPs within the metal determines the resonance condition and is given by λ/n_{eff}, where λ is the free-space wavelength and the effective mode index n_{eff} is:

$$n_{eff}(\omega) = \sqrt{1 - 4\frac{\mu_0}{\epsilon_0}\frac{1}{\sigma(\omega)^2}} \quad (1)$$

and yields, in the case of graphene, strong resonances at terahertz frequencies [5, 32, 59]. The effective mode index, among many other properties of the SPP waves, depends upon the frequency characteristics of the electrical conductivity of the metallic material $\sigma(\omega)$. Conductivity models of graphene are thus key to explore the radiation properties of graphene antennas. To model the conductivity of graphene, the main approach is to consider two approximations [5, 26]. Firstly, since we consider antennas with a size larger than 50 nm, it is possible to disregard the effects at the graphene edges. Secondly, we consider the interband contribution of the conductivity to be negligible in the frequency band of interest, which is a valid assumption when considering the terahertz band. With this, the conductivity is expressed as:

$$\sigma(\omega) = \frac{2e^2}{\pi\bar{h}} \frac{k_B T}{\bar{h}} \ln\left[2\cosh\left[\frac{E_F}{2k_B T}\right]\right] \frac{i}{\omega + i\tau^{-1}}, \tag{2}$$

where e, $\bar{h}$ and k_B are constants. Variables T, τ and E_F correspond to the temperature, the relaxation time, and the chemical potential of the graphene layer. The relaxation time is the interval required for a material to restore a uniform charge density after a charge distortion is introduced, and it highly depends upon the quality of the graphene sheet and of the underlying substrate. The chemical potential or Fermi energy E_F refers to the level in the distribution of electron energies at which a quantum state is equally likely to be occupied or empty. The chemical potential can be modified by applying a voltage to the antenna (thereby allowing to dynamically tune its radiation properties) or by means of chemical doping.

Using Eq. 2, the frequency response of the conductivity is evaluated for a fixed chemical potential and relaxation time pair. Since graphene is a one-dimensional material, the antenna can be then modeled as a patch with an equivalent surface impedance of $Z = \frac{1}{\sigma}$. Such possibility is available in commercial electromagnetic field solving simulators and allows to obtain the frequency response of the antenna upon the presence of incident electromagnetic waves. By means of this methodology, important performance aspects of the antenna can be determined as functions of graphene technological parameters (i.e., chemical potential and relaxation time), as well as the antenna design parameters (e.g., size and shape), including but not limited to:

The antenna impedance and radiation efficiency: the frequency response of the impedance and of the radiation efficiency are crucial for the design of a transceiver that will drive the antenna. The frequency and power of the input signals, as well as the characteristic impedance of the source of those signals need to be determined taking into consideration the antenna impedance and radiation efficiency. Recent works report a radiation efficiency of up to 25% for graphene patch antennas and a very high impedance in the kΩ range [59].

The antenna bandwidth: is a crucial performance metric since a high data rate potentially leading to high throughput is required. The peculiarity of graphene antennas is that bandwidth depends not only upon the shape of the antenna but also upon the technological parameters of the material. In the former case, high bandwidths can be obtained with fractal or inherently broadband structures. In the latter case, recent results state that the relaxation time of the graphene sheet has a significant impact upon the resonance bandwidth [40].

The radiation pattern: which indicates the strength of the radiated signal as a function of the radiation direction. Recent works demonstrate that the radiation pattern of graphene patches is similar to that of their metallic counterparts [40, 59], suggesting a dependence on the antenna size and shape rather than on the radiative material. For antennas based on graphene patches, the radiation efficiency is extremely low in the plane of the antenna and substantially higher in the perpendicular direction.

Graphene antennas is a thriving albeit still wide open research area. At the time of writing this book chapter, several groups are currently conducting intense research towards a further characterization of graphene patch antennas [39, 40, 59, 60]. For instance, the impact of the substrate material and thickness upon the radiation of the antenna must be taken into consideration [39]. Also, studying graphene antennas in transmission requires defining a feeding mechanism. This represents a challenge by itself, since the feeder must support the propagation of SPPs and must be matched to the antenna. The design of a matching mechanism requires, in turn, modeling the effects of the contacts between the feeder and the edge of the graphene patch.

3.1.2 Terahertz Within-Package Channel

A channel model that takes into consideration the peculiarities of the GWNoC scenario is fundamental in order to evaluate the available on-chip communication bandwidth. Mainly, the enclosed nature of chip processors causes the apparition of a large number of reflections that must be taken into consideration at the receiver. The physical landscape of a multiprocessor involves multiple dielectric/metallization layers and components printed on the chip surface, among other elements that need to be accurately described in order to model the channel [42]. Since such landscape is static, the model will be time-invariant.

In the general setting shown in Fig. 6, radiated signals reach the receiver via different paths [66]. First, surface waves propagate at the interface of the chip and the package medium. These waves show particularly low attenuation per unit of distance due to their cylindrical characteristics and are affected by the circuits printed on the chip surface. However, since graphene patch antennas show an extremely low radiation efficiency in the coplanar direction, the contribution of surface waves at the receiving end may be negligible. Second, part of the energy of patch antennas is radiated into the substrate. These waves are guided within the substrate and reach the receiver after repeated reflections upon the ground plane of the chip and the insulating layer. However, the substrate is generally lossy and introduces a very high attenuation per unit of distance. Given that surface and guided waves are highly attenuated by the antenna and the substrate, respectively, in most cases we can con-

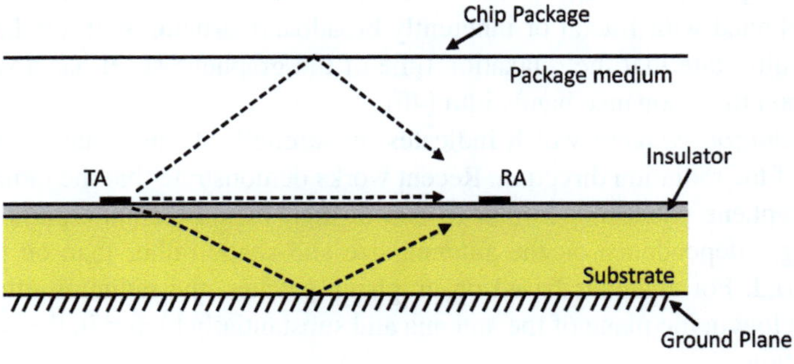

Fig. 6 Electromagnetic waves that may potentially reach the receiver

sider that communication occurs by means of a third mechanism: space waves that propagate through the medium and reflect upon the chip package and surface.

Modeling the channel implies evaluating all possible rays reaching the receiving antenna for each and every pair of antennas within the chip [42]. The result in the time domain will then be a sum of channel impulse responses: $h(t) = \sum_i \alpha_i e^{j\phi_i} g_i(\tau - \tau_i)$, where α_i, ϕ_i and τ_i are the amplitude, phase shift, and delay of the i-th ray. Note that this function is generally evaluated using Dirac deltas $\delta(t)$ instead of waveforms $g_i(t)$. In the GWNoC case, though, such ideal approach is not accurate as both propagation and reflections are frequency-dependent phenomena and antennas will radiate over a large bandwidth in the terahertz band.

Propagation: Communicating by means of terahertz waves has two main implications on propagation. First, we can assume that the far-field condition holds given both the short radiation wavelength and the fact that communication occurs by means of reflected waves. Second, the phenomenon of *molecular absorption* must be accounted for on top of typical spreading losses. Molecular absorption is the process by which part of the wave energy is converted into internal kinetic energy of the excited molecules in the medium. Molecules present in standard media have numerous resonances in the terahertz band, causing a frequency-selective attenuation of terahertz electromagnetic waves radiated by antennas [41]:

$$\alpha(f, d) = e^{k(f)d}$$

Note that the attenuation highly depends on the medium absorption coefficient k, which models the particular mixture of molecules in the medium; as well as on the transmission distance d that determines the number of molecules that the waves will find along their path. The inset of Fig. 7 exemplifies the latter dependence by representing the molecular absorption of the terahertz channel for transmission distances of 1 cm and 10 cm. Both the number of absorption peaks and their amplitude

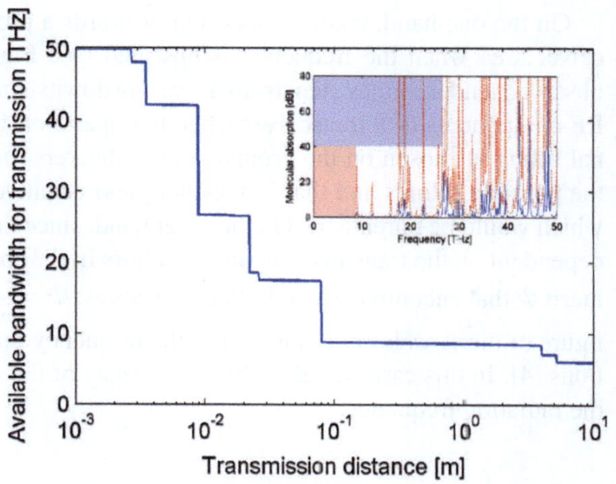

Fig. 7 Available bandwidth in the frequency band from 0 to 50 THz due to molecular absorption, as a function of the transmission distance. The *inset* shows the molecular absorption in dB and available bandwidth for two particular distances: 1 cm (*blue*) and 10 cm (*red*)

notably increase in the latter case, reducing the 10-dB bandwidth from 27 THz (top blue background) to 9 THz (bottom red background). Figure 7 shows the available 10-dB bandwidth for a range between one millimeter to ten meters [41]. In light of these results, it is concluded that molecular absorption has a limited impact on transmissions at the chip scale, fact that may lead to channel capacities over the terabit-per-second barrier [31].

Reflections: the characteristics of reflected waves depend both on the roughness of the surface and on the reflective material. The effects of the former can be neglected for conventional metallic materials in the frequency range of interest [37], whereas the latter is polarization-sensitive and given by the Fresnel coefficients of the different media [51]. The main issue here is that these coefficients are frequency-dependent and require knowing the frequency response of the materials present in on-chip environments; however, only a few materials have been characterized in the terahertz band [24, 51].

3.1.3 Transceiver

In order to enable on-chip wireless communication, it is necessary to develop a transceiver to modulate and demodulate the data and to drive the antenna. To this end, such transceiver needs to operate at the same frequency than the antenna itself. This represents a grand challenge since terahertz transceivers are still not available, even though advancements in CMOS [53] and alternative technologies based on InP [34] or graphene [25, 65] may enable their creation in the near future.

Since critical metrics such as the area and power consumed by the wireless communication unit mainly depend on the characteristics of the transceiver, accurate models are key to assess the feasibility of a GWNoC design. However, such models are not available since terahertz technologies are still in their infancy. Instead, behavioral area and energy models could be created from state-of-the-art transceiver implementations and then extrapolated to extract results in the terahertz region [4].

On the one hand, recent works point towards a promising decrease in the transceiver area when the frequency is upscaled (see Fig. 8, [4]). The reasons for the observed tendency may stem from the strong downsizing that is applied to the passive RF components of a transceiver when the operation frequency is increased. Rational fitting is chosen on the grounds that it delivers the most accurate result among the possible fittings and that it does not yield negative values for high frequencies, which would be unrealistic. On the other hand, since the transceiver energy is highly dependent on the transmission range, authors in [23] propose and discuss a figure of merit Φ that encompasses both their metrics as: $\Phi = \frac{E_{bit}}{\sqrt{d_{max}}}$. Figure 8 shows how this figure of merit scales as a function of the frequency for state-of-the-art implementations [4]. In this case, we also observe a decay of the energy per bit proportional to the radiation frequency.

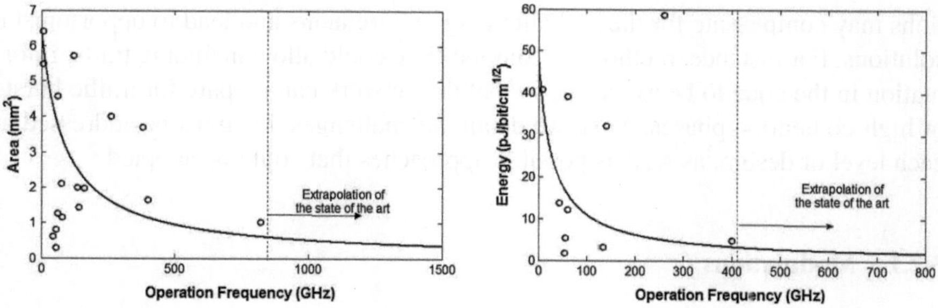

Fig. 8 Area and energy efficiency of state-of-the-art wireless transceivers as a function of their central frequency. See [4, 23, 47, 48] and references therein for more details

The results here presented confirm that terahertz circuits will likely be suitable for wireless on-chip communication purposes given the inverse relation between area, energy, and operation frequency.

3.1.4 Network Interface

As its name implies, the function of the network interface is to bridge the memory system and the network. In conventional NoCs, the network interface receives data from the memory system and creates a packet with it, to then split the packet into flow control units (flits). Finally, the network interface puts the flits in its output queue and sends them to the associated router whenever possible. At the other end, the network interface receives the flits, reconstructs the packet, and then checks that the destination address corresponds to its core address.

In our scenario, the on-chip interconnect accounts for both a wired and a wireless plane. A controller must be added to each network interface in order to determine through which plane a given piece of data should be sent. Before being sent to the transceiver, the data to be wirelessly transmitted must be packetized and serialized into a stream of bits. Finally, the inverse process is performed at the receiver: a stream of bits is received from the transceiver and interpreted. The network interface then checks the address in order to decide whether the packet must be either yielded to the core or discarded.

3.2 Design Decisions for GWNoC Protocols

The peculiarities of the GWNoC scenario require the design and development of a unique network architecture. Protocols for classical wireless networks cannot be applied to on-chip communications due to the blend of power, area limitations and stringent performance demands of multiprocessors. Luckily, some favorable conditions may compensate for these challenging requirements and lead to opportunistic solutions. For instance, methods at compile-time could allow including traffic information in the code to be executed, so that the network can prepare for traffic bursts or high-contention phases. Next, we detail the challenges that must be addressed at each level of design, as well as possible approaches that could be adopted.

3.2.1 Modulations

The area and energy figures of a transceiver not only strongly depend on the implemented modulation, but are also generally traded off against performance. Therefore, modulations are an important design step in the GWNoC scenario, as a balance between area, energy, and performance is sought. Working at terahertz speeds may allow achieving these goals provided that additional challenges are addressed. Mainly, the solution must be feasible and adapt to the terahertz components that technology progress will made available in the years to come. Jitter should also be taken into consideration with special attention, as it may become an important performance bottleneck due to the extremely fine temporal resolution needed at the receiver. Such unique features strongly limit the boundaries of the practical design space.

Within this context, Impulse Radio Ultra-Wideband (IR-UWB) techniques stand out as promising candidates for the implementation of on-chip wireless communication. The IR-UWB consists of the transmission of very short baseband pulses, the length of which determines the bandwidth of such spread spectrum signal. Academic research efforts have gone beyond commercial implementations at the 3.1–10.6 GHz band and explored frequencies up to 110 GHz [57]. Following this trend, communication in the terahertz band can be accomplished by means of the transmission and reception of picosecond long pulses. Furthermore, IR-UWB yields potential for the devising of simple and low-power systems by means of non-coherent detection. This approach advocates for the detection of the energy of the signal rather than its phase, offering simplicity at the expense of a lower performance for fixed levels of noise. Non-coherent detection eliminates the need for channel estimation and makes the system more robust against timing issues induced by jitter [63]. From an implementation standpoint, the use of power hungry components such as a phase-locked loop can be avoided in asynchronous schemes. Also, it allows to perform initial signal processing tasks in the analog domain, leading to sub-Nyquist sampling rates [7]. This aspect is critical since Nyquist sampling rates imply a need for power demanding analog-to-digital converters able to operate in the terahertz band.

Energy detection is compatible with a limited number of modulations. Among them, On-Off Keying (OOK, modulating by means of the presence/absence of pulses) is particularly suitable to the GWNoC scenario due to its simplicity and relaxed timing constraints. The probability density of OOK zeroes and ones at the energy detector are evaluated using well-known central and non-central chi-square distributions, $\chi^2(k)$ and $\chi^2(k, \mu)$ [61]. The k represents the degrees of freedom or number of samples per symbol and is generally taken as $2 \cdot TW$, where TW is the time-bandwidth product of pulses at the receiver. The non-central distribution has a non-centrality parameter μ equal to the signal to noise ratio $\gamma = hE_b/N_0$, where h accounts for the loss of energy due to jitter-induced effects. Assuming a threshold λ calculated following the *maximum a priori* criterion, the error probabilities are:

$$P(1|0) = \int_{\lambda}^{\infty} \chi^2(2TW) = \Gamma(TW, \lambda/2)/\Gamma(TW) \qquad (3)$$
$$P(0|1) = \int_{-\infty}^{\lambda} \chi^2(2TW, \gamma) = 1 - Q_{TW}(\sqrt{2\gamma}, \sqrt{\lambda})$$

where $\Gamma(\cdot,\cdot)$ corresponds to the incomplete Gamma function and $Q_u(\cdot,\cdot)$ is the generalized Marcum Q-function of order u. The error probability is then evaluated as: $P_e = P(0)P(1|0) + P(1)P(0|1)$. Upon the presence of jitter affecting the signal to noise ratio, the BER is calculated as the weighted average over the jitter probability density function: $BER = \int P_e(\varepsilon_i)f(\varepsilon_i)d\varepsilon_i$.

Next, we quantify the performance of OOK with energy detection and compare it to that of more complex options. Coherent schemes, i.e. matched filter and autocorrelator, are considered as receivers of Binary Pulse Position Modulation (BPPM), Binary Phase Shift Keying (BPSK), Transmitted Reference (TR) and differential (DIFF) schemes. On the one hand, a matched filter assumes perfect channel estimation and recovers the phase of the signal directly from the received pulse. Unlike the energy detector, this enables the demodulation of BPSK signals, the information of which is encoded within the pulse polarity. On the other hand, the autocorrelator relies upon a previously received pulse in order to estimate the channel and recover the information. In TR, a pair of pulses is sent for each symbol: the first one serves as a pilot and the second one modulates the information. In DIFF, the information is modulated differentially between each two consecutive pulses, allowing to save half of the energy per symbol.

The simulation framework assumes fixed signal-to-noise ratio (SNR) and data rate objectives, to then calculate the appropriate pulse characteristics for each scheme. Each value of jitter implies a different effective received power, which is used to evaluate the BER by using *(1)* the model explained above for the OOK case and *(2)* additional equations for the rest of cases (see [63] for more details). The BER is then averaged over all the probability density function of the jitter. Figure 9 shows the BER performance with respect to the jitter level in a system working at 100 Gbps and a nominal SNR of 18 dB. We observe that the combination of OOK and energy detector yields a performance comparable to that of coherent receivers for high levels of jitter, suggesting that the reduction in synchronization requirements of non-coherent detection makes up for its worse nominal performance.

3.2.2 Coding

As mentioned in Sect. 2, on-chip interconnects are designed to operate with a BER on the order of 10^{-15}. In light of the BER results shown above, such target will not

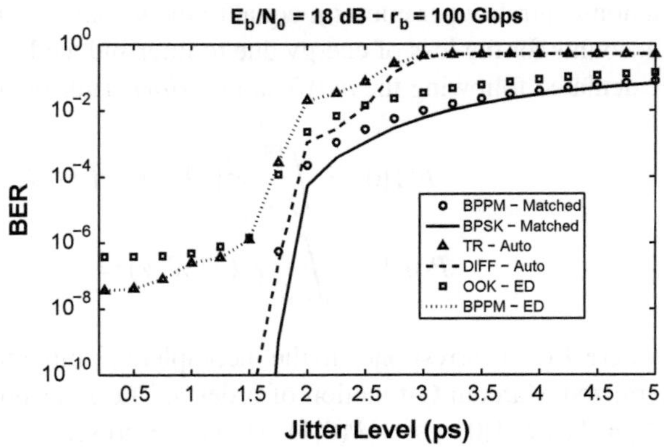

Fig. 9 Receiver performance comparison as a function of the jitter level for a SNR of 18 dB and working at 100 Gbps. The performance of the ED-OOK combination is compared to that of *(1)* ideal coherent detection (Matched) with BPPM and BPSK, as well as *(2)* autocorrelation (Auto) with TR and DIFF modulations

be likely achieved by means of increasing the signal to noise ratio. Since errors in GWNoC are due to thermal noise and will not occur in bursts, forward error correction could provide an effective way to reduce the error probability at the expense of reducing the effective data rate. Reed-Solomon (RS) with low-density parity check (LDPC) coding schemes have been proposed in the 802.15.3c standard [1], which works upon the physical layer of millimeter-wave radio for high-rate WPAN networks. These may be suitable in GWNoC environments as they are expected to provide a low-complexity implementation and high error-correcting capability. $RS(n, k)$ codes build codewords of length n, k of which are data; the remaining $2t$ bits are for parity check, allowing the correction of up to t erroneous symbols within the codeword. Assuming p to be the bit error probability considering a raw channel, the use of $RS(n, k)$ codes reduces the BER to:

$$BER = 1 - \frac{(1-p)^n}{k} - \frac{n}{k}p(1-p)^{n-1}$$

at the expense of reducing effective data rate a factor of k/n with respect to the raw data rate. In addition to codes for error correction, low-weight codes could be also employed. When combined with the OOK modulation, low-weight coding reduces the average power consumption for a given link budget.

3.2.3 Medium Access Control (MAC)

Coordinating the access to the shared medium is a huge challenge in GWNoC. All processors will be located within the same transmission range, implying the existence of one collision region accounting for hundreds or thousands of nodes. Furthermore, some applications generate large amounts of communication throughout the chip. The design of a MAC protocol that coordinates the expectedly high number of simultaneous transmissions is therefore key to guarantee that the performance requirements of GWNoC will be met. Above all, such protocol must be scalable in terms of latency since it has a critical impact in the performance of the multiprocessor. Another important aspect is that GWNoC, unlike other wireless networks, must guarantee the delivery of broadcast packets to all nodes. Acknowledgement (ACK) packets must be conveyed to the transmitter in order to avoid losses due to collisions. However, this cannot be performed through the wireless plane as the "reply storm" would saturate the medium. Instead, the wired plane could be used.

The GWNoC scenario presents a set of peculiarities that may allow the design of opportunistic solutions. For instance, hidden or exposed terminal problems are avoided as all nodes are static and within the same transmission range. Also, most of the MAC protocols are designed assuming that no prior information on the traffic is available. However, this is not the case in our scenario. The size of the messages is known as it depends upon the multiprocessor architecture and the function of broadcast transmissions. In coherence protocols, coherence requests are short messages of around 8–16 bytes and responses may include cache lines up to 128 bytes. Further, estimations on the traffic to transmit can be generated by the compiler and provided at run-time so that the MAC protocol can adapt to the instantaneous traffic load. This feature may be employed to improve fairness or to avoid saturation in high-contention phases of some parallel programs.

Existing proposals rely on channelization approaches to control the medium access. Frequency-multiplexing schemes have been evaluated [38], but their scalability is compromised by the number of channels that will be required in manycore processors. Combinations of time-multiplexing and frequency-multiplexing schemes have been also proposed, seeking to increase the number of channels [19, 43] or the available bandwidth [22]. In this case, time-multiplexing schemes introduce a latency which is proportional to the number of cores and that may not be tolerated in manycore settings. Except for the work in [19], current proposals do not offer any reconfigurability option to adapt to the time-varying needs of the application. In this scenario, MAC protocols where nodes contend for the channel are generally a better choice. A carrier sensing approach (or energy sensing approach for impulse radio) may be adopted and adapted to provide means to take advantage of the information that the compiler could provide.

3.2.4 Network Layer

Since the main aim of GWNoC is to provide one-hop broadcast communication, routing or switching strategies are not required in the wireless plane. On the contrary, switching functionalities have to be added at the network interface as it has to deliberate through which plane a message is to be sent. The decision may be simply taken depending on whether the message is broadcast or not, or depending on upper layer policies such as congestion control. Another important aspect to carefully consider is multicast addressing. Since all messages are broadcast, the network interface must be provided with means to decide whether to keep or discard a message based on the address of the packet.

4 Conclusions and Future Work

In the manycore era, the exponential increase in *one-to-many* communication requirements intensifies the need for a scalable broadcast on-chip platform. Although the concept of wireless on-chip networks has been proposed and may be suitable to this end, size constraints hinder the use of metallic antennas and require the use of nanoscale techniques. In this chapter, we presented the concept of GWNoC, wherein graphene antennas, by virtue of their downscaled size which allow per-core wireless capabilities, deliver broadcast at the core level. We analyzed the GWNoC approach from both the communication performance and protocol design perspectives, providing models and guidelines for their evaluation.

On the one hand, we analyzed the unique properties of a graphene-enabled wireless link in Sect. 3.1. We introduced a methodology based on conductivity models for the simulation of graphene antennas as a function of different technological parameters. However, further research is required in order to fully understand these antennas and develop models that will capture all the phenomena that affect radiation. In the case of the propagation channel, we presented a general model and detailed the three propagation mechanisms present in the GWNoC scenario. We conclude that communication will mainly occur by means of the space waves that propagate through air and reflect upon the chip package. Terahertz wave propagation could be challenged by molecular absorption, but we have shown that its impact becomes negligible at the chip scale; whereas reflections are frequency-dependent and will require further work in material characterization at terahertz frequencies. Finally, in the case of the transceiver, we extrapolated performance trends from the state of the art to show that area and power objectives could be met by operating in the terahertz band.

On the other hand, we discussed the main protocol design aspects in Sect. 3.2. We first qualitatively analyzed the physical layer. We proposed to employ IR-UWB modulations using subpicosecond long pulses leading to terahertz-wide signals. Seeking simplicity and energy efficiency, we both discussed the use of OOK in combination with non-coherent detection and reviewed a model for its performance evaluation. Results show a good compromise between performance and robustness in front of timing effects, but also suggest the use of coding schemes to reduce the BER to acceptable levels for on-chip communication. At the MAC layer, we analyzed the main peculiarities of the scenario and concluded that frequency- or time-multiplexing options are not suitable due to their poor scalability. Instead, protocols where nodes contend for the channel could be used due to their potential adaptability to the time-varying communication requirements of manycore processors. Furthermore, information on the traffic to be served may be available at run-time and could be used to improve the network performance. In this regard, a detailed characterization of the traffic generated by cores when running a set of benchmark applications will be a helpful tool for the design of opportunistic MAC solutions.

References

1. 802.15.3c-Part 15.3 (2009) Wireless medium access control (MAC) and physical layer (PHY) specifications for high rate wireless personal area networks (WPANs)—Amendment 2: Millimeter-wave-based alternative physical layer extension
2. Abadal S, Alarcón E, Lemme MC, Nemirovsky M, Cabellos-Aparicio A (2013) Graphene-enabled wireless communication for massive multicore architectures. IEEE Commun Mag 51(11):137–143
3. Abadal S, Cabellos-Aparicio A, Lázaro JA, Nemirovsky M, Alarcón E, Solé-Pareta J (2013) Area and laser power scalability analysis in photonic networks-on-chip. In: Proceedings of the ONDM '13
4. Abadal S, Iannazzo M, Nemirovsky M, Cabellos-Aparicio A, Alarcon E (2015) On the area and energy scalability of wireless network-on-chip: a model-based benchmarked design space exploration. IEEE/ACM Trans Netw 23(5):1501–1513
5. Abadal S, Llatser I, Mestres A, Lee H, Alarcón E, Cabellos-Aparicio A (2015) Time-domain analysis of graphene-based miniaturized antennas for ultra-short-range impulse radio communications. IEEE Trans Commun 63(4):1470–1482
6. Abadal S, Martínez R, Solé-Pareta J, Alarcón E, Cabellos-Aparicio A (2016) Characterization and modeling of multicast communication in cache-coherent manycore processors. Comput Electr Eng (51):168–183
7. Arslan H, Chen Z, Benedetto MD (2006) Ultra wideband wireless communication
8. Beausoleil RG, Kuekes PJ, Snider GS, Wang, SY, Williams RS (2008) Nanoelectronic and Nanophotonic Interconnect. Proc IEEE 96(2):230–247
9. Benini L, De Micheli G (2002) Networks on chips: a new SoC paradigm. Computer 35(1): 70–78
10. Bienia C, Kumar S, Singh JP, Li K (2008) The parsec benchmark suite: characterization and architectural implications. In: Proceedings of the PACT '08, pp 72–81. ACM
11. Binkert N, Sardashti S, Sen R, Sewell K, Shoaib M et al (2011) The gem5 simulator. ACM SIGARCH Comput Arch News 39(2):1
12. Burns J, McIlrath L, Keast C, Lewis C, Loomis A, Warner K, Wyatt P (2001) Three-dimensional integrated circuits for low-power, high-bandwidth systems on a chip. In: IEEE ISSCC Dig Tech Papers:268–269
13. Cai W, White J, Brongersma M (2009) Compact, high-speed and power-efficient electrooptic plasmonic modulators. Nano Lett 9(12):4403–4411
14. Chan J, Hendry G, Biberman A, Bergman K, Carloni LP (2010) PhoenixSim: a simulator for physical-layer analysis of chip-scale photonic interconnection networks. In: Proceedings of the DATE '10, pp 691–696
15. Chang MCF, Verbauwhede I, Chien C, Xu Z, Kim J, Ko J, Gu Q, Lai BC (2005) Advanced RF/baseband interconnect schemes for inter- and intra-ULSI communications. IEEE Trans Electron Devices 52(7):1271–1285
16. Chang MF, Cong J, Kaplan A, Naik M, Reinman G, Socher E, Tam SW (2008) CMP Network-on-chip overlaid with multi-band RF-interconnect. In: Proceedings of the HPCA '08, pp 191–202
17. David Culler AG (1999) Parallel computer architecture: a hardware/software approach
18. Deb S, Ganguly A, Pande PP, Belzer B, Heo D (2012) Wireless NoC as interconnection backbone for multicore chips: promises and challenges. IEEE J Emerg Sel Topics Circuits Syst (JETCAS) 2(2):228–239
19. DiTomaso D, Kodi A, Matolak D (2013) Energy-efficient adaptive wireless NoCs architecture. In: Proceedings of the NoCS '13, pp 1–8
20. Enright Jerger N, Peh LS, Lipasti M (2008) Virtual circuit tree multicasting: a case for on-chip hardware multicast support. In: Proceedings of the ISCA-35, pp 229–240
21. Feero BS, Pande PP (2009) Networks-on-Chip in a three-dimensional environment: a performance evaluation. IEEE Trans Comput 58(1):32–45

22. Ganguly A, Chang K, Deb S, Pande PP, Belzer B, Teuscher C (2010) Scalable hybrid wireless network-on-chip architectures for multi-core systems. IEEE Trans Comput 60(10):1485–1502
23. Gorisse J, Morche D, Jantunen J (2012) Wireless transceivers for gigabit-per-second communications. In: Proceedings of the NEWCAS '12, pp 545–548
24. Grischkowsky D, Keiding S, van Exter M, Fattinger C (1990) Far-infrared time-domain spectroscopy with terahertz beams of dielectrics and semiconductors. J Opt Soc Am 7(10): 2006–2015
25. Han SJ, Garcia AV, Oida S, Jenkins KA, Haensch W (2014) Graphene radio frequency receiver integrated circuit. Nat Commun 5
26. Hanson GW (2008) Dyadic Green's Functions for an Anisotropic, Non-Local Model of Biased Graphene. IEEE Transactions on Antennas and Propagation 56(3):747–757
27. Hennessy J, Patterson D (2012) Computer architecture: a quantitative approach
28. Hesse R, Nicholls J, Jerger NE (2012) Fine-grained bandwidth adaptivity in networks-on-chip using bidirectional channels. In: Proceedings of the NoCS '12, pp 132–141. IEEE
29. Hoskote Y, Vangal S, Singh A, Borkar N, Borkar S (2007) A 5-GHz mesh interconnect for a teraflops processor. IEEE Micro 27(5):51–61
30. Huang W, Rajamani K, Stan M, Skadron K (2011) Scaling with design constraints: predicting the future of big chips. IEEE Micro:16–29
31. Jornet JM, Akyildiz IF (2011) Channel modeling and capacity analysis for electromagnetic wireless nanonetworks in the terahertz band. IEEE Trans Wirel Commun 10(10):3211–3221
32. Jornet JM, Akyildiz IF (2013) Graphene-based plasmonic nano-antenna for terahertz band communication in nanonetworks. IEEE J Sel Areas Commun 31(12):685–694
33. Kahng A, Li B, Peh L, Samadi K (2009) Orion 2.0: A fast and accurate noc power and area model for early-stage design space exploration. In: Proceedings of the DATE '09
34. Kim M, Rieh JS, Jeon S (2012) Recent progress in terahertz monolithic integrated circuits. In: Proceedings of the ISCAS '12, pp 746–749
35. Krishna T, Peh LS (2011) Towards the ideal on-chip fabric for 1-to-many and many-to-1 communication. In: Proceedings of the MICRO-44, pp 71–82
36. Kurian G, Miller J, Psota J, Eastep J et al (2010) ATAC: A 1000-Core Cache-Coherent Processor with On-Chip Optical Network. In: Proceedings of the PACT '10
37. Kürner T, Priebe S (2013) Towards THz communications—status in research, standardization and regulation. J Infrared, Millimeter Terahertz Waves 35(1):53–62
38. Lee SB, Tam SW, Pefkianakis I, Lu S et al (2009) A scalable micro wireless interconnect structure for CMPs. In: Proceedings of the Mobicom '09, p 217
39. Llatser I, Kremers C, Cabellos-Aparicio A, Jornet JM, Alarcón E, Chigrin DN (2012) Graphene-based nano-patch antenna for terahertz radiation. Photonics Nanostruct: Fund Appl 10(4):353–358
40. Llatser I, Kremers C, Chigrin D, Jornet JM, Lemme MC, Cabellos-Aparicio A, Alarcón E (2012) Radiation characteristics of tunable graphennas in the terahertz band. Radioeng J 21(4)
41. Llatser I, Mestres A, Abadal S, Alarcón E, Lee H, Cabellos-Aparicio A (2015) Time and frequency domain analysis of molecular absorption in short-range terahertz communications. IEEE Antennas Wirel Propag Lett 14:350–353
42. Matolak D, Kaya S, Kodi A (2013) Channel modeling for wireless networks-on-chips. IEEE Commun Mag 51(6):180–186
43. Matolak D, Kodi A, Kaya S, DiTomaso D, Laha S, Rayess W (2012) Wireless networks-on-chips: architecture, wireless channel, and devices. IEEE Wireless Commun 19(5):58–65
44. Miller DAB (2009) Device requirements for optical interconnects to silicon chips. Proc IEEE 97(7):1166–1185
45. Novack A, Liu Y, Ding R, Gould M, Baehr-jones T, Li Q, Yang Y, Zhang Y, Padmaraju K, Bergmen K, Lim AEJ, Lo GQ, Hochberg M (2013) A 30 GHz silicon photonic platform. In: Proceedings of the SPIE—Integrated optics: physics and simulations, vol 8781
46. O KK, Kim K, Floyd B, Mehta J, Yoon H, Hung CM, Bravo D, Dickson T, Guo X, Li R, Trichy N, Caserta J, Yang D, Bohorquez J, Seok E, Gao L, Sugavanam A, Lin JJ, Chen J, Brewer, JE (2005) On-chip antennas in silicon ICs and their application. IEEE Trans Electron Devices 52(7):1312–1323

47. Öjefors E, Grzyb J, Heinemann B, Tillack B, Pfeiffer UR (2011) A 820 GHz SiGe chipset for terahertz active imaging applications. In: Proceedings of the ISSCC '11, pp 224–225
48. Park JD, Kang S, Thyagarajan S, Alon E, Niknejad A (2012) A 260 GHz fully integrated CMOS transceiver for wireless chip-to-chip communication. In: Proceedings of the VLSIC '12, pp 48–49
49. Pande P, Grecu P, Jones C, Ivanov M, Saleh A (2005) Performance evaluation and design trade-offs for network-on-chip interconnect architectures. IEEE Trans Comput 54(8):1025–1040
50. Rodrigo S, Flich J, Duato J, Hummel M (2008) Efficient unicast and multicast support for CMPs. In: Proceedings of the MICRO-41 pp 364–375
51. Ronne C, Thrane L, AŁstrand PO, Wallqvist A et al (1997) Investigation of the temperature dependence of dielectric relaxation in liquid water by THz reflection spectroscopy and molecular dynamics simulation. J Chem Phys 107(14):5319
52. Schinkel D, Mensink E (2009) Low-power, high-speed transceivers for network-on-chip communication. IEEE Trans VLSI Syst 17(1):12–21
53. Seok E, Shim D, Mao C, Han R, Sankaran S, Cao C, Knap W (2010) Progress and challenges towards terahertz CMOS integrated circuits. IEEE J Solid-State Circuits 45(8):1554–1564
54. Shacham A, Bergman K, Carloni LP (2008) Photonic networks-on-chip for future generations of chip multiprocessors. IEEE Trans Comput 57(9):1246–1260
55. Socher E, Chang MCF (2007) Can RF Help CMOS processors? IEEE Commun Mag 45(8):104–111
56. Soteriou V, Wang H, Peh LS (2006) A statistical traffic model for on-chip interconnection networks. In: Proceedings of the MASCOTS '06
57. Stallo C, Mukherjee S (2010) IR-UWB for high bit rate communications beyond 60 GHz. In: Proceedings of the PIMRC '10
58. Sun C, Chen C, Kurian G (2012) DSENT—a tool connecting emerging photonics with electronics for opto-electronic networks-on-chip modeling. In: Proceedings of the NoCS '12, pp 201–210
59. Tamagnone M, Gomez-Díaz JS, Mosig JR, Perruisseau-Carrier J (2012) Analysis and design of terahertz antennas based on plasmonic resonant graphene sheets. J Appl Phys 112:114, 915
60. Tamagnone M, Gomez-Díaz JS, Mosig JR, Perruisseau-Carrier J (2012)Reconfigurable terahertz plasmonic antenna concept using a graphene stack. Appl Phys Lett 101(21), 214, 102
61. Urkowitz H (1967) Energy detection of unknown deterministic signals. Proc IEEE 55(4)
62. Vantrease D, Schreiber R, Monchiero M, McLaren M, Jouppi N, Fiorentino M, Davis A, Binkert N, Beausoleil R, Ahn J (2008) Corona: system implications of emerging nanophotonic technology. ACM SIGARCH Comput Architect News 36(3):153–164
63. Witrisal K, Leus G, Janssen GJM, Pausini M, Troesch F, Zasowski T, Romme J (2009) Noncoherent ultra-wideband systems. IEEE Signal Process Mag 26(4):48–66
64. Woo S, Ohara M, Torrie E, Singh J (1995) The SPLASH-2 programs: characterization and methodological considerations. In: Proceedings of the ISCA-22, vol 23, issue no 2, pp 24–36
65. Wu Y, Farmer DB, Xia F, Avouris P (2013) Graphene electronics: materials, devices, and circuits. Proc IEEE 101(7):1620–1637
66. Zhang YP, Chen ZM, Sun M (2007) Propagation mechanisms of radio waves over intra-chip channels with integrated antennas: frequency-domain measurements and time-domain analysis. IEEE Trans Antennas Propag 55(10):2900–2906

Energy Harvesting in Nanonetworks

Shahram Mohrehkesh, Michele C. Weigle and Sajal K. Das

Abstract The goal of this chapter is to review the process, issues, and challenges of energy harvesting in nanonetworks, composed of nanonodes that are nano to micro meters in size. A nanonode consisting of nan-memory, a nano-processor, nano-harvesters, ultra nano-capacitor, and a nano-transceiver harvests the energy required for its operations, such as processing and communication. The energy harvesting process in nanonetworks differs from traditional networks (e.g. wireless sensor networks, RFID) due to their unique characteristics such as nanoscale, communication model, and molecular operating environment. After reviewing the energy harvesting process and sources, we introduce the communication model, which is the main source of energy consumption for nanonodes. This is followed by a discussion on the models for joint energy harvesting and consumption processes. Finally, we describe approaches for optimizing the energy consumption process, which includes optimum data packet design, optimal energy utilization, energy consumption scheduling, and energy-harvesting-aware protocols.

1 Introduction

The advancement of nanotechnology promises to provide a significant rise in small scale communication. Wireless nanonetworks [1, 2] are a next generation of net-

S. Mohrehkesh (✉)
Department of Computer and Information Sciences, Temple University,
Philadelphia, PA, USA
e-mail: shahram@temple.edu

M.C. Weigle
Department of Computer Science, Old Dominion University,
Norfolk, VA, USA
e-mail: mweigle@cs.odu.edu

S.K. Das
Department of Computer Science, Missouri University of Science and Technology,
Rolla, MO, USA
e-mail: sdas@mst.edu

J. Suzuki et al. (eds.), *Modeling, Methodologies and Tools for Molecular and Nano-scale Communications*, Modeling and Optimization in Science and Technologies 9, DOI 10.1007/978-3-319-50688-3_14

works at nano scale. Nanonodes in such a network are composed of nano antennas, nano-memory, nano-processor, nano-sensors, nano-sclae energy storage, and so on. Each nanonode is in the range of nano to micro meters in size. The nanoscale property of nanonodes opens the door for exciting new applications. For example, nanosensors could detect chemical compounds at the molecular level or the presence of different infectious agents, such as viruses or harmful bacteria [2]. Many other applications can also be imagined in the biological, medical, chemical, environmental, military, and industrial domain [2]. For example, nanosensors could be used to develop new touch surfaces or be added to standard office products (e.g., pens, papers, etc.), thus making the idea of smart offices a reality.

The functionalities of nanonodes are realized only through communication. Nanosensors will collect useful information that must be sent outside of their sensing environment for storage and additional processing. Nanonodes need to communicate to control or actuate an action, or similarly monitor a phenomena. In other words, they will need to communicate between themselves as well as with nodes in other networks, e.g., local area network. Among all possible models of communication in nanonodes (e.g. electrical, molecular, optical, or acoustic), studies [1] show that electromagnetic communication in the 0.1–10.0 terahertz (THz) frequency band is a promising approach for communication in nanonetworks. Therefore, we focus on the THz communication mechanism, in which nanosensors can consume low energy while having connectivity at the nano scale. The energy for communication is the main part of energy consumption for nanonodes [23].

Due to the size limitation of nanonodes, only a limited energy storage can be considered, where a nanonode harvests and stores energy from ambient sources. Various sources, e.g., vibration, heat, and light, exist for energy harvesting. The use of each type of source corresponds to the particular environment or application. For example, where light is not available, heat can be used as the source of energy. The energy availability of most sources follows stochastic processes.

In addition to the size limitation of energy storage and the stochastic energy harvesting process, the harvester size also needs to be adapted to nanoscale for integration with nanonodes. For example, piezoelectric nanogenerators enable a high energy harvesting rate through compress-release cycles of the nanowires on an ultra-nanocapacitor [23]. Similarly, nano-carbon-based and nano-ceramics-based photovoltaics show promising light absorption properties and can be integrated with nanonodes. The use of biofuel cells to harvest energy from various materials such as blood sugar is another recent advancement in harvesters for nanonodes.

The new communication model, size limitation of nanonodes (which result in low processing capabilities and low harvest rate), and stochastic properties of energy harvesting process create a paradigm of new challenges (in modeling and optimization of energy consumption) to solve for the realization of nanonetworks. Because the energy is expected to be renewed, it is important to achieve the maximum utilization of this energy while keeping a nanonode operational. This differs from traditional energy-saving models (e.g., duty cycles, balancing energy consumption among all nodes, data compression, data aggregation, etc.) although some of these may still be applicable to nanonetworks. Because energy harvesting in nanonetworks is in its early stage of research, we briefly study the energy harvesting in wireless sensor and RFID networks with two goals: (I) introducing the ideas in these networks that may be adapted for nanonetworks; and (II) highlighting both their limitations that prevent them from being used in nanonetworks, and new approaches that could be taken.

Typically, communication is the most energy-consuming operation for nanonodes, implying that the models should focus on energy harvesting in combination with energy consumption for communication. We introduce some methods for optimizing the consumption of energy, including finding the optimum packet size, packet scheduling, and energy harvesting-aware techniques in the realm of nanonetworks. More importantly, the effect of optimizing energy consumption at each nanonode on the overall performance of the nanonetwork should be evaluated. For example, while a nanonode should not be a no-energy state in order to avoid unsuccessful transmission of packets, it should also not remain in a full energy state in order to ensure that available energy is harvested and network utilization is maximized.

The remainder of this chapter is organized as follows. In Sect. 2, we give an overview of the taxonomy and properties of energy harvesting. Moreover, we introduce various sources of energy as well as nano scale harvesters for nanonodes. Section 3 introduces the communication model between nanonodes. Section 4 discusses the techniques for modeling energy harvesting and consumption processes. Section 5 analyzes the optimizing factors of energy consumption based on the properties of the pulse-based communication model. Finally, the chapter is concluded in Sect. 6 with open research issues in the energy harvesting process for nanonetworks. Particularly, open questions related to optimum energy consumption and energy harvesting-aware protocols are discussed.

2 Energy Harvesting

Research in energy harvesting has attracted attention in recent years due to the availability of devices that can harvest solar energy. However, solar energy is limited to specific times and locations. Therefore, researchers have investigated new methods of energy harvesting such as ambient vibration or heat. Independent of the type of source for energy harvesting, energy sources mainly share a common property: the availability and quantity of energy follows a stochastic process.

Energy sources are categorized broadly into (I) ambient energy sources such as, solar, wind, radio frequency (RF), and vibration; and (II) human power [56], which in turn could be passive (i.e., uncontrollable) such as blood pressure, body heat, heartbeat and breath, or active (i.e., controllable) such as finger motion, paddling, and walking. In the following we present a taxonomy of energy harvesting.

2.1 Taxonomy of Energy Harvesting

There are three main metrics for the evaluation of harvesting methods [4, 10, 56]:

- ***Conversion Efficiency***: This is the amount of energy harvested as compared to the amount of available energy.
- ***Energy Harvest Rate***: This specifies how fast the energy can be harvested and depends on the type of energy, among other factors. For example, in the case of solar energy, the size of the solar panel or weather conditions (e.g., sunny or cloudy) affect this parameter. On the other hand, in vibrant energy harvesting, the rate of vibration affects the rate of energy harvesting.
- ***Power Density***: This indicates the amount of power (time rate of energy transfer) per unit volume, measured in $\frac{Watt}{m^3}$, that a device (harvester or battery) can offer.

The harvested energy is used in two ways:

- **Harvest-Use**: In this method when the energy is produced, it is used immediately. An example of this method is pushing a key/button. Pushing produces some energy that can be used to transfer an electronic signal.
- **Harvest-Store-Use**: In this method, energy is harvested whenever possible, and is stored for future use. This method is more useful since there is always some energy available if it is consumed wisely. The limitation comes only from the capacity of storage. Most of the studies in the domain of networking use the harvest-store-use method. In these situations, two approaches are taken: (I) finding the required capacity of storage to meet the application requirements; and/or (II) trying to optimize usage of this energy.

More detailed studies describing the energy harvesting taxonomy are available [4, 50, 56]. In the following, we focus on possible energy sources and models of energy harvesting for nanonetowrks.

2.2 Sources of Energy for Nanonodes

Energy harvesting plays a major role in the realization of nanonetworks. Due to the limited size of nanonodes as well as some of their applications in environments with no light (e.g. inside body, in liquids), other energy sources are considered. Mechanical energy (from vibration and motion) and chemical energy are the two main sources of energy nanonodes, especially in biological environments. Thermal energy is not efficient and has downsizing limitations. In the following, we discuss the state-of-art in energy harvesting for nanonodes.

2.2.1 Mechanical-Energy Harvesting

Vibration-based mechanical energy is ubiquitous in environments where solar and thermal energy are not available or accessible. Mechanical vibrations exist in a wide range of frequencies, from a few hertz to several kilohertz, which result in power densities ranging from a few microwatts to milliwatts per cubic centimeter [49].

The harvesting of mechanical energy by piezoelectric materials is the main approach to directly convert mechanical energy into electricity. Traditionally, lead zirconate titanate, or PZT, has been the material mostly used for mechanical energy harvesting. However, issues such as the reliability, durability, and safety of these materials limit their usage for long-term operations. Recently, piezoelectric nanowires (NWs) have shown promise in the harvesting of mechanical energy at nanoscale [18].

Nano-generators (NGs) based on NWs can be fabricated on various substrates, including polymers, semiconductors, and metals, and even on unconventional non-planar substrates, such as fibers [58]. By weaving bundles of such fibers into fabrics, potential applications such as smart clothes can be imagined. Fiber NGs based on similar configurations have been developed to harvest low-frequency vibrations induced by air, liquid flow, exhalation and the heartbeat of a human body [32]. The human body contains other significant mechanical energy induced by continuous activities, such as breathing and the beating of the heart, and discontinuous motions, such as walking and muscle stretching.

Mechanical energy from vibration and motion is available in many other environments, which makes it an invaluable source of energy in many medical as well as indoor industrial applications for nanonetworks. Table 1 represents some of the potential sources for harvest energying from vibration.

Table 1 Peak frequency and acceleration amplitude for various vibration sources [11, 48, 51]

Vibration source	Peak frequency (Hz)	Acceleration amplitude ($\frac{m}{s^2}$)
Refrigerator	240	0.1
Car engine compartment	200	12
Door frame just after door closes	125	3
Kitchen blender casing	121	6.4
Clothes dryer machine	121	3.5
Small microwave oven	121	2.25
Washing machine	109	0.5
External windows next to a busy street	100	0.7
Second story of wood frame office building	100	0.2
HVAC in office buildings	60	0.2–1.5
Vehicles	5–2000	0.5–110
Person nervously tapping their heel	1	3

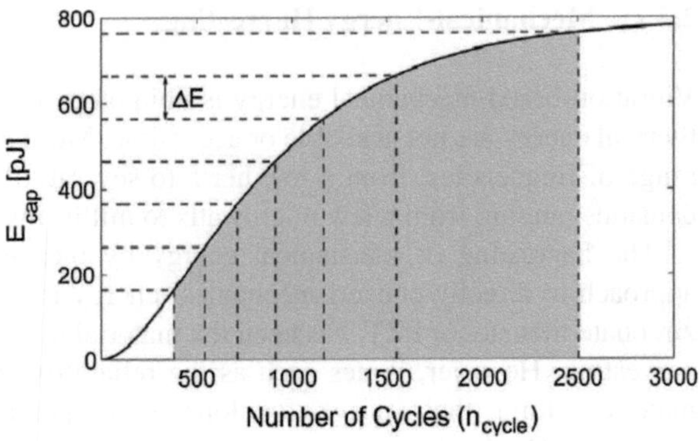

Fig. 1 Energy harvesting model [23]

New generations of piezoelectric-nanowire are sensitive to very low acceleration [60]. Therefore, the main parameter that affects the amount of energy harvesting is the frequency. This means that the variation in the vibration rate will result in a stochastic model for available energy for a nanonode at different times and different locations. Moreover, energy storage in ultra nano-capacitors is a non-linear process. Therefore, the first issue is to understand and model the energy harvesting process where the stochastic and nonlinear behavior of harvesting is included.

In [23], an energy harvesting model has been proposed for storing energy process in an ultra-nanocapacitor by piezoelectric nanogenerators, as shown in Fig. 1. In this model, energy is harvested through vibrations, which produce compress-release cycles of the nanowires on a nano-capacitor. For a specific ultra nano-capacitor, the stored energy is specified by the number of cycles. The energy-harvesting rate (Joules/second) is defined as

$$\lambda(E_{cur}, \Delta E) = \frac{1}{t_{cycle}} \cdot \frac{\Delta E}{n_{cycle}(E_{cur} + \Delta E) - n_{cycle}(E_{cur})} \tag{1}$$

where t_{cycle} is the time between cycles, $n_{cycle}(E)$ is the number of cycles required to generate E Joules, E_{cur} is the current energy level, and ΔE is the amount of energy increase. We note that if every vibration generates one cycle, then the inverse of t_{cycle} is the vibration rate.

The amount of power that can be harvested through vibration is compared with other sources of energy in Table 2 in terms of power density. Power density is the amount of power (energy transfer per unit time) per unit volume or surface. Depending on the power source, the power is measured [50] for volume or surface, respectively called volume power density, which is expressed as $\mathrm{W/m^3}$, and surface power density, which is expressed as $\mathrm{W/m^2}$.

As can be seen, piezoeletric nanowire provides a significant amount of power density. The limitations in harvesting the energy comes from size limitations for nanonodes (scale of nano to micro meters) as well as the availability of the vibration source. For example, as can be seen from arm motion, power density up to 330 $\mu\mathrm{W/cm^3}$ can be extracted.

Table 2 Comparison of power density for various harvesting sources and technologies [51, 60, 61], * = μW/cm²

Source/technology	Power density (μW/cm³)
Solar (outdoor)	15,000 direct sun, 150 cloudy day
Solar (indoor)	6 office desk
Vibration (piezoelectric conversion)	250
Vibration (electrostatic conversion)	50
Acoustic noise	0.003 at 75 dB, 0.96 at 100 dB
Temperature gradient	10–60*, depends on temperatures and difference, known as Carnot efficiency
Shoe inserts (pizoelectric vibration)	330 [54, 55]
Vibration (small microwave oven)	116
Batteries (non-rechargeable lithium)	45
Batteries (rechargeable lithium)	7
Arm motion	330
Piezoelectric-nanowire [60]	2800
Running	Max from kinetic 300 [39]
Walking	Max from kinetic 30 [39]
Light	Outdoor at night 25, indoor 100* [12]
RF	0.02–40*, depends on source and distance

2.2.2 Biofuel Cells (BFCs) for the Harvesting of Chemical and Biochemical Energy

A fuel cell converts the chemical energy of a fuel, such as hydrogen or methanol, into electricity through a chemical reaction with an oxidizing agent, such as oxygen or air [16]. In contrast to batteries, in which chemical materials are used to store electrical energy, fuel cells extract chemical energy from reactants and convert the extracted chemical energy into electricity as long as the reactants are available. Although it is a mature technology that has been used extensively at macroscale applications, conventional fuel cell technology has several inherent disadvantages such as the materials used, the fabrication cost, and size restrictions, for the cost-effective solution at the micro and nano scale applications such as implanted biomedical sensors. Therefore, a biofuel cell (BFC) is introduced where it simply uses biological enzymatic substances, rather than metals, to catalyze the anode and/or cathode reactions.

BFCs can be classified into two categories: (I) microbial fuel cells (MFCs), where the catalytic enzymes involved are in living cells; and (II) enzymatic BFCs, where the catalytic enzymes involved are located outside of living cells [19]. MFCs demonstrate unique features such as long-term stability and fuel efficiency. However, the power densities associated with MFCs are typically lower than BFCs [19]. Thus, the application of MFCs at the micro and nano scale is limited. On one hand, enzymatic BFCs are biocompatible and can provide efficient power on order of sub-mWcm^{-2},

Table 3 Comparison of energy-harvesting techniques for biological environments

Energy source	Harvesting principle	Power density	Advantages	Disadvantages
Mechanical	Piezoelectric	1–10 $mWcm^{-2}$	Ubiquitous and abundant	Low efficiency, stochastic availability
Biochemical	chemical reactant	0.1–1 $mWcm^{-2}$	Biocompatible, inexpensive, abundant in biological environment	Low power output, poor reliability, limited lifetime

which makes them applicable in micro and nanoscale applications such as in vivo biochemical/biomedical applications through the harvesting of biochemical energy directly from the human body. On the other hand, current enzymatic BFCs normally suffer poor stability. Table 3 summarizes the two main approaches of energy harvesting in biological environments.

2.2.3 Hybrid Cells for the Harvesting of Biomechanical and Biochemical Energy

Most energy harvesting methods, e.g., biomechanical or biochemical, are developed based on the existence of a certain type of energy source, while the other types of energy were wasted. Moreover, as illustrated in Table 3, the properties of biomechanical and biochemical energy harvesting methods are complements of each other. Therefore, new research directions try to develop innovative approaches for concurrent harvesting of energy from multiple types of sources through the use of integrated structures/materials [58]. This will help the energy harvesting process because at nano scale, the temporal/spatial distribution and availability of energy sources vary drastically [58].

Mechanical and biochemical energy are abundant in the biological environment due to body motion, muscle stretching, and metabolic processes. Therefore, hybrid solutions of these two energy sources are emerging as a new approach for energy supply in biological environments. A hybrid energy scavenger [15] was developed recently, which consists of a piezoelectric nanofiber NG for harvesting mechanical energy, such as from respiration and blood flow in the vessels, integrated with a flexible enzymatic BFC for harvesting the biochemical energy from the chemical processes between glucose and O_2 in biofluids. These two energy harvesting approaches, integrated within one single device, can work individually as well. Studies [15, 44] have demonstrated the feasibility of applying these energy harvesters in building self-powered nanodevices for in vivo biomedical applications to power nanosensors.

2.3 *Future*

More advancement in energy harvesting downscaling is required to integrate the harvesters from sources such as solar, light and thermal into nanonodes. Currently, energy harvesting from mechanical or biochemical sources are the main approaches to supply energy for nanodevices. These are also applicable for in vivo medical applications. New sources of energy for biochemical energy harvesting are emerging every day. For example, energy harvesting from blood sugar by biofuel cells [37] or from electrical differences in the inner ear [38] are new sources of energy. Moreover, advancements in nanodevices can be helpful in the production of nanoscale RF energy harvesters. Currently, RF energy harvesters are widely used for wireless sensor or RFID networks [45]. With the help of nanotechnology, this could be a significant source of energy, which is also controllable. Moreover, inductive charging [6], which is currently deployed for many medical applications in body area networks, could be investigated. Again, the size limitation is likely the main barrier for its usage at nanoscale.

3 Communication

As the energy for communication is the main part of energy consumption for nanonodes, in this section, we briefly describe the communication model for nanonodes. Two major possible mechanisms are envisioned [1] for communications among nanonodes: molecular communication and electromagnetic communication. The molecular communication is mainly based on the chemical and physical interactions, which have different consumption models and are not yet known completely [46]. Therefore, we focus only on the electromagnetic communication.

3.1 *Electromagnetic Communication*

Electromagnetic (EM) communication has been proposed [1, 2] as a communication method for nanonetworks. More specifically, pulse based communication in the 0.1–10 THz has been studied. There are several limitations in existing silicon-based manufacturing techniques (e.g., silicon atom size, heating and current leakage) that make the downscaling of existing EM transceivers infeasible [36]. Alternatively, nanomaterials are envisioned to solve part of building a new generation of electronic components that overcome shortcomings of current technology [3]. Carbon Nanotubes (CNTs) and Graphene Nanoribbons (GNRs) among other graphene based materials are expected to be the silicon of the 21st century [30]. The EM properties on these nanomaterials should be evaluated in terms of bandwidth for EM emission, the time lag of the emission, and the magnitude of the emitted power for a given input energy, among others. Ongoing research on the EM emission on graphene are

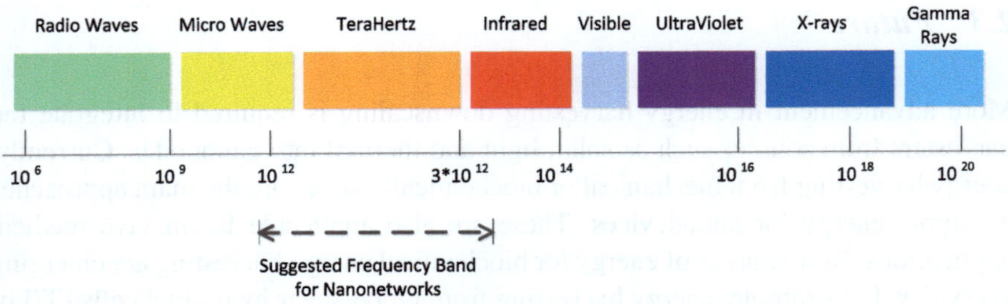

Fig. 2 Frequency bands 0.1–10 THz is suggested for nanonetworks

Fig. 3 Path loss in terahertz [21]

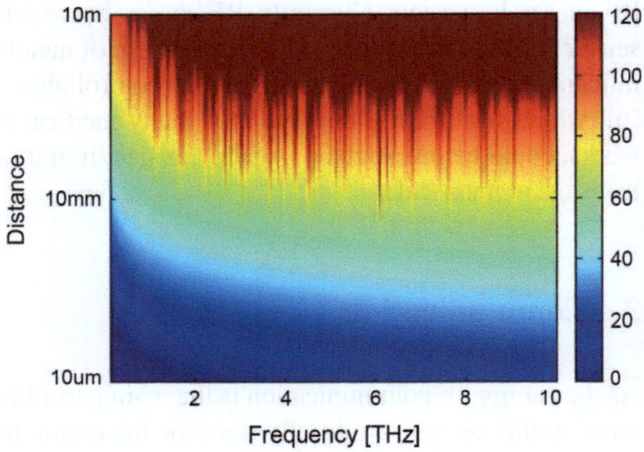

indicating the 0.1–10 Terahertz band (Fig. 2) as the expected frequency range of operation of future nano EM transceivers [24, 52]. In particular, it is determined that a 1 μm long graphene-based nano-antenna can only efficiently radiate in the Terahertz range. This matches the initial predictions for the operation frequency of graphene-based RF transistors [33].

Communication in terahertz is very sensitive to communication distance. Figure 3 illustrates the path loss at various distances in the THz band. For distances larger than one meter in a gaseous environment with 10% water vapor, path loss exceeds 100 dB. The path loss for 1 cm distance is around 50 dB. As can be seen, the path loss depends significantly on both the distance and frequency. Therefore, the power requirement for various distance and frequencies would vary significantly and should be considered in any communication design scheme.

3.2 *Pulse-Based Communication Model for Nanonetworks*

Pulse-based communication [8] is a known method in Ultra-Wide-Band (UWB) networks as Impulse Radio Ultra-Wide-Band (IR-UWB) systems. The pulse-based communication model for nanonodes, based on the model proposed in [2, 23], oper-

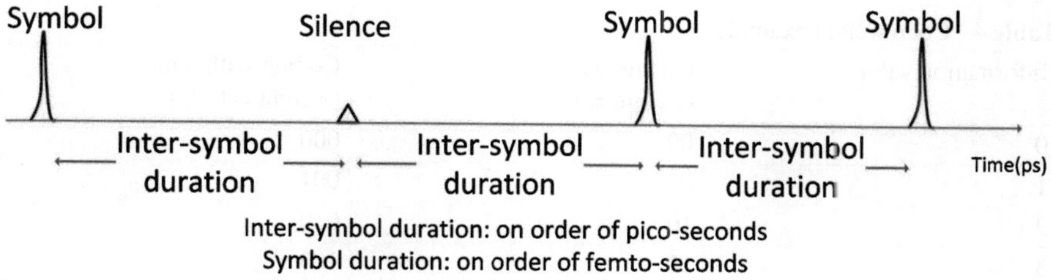

Fig. 4 RD TS-OOK modulation for transfer of `1011`

ates in the 0.1–10 THz frequency band, which results in a micro to millimeter communication range [2, 20]. The main reason for this frequency selection is due to the limitation in antenna size that a nanonode can afford, e.g., 1 μm. Communication in the THz band presents new channel properties: molecular absorption, thermal effects, and so on [20].

Nanonodes use the pulse-based communication and Rate Division Time Spread On-Off Keying (RD TS-OOK) [25] as the modulation mechanism (Fig. 4). A logical `1` is transmitted as a femto-second long pulse, and a logical `0` is transmitted as silence. RD TS-OOK is a simple modulations, but it is envisioned because more complex modulation (e.g., pulse amplitude, pulse width, pulse rate) used in IR-UWB are not applicable in nanonodes due to the capabilities limitations of nanonodes [25].

The duration of each pulse is T_p and the time between two symbols is T_s, producing a symbol rate of $\beta = \frac{T_s}{T_p}$. The selection of optimal β is still an open question. It certainly will depend on the hardware capabilities of the transmitter and receiver. Assuming there is no limitation in hardware capabilities, the existence of several flows of symbols from neighbor nanonodes will result in the collision of symbols. Moreover, energy availability is another factor that can affect the design of β. Currently, it is assumed that β takes values on the order of thousands.

In OOK modulation, since silence does not consume energy, any scheme that could produce fewer `1`s is preferred. For example, using *code weight* [22] has been proposed [23] to reduce energy consumption. The code weight basically reduces the number of `1`s by adding extra bits so that data is coded in a way that a fewer number of `1`s are present in the coded bits. This results in less energy in transmission and higher energy in reception. Reception of either a `0` or `1` costs the same energy, so sending more bits results in higher energy consumption for the receiver. Energy savings could happen only if the energy for reception is lower than transmission, which is the typical case in wireless transmission and has been shown in [23, 25]. Moreover, the code weight will not necessarily save energy in broadcast or multicast scenarios. Not only should the optimum value for this trade-off be identified, but other methods of coding information regarding the limitation of nanonodes are also of interest.

Since the transmission of `0`s in RD TS-OOK pulse-based modulation is equal to silences which do not consume energy, the lower code weight can reduce the energy

Table 4 Code weight example

Information value	Coding with 2 bits (weight = 0.5)	Coding with 3 bits (weight = 0.25)
0	00	000
1	01	001
2	10	010
3	11	100

consumption. Moreover, the code weight can lower the collisions since fewer `1`s, which are the only pulses that can face collision, are transmitted. A code weight of 0.5 means that, on average, there are an equal number of `1`s and `0`s in the packets. A lower weight, such as 0.4, means that there are fewer `1`s. However, it also means that more bits should be used to send the same of amount of information. For example, Table 4 shows how the number of `1`s for sending two bits of information could be reduced by using three bits. The code weight in this example is decreased from 0.5 to 0.25.

As a more realistic example, for sending $n = 64$ bits of information with a code weight of 0.4, at least $a = 6$ more bits will be added to each packet. In this case, the total number of encoded bits would be $m = 70$ and the number of `1`s, denoted as u, is less than or equal to 28.

To make sure that, for a target code weight, there are at most u `1`s independent of the original bit values, for n bits of information, the $\frac{m!}{(m-u)!u!} \geq 2^n$ condition must be satisfied [22], where m is the total number of bits, a is the number of additional bits, and $m = n + a$.

The method to determine the additional number of required bits is as follows. First, for a specific code weight W, u is specified as

$$u = \lceil W \cdot m \rceil, \tag{2}$$

and the following condition must be satisfied with the minimum a, where $m = n + a$.

$$\frac{m!}{(m-u)!u!} \geq 2^n \tag{3}$$

Note that sending fewer `1`s consumes less energy in the sender while it consumes more energy in the receiver. Energy is consumed when receiving any bit, `0`s or `1`s. Decreasing the code weight necessarily increases the packet size, increasing the cost to the receiver. Depending on the packet length and the ratio of energy required for reception to that for transmission of a pulse, named as α, the code weight may or may not save energy in total. Here, the assumption is that α is small, say 0.1; therefore, the aim is to find the optimum values for *packet length* and *code weight*, which we address in Sect. 5.

The probability of collision between symbols is extremely low due to the fact that there can be no collisions for `0` symbols (silences) and that the length of T_s is much longer than T_p (typically 1000 times larger). However, unlike other frequency ranges of electromagnetic signals, there is molecular absorption noise, for example 10^{-4} bit error rate (BER) for 10% water vapor. To mitigate the effect of these problems, repetition and code weight techniques have been proposed in [22, 23].

Repetition is a simple mechanism for error detection and correction. In this method, the sender simply repeats the symbol several times, typically 1 to 9 times. For example, in 3-repetition, a `1` would be transmitted as `111`. In this case, if one or two of these `1`s were not received, the problem could be detected at the receiver, and the information (i.e., a bit of `1`) would still be received. Although it is not the most efficient method, it is the simplest. Other methods for coding and error detection and correction are being investigated [5, 31].

4 Modeling of Energy Harvesting and Consumption

Modeling of energy harvesting and consumption has been the topic of research in other networks such as sensor and RFID networks. In this section, we first give an overview of existing models. Then we discuss the lack of models for nanonetworks. A recent model for nanonetworks that incorporates some of the properties of nanonetworks is introduced at the end of this section.

4.1 Models for Other Networks

There has been much previous works on modeling energy harvesting (see for example [12, 53]). In [53], energy harvesting and energy consumption is modeled as a queuing system, and based on stationary analysis, a transmission strategy is proposed to optimize the throughput of a sensor node. This model considers only one node and the energy required for transmission. It also assumes that the data buffer and energy storage are infinite, which may not be the case in many situations such as nanoscale nodes. In [12], the best spending rate of energy consumption for a node/link is derived through optimization and lexicographic frameworks. The authors developed an algorithm for predictable energy inputs as well as stochastic models. The model has been evaluated in a network of RFID active tags.

Models and algorithms for energy harvesting and consumption could be categorized in various aspects as follows [12]:

- *energy model profile*: Several parameters such as energy source (e.g., solar, vibration, RF) and environment (e.g., indoor/outdoor, vibration rate, temperature) can produce different energy model profiles. *Predictable*, *partially predictable*, *stochastic*, and *model free* are known categories that have been identified and studied [4, 56].

- *ratio of energy storage capacity to energy harvested*: This parameter specifies how fast the energy storage is filled. It depends both on the capacity of energy storage and the availability of energy. In other words, it connects the *energy harvesting rate* to *power density*.
- *time granularity*: This specifies the timescale of decision making and design schemes, algorithms, and protocols. The timescale can be in the range of seconds to days. It is related to the *storage-harvesting ratio* as well as the *energy profile model*. The higher the time granularity, the more accuracy is required of a design. This is important in applications where there are QoS requirements for data transfer.
- *problem size*: When solving any problem for efficient energy harvesting, the design can be evaluated in the domain of a node, pairwise nodes (link), or network wide (e.g., routing).

In the following, we describe some literature that model the energy harvesting process. Table 5 compares the works based on various design aspects. These models can be categorized into two general types: lexicographic and stochastic.

Lexicographic[1]: In [12] a solar power model for active tag RFID nodes is proposed, which operates based on various time fair energy allocations for both predictable energy inputs as well as stochastic inputs. Here, the authors provide some real environment measurements, and develop a prediction model for energy arrival. Next, they use the lexicographic maximization and utility maximization framework for modeling the energy spending rate, and achieve fair allocation of resources among nodes over a one day duration. Considering a stochastic energy arrival, the authors claim that, based on a developed Markov Decision Process, an optimal energy spending policy can be achieved for a single node or link. In [34], a fair and high throughput data extraction as well as a routing path solution among all nodes is designed, where the energy model is developed for solar power. They develop a centralized solution and two distributed solutions. The main idea is to adapt the extraction rate (sensing and sending rate of information) based on the available energy. A rate assignment for data transfer is found through lexicographical optimization. Even though the strength of the scheme is that it is independent of the energy arrival profile, the optimization solution works only on a large time scale, i.e., a day.

Stochastic: The energy arrival and consumption as a $G/G/1$ queue is modeled in [53]. After finding the stationary state of the model in some specific conditions, this model attempts to find the optimum throughput (largest possible data rate of packets) based on energy management policy, and also minimizes the delay of packets in the buffer. The optimization model is called α-discount optimal and is developed based on the stationary state of Markov model. The main weakness of this model

[1]Lexicographic optimization is a form of multi-criteria (multi-objective) optimization in which the various objectives under consideration cannot be quantitatively traded off between each other, at least not in a meaningful and numerically tractable way. The lexicographic method assumes that the objectives can be ranked in the order of importance. It can be assumed, without loss of generality, that the k objective functions are in the order of importance so that f_1 is the most important and f_k the least important to the decision maker. Then, the lexicographic method consists of solving a sequence of single objective optimization problems [47].

Table 5 Comparison of energy harvesting and consumption modeling

References	Energy model profile	Time granularity	Problem size	Solution method	Network	Energy source
[12]	Predictable and stochastic	Day	Node, link	Lexicographical	RFID	Light
[34]	Almost independent	Day	Node, network	Lexicographical	Sensor network	Solar
[53]	Independent (general arrival)	Seconds-minutes	Node, partially network	Queueing	Sensor network	Any
[29]	Almost independent (general arrival)	Seconds-minutes	Node, network	Time discrete	Sensor network	Any
[59]	Independent	Seconds	Network	Stochastic network calculus	N/A	N/A

is that it assumes that the energy buffer and the data buffer are infinite. The goal of the scheme developed in [29] is to achieve the highest data rate that results in a long term optimal solution. The advantage of the scheme is that it requires no explicit knowledge of the energy harvesting profile or traffic generation process. In fact, it is a learning system that adapts itself based on the environment (i.e., available energy) and network circumstances. This scheme works at the node level as well as at the network layer. This work is limited to analysis, with no simulation or test-bed results. In [35], on the other hand, an optimized training model is developed for a transmission policy that specifies the energy spending based on channel state information (CSI). This model also assumes an infinite buffer level. Finally, in [59], a model is proposed for evaluating the stochastic properties of energy harvesting while evaluating the network performance, such as satisfying a soft QoS. However, it is not clear how efficient the model would be.

There is additional works in the literature involving stochastic modeling of energy consumption that focus on other aspects of energy harvesting. For example, a stochastic optimization framework is proposed in [9] for modeling the stochastic behavior of the channel while achieving the best policy on transmission and energy consumption. In [28] the duty cycle of sensor nodes is modeled, assuming that nodes cannot harvest energy and communicate simultaneously. Therefore, nodes are switched between active and passive states. The goal is to optimize the timing of sleep/awake to maximize a utility function, such as throughput.

4.2 Model for Nanonetworks

As discussed in Sect. 4.1, many models have been developed for energy harvesting and consumption in other networks. However, there are special characteristics of

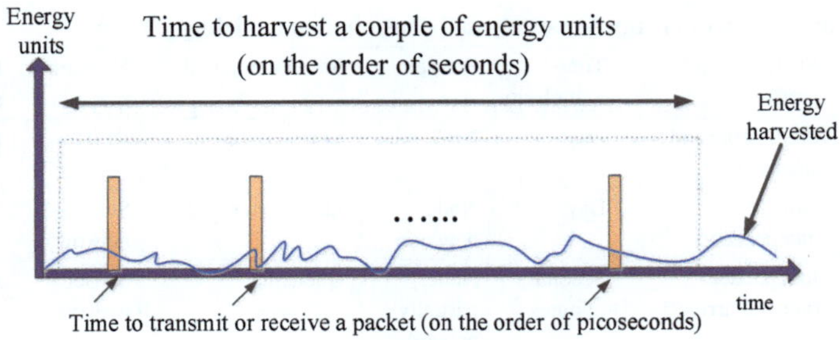

Fig. 5 Comparison of timescales between harvesting and consumption of energy

nanonetworks that necessitate the development of new models for the evaluation of energy consumption and harvesting processes. More specifically, unlike other networks, the granularity of the energy harvesting rate in nanonetworks is slower than the energy consumption rate (Fig. 5). In other words, it means that the energy that it takes a couple of seconds to harvest can be consumed in a couple of picosconds. For example, it can take up to 5 min to harvest energy to transmit only a small packet [23]. Moreover, new harvester elements such as nanowires present different behaviors than previously studied models, such as photovoltaic or electrostatic cells. In addition, new sources of energy are emerging. For example, energy harvesting from blood sugar by biofuel cells [37] or from electrical difference in the inner ear [38] are new sources of energy with unique properties. Furthermore, nanocapacitors represent a nonlinear behavior as compared to most battery-based models. All such properties mandate the need for novel models of energy harvesting and consumption for nanonodes.

A model for the joint evaluation of energy consumption and harvesting in nanonetworks has been proposed in [23]. The model can be used to determine the energy status of nanonodes, and consequently, to evaluate the performance of nanonetworks. A continuous time Markov process, as illustrated in Fig. 6, is developed, where states represent the level of energy, λ_is represent the harvesting rates, and μ_is represent the consumption rates. In the first state, the nanonode does not have any energy to receive or transmit a packet, and in the last state, the nanonode's energy storage is full. The model considers Poisson models for the energy arrival and consumption. Energy is harvested by a nanogenerator through the vibrations of nanowires and is stored in an ultra nano-capacitor with a nonlinear storage behavior.

The nanonodes are assumed to be in a grid, and transmit the received packets from their neighbors in addition to their own generated packets. The authors model the joint process as a continuous time Markov process, where the steady states of the system represent the energy level probability distribution of each nanonode. This metric is later used to evaluate the delay, throughput, and the probability of successful delivery of packets. However, this model does not evaluate the effect of various parameters to find the optimal performance. To be specific, several parameters are introduced in the modeling of energy harvesting and consumption that can affect

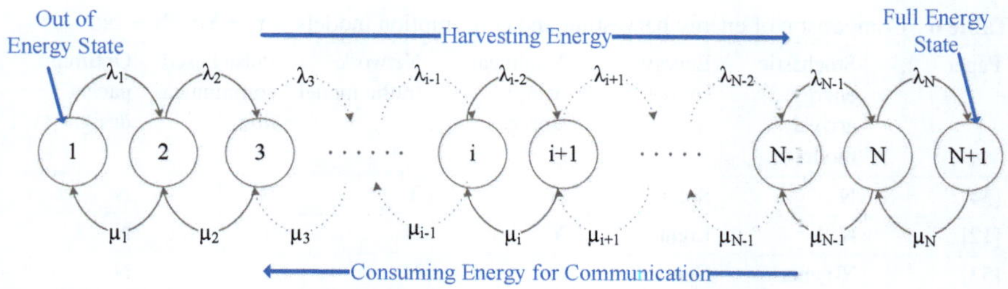

Fig. 6 Markov model for the joint process of energy harvesting and consumption

the optimum energy utilization of a nanonode. The authors in [23] argue that using code weight can save energy in transmission since the lower the code weight, the lower the energy for transmission. However, a deeper look shows that the selection of optimal code weight depends on the energy ratio of reception to transmission of pulses as well as topology (i.e., number of neighbors). Moreover, the optimum packet size is dependent on code weight. Finally, the effect of code weight and repetition on communication reliability in combination with energy consumption needs to be considered. Section 5 shows how to find an optimum combination of these variables through solving an optimization problem. Moreover, the scheduling of packet transmission, which is assumed to be Poisson, can be modified to utilize the energy more efficiently. We describe this in more detail in Sect. 6.

4.3 Summary

To summarize, most energy harvesting models, developed so far in other domains such as sensor networks, are not applicable for many reasons. First of all, each of the stochastic-based models has assumptions such as unlimited energy buffer that cannot be used in nanonetwork scenarios. Second, they mainly assume a linear model for charging their energy storage, while energy storage for nanonodes follows non-linear models. Third, new energy sources such as ambient vibration are less studied. Also, models that are independent of energy sources are not applicable due to their very generic modeling. Moreover, most models do not include consumption and harvesting at the same time. Even if they do, they are not built on pulse-based communication characteristics. Therefore, it is not possible to evaluate the model for different parameters such as packet length, traffic model, etc. Table 6 summarizes the differences between previous work and the two models for nanonetworks [23, 41]. As can be viewed, the model in [23] has many limitations in the stochastic energy arrival model and network traffic model, among others. The model in [41] is a more general model and addresses optimization issues also. This model will be discussed in more details in Sect. 6.

Table 6 Comparison of energy harvesting and consumption models—Y = Yes, N = No

Paper	Stochastic energy arrival model	Energy source	Nonlinear energy storage	Network traffic model	Pulse based communication	Optimum packet design
[34]	N	Solar	N	Y	N	N
[12]	N	Light	Y	N	Y	N
[53]	Y-generic	General	N	N	Y	N
[59]	Y-generic	General	N	Y	N	N
[23]	Only Poisson	Vibration	Y	Partially	Y	Partially
[41]	Y-generic	General	Y	Y	Y	Y

5 Optimizing Energy Consumption Factors

As described in Sect. 4, various parameters can affect the model of energy harvesting and consumption. Particularly, the packet size, code weight, and repetition can affect the amount of energy consumed. Repetition and code weight should be selected in a way that provides an efficient bit rate. Therefore, finding the optimum design point between the energy usage efficiency and bit rate efficiency is the challenge that is addressed in this section. We first provide an overview of a model, previously developed in [40], that can find the best combination of these parameters. Then, we show how the best answer could be selected among a list of candidates when traffic load and utilization are taken into account. More details about the model and results can be found in [40].

5.1 Optimization Model

Multi-Objective Combinatorial Optimization (MOCO) is a special form of Multi-Objective Optimization (MOP) [7], where variables can take discrete values. In a MOP/MOCO problem, several functions need to be optimized at the same time. In our problem, the functions to be optimized at the same time for the packet size (N), repetition (R), and code weight (W) variables are defined as follows.

$$\begin{aligned} &Min_v\ [f_1(v), f_2(v), f_3(v), f_4(v), f_5(v)] \\ &s.t. \\ &g_1(v) \leq 0, \\ &g_2(v) \leq 0, \\ &v = [N, R, W], \end{aligned}$$

$$W \in (0.15 : 0.05 : 0.5),$$
$$R \in (1, 3, 5),$$
$$1 \leq N \leq 2500\,.$$

The first function is energy consumption, that is, the energy consumed for transmission ($E_{packet-tx}$) plus reception of a packet ($E_{packet-rx}$) by all the G neighbors with N bits data.

$$f_1 = \frac{E_{packet-tx} + G \cdot E_{packet-rx}}{N}. \tag{4}$$

The energy required for the transmission and reception of a packet can be computed as follows [23]. For a packet of size N bits, the energy consumed when transmitting and receiving a packet with code weight W are respectively given by

$$E_{packet-tx} = N \cdot W \cdot E_{pulse-tx}, \tag{5}$$

$$E_{packet-rx} = N \cdot E_{pulse-rx}, \tag{6}$$

where $E_{pulse-tx}$ and $E_{pulse-rx}$ are the energy consumed in the transmission and in the reception of a pulse, respectively.

After substituting (5) and (6) in (4), we can write:

$$f_1 = \frac{E_{packet-tx} + G \cdot E_{packet-rx}}{N} = \frac{m' \cdot W \cdot E_{pulse-tx} + G \cdot m' \cdot \alpha \cdot E_{pulse-tx}}{N}$$
$$= \frac{m' \cdot E_{pulse-tx}}{N} \cdot (W + G \cdot \alpha),$$

where α is the ratio of energy for pulse reception to transmission, G is the number of neighbors, W is the code weigh. The value of m' is equal to $N + a$, where a is the number of additional bits added to N that enables coding with code weight W.

The value of $E_{pulse-tx}$ is set to 1 picoJoule (pJ). We developed the model in the general form that there are G neighbors. Therefore, it would cover most unicast or broadcast scenarios where the packet will be received by one, some, or all of the neighbors. Moreover, a preamble or handshake method could be deployed to avoid reception of packets by all neighbors when it is not targeted for them. This objective function is set to be minimized, which means that the total energy that is consumed for transmission and reception per bit of information should be minimized.

The second objective function concerns delay. Since N is larger than the information generation rate, the packet would contain several pieces of information together to avoid the overhead of packet transmission. However, this increases the delay in transmission of information. For example, if information is generated at 10 bits per second and the packet size is 1000 bits, it means that it will take 100 seconds to prepare a packet. This may be acceptable for non-real time applications, or when the rest of the packet can be filled with neighbors' forwarding data, or can just be left empty. However, in our model, we are assuming that packets only contain information generated from one node. The simplest way to define the delay function is to model it in a linear relation with packet length, N. However, if delay has higher importance, the function could be modeled as a higher degree polynomial function of N.

$$f_2 = N\,. \tag{7}$$

This function is set to be minimized.

The third objective function associates the chance of bit error rate with code weight. A lower code weight means the transmission of fewer `1`s, which results in a lower probability of absorption as well as collision between `1`s.

$$f_3 = W\,. \tag{8}$$

This function is set to be minimized.

The optimization problem can be formulated with only the f_1, f_2 and f_3 functions, if repetition is not required to be considered as a variable. This could be the case if it is known that the environment would not affect the pulses significantly and it is better to repeat the entire packet in case of error rather than consume energy with the repetition of symbols. However, we define two objective functions for repetition in order to have a comprehensive model.

The following function shows the effect of repetition. The higher the repetition, the higher the chance of error detection and recovery.

$$f_4 = \lfloor \frac{\frac{R-1}{2}}{R} \rfloor\,. \tag{9}$$

On the other hand, lower repetition means fewer bits and less energy consumption.

$$f_5 = \frac{N}{R}\,. \tag{10}$$

The function f_5 shows the efficient bit rate when repetition is used, and it should be maximized.

The constraint functions would be

$$g_1 = m' \cdot W \cdot E_{pulse-tx} - E_{max} \leq 0\,,$$
$$g_2 = m' \cdot E_{pulse-rx} - E_{max} \leq 0.$$

This means that the energy for transmission or reception of one packet cannot exceed the maximum energy capacity of the node, E_{max}.

As stated earlier, the bounds on the variables of the problem are defined as follows:

$$\begin{aligned} &W \in (0.15 : 0.05 : 0.5)\,, \\ &R \in (1,3,5) \text{ and} \\ &1 \le N \le 2500\,. \end{aligned}$$

Since this is a combinatorial problem, the bounds are actually the set of valid values that can be assigned to variables, i.e., W and R. For N, in addition to the bounds, the values should be discrete.

The output of a MOCO would be a set of Pareto optimal points. Typically, the selection of one point depends on the application and context that a decision maker is facing.

5.2 Simulation

We solved the above MOCO problem with the optimization toolbox of MATLAB. We ran the optimization with different values for α, G and *repetition* to show the effect of these parameters on the points that are selected as optimum. The results for two scenarios are presented in the following subsections. The results of additional scenarios can be found in [40]. Note that Pareto optimal points are not unique and can even be different in several runs. However, the results that are presented here have a similar pattern for all runs, and different runs give only non-significant bit differences in packet size.

5.3 Scenario 1 ($G = 1$, $\alpha = 0.1$, *Repetition* = 1)

In this scenario, we set G to 1 and use no repetition. This scenario will evaluate the case of transmission between two adjacent nodes when broadcast will result only in reception by one neighbor. The value of α is set to 0.1, based on the numerical values in [25] and modeling in [23].

Figure 7 shows the Pareto optimal points that are selected. This scatter plot represents the value of first and second objective functions for each of the Pareto points. The code weight and the packet length for each of the points is presented in the legend. Recall that the first objective function tries to minimize the amount of consumed energy per bit, and, the second function, minimizing delay, is related to the packet length. Each of these points dominates another in one of the two objective functions. Therefore, depending on design priority, any of these points can be selected as the optimal solution. For example, if the priority is energy consumption, one of the

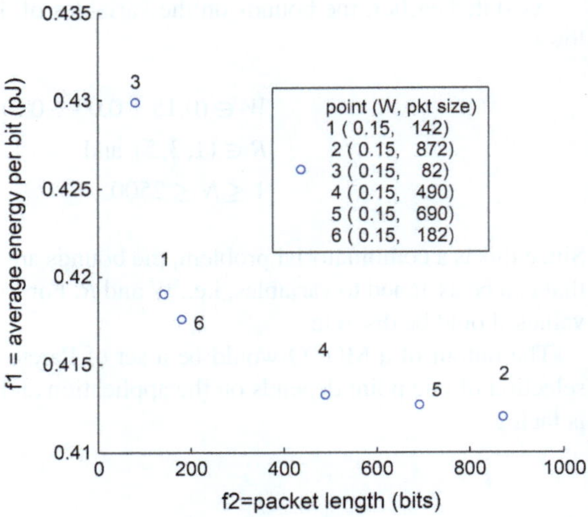

Fig. 7 Pareto point and function values for scenario 1

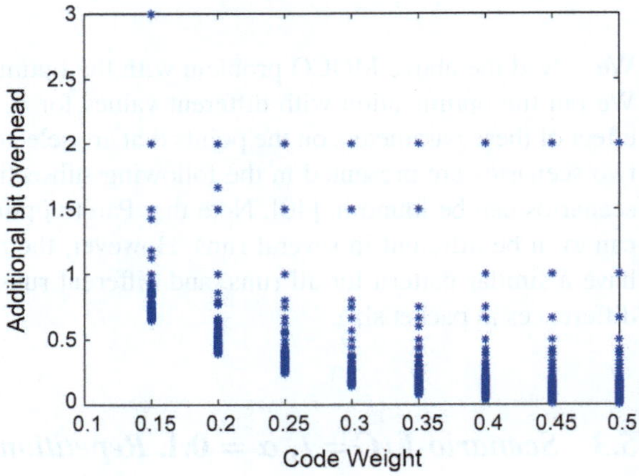

Fig. 8 Additional bit overhead for various code weights

points in lower-right of the chart could be selected. If delay has priority, one of the points in the left side of the chart would be the choice.

Figure 7 illustrates that various packet lengths are selected. A deeper look at the selected code weight for these points shows that all of them are equal to 0.15, which is the minimum code weight. It means that with this setting for G and α, it is better to choose the minimum code weight that is available.

Figure 7 also shows that the difference in terms of efficient energy per actual information bit, f_1, is not significantly different among all the optimal points. This observation can be confirmed by the fact that for a selected code weight, usually the ratio of additional bit rate to actual bits, i.e. $\frac{a}{N}$, illustrated in Fig. 8, is almost the same for each code weight independent of N.

Figure 8 also shows that the overhead from code weight generally does not depend on the length of data. The figure illustrates data lengths that are in the range [1..1000]

Fig. 9 Pareto point and function values for scenario 2

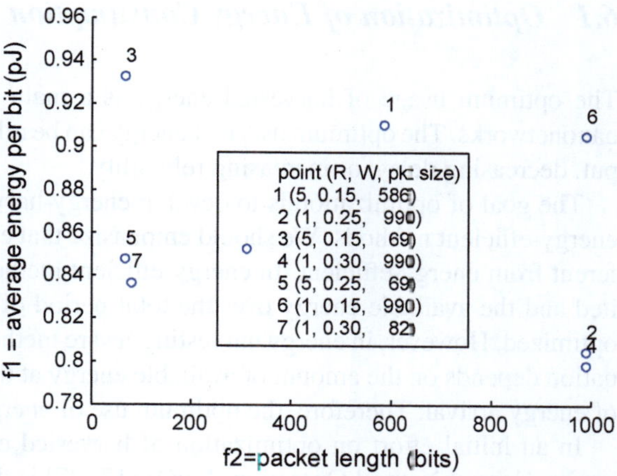

bits. Outliers occur when the number of original bits are very small, i.e., less than 10 bits, but these short data lengths are not applicable for packet transmission.

5.4 Scenario 2 (G = 4, α = 0.1, Repetition = 5)

This scenario evaluates the effect of repetition in combination with a higher number of neighbors (from one to four). The maximum repetition and α are set to 5 and 0.1, respectively. In this scenario, the optimal points, as illustrated in Fig. 9, are selected from almost all ranges of code weight and repetition. However, packet length values are mainly chosen from very short or very large packet sizes. The reason is that when a short packet size is selected, the energy bit efficiency and delay will be the dominant functions. On the other hand, for large packet sizes, code weight will be the dominant factor that leads to lower average energy per bit.

6 Open Issues and Challenges

As the harvester's size is reduced, and possibly energy is harvested from new sources, there are other issues and challenges to be investigated in the domain of energy harvesting for nanonetworks. First of all, as shown in Sect. 4, there is a need for new models of energy harvesting and consumption, based on the characteristics of nanonetworks. In addition, the optimization of energy consumption remains an open issue in several aspects. Furthermore, protocol design in an energy harvesting context and with nano scale properties is an open question. We describe these issues in the following subsections.

6.1 Optimization of Energy Consumption

The optimum usage of harvested energy is a main challenge to be addressed in nanonetworks. The optimum usage of energy can be related to increasing the throughput, decreasing delay, or increasing reliability.

The goal of optimization is to develop energy-harvesting-aware [17] rather than energy-efficient methods. We should emphasize that energy harvesting-aware is different from energy-efficient. In energy-efficient methods, the energy budget is limited and the available energy over the total period of problem modeling should be optimized. However, in energy harvesting-aware methods, the decision about the situation depends on the amount of available energy at the moment, and the prediction of energy arrival. Therefore, the optimum use of energy needs a different model.

In an initial effort on optimization of harvested energy in wireless sensor networks, Energy-Neutral Operation (ENO) [17, 27] is defined as how to operate such that the amount of energy used is always less than the amount of energy harvested. This concept estimates battery size based on an average approach for the rates of energy harvesting and consumption, where energy storage is not 100% efficient and there is energy leakage. Also a power management system is developed to optimize the harvested energy. An exponentially weighted moving-average (EWMA) filter is used to predict the arrival of energy at each time slot based on previous time slots, and then compute the consumption rate based on the prediction. In the next time slot, the prediction is adjusted based on real values.

Additional literature in wireless sensor networks and RFID networks (e.g., [12, 13, 34, 43]) follow the idea that the optimization of energy consumption in perpetual networks is different from typical battery-based networks. However, they do not address the problem in a way that is suitable for nanonetworks. In [34], the authors focus on consumption for data collection, not energy consumption for communication. In [43] the problem of optimization is described, when energy arrivals are stochastic. However, the authors develop their solution based on a historical prediction model of energy arrival, not an exact probability distribution function. In [12, 13], a stochastic model is considered which maximizes the data rate and smoothing consumption for a discrete distribution of energy arrival. Moreover, the model does not behave based on the stochastic arrival of energy. Therefore, nodes can be out of energy for unknown amount of time. Therefore, these methods are not applicable for nanonetworks.

A model for optimum energy consumption for nanonetworks has been proposed in [41]. Knowing the model for energy harvesting, the problem of finding the optimal usage of the harvested energy can be investigated. In fact, with optimum rates, a better data rate can be achieved as compared to fixed consumption rates. Intuitively, it is better to consume more if more energy is harvested and vice versa. Energy should be always available in order to avoid node failure and consequently lack of communication. In this case, the question is: what are the optimal rates of energy consumption? If the harvested energy is not consumed optimally, a node will miss some energy that it could have harvested. This, for example, can occur if a conservative policy (i.e., minimum consumption rate) is used. On the other hand, an aggressive strategy will create nodes with low energy levels which will lead to many failures in packet transmission. A comparison of various strategies can be found in [41]. Moreover, an optimal model is defined. The current amount of energy and the harvesting model determine the optimal energy consumption policy. It is a challenging problem since energy arrival follows a stochastic process. Moreover, finding the packet size and/or the feasible transmission rates to satisfy the optimal rates make the problem more difficult. Using a variable packet size, which has some overhead, is another option that can be investigated.

An optimal energy allocation policy should consider these requirements. First, the energy that is consumed cannot be more than the harvested energy. Since the amount of harvested energy follows a stochastic process, the consumption process should consider it when the optimal policy is designed. Second, a conservative policy is not an optimal policy since it is not acceptable to have energy harvested that cannot be stored due to a full battery. Therefore, it is more important to make the best use of harvested energy rather than to minimize the energy consumption.

In the current state-of-the-art for energy harvesting at nanoscale, the rate of energy consumption is much higher than the rate of energy harvesting. Moreover, limited energy storage (ultra-nanocapacitor/nanobattery) capacity as well as limited queuing space for packets makes the problem more challenging. Thus, these constraints should be taken into account for modeling and optimization of the energy harvesting processes.

Given that nanonodes may be in unknown environments, at least in terms of available energy for harvesting, they need to understand their environment. Moreover, optimization models to maximize the utilization of harvested energy typically result in computationally expensive schemes. Since the processing and memory resources are limited at nanonodes, the optimum solutions should be designed as offline solutions. Another approach to consider is to develop light-weight heuristic methods with near-optimal performance [41].

6.2 *Energy Harvesting-Aware Protocols*

After the optimum energy consumption design, energy harvesting-aware protocols need to be developed. Medium access design is the main issue that needs to be addressed. Not only do the MAC protocols for nanonodes need to be harvesting-aware, but they also have to be designed based on other characteristics of nanonodes, e.g., limited capabilities, pulse-based communication, and the large scale of the nanonetwork. Some pulse-based MAC protocols have been developed for UWB networks [14] that may have the potential to be used in nanonetworks. However, characteristics of the THz band, as well as the limited processing capabilities of nanodevices, are the major factors that can lead to the need for the redesign of protocols for the networking of nanonodes.

Similar to reasoning about the need for new models of energy harvesting in nanonetworks, energy harvesting-aware protocols need to be customized and may even be newly created. Pulse-based communication in the THz band, the unique properties of energy harvesting and consumption for nanonodes, and the capability limitations due to size constraints are the main factors that mandate the development of novel energy harvesting-aware protocols.

Recently, some energy harvesting-aware MAC protocols have been proposed for nanonetworks. One such proposal [26] exploits the benefits of novel low-weight coding and chooses the optimal value of code weight and repetition. The performance of the proposed protocol is analytically studied in terms of energy consumption, delay, and achievable throughput, using models of the Terahertz channel (path-loss and molecular absorption noise) and interference. However, the feasibility of the protocol implementation and an energy efficiency evaluation of the method are still open issues. Later, in [57] an energy harvesting-aware and light-weight MAC protocol has been proposed. The aim of the protocol is to achieve fair throughput and optimal channel access among nanosensors which are controlled by a nano-controller. Towards this end, the critical packet transmission ratio is defined, which is the maximum allowable ratio between the transmission time and the energy harvesting time, below which a nanosensor can harvest more energy. However, the focus of the work is on the scheduling of packet transmissions by the nanocontroller, rather than energy aspects.

An energy harvesting-aware MAC protocol (RIH-MAC) is proposed in [42], where nanonodes communicate based on a receiver-initiated mechanism. The advantage of the protocol is that through a receiver-initiated mechanism, harvested energy is utilized more efficiently because the transmitter and receiver spend their energy wisely to maximize the probability that both the transmitter and receiver will have energy for communication. RIH-MAC can be used either in a centralized network topology or in an ad hoc formation of nanonodes, i.e., a distributed network topology. Distributed RIH-MAC protocol exploits a distributed edge-graph coloring scheme to facilitate a coordination among nanonodes to access the medium. Furthermore, RIH-MAC adapts to various energy harvesting rates. These protocols are the first steps towards energy harvesting-aware protocols for nanonetworks. More protocols for the MAC layer as well as upper layers remain open for further investigation.

7 Summary

In this chapter, we introduced a taxonomy of energy harvesting. Recent advances in nanomaterials have enabled the development of nanoscale harvesters such as nanogenerators. Nanonodes are expected to harvest their required energy mainly from mechanical and chemical sources. Harvested energy is consumed for communication in the THz band among nanonodes. Modeling the joint process of energy harvesting and energy consumption is required to understand the special characteristics of this process due to the nanoscale properties of harvesting process, a new communication model, and limited energy storage. Next, the optimization of the consumption of the harvested energy needs to be studied. Development of energy harvesting-aware protocols and maximizing the utilization of energy are the main approaches to optimize energy consumption.

References

1. Akyildiz IF, Brunetti F, Blazquez C (2008) Nanonetworks: a new communication paradigm. Comput Netw 52(12):2260–2279. doi:10.1016/j.comnet.2008.04.001
2. Akyildiz IF, Jornet JM (2010) Electromagnetic wireless nanosensor networks. Nano Commun Netw 1(1):3–19. doi:10.1016/j.nancom.2010.04.001
3. Avouris P (2009) Carbon nanotube electronics and photonics. Phys Today 62, 3440
4. Chalasani S, Conrad J (2008) A survey of energy harvesting sources for embedded systems. In: IEEE Southeastcon, pp 442–447 (2008). doi:10.1109/SECON.2008.4494336
5. Chi K, Zhu Y, Jiang X, Tian X (2013) Optimal coding for transmission energy minimization in wireless nanosensor networks. Nano Commun Netw. doi:10.1016/j.nancom.2013.07.001
6. Christ A, Douglas M, Roman J, Cooper E, Sample A, Waters B, Smith J, Kuster N (2013) Evaluation of wireless resonant power transfer systems with human electromagnetic exposure limits. IEEE Trans Electromag Compatib 55(2):265–274. doi:10.1109/TEMC.2012.2219870
7. Deb K (2005) Multi-objective optimization. In: Burke, E Kendall G (eds) Search methodologies, pp 273–316. Springer US. doi:10.1007/0-387-28356-0_10
8. Fontana R (2004) Recent system applications of short-pulse ultra-wideband (UWB) technology. IEEE Trans Microw Theory Techn 52(9):2087–2104. doi:10.1109/TMTT.2004.834186
9. Gatzianas M, Georgiadis L, Tassiulas L (2010) Control of wireless networks with rechargeable batteries. IEEE Trans Wirel Commun 9(2):581–593. doi:10.1109/TWC.2010.080903
10. Gilbert J, Balouchi F (2008) Comparison of energy harvesting systems for wireless sensor networks. Int J Autom Comput 5:334–347. doi:10.1007/s11633-008-0334-2
11. Gorlatova M, Sarik J, Cong M, Kymissis I, Zussman G (2013) Movers and shakers: kinetic energy harvesting for the internet of things. http://arxiv.org/pdf/1307.0044v1.pdf
12. Gorlatova M, Wallwater A, Zussman G (2011) Networking low-power energy harvesting devices: measurements and algorithms. In: Proceedings of IEEE INFOCOM, pp 1602–1610. doi:10.1109/INFCOM.2011.5934952
13. Gorlatova M, Wallwater A, Zussman G (2012) Networking low-power energy harvesting devices: measurements and algorithms. IEEE Trans Mobile Comput 12(9) 1853–1865. doi:10.1109/TMC.2012.154
14. Gupta A, Mohapatra P (2007) A survey on ultra wide band medium access control schemes. Comput Netw 51(11):2976–2993. doi:10.1016/j.comnet.2006.12.008
15. Hansen BJ, Liu Y, Yang R, Wang ZL (2010) Hybrid nanogenerator for concurrently harvesting biomechanical and biochemical energy. ACS Nano 4(7):3647–3652
16. Hoogers G (2002) Fuel cell technology handbook. Handbook series for mechanical engineering. Taylor & Francis
17. Hsu J, Zahedi S, Kansal A, Srivastava M, Raghunathan V (2006) Adaptive duty cycling for energy harvesting systems. In: Proceedings of the international symposium on low power electronics and design, pp 180–185. doi:10.1109/LPE.2006.4271832
18. Hu Y, Zhang Y, Xu C, Zhu G, Wang ZL (2010) High-output nanogenerator by rational unipolar assembly of conical nanowires and its application for driving a small liquid crystal display. Nano Lett 10(12):5025–5031
19. Ivanov I, Vidakovi-Koch T, Sundmacher K (2010) Recent advances in enzymatic fuel cells: experiments and modeling. Energies 3(4):803–846

20. Jornet J, Akyildiz I (2010) Channel capacity of electromagnetic nanonetworks in the terahertz band. In: IEEE international conference on communications (ICC), pp 1 –6. doi:10.1109/ICC.2010.5501885
21. Jornet J, Akyildiz I (2011) Channel modeling and capacity analysis for electromagnetic wireless nanonetworks in the terahertz band. IEEE Trans Wirel Commun 10(10):3211–3221. doi:10.1109/TWC.2011.081011.100545
22. Jornet J, Akyildiz I (2011) Low-weight channel coding for interference mitigation in electromagnetic nanonetworks in the terahertz band. In: IEEE international conference on communication (ICC), pp 1–6. doi:10.1109/icc.2011.5962987
23. Jornet J, Akyildiz I (2012) Joint energy harvesting and communication analysis for perpetual wireless nanosensor networks in the Terahertz band. IEEE Trans Nanotechnol 11(3):570–580. doi:10.1109/TNANO.2012.2186313
24. Jornet JM, Akyildiz IF (2010) Graphene-based nano-antennas for electromagnetic nanocommunications in the terahertz band. In: Proceedings of the European conference on antennas and propagation
25. Jornet JM, Akyildiz IF (2011) Information capacity of pulse-based wireless nanosensor networks. In: Proceedings of IEEE SECON, pp 80–88
26. Jornet JM, Pujol JC, Pareta JS (2012) PHLAME: a physical layer aware MAC protocol for electromagnetic nanonetworks in the terahertz band. Nano Commun Netw 3(1):74–81. doi:10.1016/j.nancom.2012.01.006
27. Kansal A, Hsu J, Zahedi S, Srivastava MB (2007) Power management in energy harvesting sensor networks. ACM Trans Embed Comput Syst 6(4). doi:10.1145/1274858.1274870
28. Kar K, Krishnamurthy A, Jaggi N (2006) Dynamic node activation in networks of rechargeable sensors. IEEE/ACM Trans Netw 14(1):15–26. doi:10.1109/TNET.2005.863710
29. Khouzani M, Sarkar S, Kar K (2011) Optimal routing and scheduling in multihop wireless renewable energy networks. In: Proceedings of sixth information theory and applications workshop (ITA)
30. Kim P (2008) Toward carbon based electronics. In: Proceedings of device research conference, p 9. doi:10.1109/DRC.2008.4800712
31. Kocaoglu M, Akan O (2012) Minimum energy coding for wireless nanosensor networks. In: Proceedings of IEEE INFOCOM, pp 2826–2830. doi:10.1109/INFCOM.2012.6195709
32. Li Z, Wang ZL (2011) Air/liquid-pressure and heartbeat-driven flexible fiber nanogenerators as a micro/nano-power source or diagnostic sensor. Adv Mater 23(1):84–89
33. Lin YM et al (2010) 100-GHz transistors from wafer-scale epitaxial graphene. Science 327:662
34. Liu RS, Fan KW, Zheng Z, Sinha P (2011) Perpetual and fair data collection for environmental energy harvesting sensor networks. IEEE/ACM Trans Netw 19(4):947–960. doi:10.1109/TNET.2010.2091280
35. Luo Y, Zhang J, Letaief KB (2012) Training optimization for energy harvesting communication systems. In: Proceedings of IEEE Globecom
36. Luryi S, Xu J, Zaslavsky A (2007) Future trends in microelectronics: up the nano creek. Wiley, IEEE
37. MacVittie K, Halamek J, Halamkova L, Southcott M, Jemison WD, Lobel R, Katz E (2013) From "cyborg" lobsters to a pacemaker powered by implantable biofuel cells. Energy Environ Sci 6:81–86. doi:10.1039/C2EE23209J
38. Mercier PP, Lysaght AC, Bandyopadhyay S, Stankovic APCKM (2012) Energy extraction from the biologic battery in the inner ear. Nat Biotechnol 30:1240–1243
39. Mitcheson P (2010) Energy harvesting for human wearable and implantable bio-sensors. In: IEEE engineering in medicine and biology society, pp 3432–3436 (2010)
40. Mohrehkesh S, Weigle MC (2013) Optimizing communication energy consumption in perpetual wireless nanosensor networks. In: Proceedings of the IEEE Globecom, Atlanta, GA, pp 545–550
41. Mohrehkesh S, Weigle MC (2014) Optimizing energy consumption in terahertz band nanonetworks. To appear in IEEE JSAC: molecular, biological, and multi-scale communications series

42. Mohrehkesh S, Weigle MC (2014) RIH-MAC: receiver-initiated harvesting-aware MAC for nanonetworks. In: Proceedings of the first ACM annual international conference on nanoscale computing and communication (NANOCOM), pp 6:1–6:9
43. Noh DK, Abdelzaher TF (2012) Efficient flow-control algorithm cooperating with energy allocation scheme for solar-powered WSNs. Wirel Commun Mobile Comput 12(5):379–392. doi:10.1002/wcm.965
44. Pan C, Li Z, Guo W, Zhu J, Wang ZL (2011) Fiber-based hybrid nanogenerators for/as self-powered systems in biological liquid. Angewandte Chemie 123(47):11388–11392
45. Parks A, Sample A, Zhao Y, Smith J (2013) A wireless sensing platform utilizing ambient RF energy. In: IEEE topical conference on wireless sensors and sensor networks (WiSNet), pp 127–129. doi:10.1109/WiSNet.2013.6488656
46. Pierobon M, Akyildiz I (2013) Capacity of a diffusion-based molecular communication system with channel memory and molecular noise. IEEE Trans Inf Theory 59(2):942–954. doi:10.1109/TIT.2012.2219496
47. Rentmeesters M, Tsai W, Lin KJ (1996) A theory of lexicographic multi-criteria optimization. In: Proceedings of second IEEE international conference on engineering of complex computer systems, pp 76–79. doi:10.1109/ICECCS.1996.558386
48. Roundy S (2005) On the effectiveness of vibration-based energy harvesting. J Intell Mater Syst Struct 16(10):809–823. doi:10.1177/1045389X05054042
49. Roundy S, Leland E, Baker J, Carleton E, Reilly E, Lai E, Otis B, Rabaey J, Wright P, Sundararajan V (2005) Improving power output for vibration-based energy scavengers. IEEE Pervasive Comput 4(1):28–36
50. Roundy S, Wright P, Rabaey J (2004) Energy scavenging for wireless sensor networks: with special focus on vibrations. Kluwer Academic Publishers
51. Roundy S, Wright PK, Rabaey J (2003) A study of low level vibrations as a power source for wireless sensor nodes. Comput Commun 26(11):1131–1144. doi:10.1016/S0140-3664(02)00248-7
52. Sensale-Rodriguez B, Yan R, Kelly MM, Fang T, Tahy K, Hwang WS, Jena D, Liu L, Xing HG (2012) Broadband graphene terahertz modulators enabled by intraband transitions. Nat Commun
53. Sharma V, Mukherji U, Joseph V, Gupta S (2010) Optimal energy management policies for energy harvesting sensor nodes. IEEE Trans Wirel Commun 9(4):1326–1336. doi:10.1109/TWC.2010.04.080749
54. Shenck N, Paradiso J (2001) Energy scavenging with shoe-mounted piezoelectrics. IEEE Micro 21(3):30–42. doi:10.1109/40.928763
55. Starner T (1996) Human-powered wearable computing. IBM Syst J 35(34):618–629. doi:10.1147/sj.353.0618
56. Sudevalayam S, Kulkarni P (2011) Energy harvesting sensor nodes: survey and implications. IEEE Commun Surv Tutor 13(3):443–461. doi:10.1109/SURV.2011.060710.00094
57. Wang P, Jornet JM, Malik MA, Akkari N, Akyildiz IF (2013) Energy and spectrum-aware MAC protocol for perpetual wireless nanosensor networks in the terahertz band. Ad Hoc Netw 11(8):2541–2555. doi:10.1016/j.adhoc.2013.07.002
58. Wang ZL, Wu W (2012) Nanotechnology-enabled energy harvesting for self-powered micro-/nanosystems. Angewandte Chemie International Edition 51(47):11700–11721
59. Wu K, Jiang Y, Marinakis D (2012) A stochastic calculus for network systems with renewable energy sources. In: Proceedings of IEEE conference on computer communications workshops (INFOCOM Workshops), pp 109–114. doi:10.1109/INFCOMW.2012.6193470
60. Xu S, Hansen BJ, Wang ZL (2010) Piezoelectric-nanowire-enabled power source for driving wireless microelectronics. Nat Commun 1
61. Zungeru AM, Ang LM, Prabaharan S, Seng KP (2012) Chapter 13. Radio frequency energy harvesting and management for wireless sensor networks. CRC Press

42. Mohrehkesh S, Weigle MC (2014) RIH-MAC: receiver-initiated harvesting-aware MAC for nanonetworks. In: Proceedings of the first ACM annual international conference on nanoscale computing and communication (NANOCOM), pp 6:1–6:9
43. Noh DK, Abdelzaher TF (2012) Efficient flow-control algorithm cooperating with energy allocation scheme for solar-powered WSNs. Wirel Commun Mobile Comput 12(5):379–392. doi:10.1002/wcm.965
44. Pan C, Li Z, Guo W, Zhu J, Wang ZL (2011) Fiber-based hybrid nanogenerators for/as self-powered systems in biological liquid. Angewandte Chemie 123(47):11388–11392
45. Parks A, Sample A, Zhao Y, Smith J (2013) A wireless sensing platform utilizing ambient RF energy. In: IEEE topical conference on wireless sensors and sensor networks (WiSNet), pp 127–129. doi:10.1109/WiSNet.2013.6488656
46. Pierobon M, Akyildiz IF (2013) Capacity of a diffusion-based molecular communication system with channel memory and molecular noise. IEEE Trans Inf Theory 59(2):942–954. doi:10.1109/TIT.2012.2219496
47. Rentmeesters M, Tsai W, Lin KJ (1996) A theory of lexicographic multi-criteria optimization. In: Proceedings of second IEEE international conference on engineering of complex computer systems, pp 76–79. doi:10.1109/ICECCS.1996.558386
48. Roundy S (2005) On the effectiveness of vibration-based energy harvesting. J Intell Mater Syst Struct 16(10):809–823. doi:10.1177/1045389X05054042
49. Roundy S, Leland E, Baker J, Carleton E, Reilly E, Lai E, Otis B, Rabaey J, Wright P, Sundararajan V (2005) Improving power output for vibration-based energy scavengers. IEEE Pervasive Comput 4(1):28–36
50. Roundy S, Wright P, Rabaey J (2004) Energy scavenging for wireless sensor networks: with special focus on vibrations. Kluwer Academic Publishers
51. Roundy S, Wright PK, Rabaey J (2003) A study of low level vibrations as a power source for wireless sensor nodes. Comput Commun 26(11):1131–1144. doi:10.1016/S0140-3664(02)00248-7
52. Sensale-Rodriguez B, Yan R, Kelly MM, Fang T, Tahy K, Hwang WS, Jena D, Liu L, Xing HG (2012) Broadband graphene terahertz modulators enabled by intraband transitions. Nat Commun 3
53. Sharma V, Mukherji U, Joseph V, Gupta S (2010) Optimal energy management policies for energy harvesting sensor nodes. IEEE Trans Wirel Commun 9(4):1326–1336. doi:10.1109/TWC.2010.04.080749
54. Shenck N, Paradiso J (2001) Energy scavenging with shoe-mounted piezoelectrics. IEEE Micro 21(3):30–42. doi:10.1109/40.928763
55. Starner T (1996) Human-powered wearable computing. IBM Syst J 35(3.4):618–629. doi:10.1147/sj.353.0618
56. Sudevalayam S, Kulkarni P (2011) Energy harvesting sensor nodes: survey and implications. IEEE Commun Surv Tutor 13(3):443–461. doi:10.1109/SURV.2011.060710.00094
57. Wang P, Jornet JM, Malik MA, Akkari N, Akyildiz IF (2013) Energy and spectrum-aware MAC protocol for perpetual wireless nanosensor networks in the terahertz band. Ad Hoc Netw 11(8):2541–2555. doi:10.1016/j.adhoc.2013.07.002
58. Wang ZL, Wu W (2012) Nanotechnology-enabled energy harvesting for self-powered micro-/nanosystems. Angewandte Chemie International Edition 51(47):11700–11721
59. Wu K, Jiang Y, Marinakis D (2012) A stochastic calculus for network systems with renewable energy sources. In: Proceedings of IEEE conference on computer communications workshops (INFOCOM Workshops), pp 109–114. doi:10.1109/INFCOMW.2012.6193470
60. Xu S, Hansen BJ, Wang ZL (2010) Piezoelectric-nanowire-enabled power source for driving wireless microelectronics. Nat Commun 1
61. Zungeru AM, Ang LM, Prabaharan S, Seng KP (2012) Chapter 13, Radio frequency energy harvesting and management for wireless sensor networks. CRC Press

Nanoscale Communications Based on Fluorescence Resonance Energy Transfer (FRET)

Murat Kuscu and Ozgur B. Akan

1 Introduction

Nanoscale communication is a novel and quite interdisciplinary research area which aims to design and develop communication networks among nano-size machines to extend their limited capabilities for groundbreaking biomedical, industrial and environmental applications [1]. In this chapter, we thoroughly investigate a novel nanoscale communication method based on a physically realizable phenomenon, Förster Resonance Energy Transfer (FRET) which is the non-radiative transfer of excited state energies, i.e., excitons, between two fluorescent molecules, i.e., fluorophores, such as fluorescent proteins, organic dyes, and semiconductor nanoparticles which have spectral similarities and are located in a close proximity such as 0–10 nm [15]. FRET-based nanoscale communication is a promising paradigm that allows future molecular-size machines to communicate with each other over distances up to tens of nm.

FRET is a quantum mechanical phenomenon based on the dipole-dipole interactions of flurophores, and observed in some biological systems, such as photosynthesis. The phenomenon has been widely used in studies of biotechnological research including fluorescence microscopy, molecular biology and optical imaging, since it provides a significant amount of structural and spatial information about molecules by means of optical signals with nanoscale resolution [6]. Also in nanomedicine, exploiting FRET, Quantum Dots (QDs) have been employed as the photosensitizing agents for photodynamic therapy (PDT) of cancer [17]. Furthermore, the quantum coherence behavior of FRET in short distances has been widely studied showing the potential of FRET for future quantum computer designs [18]. Low dependence

M. Kuscu (✉) · O.B. Akan
Koc University, Istanbul, Turkey
e-mail: mkuscu@ku.edu.tr

O.B. Akan
e-mail: akan@ku.edu.tr

J. Suzuki et al. (eds.), *Modeling, Methodologies and Tools for Molecular and Nano-scale Communications*, Modeling and Optimization in Science and Technologies 9, DOI 10.1007/978-3-319-50688-3_15

on the environmental factors, controllability of its fundamental parameters, and relatively wide transfer range make FRET also a promising candidate to be used for high-rate nanoscale wireless communications.

In this chapter, we first present the communication theoretical model of FRET-based nanocommunication channel between a donor fluorophore as the transmitter nanomachine (TN) and an acceptor fluorophore as the receiver nanomachine (RN) [9, 10]. In this model, a single exciton is considered to be the information carrier. Assuming that a remote optical information source is available to excite the donor fluorophore, i.e., TN, and employing binary ON/OFF Keying (OOK) modulation scheme, we investigate the probability of successful transmission of information from TN to RN. We information theoretically derive analytical expressions for the mutual information between TN and RN, and the channel capacity. The dependence of the channel capacity on the fundamental parameters of FRET, e.g., intermolecular distance, spectral similarity, is demonstrated through numerical simulations.

We also investigate another method for FRET communication based on multi-step FRET employing identical relay fluorophores between TN and RN, and utilizing multi-exciton transmission scheme which further improves the spatial range and the achievable transmission rate of the communication [11, 13]. The relays are considered to be randomly deployed in a three dimensional aqueous medium and expected to transfer the information between TN and RN through multiple energy transfers. We simulate the communication through the proposed channel following a realistic algorithm based on the competitive behavior of the multiple excitons and concerning many sources of randomness intrinsic to the phenomenon. Following a Monte Carlo approach, we evaluate the performance of the channel by means of information theoretical capacity and interference probability between successive transmissions, then we derive the maximum achievable data transmission rate.

We finally explore network of mobile nanomachines communicating through FRET. We focus on FRET-based mobile molecular sensor/actor nanonetwork (FRET-MSAN) which is a distributed system of mobile fluorophores acting as sensor or actor node [12, 14]. We present the model of single message propagation based on modified birth-death process with continuous time Markov chain. We evaluate the performance of FRET-MSAN in terms of successful transmission probability, mean extinction time of the messages, system throughput, and mutual information between transmitted and received alphabets.

The remainder of this chapter is organized as follows. In Sect. 2, we review the FRET theory. In Sect. 3, we present the information theoretical model of the single-pair FRET-based nanocommunication channel. The multi-step FRET-based nanocommunications is investigated in Sect. 4. Section 5 introduces the FRET-based mobile molecular nanonetworks. Finally, the concluding remarks are given in Sect. 6.

2 Theory of FRET

FRET is a fundamental excited state process observed among fluorescent molecules. It is the transfer of excited state energy in the form of exciton from an excited molecule, called the donor, to a ground-state molecule, called the acceptor, in close proximity of each other. The theory of FRET between immobile fluorophores is established in Theodor Förster's seminal work [7], and then, fundamentally validated by an extensive number of experimental studies [19].

When there exists no quenching mechanism such as molecular collisions or FRET, the excited-state donor is expected to return to ground-state after a randomly short time by fluorescing, i.e., releasing a photon. The excited state lifetime τ, i.e., the time interval between the excitation and relaxation of a molecule, is an exponential random variable of which mean is generally in the range between 2 ns and 1μs [15]. The reciprocal of the mean excited-state lifetime μ_0 when the fluorescence is the only quenching mechanism gives the natural fluorescence rate k_0 of a fluorophore, i.e., $k_0 = 1/\mu_0$. k_0 is the number of photons released at a unit time through fluorescence. The existence of a quenching mechanism provides additional pathways for the excited-state donor to relax to the ground-state, and thus, significantly reduces the lifetime of excitons by increasing the relaxation rate of the donor.

As a fundamental quenching mechanism, FRET can be realized if the following conditions are satisfied: (i) the excited fluorophore must be in close proximity (0–10 nm) with at least one ground-state acceptor; (ii) the emission spectrum of the donor and the absorption spectrum of the acceptor must overlap; and (iii) the relative orientation of transition dipole moments of the donor and the acceptor fluorophores must not be orthogonal.

If all requirements are met, the rate of the energy transfer as a function of the natural fluorescence rate of the donor k_0 and in terms of number of excitons transferred per unit time is given as

$$k_t = k_0(\frac{R_0}{R})^6 \tag{1}$$

where R is the intermolecular distance, and R_0 is the Förster radius which incorporates the effects of some intrinsic and environmental parameters Since information is encoded into the excitons, the exciton transfer rate k_t is the most crucial parameter for the performance of FRET-based communications. It is a measure of how fast the information can be transferred between fluorophore-based nanonodes. Förster radius R_0 can be expressed by

$$R_0 = (8.8 \times 10^{22} \kappa^2 n^{-4} Q_D J)^{-\frac{1}{6}} \tag{2}$$

where κ^2 is the relative orientation factor, Q_D is the quantum yield of the donor, n is the refractive index of the medium, and J is the degree of the spectral overlap. R_0 ranges between 4 and 10 nm [15].

The mean excited state lifetime of the donor is shortened when there exists different pathways to relax including FRET. When FRET is a possible relaxation pathway for the donor, the mean lifetime is given as

$$\mu_{FRET} = \frac{1}{k_0 + k_t} \tag{3}$$

At the same conditions, the probability of FRET for an excited donor fluorophore can be given in terms of process rates as follows

$$P_{FRET} = \frac{k_t}{k_0 + k_t} \tag{4}$$

The degree of the overlap between the emission spectrum of the donor and the absorption spectrum of the acceptor J is formulated as

$$J(\lambda) = \int_0^{\infty} f_D(\lambda)\varepsilon_A(\lambda)\lambda^4 \mathrm{d}\lambda \tag{5}$$

where $f_D(\lambda)$ is the normalized fluorescence emission intensity and $\varepsilon_A(\lambda)$ is the acceptor molar absorptivity.

The orientation factor κ^2 can be given as follows

$$\kappa^2 = (\cos\theta_T - 3\cos\theta_D\cos\theta_A)^2 \tag{6}$$

where θ_T, θ_D and θ_A are the angles determined by the emission and absorption transition dipoles of the fluorophores [15] as shown in Fig. 1. Assuming isotropic and unrestricted distributions for all three angles, the distribution of κ^2 is as follows [3]

$$p_{\kappa^2}(\kappa^2) = \begin{cases} \frac{1}{2\sqrt{3\kappa^2}} \ln(2+\sqrt{3}) & 0 \le \kappa^2 \le 1 \\ \frac{1}{2\sqrt{3\kappa^2}} \ln\left(\frac{2+\sqrt{3}}{\sqrt{\kappa^2}+\sqrt{\kappa^2-1}}\right) & 1 \le \kappa^2 \le 4 \end{cases} \tag{7}$$

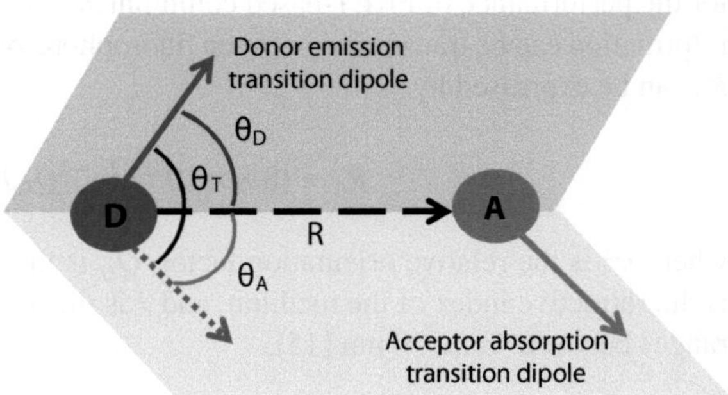

Fig. 1 Critical angles of relative orientation between donor and acceptor fluorophores

In most studies, κ^2 is taken as its mean value which is 2/3 assuming isotropically free molecules.

In the case of mobile fluorophores, the situation is radically different in the sense that during the excited state lifetime of the donor, the intermolecular distances and the relative orientation of the dipole moments of fluorophores are not constant. Furthermore, the excited donor fluorophore can get in close proximity with a varying number of acceptors if the donor lifetime or the diffusion coefficient of the fluorophores is sufficiently long. Considering that the excitons randomly walk in a random lattice consisting of diffusing fluorophores, giving a closed form expression for the transfer rate requires some assumptions. Stryer et al. postulated the governing rate equations for the energy transfer from a single excited donor to a single ground-state acceptor in a three dimensional environment assuming that the fluorophores are in the rapid-diffusion limit [20]:

$$k_{rd} = \frac{4\pi k_0 R_{0,d-a}^6}{3Va_{d-a}^{-3}} \tag{8}$$

where $R_{0,d-a}$ is the Förster radius between the donor and the acceptor, and V is the volume of the three dimensional medium. a_{d-a} is the possible closest distance between the centers of the donor and the acceptor. When there are more than one acceptor molecules, the total FRET rate between the donor and the acceptors becomes

$$k_t = k_{rd} N_a \tag{9}$$

where N_a is the number of available, i.e., ground-state, acceptor molecules in the environment. The rapid-diffusion criterion is given as

$$\frac{D\tau_0}{s^2} \gg 1 \tag{10}$$

where D is the sum of diffusion coefficients of the donor and the acceptor, τ_0 is the natural excited state lifetime of the donor, i.e., $\tau_0 = 1/k_0$, and s is the mean intermolecular distance between donor-acceptor fluorophores [20]. The rapid diffusion limit can be achieved by using fluorophores with moderate diffusion coefficients and long lifetimes such as 1–2 μs.

The excited-state energy of fluorescent molecules is transferred at times in the range of picoseconds to nanoseconds via FRET, therefore, a nanocommunication method to be developed based on the idea of encoding information into the existence of excitons can provide significantly higher communication rates compared to other molecular communication techniques which are generally originated from very slow diffusion process of molecules. Moreover, high-level controllability of almost all of the main system parameters, and the intrinsic macro-nano interface resultant from the possibility of transferring information optically between the flurophore-nanonetwork and a human-controlled laser-photodetector system make FRET-based nanocommunications incomparably more practical and feasible. Furthermore, the abundance of both theoretical and empirical studies about FRET in the literature and commercially availability of its experimental setups provide the opportunity of making improvements, validating theoretical models based on experiments. Hence, unlike most of the existing molecular approaches in the literature, we introduce an already analyzed and experimented, therefore, a physically realizable, and hence, clearly realistic solution to the problem of nanoscale molecular communications.

3 FRET-Based Point-to-Point Nanoscale Communication Channel with Single Exciton Transmission

In this section, the simplest form of the FRET-based molecular communication channel comprising a single transmitter-receiver nanomachine pair is modeled. Furthermore, using the information theoretical approach, the capacity of the communication channel is derived.

3.1 Channel Model

The point-to-point FRET-based nanoscale communication system consists of three main parts; a Transmitter Nanonode (TN) which is a single donor fluorophore, a communication channel, and a Receiver Nanonode (RN) which is a single acceptor fluorophore. The nodes are assumed to be located at fixed locations separated by a reasonable distance R as shown in Fig. 2.

The system utilizes a pulsed laser as the Information Source (IS) that can generate picosecond-duration excitation pulses, and implements ON-OFF keying (OOK) modulation with two bits available; bit-0 and bit-1. When IS intends to transfer bit-1,

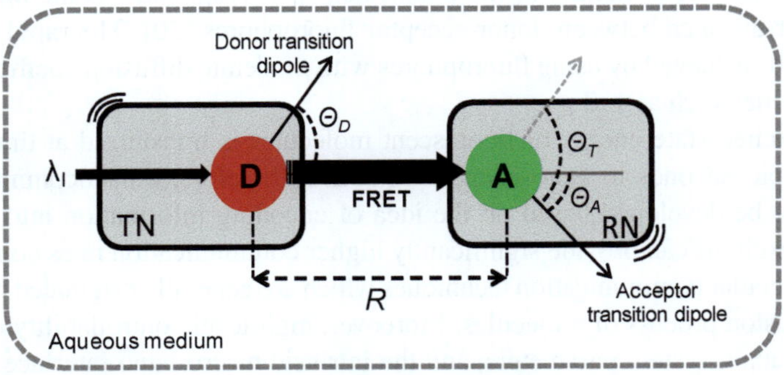

Fig. 2 Point-to-point FRET-based molecular communication channel model with single TN and RN communicating via FRET

a single laser pulse which has a wavelength near to the excitation maximum of the donor is sent to TN at the beginning of a fixed-duration time slot. The donor fluorophore acting as TN is excited by this pulse and after some time it relaxes through either fluorescence or FRET by transferring its excited energy to a nearby acceptor fluorophore, i.e., RN, and making the acceptor excited. If FRET occurs following the excitation of the donor, RN detects bit-1. In the bit-0 case, IS does not send any pulse to TN and keeps it silent in a time slot duration. Therefore, RN also stays in the ground state, detects bit-0. The time slot determines the bit interval, i.e., bit period, T_b.

A fluorophore in its excited state cannot be re-excited until it relaxes to the ground state [15]. If slot durations are not arranged properly, this fact results in Inter Symbol Interference (ISI) such that when IS sends successive bit-1's to TN, some of the bits cannot be transmitted through the channel if the donor or acceptor is in the excited state due to the preceding bit-1 transmission. An exciton generated by IS stays on the donor for a time of τ_{FRET}, which is the excited state lifetime of the donor when there is a ground-state acceptor in its vicinity, and then, it is transferred to the acceptor where it stays for a time of τ_A, which is the excited state lifetime of the acceptor. Therefore, after a sojourn time of $T_e = \tau_{FRET} + \tau_A$, the exciton is removed from the system. Since T_e is the sum of two independent exponential variables with the mean values of μ_{FRET} and μ_A, its cumulative distribution function can be written as

$$F(T_e) = 1 + \frac{1/\mu_{FRET}}{1/\mu_A - 1/\mu_{FRET}} e^{-1/\mu_A T_e} - \frac{1/\mu_A}{1/\mu_A - 1/\mu_{FRET}} e^{-1/\mu_{FRET} T_e} \quad (11)$$

If we set to slot duration as $T_b = 4 \times (\mu_{FRET} + \mu_A)$, the probability that the exciton removal time is greater than the slot duration, which is the case that results in ISI, becomes as low as 10^{-3} that might be acceptable for nanonetwork applications.

If the bit interval satisfies the no-ISI criteria, assuming that the donor is guaranteed to be excited by each laser pulse, and neglecting the direct excitation of the acceptor by the laser pulse, the successful transmission probability of bit-1, i.e., p_1, is equal to P_{FRET}. Assuming there is no another excitation source that can excite the acceptor independently, the successful transmission probability of bit-0 is equal to 1. Therefore, the channel is modeled information theoretically as a Z-channel.

The transition matrix of the Z-channel considering X as the transmitted bit by TN, and Y as the received bit by RN is given as

$$P(Y|X) = \begin{bmatrix} 1 & 0 \\ \frac{k_0}{k_0+k_t} & \frac{k_t}{k_0+k_t} \end{bmatrix}$$

When the input alphabet X is Bernoulli distributed with probability P_F, the mutual information $I(X;Y)$ between X and Y can be inferred from the transition matrix as

$$I(X;Y) = H(P_F \frac{k_t}{k_0 + k_t}) - P_F\, H(\frac{k_0}{k_0 + k_t}), \quad (12)$$

Table 1 Simulation parameters for point-to-point single-exciton FRET-based nanoscale communication channel

Donor-acceptor pair	ECFP-EYFP
Intermolecular distance (R)	(3–6) nm
Refractive index (n)	1 (vacuum), 1.3342 (water at 25 °C), 1.5185 (silicon oil at 25 °C)
Orientation factor (κ^2)	2/3 (rapid randomization), 4 (parallel dipole moments)

where $H(.)$ denotes the binary entropy. Therefore, the capacity of the FRET channel C_F can be given as the maximum mutual information over the input distribution as follows

$$C_F = \max_{P_F} I(\mathbf{X}; \mathbf{Y}) \tag{13}$$

3.2 Numerical Analysis

In this section, we present the numerical analysis performed over the mutual information expression given in (12) to show how the information theoretical capacity of the FRET-based nanocommunication channel varies according to some system parameters. The simulation parameters are given in Table 1

3.2.1 Effect of Intermolecular Distance

For the first analysis, we investigate the effect of the intermolecular distance R on the capacity of FRET-based communication channel. The analysis is carried out assuming that a commonly used fluorophore pair ECFP-EYFP is employed as the donor-acceptor pair. We also assume rapid randomization of the relative orientation of the fluorophores in a medium of water at 25 °C. Under these assumptions, Förster radius for ECFP-EYFP pair is calculated as $R_0 = 4.92$ nm [16].

Figure 3a demonstrates the mutual information for varying excitation probability of the donor (P_F) with different intermolecular distances. For R values higher than the R_0, the probability of FRET in the case of donor excitation, i.e., p_1, significantly decreases. As a result, the transmission of bit-1 becomes very erroneous. Therefore, the capacity decreases for higher R. The mutual information is maximized when $R = 3$ nm and $P_F = 0.474$. The resultant capacity for this arrangement is observed as 0.86 bit/use.

3.2.2 Effect of Medium

In the second analysis, we investigate the channel capacity for different media. ECFP-EYFP pair as TN-RN pair is supposed to be located in different media and separated

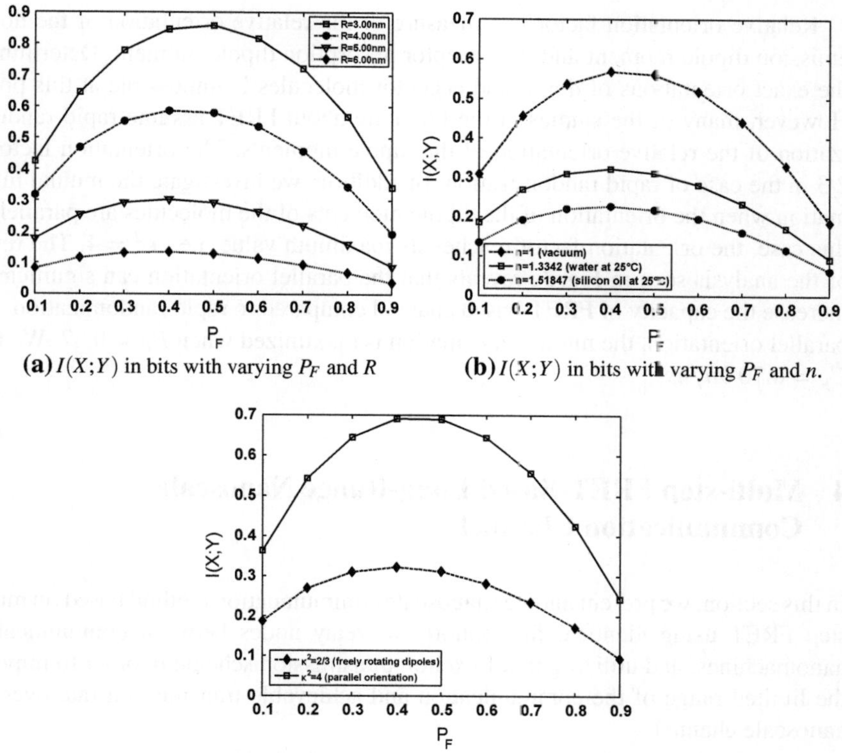

(a) $I(X;Y)$ in bits with varying P_F and R

(b) $I(X;Y)$ in bits with varying P_F and n.

(c) $I(X;Y)$ in bits with varying P_F and κ^2

Fig. 3 Mutual information for varying system parameters

by a distance of $4nm$ assuming rapid randomization of relative orientation of the molecules.

The mutual information $I(X;Y)$ is shown for varying excitation probability P_F with different media, i.e., different refractive indices, in Fig. 3b. As the refractive index of the medium decreases, the Förster radius increases. Therefore, the probability of FRET, i.e., successful transmission probability of bit 1 increases. As a consequence, the mutual information between TN and RN increases with decreasing refractive index of medium, and it is maximized for vacuum when $P_F = 0.43$. We find $C_F = 0.57\ bit/use$.

3.2.3 Effect of Relative Orientation Factor

The last analysis investigates the effect of relative orientation factor (κ^2) on the channel capacity using ECFP-EYFP as the TN-RN pair which are assumed to be located in a medium of water at 25 °C and separated by a distance of 4 nm.

Relative orientation factor is a measure of the relative orientation of the donor emission dipole moment and the acceptor absorption dipole moment. Determining the exact orientations of donor and acceptor molecules is impossible at this point. However, many of the studies in the literature about FRET assume rapid randomization of the relative orientation of the dipole moments. The orientation factor is 2/3 in the case of rapid randomization. In addition, we investigate the mutual information when the orientation of the dipole moments of the molecules are parallel. In this case, the orientation factor reaches its maximum value, i.e., $\kappa^2 = 4$. The result of the analysis seen in Fig. 3c reveals that the parallel orientation can significantly increase the capacity of FRET-based channel compared to rapid randomization. For parallel orientation, the mutual information is maximized when $P_F = 0.47$. We find $C_F = 0.70\ bit/use$.

4 Multi-step FRET-Based Long-Range Nanoscale Communication Channel

In this section, we present another nanoscale communication method based on multi-step FRET using identical fluorophores as relay nodes between communicating nanomachines, and utilizing multi-exciton transmission scheme in order to improve the limited range of the communication and achievable transmission rate over the nanoscale channel.

4.1 Principles of Multi-step FRET-Based Communications

Multi-step FRET defines the sequential transfer of excitons, i.e., excited state of fluorophores, through more than one fluorophore. The excitons generated on the donor and transferred to the acceptor molecule might be transferred to another fluorophore that is spectrally similar and spatially proximal to the acceptor. Multi-step FRET has been studied extensively in order to improve the spatial range of FRET over 10 nm [8]. Based on the multi-step FRET, the basic model presented in Sect. 3 is extended employing more than one identical fluorophore as the relay nodes between TN and RN, and utilizing multi-exciton transmission with ns-duration pulses.

4.1.1 Communication System Model

The multi-step FRET-based nanoscale communication system is composed of four main parts similar to traditional communication systems: an information source, a transmitter nanomachine, a communication channel, and a receiver nanomachine:

- ***Information Source***: The main information source of the system can be an external excitation source such as commercial laser that aims to remotely control the operation of a nanonetwork by sending optical pulses to TN. Utilizing OOK modulation, the information is encoded into two bits: bit-1 and bit-0. We assume that IS sends optical or electrical excitation pulses to TN with T_{pulse}-duration at the beginning of a bit interval T_b in order to represent bit-1, and keeps TN silent during T_b to represent bit-0.
- ***Transmitter Nanomachine***: The transmitter is assumed to be a single donor fluorophore that receives the excitation pulses sent from IS and generates excitons to the system. The emission spectrum of the donor molecule is assumed to overlap with the absorption spectra of the channel molecules, i.e., relay fluorophores, and the donor is assumed to be in close proximity of the relays. On the contrary, we neglect the back transfer of the excitons from the relay fluorophores to the donor assuming the emission spectra of the relays and the absorption spectrum of the donor do not overlap.
- ***Communication Channel***: The communication channel is composed of relay fluorophores that are randomly located and oriented in a three dimensional virtual lattice over aqueous medium and assumed to undergo random movements following Brownian statistics. The excitons are transferred between identical relay fluorophores via homoFRET. The underlying mechanism of homoFRET is the same as the one of FRET, except that it occurs between identical molecules, therefore, it requires that the emission and absorption spectra of the employed molecules must have a significant overlap.
- ***Receiver Nanomachine***: The receiver is assumed to be a single acceptor fluorophore that is the final destination of the excitons. The absorption spectrum of the acceptor and the emission spectra of the relays are assumed to overlap. However, the emission spectrum of the acceptor and the absorption spectra of the relays do not overlap in order to avoid back transfer from the acceptor to the relays. RN is supposed to be time-synchronized with TN. When RN is excited via FRET during T_b, it decides bit-1. If RN does not receive any exciton during the bit interval, it decides bit-0.

4.1.2 Principles of Exciton Activity

During the transmission of bit-1, the generated excitons occupy fluorophores for a random time, then they are transferred to another fluorophore via FRET or removed from the system via fluorescence. These processes are detailed as follows:

- ***Exciton generation***: When IS sends an excitation pulse to TN i.e., the donor, it is immediately excited, i.e., it generates an exciton, assuming the absorption coefficient of the donor is 1. Once the exciton is generated, it stays on the donor for a random time, then the donor relaxes through either fluorescence or FRET. In the case of FRET, the generated exciton is transferred to a proximal fluorophore. If T_{pulse} is large enough, after the first relaxation, the donor is expected to undergo

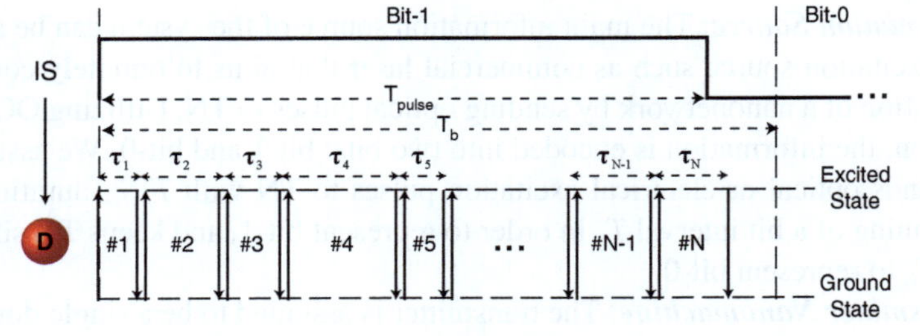

Fig. 4 Representation of bit-1 and bit-0 by IS, and sequential generation of N excitons on Donor (D) with random inter-generation times

many excitation and relaxation cycles during T_{pulse}, i.e., we expect more than one exciton to be generated by one pulse as demonstrated in Fig. 4. However, the number of generated excitons by a single pulse is a random variable, since the inter-generation times, i.e., the occupation times of generated excitons on donor, are random.

- ***Exciton occupation***: An exciton occupies a fluorophore when it is generated on or transferred to that molecule. An occupied fluorophore is not available for the occupation of another exciton until it relaxes. The interval between the excitation and the relaxation of a fluorophore τ gives the occupation time of that exciton. τ is an exponential random variable with a mean μ_τ which depends on the FRET rate of the donor molecule to the proximal and available, i.e., ground-state, fluorophores, and the intrinsic lifetime of the donor molecule. Assuming there are k fluorophores available for energy transfer in the range of the donor, the mean occupation time of the ith exciton can be expressed in terms of the process rates:

$$\mu_{\tau_i} = (k_0 + \sum_{j=1}^{k} k_{t,j,i})^{-1} \tag{14}$$

where $k_{t,j,i}$ is the FRET rate between the excited fluorophore and the jth proximal fluorophore for the ith exciton. k_0 is the intrinsic fluorescence rate of the excited fluorophore. $k_{t,j,i}$ depends on intrinsic and environmental parameters [15] and can be expressed by:

$$k_{t,j,i} = 8.8 \times 10^{22} \kappa_{j,i}^2 n^{-4} J_j(\lambda) \frac{k_0}{R_{j,i}^6} \tag{15}$$

where $\kappa_{j,i}^2$ is the orientation factor of the occupied and the jth proximal fluorophore during the occupation of the ith exciton. J_j is the degree of the spectral overlap between the emission spectrum of the occupied fluorophore and the absorption spectrum of the jth proximal fluorophore. $R_{j,i}$ is the intermolecular distance between the occupied fluorophore and the jth proximal fluorophore for the occupation time of the ith exciton.

- ***Exciton transfer***: After a random occupation time, the exciton leaves the fluorophore by either fluorescence or FRET to another proximal fluorophore. The exciton randomly selects the next pathway from a set of possible pathways. Assuming that k proximal fluorophores are available for the energy transfer, the probability of FRET to a specific fluorophore can be expressed by

$$P_{FRET\,j,i} = \frac{k_{t,j,i}}{k_0 + \sum_{l=1}^{k} k_{t,l,i}} \tag{16}$$

- ***Exciton removal***: An exciton is removed from the system by exciton recombination, i.e., the recombination of electron and hole. The prevalent recombination mechanism among dyes and fluorescent proteins is radiative recombination that results in radiation of a photon. The excitonic energy is converted to a photon with a wavelength dependent on the emission spectrum of the occupied fluorophore. For an exciton that occupies a fluorophore with k available neighbor fluorophores, the probability of fluorescence is given as

$$P_{Fluo,i} = \frac{k_0}{k_0 + \sum_{l=1}^{k} k_{t,l,i}} \tag{17}$$

4.2 *Multi-step FRET-Based Communication Channel with Disordered Relays*

Figure 5 demonstrates the multi-step FRET based communication system comprising identical fluorophores located randomly in a 3-dimensional aqueous environment. The fluorophores are expected to move uniformly in each direction following

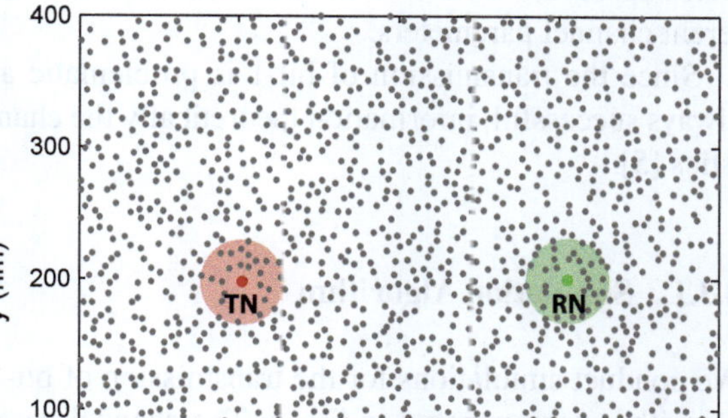

Fig. 5 Two dimensional demonstration of randomly deployed relays throughout the lattice surrounding TN and RN. The excitons released by TN undergo random jumps through the relay nodes, and some of them reach RN

Brownian statistics, therefore, at any time instant, the spatial distribution of the fluorophores is assumed to be uniform throughout the environment. Assuming isotropic and unrestricted distributions for all three angles, the individual relay fluorophores freely rotate. Following the assumption, the orientation factor for each pair of the relays is subject to the probability density function given in (7).

The donor, acceptor and relay fluorophores are considered as spherical molecules with diameter of 1.2 nm. There are just one donor and one acceptor in the considered lattice. We assume that there is no collisional quencher which removes the excitons when it collides with an excited fluorophore. Therefore, there are only two processes that an exciton can undergo: fluorescence or FRET to a nearby fluorophore.

4.3 *Information Theoretical Analysis*

In this section, we analyze the information theoretical capacity of the multi-step channel. Additionally, we investigate the interference between successively transmitted bits, and derive the maximum achievable data transmission rate in the case of OOK modulation.

For each communication system, we utilize the most basic modulation technique: binary OOK modulation with two bits available, bit-1 and bit-0. IS sends a T_{pulse}-duration optical pulse to TN at the beginning of a T_b-duration time slot in order for TN to transmit bit-1. For bit-0, IS does not send any pulse and keeps TN silent during T_b. Bit-1 is successfully transmitted if RN is excited in the relevant time slot. Bit-0 is transmitted successfully if RN stays in ground-state during T_b.

Assuming that there is no excitation-source except from IS, the transmission of bit-0 is always successful, i.e., $p_0 = 1$, if we neglect a probable ISI situation. However, the transmission of bit-1 is ambiguous because of the high degree of randomness and the correlated behavior in the motion of excitons. It is very difficult to derive an analytical solution for the successful transmission probability of bit-1, i.e., p_1. For that reason, we simulate the transmission of bit-1 in Matlab. Following a Monte Carlo approach, we derive the successful transmission and ISI probabilities for different channel parameters.

Since the transmission of bit-1 is problematic and the transmission of bit-0 is always successful, information theoretically the channels show Z-channel characteristics [5].

4.3.1 Simulation Algorithm

We conduct simulations for the transmission of bit-1 on each channel based on the algorithm demonstrated in Fig. 6. The channel parameters used in the simulations are given in Table 2.

The algorithm used in channel simulation is based on the competitive behavior of the excitons, and operates through the following steps:

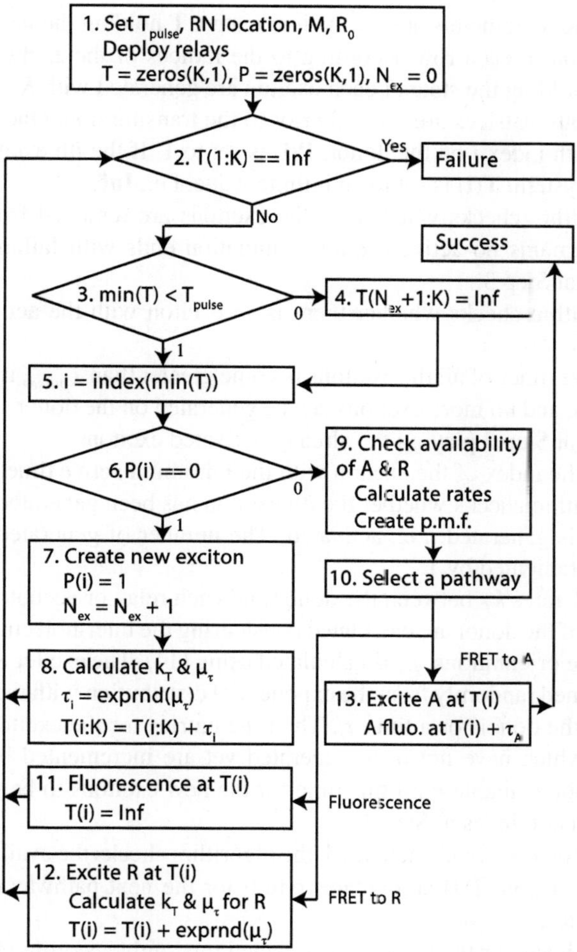

Fig. 6 Monte Carlo algorithm for the simulation of bit-1 transmission through the multi-step FRET-based communication channel

Table 2 Simulation parameters for multi-step FRET-based communication channel with disordered relays

Förster radius (R_0)	3–7 nm
Pulse length (T_{pulse})	10, 50, 100, 500 ns
Lattice size	$400 \times 400 \times 400$ nm
TN location	(100, 100, 100) nm
RN location	(103, 100, 100)–(301, 100, 100) nm
Relay concentration (M)	480, 960, 1920 mol/m^3
Fluorescence rates (k_D, k_R, k_A)	5×10^8 1/s
Molecular radius (r)	1.2 nm
Transfer range	$2 \times R_0$

1. The channel parameters are set. A time matrix **T** holding the last active time of each exciton in each row according to the indices of the excitons, and a state matrix **P** holding the state of each exciton are generated with K rows. All of the rows of both matrices are set to 0 prior to the transmission. Once the exciton is created with index i on the donor, **P(i)** is set to 1. If the ith exciton is removed from the system, **T(i)** is set to an infinite value, i.e., **Inf**.
2. The algorithm checks whether all the excitons are removed from the system. If there remains no active exciton, simulation ends with failure, otherwise it continues at Step 3.
3. The algorithm checks whether there is an exciton with the active time is less than T_{pulse}.
4. If the active times of all the excitons become greater than T_{pulse}, it means that the pulse ends, and no more excitons can be generated on the donor. The simulation continues at Step 5 playing the already-activated excitons.
5. i is set to the index of the exciton with the minimum active time.
6. The algorithm checks whether the ith exciton has been generated before.
7. Exciton i is generated, i.e., activated. The number of generated excitons, i.e., N_{ex}, is incremented by 1.
8. The FRET rates k_T between the donor and each relay or acceptor molecules in the range of the donor are calculated considering the intermolecular distance and the relative orientations. μ_τ is calculated using (14). The excited state lifetime τ_i is determined randomly from the exponential distribution with mean μ_τ. Exciton i stays at the donor for a time τ_i. The time entries for the exciton i and for the excitons which have not been generated yet are incremented by τ_i, since the donor is not available until this time for the new excitons to be generated. The simulation continues at Step 2.
9. If the exciton i is already activated, the algorithm checks the available molecules for FRET at time **T(i)** and creates p.m.f. for the next pathway calculating the process rates.
10. A new pathway for the exciton i is selected randomly according to the p.m.f.
11. If exciton i results in fluorescence at time **T(i)**, it is removed from the system by setting **T(i)** = **Inf**. The occupied molecule is relaxed and becomes available for the new excitons. The simulation continues at Step 2.
12. If exciton i occupies a relay molecule through FRET, the process rates are calculated checking the available molecules at time **T(i)**. μ_τ is calculated accordingly. The exciton i stays at the relay molecule for a time τ_i which is determined randomly from the exponential distribution with mean μ_τ. The simulation continues with Step 2.
13. If exciton i is transferred to an acceptor molecule at time **T(i)**, receiver detects bit-1 when the acceptor fluoresces at time $\mathbf{T(i)} + \tau_\mathbf{A}$. Since, the only way for the acceptors to relax is to fluorescence, τ_A is determined randomly from the exponential distribution with mean μ_{τ_A} regardless of the available molecules in the range. The simulation ends with the successful transmission of bit-1.

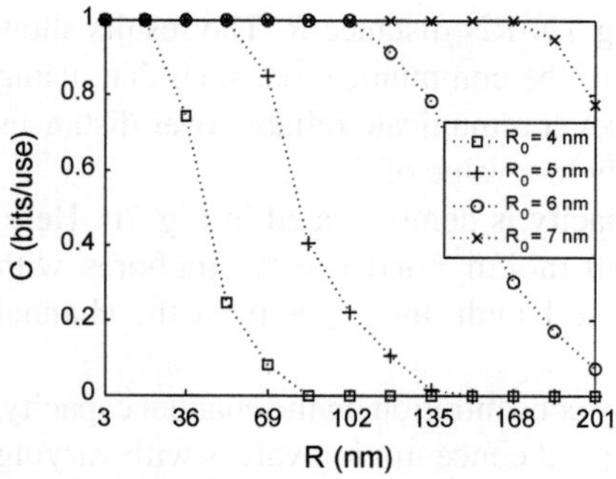

(a) Capacity for $M = 960\text{mol/m}^3$, $T_{pulse} =$ 50ns, and varying R_0.

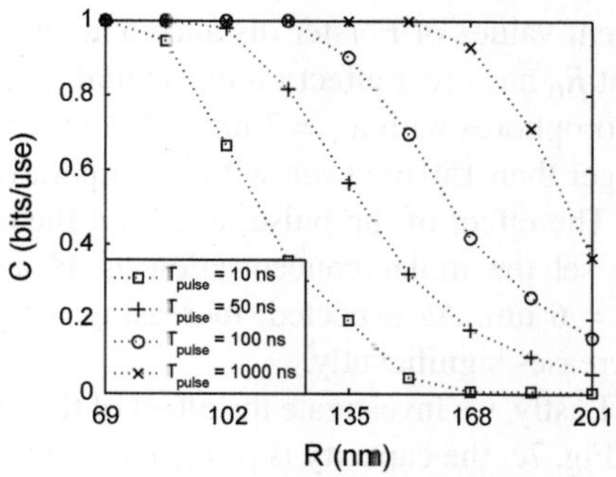

(b) Capacity for $M = 960\text{mol/m}^3$, $R_0 =$ 6nm, and varying T_{pulse}.

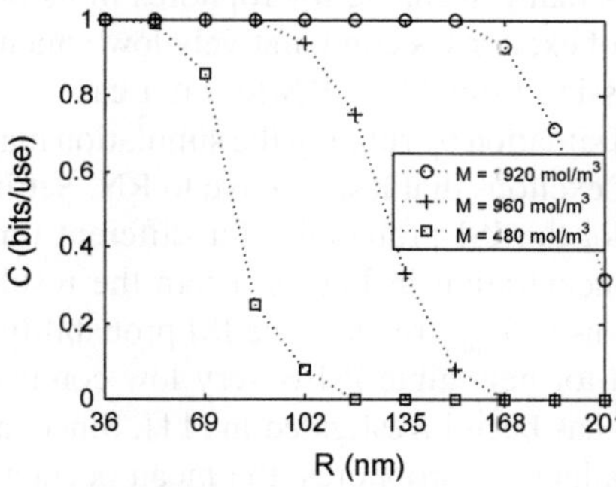

(c) Capacity for $T_{pulse} = 50\text{ns}$, $R_0 = 6\text{nm}$, and varying M.

Fig. 7 Information theoretical capacity of disordered channel for different channel parameters with varying TN-RN distance R

4.3.2 Analysis

Here, we investigate the capacity of the channel. The transmission probability of bit-1 is simulated following the algorithm described in Fig. 6, and using the parameters given in Table 2. The simulations are repeated until p_1 converges to a finite value. Note that, for each run of the simulation, the relay fluorophores are deployed again in random locations following a uniform distribution. The relative orientation of each pair of fluorophores is randomly selected according to the distribution given in (7) for each exciton occupation in order to imitate the isotropic and unrestricted rotations of fluorophores. Using the converged value of p_1, the capacity is derived as the maximum mutual information between the transmitted and received bits over all input distributions.

Setting the molar concentration of relay fluorophores in the environment as 960 mol/m^3, and using excitation pulses with duration $T_{pulse} = 50$ ns to represent bit-1, the information theoretical capacity of the channel is plotted in Fig. 7a for different values of Förster distance and varying TN-RN distance R. The results show that R_0 has great effect on the spatial range of the communication, such that, using fluorophores with $R_0 = 7$ nm, TN and RN can communicate reliably over distances larger than 150 nm even with a comparatively low value of T_{pulse}.

The effect of the pulse length on the capacity is demonstrated in Fig. 7b. Here, we set the molar concentration as $M = 960$ mol/m^3, and use fluorophores with $R_0 = 6$ nm. As expected, increasing the pulse length, the capacity of the channel increases significantly.

Lastly, we investigate the effect of the relay concentration on the channel capacity. In Fig. 7c, the capacity is plotted for three typical concentration values with varying distance between TN and RN. As is seen, increasing the molecular concentration improves the communication range. For relatively denser concentrations, an individual relay fluorophore has many available fluorophores in its proximity. As a result, the removal probability of excitons is comparatively low which increases the probability of exciton transmission from TN to RN for bit-1 case.

We investigate the ISI situation by running the simulation many times, and recording the removal times of excitons that lastly arrive to RN. Setting $T_{pulse} = 10$ ns and $T_b = T_{pulse} + T_{off}$, we plot the ISI probability for different time offsets with varying R and molecular concentration in Fig. 8a. From the results, we conclude that adding a time offset of 2 ns to T_{pulse} reduces the ISI probability to negligible values. The required time offset for negligible ISI is very low compared to that of zeolite L based channels which has been investigated in [11]. Since, a relay fluorophore is surrounded by many available fluorophores, the mean occupation time of excitons on relay fluorophores decreases to very low values, e.g., 1–10 ps for very dense environments. As a result, the transit time of excitons from TN to RN is also reduced. The decrease in the ISI probability with increasing molecular concentration also justifies this reasoning.

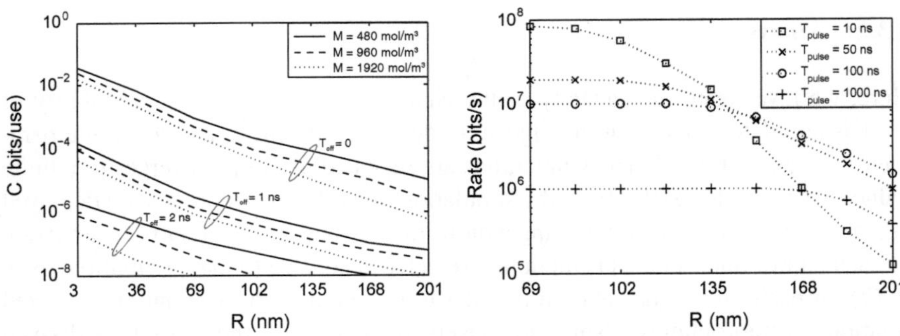

(a) ISI probability for disordered channel with varying T_{off}, M and R.

(b) Maximum achievable rate for disordered channel with varying T_{pulse} and R.

Fig. 8 Results of ISI and achievable rate analysis for disordered channel

Setting $T_{off} = 2$ ns and neglecting ISI, in Fig. 8b, we plot the achievable rates with varying R. It is demonstrated that mobile TN and RN can communicate at a rate over 80 Mbps if they are in approximately 70 nm-proximity of each other, and with 200 nm-proximity, the reliable communication can be realized at a rate over 1Mbps.

5 FRET-Based Mobile Molecular Nanonetworks

In the previous sections, FRET-based nanocommunication channel previously has been modelled with different configurations assuming that the communicating nanomachines are immobile during the communication. However, most of the applications, especially in-body applications, that nanonetworks promise, require for nanomachines to be mobile. Following this motivation, in this section, we focus on a network of mobile nanomachines communicating through FRET, i.e., FRET-Based Mobile Molecular Sensor/Actor Network (FRET-MSAN). We model the single message propagation in FRET-MSAN based on Continuous Time Markov Chains (CTMCs) assuming that the nanonetwork nodes are freely and randomly diffusing in a three dimensional aqueous environment satisfying the rapid-diffusion criterion. The performance of the network is investigated based on the results of the Monte Carlo simulations on Markov chain models.

5.1 FRET-Based Mobile Molecular Sensor/Actor Network (FRET-MSAN)

Bioluminescent molecules define a class of fluorescent molecules which are excited upon binding a target molecule [15]. Since they do not need a remote excitation source, e.g., optical laser, they are extensively used in biotechnological research as biomolecular sensors optically indicating the presence of a certain kind of molecule [4]. For example, *aequorin*, a bioluminescent protein, reacts with calcium ions, and relaxes through releasing a photon, thus, it is extensively used to measure Ca^{2+} concentration [2].

Fluorescent molecules also find applications in photodynamic therapy (PDT) of cancer as actuators. In QD-based PDT, QDs are excited by optical energy from a remote source and then transfer its exciton to the conjugated photosensitizing agent which synthesizes a reactive singlet oxygen via energy transfer [17]. The produced singlet oxygen initiates the apoptosis of nearby cancer cells. However, the reactive singlet oxygen is also harmful for normal cells, therefore, the spatial precision of the activation of singlet oxygen is crucial.

Fig. 9 Information flow in FRET-MSAN

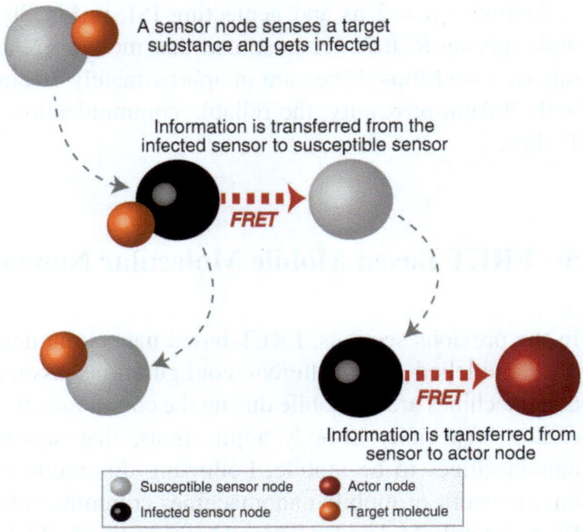

FRET-MSAN is composed of mobile bioluminescent sensors and fluorophore-based actors that can collect the information from the sensors and perform an appropriate action upon the environment. The investigated scenario in this section can pave the way for designing autonomous networks of nanomachines which are able to collaboratively sense the presence of tumor cells, and act precisely for the apoptosis of them. The information flow in FRET-MSAN is demonstrated in Fig. 9.

The generated message by the nanosensors is a one-bit binary message which is an exciton indicating whether the nanosensors bind to a target molecule or not. More than one nanosensor can detect the same target at the same time, and therefore, more than one excitons can be generated on the network. We assume that the nanoactors are not capable of the detecting the target and generating excitons individually. The initially generated excitons occupy on the detecting nanosensors during an exponentially random time, and then transferred to the other sensors or actors in the ground-state, i.e., randomly walk on a random lattice, or they can be removed from the system by fluorescence.

We assume that the detection is realized with a varying number of nanosensor nodes at the same, and during the transmission of the detection message, no other detection occurs. Therefore, the initially generated excitons set an upper bound on the number of excitons on the network, and the number of excitons is a monotonically decreasing function. During this discrete fade away of the excitons, if at least one nanoactor receives an exciton from an excited sensor node, the message is successfully transferred. These characteristics of the network imply that the single message transmission resembles a modified death process which can be modelled with a continuous time Markov chain which is demonstrated in Fig. 10.

In the model, the states demonstrate the number of excitons that are available on the network. The transition rates depend on the state of the network, i.e., number

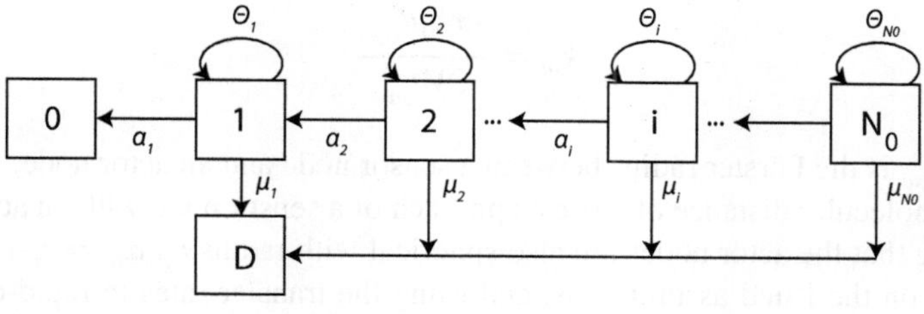

Fig. 10 Markov chain model of FRET-MSAN

of excitons that randomly walk throughout the network. There are two absorbing states in this Markov chain, i.e., state-0 and state-D. The state-0 represents the complete extinction of the excitons in the network. Detection cannot be realized once the extinction occurs since no other exciton is generated. If none of the nanoactors receive an exciton until the extinction, the message is lost, and the transmission fails. The state-D represents the successful detection of the message which occurs when a nanoactor receives an exciton. All of the states except the state-0 are connected to the state-D. The model is based on the following assumptions:

- Initially N_0 number of sensors are in the excited state.
- No additional exciton is generated during the message propagation.
- The number of sensor nodes, N_s, and the number of actor nodes, N_a, and the total number of the network nodes, $N = N_s + N_a$, are constant.
- An excited sensor node gets rid of the excitation and returns to the ground-state without an exciton transfer at a rate of k_0. Note that, this is the natural fluorescence rate of the bioluminescent fluorophores used as molecular sensors.
- An excited sensor node transfers the infection to a ground-state sensor node making it excited while returning to the ground-state with a rate of k_{ss}.
- An excited sensor node transfers its excitation to an actor, and return to the ground-state with a rate of k_{sa}.
- The exciton transfer is pairwise, i.e., an excited nanonode can transfer its excitation to only a single nanonode at a time.

The information transfer rate between a rapidly-moving excited sensor node to a ground-state, i.e., available, sensor node in the environment, i.e., k_{ss}, is given using (8) as follows

$$k_{ss} = \frac{4\pi k_0 R_{0,ss}^6}{3Va_{ss}^{-3}} \tag{18}$$

where $R_{0,ss}$ is the Förster radius between sensor nodes, and a_{ss} is the intermolecular distance of closest approach of two sensor nodes. We assume that the bioluminescent sensors are spherical with radius r_s, therefore, $a_{ss} = 2r_s$. Similarly, the rate of the information transfer from an excited sensor node to an actor node is given by

$$k_{sa} = \frac{4\pi k_0 R_{0,sa}^6}{3Va_{sa}^{-3}} \tag{19}$$

where $R_{0,sa}$ is the Förster radius between a sensor node and an actor node, and a_{sa} is the intermolecular distance of closest approach of a sensor node with an actor node. Assuming that the actor nodes are also spherical with radius r_a, $a_{sa} = r_s + r_a$.

Based on the listed assumptions, and using the transfer rates in rapid-diffusion, we derive the transition rates of the Markov chain. The state-dependent death rate, α_i, is the overall fluorescence rate on the network, and it can be given by

$$\alpha_i = ik_0 \tag{20}$$

where i is the number of excitons, i.e., number of excited sensor nodes, on the network, and k_0 is the natural fluorescence rate of the fluorophores which we assume identical for each fluorophore. The exciton transfers between sensor nodes do not alter the state of the network, since the number of excitons does not change. The overall transfer rate between nanosensors is given in terms of number of excited state sensors and ground-state sensors as

$$\theta_i = i(N_s - i)k_{ss} \tag{21}$$

where N_s is the total number of sensor nodes. The rate of transfer from the excited sensor nodes to the actor nodes can be given by

$$\mu_i = iN_a k_{sa} \tag{22}$$

where N_a is the number of actor nodes. Considering the infeasibility of deriving analytical expressions for the performance metrics, we simulate the model by dividing time into small intervals, i.e., Δt, during which we assume only one event can occur. For this time interval, the transition probabilities can be given as

$$P_{i,i-1} = P(S_{t+\Delta t} = i-1 | S_t = i) \approx \alpha_i \Delta t \tag{23}$$

$$P_{i,i+1} = P(S_{t+\Delta t} = i+1 | S_t = i) = 0 \tag{24}$$

$$P_{i,D} = P(S_{t+\Delta t} = D | S_t = i) \approx \mu_i \Delta t \tag{25}$$

$$P(S_{t+\Delta t} = i | S_t = i) \approx 1 - P_{i,i-1} - P_{i,D} \tag{26}$$

where $i = 1, 2, ...N_0$, and S_t demonstrates the state of the network at time t. Using the transition probabilities and setting a sufficiently small time interval Δt, we conduct a Monte Carlo simulation in Matlab to obtain the successful detection probability of a single message, $Pr(success)$ and average message delivery delay, $E[T_d]$ for different system parameters. The detection probability is calculated as the number of successful transmissions to the number of extinctions without detection. The simulation is run until the detection probability converges to a finite value. We also obtain the average number of actor nodes that receive an exciton for each cycle of message transmission, i.e., n, and we derive the average system throughput as

$$T_{avg} = \frac{n}{E[T_d]} \tag{27}$$

5.2 *Performance Analysis of FRET-Based Mobile Molecular Nanonetworks*

We conduct Monte-Carlo simulations to obtain the successful detection probability of a single message and the mean extinction time of excitons in FRET-MSAN with varying network and node parameters to understand the effect of each parameter on the network performance and gain insight on the feasibility of the FRET-based mobile nanonetworks.

5.2.1 Performance Analysis of FRET-MSAN

In this section, we present the performance of the FRET-MSAN based on the simulation results for single message propagation. We analyze the single-message transmission probability, mean transmission delay as the mean extinction time of excitons and average system throughput for varying number of nanoactors and initially excited nanosensors, varying Förster radius between nanosensors and nanoactor, and varying size of nanomachines. The default values of the parameters used in the numerical simulations are presented in Table 3.

Table 3 Simulation parameters for FRET-MSAN

Radius of nanosensors (r_s)	0.25 (nm)
Radius of nanoactors (r_a)	0.25 (nm)
Natural fluorescence rate of sensor and actor nodes (k_0)	10^6 (s^{-1})
Volume (V)	1 (μm^3)
Number of initially excited sensor nodes (N_0)	5
Total number of sensor nodes (N_s)	100
Number of actor nodes (N_a)	20
Förster radius of nanosensor-nanosensor pair ($R_{0,ss}$)	8 (nm)
Förster radius of nanosensor-nanoactor pair ($R_{0,sa}$)	8 (nm)

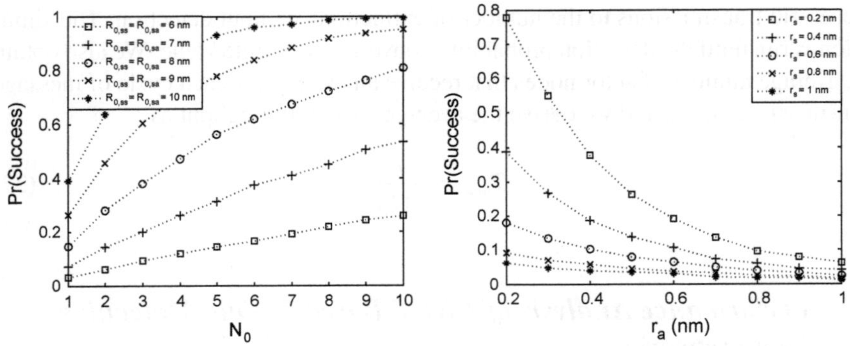

(a) $Pr(success)$ with varying N_0 for different $R_{0,ss} = R_{0,sa}$.

(b) $Pr(success)$ with varying r_a for different r_s.

Fig. 11 The successful detection probability for one-bit message transmission in FRET-MSAN

Probability of Successful Detection

The effect of the number of initially infected nanosensors on the detection probability is shown in Fig. 11a with varying number of nanoreceivers. Since we assume that no other external infection occurs, and the infection is only transferred between the network nodes, the number of infected nanosensors is a decreasing function of time. Therefore, the initial number of infection is an important parameter for the detection performance. It is observed that the detection probability significantly increases even with a small increase in the number of initially infected nanosensors.

The size of nanomachines also significantly affects the successful detection probability, since the extent of the distance of closest approach has a direct effect on the energy transfer rate between nanonodes. The detection probability for varying radii of nanosensors and nanoactors is shown in Fig. 11b. The selected values for radii are in the range of size of common fluorophores, e.g., fluorescent dyes. We observe that, using small-size nodes significantly increases the detection probability.

Mean Extinction Time of Excitons

Mean extinction time of the excitons determines the average time that the generated message (bit-1) propagates through the network. Since the actor nodes can get the excitation even when there is a single exciton existing in the network, the mean extinction time, $E[T_d]$ can be regarded as the mean delay for the message transmission.

In Fig. 12a, the mean extinction time is demonstrated with varying N_0 for different Förster radii. As expected, when the message is encoded into higher number of initial excitons, the extinction time increases, since the initial state of the network is further from the extinction state, i.e., state-0, when there are more initial excitons.

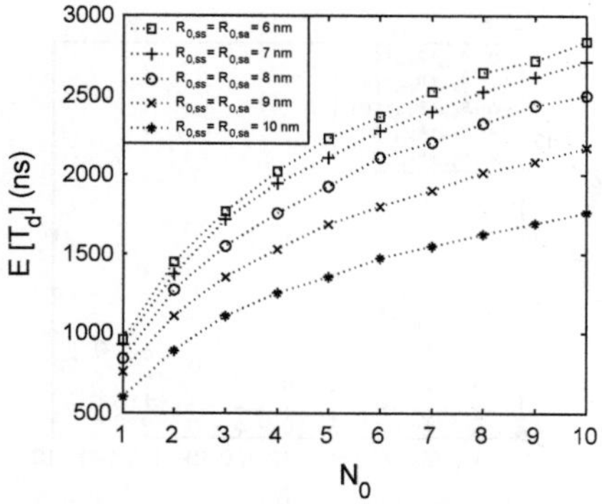

(a) $E(T_d)$ with varying N_0 for different $R_{0,ss} = R_{0,sa}$.

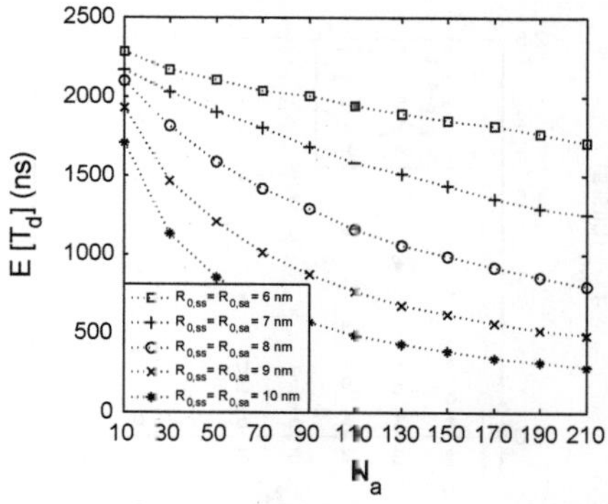

(b) $E(T_d)$ with varying N_a for different $R_{0,ss} = R_{0,sa}$.

Fig. 12 Mean extinction time of excitons for one-bit message transmission in FRET-MSAN

Förster radius between sensor nodes, $R_{0,ss}$, has no effect on the average decay time of the excitons, since it only has a contribution in the rate of the exciton transfer between sensor nodes which does not alter the state of network, i.e., the number of excitons. However, Förster radius between sensor and actor nodes, $R_{0,sa}$ has a significant effect on the extinction time. This is because, increasing $R_{0,sa}$ increases the exciton transfer rate from sensor nodes to actor nodes, i.e., k_{sa}. Since an exciton that is transferred to an actor node is assumed to be lost during the decoding process, the number of excitons decreases more rapidly, when k_{sa} increases. This is directly related to the result presented in Fig. 12b which demonstrates that the mean extinction time decreases with increasing number of actor nodes which also increases k_{sa}.

System Throughput

System throughput is a significant factor which determines how many actor nodes can receive the one-bit message per unit time. To calculate the system throughput, the number of actor nodes that achieve to receive excitation until the extinction of the excitons, is observed in the Monte Carlo simulations. The average system throughput that is calculated using (27) is demonstrated in Fig. 13a, b with varying N_0, N_a and Förster radii.

When the initial number of excitons increases, the extinction time of the excitons also increases, therefore it is more likely that more actor nodes get into close proximity of excited sensor nodes and get an excitation. Increasing the number of actor nodes and the Förster radius between sensor and actor nodes, the exciton transfer rate from excited sensor nodes to the actor nodes increases. This results in a higher detection probability at any time interval. Therefore, excitons can be transferred to more actor nodes during a single message transmission.

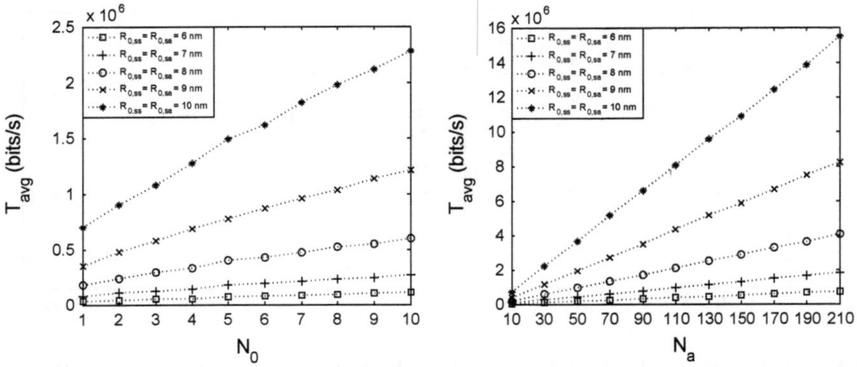

(a) T_{avg} with varying N_0 for different $R_{0,ss} = R_{0,sa}$.

(b) T_{avg} with varying N_a for different $R_{0,ss} = R_{0,sa}$.

Fig. 13 Average system throughput for one-bit message transmission in FRET-MSAN

6 Conclusion

In this chapter, we review the nanoscale molecular communication method which is based on a well-known phenomenon FRET. First, we investigate FRET-based point-to-point nanoscale communication channel from the information theoretical perspective employing ON/OFF Keying modulation with single-exciton transmission scheme. Later, we present the long-range nanoscale communication channel based on multi-step FRET applying multi-exciton transmission scheme which extends the communication range of the FRET-based communications. Furthermore, we investigate FRET-based mobile molecular nanonetworks, and present continuous time Markov chain models for FRET-based mobile molecular sensor and actor networks (FRET-MSAN) benefiting from the modified birth-and-death processes. As with the further advances in nanotechnology, we believe that FRET-based nanocommunication and nanonetworking concepts which are reviewed in this chapter, will pave the way for the design of efficient and reliable nanonetworks to be used for groundbreaking applications.

References

1. Akyildiz IF, Jornet JM, Pierobon M (2011) Nanonetworks: a new frontier in communications. Commun ACM 54(11):84–89
2. Allen DG, Blinks JR (1978) Calcium transients in aequorin-injected frog cardiac muscle. Nature 273(5663):509–513
3. Badali D, Gradinaru CC (2011) The effect of Brownian motion of fluorescent probes on measuring nanoscale distances by Förster resonance energy transfer. J Chem Phys 134(22):225102
4. Contag CH, Bachmann MH (2002) Advances in in vivo bioluminescence imaging of gene expression. Ann Rev Biomed Eng 4:235–260

5. Cover TM, Thomas JA (1991) Elements of information theory. Wiley, New York
6. Didenko VV (2001) DNA probes using fluorescence resonance energy transfer (FRET): designs and applications. Biotechniques 31(5):1106–1121
7. Förster T (1948) Zwischenmolekulare energiewanderung und fluoreszenz. Ann Phys 437(1–2):55–75
8. Heilemann M, Tinnefeld P, Mosteiro Parajo MG et al (2004) Multistep energy transfer in single molecular photonic wires. J Am Chem Soc 126:6514–6515
9. Kuscu M, Akan OB (2011) A nanoscale communication channel with fluorescence resonance energy transfer (FRET). In: Proceedings of 1st IEEE international workshop molecular nano scale communication/IEEE conference on computer communication workshops, Shanghai, China, April 2011
10. Kuscu M, Akan OB (2012) A physical channel model and analysis for nanoscale molecular communications with Förster resonance energy transfer (FRET). IEEE Trans Nanotechnol 11(1):200–207
11. Kuscu M, Akan OB (2013) Multi-step FRET-based long-range nanoscale communication channel. IEEE J Sel Areas Commun 31(12):715–725
12. Kuscu M, Akan OB (2014) A communication theoretical analysis of FRET-based mobile ad hoc molecular nanonetworks. IEEE Trans Nanobiosci 13(3):255–266
13. Kuscu M, Akan OB (2014) FRET-based nanoscale point-to-point and broadcast communications with multi-exciton transmission and channel routing. IEEE Trans Nanobiosci 13(3):315–326
14. Kuscu M, Akan OB (2014) Coverage and throughput analysis for FRET-based mobile molecular sensor/actor nanonetworks. Nano Commun Netw J (Elsevier) 5(1–2):45–53
15. Lakowicz JR (2006) Principles of fluorescence spectroscopy. Springer, Baltimore
16. Patterson GH, Piston DW, Barisas BG (2000) Förster distances between green fluorescent protein pairs. Anal Biochem 284(2):438–440
17. Samia ACS, Chen X, Burda C (2003) Semiconductor quantum dots for photodynamic therapy. J Am Chem Soc 125(51):15736–15737
18. Sekatskii SK, Chergui M, Dietler G (2003) Coherent fluorescence resonance energy transfer: construction of nonlocal multiparticle entangled states and quantum computing. Europhys Lett 63:21
19. Stryer L (1978) Fluorescence energy transfer as a spectroscopic ruler. Ann Rev Biochem 47:819–846
20. Stryer L (1982) Diffusion-enhanced fluorescence energy transfer. Ann Rev Biophys Bioeng 11:203–222

Part IV
Nanomaterial and Nanostructure

Ultrasonics—An Effective Non-invasive Tool to Characterize Nanofluids

M. Nabeel Rashin and J. Hemalatha

Abstract Nanofluids are smart colloidal suspensions of fine nanomaterials in the size range of 1–100 nm in base fluids. For the last few years, nanofluids have been an important focus of research, due to their superior thermo physical properties and promising heat transfer applications. Regardless of various experimental studies, it is still unclear whether the thermal conductivity enhancement in nanofluids is anomalous, or lies within the predictions of theoretical models. Moreover, most of the reported values on their thermo physical properties are inconsistent, due to the complexity associated with the surface chemistry of nanofluids. In this chapter, the versatility of ultrasonics, as an effective non-invasive tool in characterizing nanofluids, is discussed. The chapter encompasses the significance and measurement methods of various ultrasonic parameters. The ultrasonic investigations, being non-invasive in nature, highly efficient and relatively cheap, can provide a powerful means to explore complex colloidal systems, like nanofluids and ferrofluids.

1 Introduction

Technological advances led to the miniaturization of gadgets and increased operating speeds, which in turn, augmented heat dissipation that demands a novel and innovative perception for cooling with improved performance. Heat transfer becomes a crucial issue to sustain and maintain an enhanced, reliable performance of a wide variety of products like the ultrahigh heat flux optical devices, car engines, exhaust gas regulators, high powered lasers, X-rays, computers, power electronics etc. Thus, the challenge of attaining a good cooling technology is faced

M. Nabeel Rashin · J. Hemalatha (✉)
Advanced Materials Lab, Department of Physics, National Institute of Technology, Tiruchirappalli 620 015, Tamilnadu, India
e-mail: hemalatha@nitt.edu

M. Nabeel Rashin
e-mail: nabeelrashin@gmail.com

J. Suzuki et al. (eds.), *Modeling, Methodologies and Tools for Molecular and Nano-scale Communications*, Modeling and Optimization in Science and Technologies 9, DOI 10.1007/978-3-319-50688-3_16

by almost all sectors like the transport industry, electronic industry, metrology, medicine, and defense [1, 2].

The insignificant thermal conductivity of conventional heat transfer fluids is a serious constraint in achieving effective performance. By considering the fact that metallic solids possess higher thermal conductivity than fluids, fluids containing suspended solid particles are expected to exhibit significantly higher thermal conductivity than conventional heat transfer fluids [1]. The conventional solid-liquid suspensions have millimeter or micrometer sized particles in the heat transfer fluids. These are the old and traditional materials employed for transferring heat. But these suspensions could not meet the demands of the industry because of their rapid settling. To resolve this problem, several attempts have been made to circulate the suspensions through the pipes or channels. Unfortunately, such particulate suspensions corrode and wear out the pipes, pumps, channels and the bearings. Moreover, these fluids are not applicable to Microsystems, as they clog the microchannels. Another major problem faced is the requirement of a large number of particles to be added to make up the suspension. This creates a pressure drop and reduces the pumping power. The thermal conductivity of these conventional fluids is very poor and this imposes a fundamental limit on the heat transfer.

Very few nanoparticles, when dispersed uniformly and suspended stably in host fluids, can provide dramatic improvements in their thermal properties. Such smart colloidal suspensions of nanoparticles in carrier liquids are called nanofluids. The term nanofluid is introduced by Choi in 1995 for describing this new class of nanotechnology—based heat transfer fluids that exhibit thermal properties superior to those of conventional particle–fluid suspensions [3]. The enhancements in the thermal conductivity of nanofluids have been ascribed to a variety of mechanisms, including Brownian motion, microconvection, clustering, and agglomeration or the combination of all these, which are specific to a system and test conditions.

Nanofluids have promising heat transfer applications, which are of major importance to industrial sectors including transportation, power generation, micro-manufacturing, thermal therapy, heating, cooling, ventilation, air conditioning, MEMS, etc. Despite numerous investigations, it is still ambiguous whether the enhancement in the thermophysical properties of nanofluids is anomalous or within the predictions of classical theory. The ever growing technological developments and rapid increase in energy needs, demand the creation of a novel heat transfer fluid with favorable rheological behavior and high stability, which can strengthen the heat transfer processes and trim down energy losses. Nanofluids provide such a potential heat transfer medium, as they can surpass conventional heat transfer liquids in thermal conductivity, convective heat transfer coefficient, critical heat flux, viscosity, wettability etc. Besides high thermal conductivity and stability, they can offer numerous benefits, such as microchannel cooling without clogging, the possibility of miniaturizing systems scaling, reduction in pumping power etc. Thus, they act as the most efficient industrial coolants in microsystems, and in operating the high efficiency engines with significantly higher thermal loads.

Nanofluids have received significant attention, after the first report of unusually large thermal conductivity enhancement in Cu nanofluids at very low particle loading [4]. Numerous theoretical and experimental studies are now available on the thermal conductivity of nanofluids [1, 5–7].

Thermal conductivity enhancements in nanofluids are normally predicted using the effective medium theory (EMT) of Maxwell for low concentrations (<10 vol. %). The theory is initially developed by Maxwell [8] for non–interacting spheres and later extended by Hamilton and Crosser [9] for nonspherical particle shapes and is given by the following expression.

$$k_{nf} = \left[\frac{k_p + (n-1)k_{bf} + (n-1)(k_p - k_{bf})\phi}{k_p + (n-1)k_{bf} - (k_p - k_{bf})\phi}\right] k_{bf} \tag{1}$$

where, k_p, k_{nf} and k_{bf} are the thermal conductivities of the nanoparticle, the nanofluid and the base fluid respectively. ϕ is the volume fraction and n is the empirical shape factor given by $n = 3/\psi$. ψ is the sphericity, defined as the ratio of the surface area of a sphere to the surface area of the particle having, the same volume as that of the sphere.

However, there are reports [10–22] which show that the thermal conductivity enhancements are considerably higher than those predicted by using the effective medium theory for nanofluids of very low particle concentrations, whereas, recent reports show the enhancement lying within the EMT predictions [23–30]. It is commonly agreed [31] that the initial reports on thermal conductivity enhancements are unrepeatable. Despite numerous experimental and theoretical studies, it is still debated whether the thermal conductivity enhancements in nanofluids are anomalous or within the predictions of the effective medium theory. The fundamental understanding of the exact mechanisms behind the anomalous behavior is unclear, because of the lack of molecular level understanding of the ultrafine particles dispersed in the liquid medium.

It is easy to understand from the history of colloid science, that researchers have been trying to explore the internal structure and interactions of nanofluids through light aided equipments. Light has been used as an important tool from the time of the microscopic observations of Brownian motion to the light scattering techniques. On the other side, sound remains unpopular in colloid science, despite a little amount of work in the field of acoustics in the past decade. The roots of the current understanding of sound starts from more than 300 years back, when Newton suggested the first theory for calculating the speed of sound [32]. Newton proposed that sound propagated while maintaining a constant temperature, which is, an isothermal case. Later, Laplace corrected this misinterpretation by showing that it was actually adiabatic in nature [32, 33].

Even though ultrasonics is not explicitly used for nanofluids, it has been used as a powerful tool to learn about the structure of pure liquids, the dissolution mechanisms in polymer solutions, the molecular interactions in solutions and the nature of chemical reactions in liquids. Ultrasonic velocity depends upon the relation between elasticity and density in the medium through which it is transmitted [33].

A major drawback of traditional light based characterization techniques, is that light cannot propagate through concentrated and opaque liquid systems. They do not characterize the systems as they are, since they require specific sample preparation, including super–dilution which can destroy aggregates or flocs, and hence, the corresponding measured parameters would not be as exact as the original.

Ultrasonic investigation can be applied to any fluid system that is Newtonian or non-Newtonian. The diversity in application ranges from pure solvent to concentrated slurries, which eliminate the need for dilution; i.e., aggregative stability, cosmetic emulsions, ceramics, chemical-mechanical polishing, coal slurries, coatings, environmental protection, flotation, ore enrichment, food products, latex, cement slurries, emulsions and micro emulsions, mixed dispersions, nanosized dispersions, nonaqueous dispersions, paints, and photo materials.

The ultrasonic technique is very robust, accurate and relatively fast. Also, it is simple because the process involves only a single field, the mechanical stress. Thus, any concentrated and opaque liquid system, such as nanofluids and ferrofluids can be characterized as they are. All these peculiarities make ultrasonics as an inevitable, effective and attractive charactering tool for nanofluids and other emerging liquids.

From the literature it is found that there are only a few reports available on the ultrasonic properties of nanofluids [34–41]. Reports on magnetic nanofluids [42–44] prove their tunable optical, rheological and thermal properties, and also show the dependence of ultrasonic velocity on the clustering structure of the magnetic fluid. A big deviation of the experimental values of velocity and attenuation from the theoretical predictions are also reported [45–47].

2 Versatility of Ultrasonics in Nanofluid Characterization

The ultrasonic technique allows us to determine the significant parameters through which various thermo physical properties of fluids can be explored. The most important and widely used among them are listed in Table 1.

Table 1 Ultrasonic parameters and their significance in nanofluid research

Parameter	Unit	Significance
Sound velocity (v)	m/s	Velocity of ultrasound in a medium which is the quantum of molecular vibration that is very sensitive to temperature but not to the frequency
Acoustic impedance (Z)	Ns/m^3	It is the ratio between the acoustic pressure and the fluid velocity
Adiabatic compressibility (β)	m^2/N	It is a measure of liquid elasticity and is a thermodynamic parameter. It gives information regarding the bulk volume and liquid structure
Bulk modulus (K)	Pa	It is the measure of the rigidity of the fluid medium
Attenuation (α)	Np/m	It is a measure of the damping of acoustical energy due to absorption and scattering. It is dependent on frequency, and independent of temperature

2.1 Measurement Techniques

Ultrasound is a versatile tool for characterizing materials, especially liquid-based systems including complex colloids. For a long time, many useful fluid properties have been computed from ultrasonic measurements. However, there are a number of other techniques that have been found useful for characterizing colloidal systems including nanofluids. Regarding ultrasonic velocity and attenuation, these techniques are classified into two categories: transmission and interferometric. The interferometric technique uses standing waves, while the transmission method exploits the transverse waves [33].

2.1.1 Interferometry

Interferometry is based on the generation of standing waves inside the sample chamber. It can be achieved by placing an acoustic reflector in front of the ultrasound transducer (mostly piezoelectric) at a distance across the chamber. The reflected wave interferes with the incident wave generated by the transducer, which forms a standing wave. Such a standing wave depends on both the wavelength and the distance between the transducer and the reflector. In this setup, the transmitting transducer works as the receiver as well. It monitors the intensity of the ultrasound at its surface. It is possible to change this amplitude by varying either the wavelength or the distance. The former is known as swept-frequency interferometry [48] and the latter as swept-distance interferometry [49–51].

The intensity of the ultrasound goes through a successive minima and maxima, when either the wavelength or the distance changes. The distance between the successive maxima is equal to the ultrasound wavelength in the nanofluid. Thus, ultrasonic velocity can be found using the following equation:

$$v = f\lambda \quad (2)$$

where, v is the ultrasonic velocity, f is the frequency, and λ is the measured wavelength in the nanofluid. It is an accurate and precise method for ultrasonic velocity measurement.

It is also possible to measure attenuation using interferometry. The amplitude of the maxima is a function of the distance between the transducer and the receiver, and also the amplitude decays with increasing distance. Hence, the attenuation coefficient can be derived as a measure of this maximum amplitude-distance dependence. Nevertheless, Sette [52] suggested that the transmission technique is far better and well-matched for attenuation measurements, due to its ability to handle a much larger dynamic range.

2.1.2 Transmission Technique

The transmission technique is primarily of two categories, i.e., based on two basic principles: the use of a pulse technique, and the use of a sensor with a variable acoustic path length (variable gap). The general principles of this technique are originally formulated by Pellam and Galt [53] and independently by Pinkerton [54]. In each case, the velocity measurement is taken by determining the distance; the transducer must be moved to delay the received echo by a specified increment [49–51]. Likewise, the attenuation is found by determining the attenuation that needs to be added or subtracted in order to keep the received signal constant as the transducer is moved. The velocity is obtained directly from the slope of the measured pulse delay versus the distance travelled. The attenuation is measured from the slope of the compensating attenuation versus distance. There is no need to know the exact distance traveled by the sound because the measured parameters depend only on the difference in the acoustic path. These instruments work only at a single frequency, by means of one transducer in a pulse-echo mode. This method is significantly improved by Andrea et al. [55]. They extended the Pellam-Galt method to incorporate measurements at multiple frequencies. Instead of a single transducer as in a pulse-echo mode, they used separate transducers for transmitting and receiving [33]. It is an accurate and precise method for ultrasonic attenuation measurement.

2.2 *Ultrasonic Parameters*

There are two key acoustic parameters: the ultrasonic velocity and the attenuation coefficient, which are measured experimentally. On the other hand, the theoretical treatment of the measured data is the main reason for ultrasound's versatility for characterization purposes. i.e., the raw data should be theoretically treated, so that a multitude of other properties can be extracted. It turns out that there are various significant calculated parameters that can be dug out from the experimental raw data, through which various thermo physical properties of fluids can be explored, depending on the degree of our a priori knowledge about the system. The ultrasonic technique allows us to determine various significant parameters through which various thermo physical properties of nanofluids can be explored. Most essential and widely used among them are discussed in this session.

2.2.1 Ultrasonic Velocity

The velocity of ultrasound v is the most important ultrasonic parameter, especially for characterizing nanofluids. It is the quantum of the molecular vibration in the medium through which the wave passes. It is very sensitive to temperature, but not to frequency. The magnitude of this temperature dependence is usually several meters per second per degree. For example, the ultrasonic velocity of water changes, 2.4 m/s per degree Celsius, at room temperature. For simple Newtonian liquids, the ultrasonic velocity is almost independent of frequency [33]. In addition, the ultrasonic velocity through rigid particles is much higher than that of fluids (for instance the ultrasonic velocity of water is 1496 m/s at 25 °C) and is typically close to 6000 m/s.

The variation of the ultrasonic velocity of Fe_3O_4—water nanofluid [56] with concentration at various temperatures ranging from 308 to 318 K is shown in Fig. 1. It is clear that, the velocity decreases with the increase in nanoparticle concentration. This decrease in velocity with increase of concentration is ascribed to the particle–fluid interactions, and it further ensures the dominance of the intermolecular interactions over the intramolecular interactions [57].

It can be further understood as follows. Typically, in any complex colloidal liquid, there will be a trivial probability for particle–particle interaction as also a particle–fluid interaction. The particle–particle interactions necessarily induce aggregation and flocculation. The aggregation can create percolation which, in turn, makes a mean free path for ultrasonic wave and thereby, accelerates the ultrasonic velocity. On the other hand, as the particle–fluid interaction dominates, the possibility of interparticle aggregation will be less, and hence there is less likelihood of percolation. Moreover, with the particle loading there is a possibility for the decreasing rate of occurrence of Brownian motion of the fluid molecule, along with the formation of a resistive surface layer that can cause a decrease in ultrasonic velocity. Besides, with the increase in nanoparticle concentration, there is an

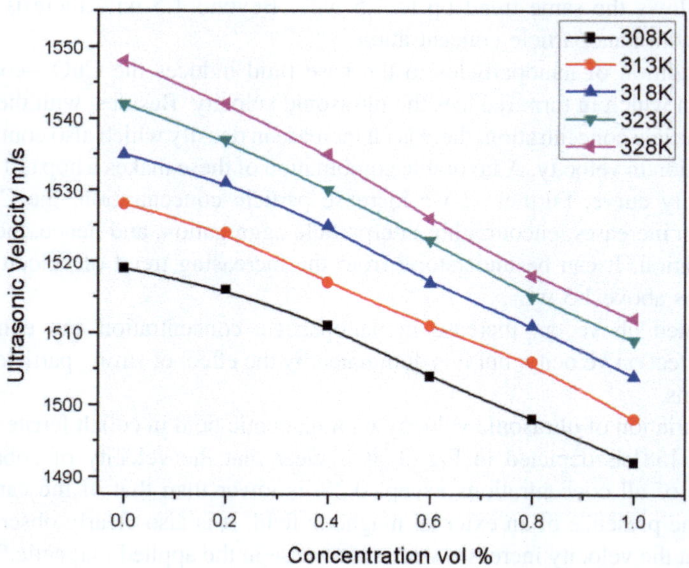

Fig. 1 Ultrasonic velocity versus concentration of Fe_3O_4—water nanofluid [56]

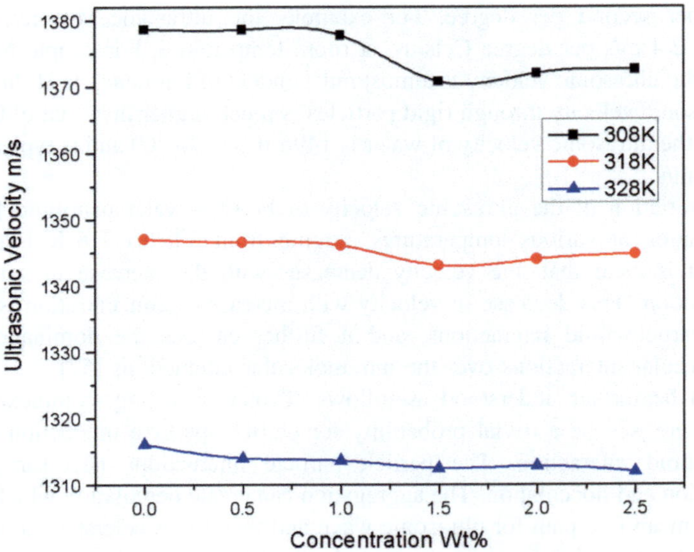

Fig. 2 Variations of ultrasonic velocity with concentration in CuO—coconut oil nanofluid [57]

increase in density, which also contributes to the reduction in velocity. This can be further understood from the trend of adiabatic compressibility with concentration.

The ultrasonic velocity in the CuO—coconut oil nanofluid [57] shows a progressive change with the concentration of the CuO nanoparticle, as shown in Fig. 2. The velocity decreases with the increase of particle loading at lower concentrations and it follows the same trend up to 1.5 wt%. Beyond 1.5 wt% there is a rise in velocity with nanoparticle concentration.

The addition of nanoparticles to the base fluid induces the CuO—coconut oil interaction which in turn, reduces the ultrasonic velocity. Besides, with the increase in nanoparticle concentration, there is an increase in density which also contributes to the reduction in velocity. A favorable combination of these makes a hop at 1.5 wt% in the velocity curve. Further, if we increase particle concentration, the CuO–CuO interaction increases, encouraging interparticle aggregation, and hence, the chances of percolation. It can be understood from the increasing trend of velocity at concentrations above 1.5 wt%.

As stated above, an increase in nanoparticle concentration can enhance the density effect on velocity, but it is dominated by the effect of strong particle–particle interactions.

The variation of ultrasonic velocity with magnetic field in cobalt ferrite magnetic nanofluid [58] is depicted in Fig. 3. It is clear that the velocity of cobalt ferrite nanofluid of all concentrations except 0.2% is lower than that of the carrier fluid even in the presence of an external magnetic field. It is also clearly observed from Fig. 3 that the velocity increases with an increase in the applied magnetic field. This increase of velocity with the magnetic field is the sign of clustering of the magnetic

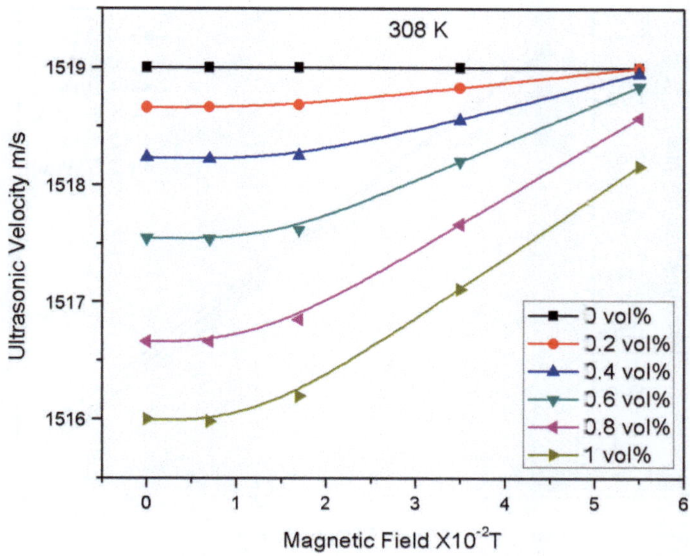

Fig. 3 Plots of ultrasonic velocity versus magnetic field for cobalt ferrite magnetic nanofluids [58]

particles. As the applied field sets the magnetic moments along the direction of the magnetic field forming the clusters, the rigidity of the system increases, and hence, the velocity increases. When the magnetic field is turned off, the velocity returns to its initial value without exhibiting any hysteresis. Because, as the particles are sterically stabilized, no permanent aggregation by means of van der Waals attraction occurs, and the aggregation phenomenon observed in the magnetic field is perfectly reversible [6, 18, 19].

It can also be understood from Fig. 3 that at lower concentrations, the enhancement in velocity with increasing external magnetic field is not so significant, indicating the slightest influence of the magnetic field on the ultrasonic velocity. But the magnetic nanofluid concentration of 1% shows considerably higher variations in the velocity with respect to the change in the magnetic field. It points out that at 1% the concentration of cobalt ferrite is adequately high, and the intensity of the magnetic field is strong enough to stimulate strong interparticle interactions.

The effect of temperature on ultrasonic velocity in CuO—coconut oil nanofluid is shown in Fig. 4. It elucidates the significant effect of temperature on the ultrasonic velocity of nanofluids at various concentrations. The decrease of ultrasonic velocity with increasing temperature, shown in Fig. 4, proves the nanofluids follow the common behavior of non-aqueous liquids.

As the temperature rises, the fluid tends to expand; this in turn, induces a reduction in density, which can provide a consequent increase in velocity. At the same time, with the increase of fluid temperature, the average speed of the molecules increases, and leads to the weakening of the interparticle and intermolecular

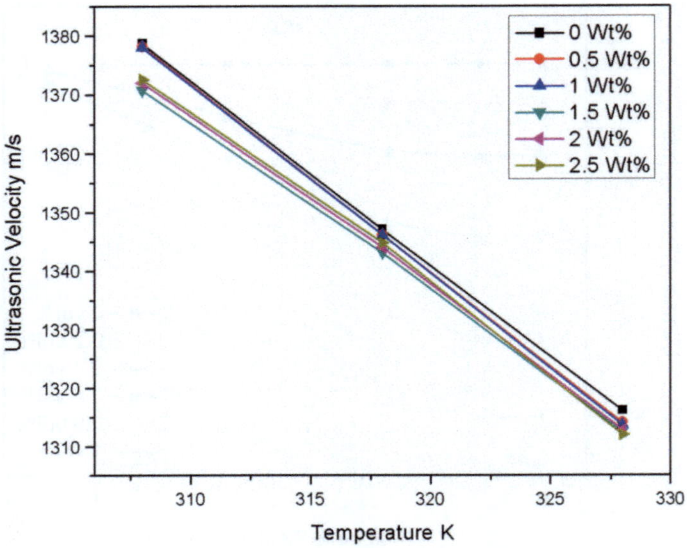

Fig. 4 Plots of ultrasonic velocity versus temperature in CuO—coconut oil nanofluid [57]

adhesion forces [59], thereby improving the compressibility. In fact, the effect of compressibility dominates that of density on velocity data, and ensures a consequent decrease in the magnitude.

Conversely, a water-based magnetic nanofluid (Fig. 1) shows an exactly reverse dependence of velocity on temperature [56]. It can easily be understood that magnetic nanofluids follow the common behavior of water, showing an increase in velocity with an increase of temperature, which can be elucidated using the open and close packed structure of water. Water consists of hydrogen bonded clusters and unbounded water molecules. The molecules in the interior clusters are quadruply bonded, and the unbounded water molecules are supposed to occupy the space between the clusters. The clusters are sometimes referred to as open structure water, and the dense monomeric fluid is referred to as closed structure water. In water, the rise in temperature causes the thermal rupture of the open packed water structure, which in turn, enhances the cohesion of the water molecules, and the less compressible closed packed structure, leading to an increase in the ultrasonic velocity. It further appears that the cohesion factor dominates over the thermal expansion factor with an increase in the temperature. The same explanation holds good for the increase in velocity, and the corresponding increase in acoustic impedance of water based magnetic nanofluids observed at elevated temperatures [56, 58].

2.2.2 Acoustic Impedance

In general, impedance can be described in terms of the linear relationship between the driving force and the corresponding flow. For an AC field, it is electric voltage and electric current. The acoustic impedance, Z is the coefficient of the proportionality between the pressure, P and the velocity of the particles,v_p in the sound wave, and is given by the following expression,

$$Z = \frac{P}{v_p} \tag{3}$$

Like electric impedance, acoustic impedance is a complex number. It depends on the acoustic properties of the medium. It is also widely used in acoustics and it is defined [33] as,

$$Z = \rho v \left(1 - j\frac{\alpha\lambda}{2\pi}\right) \tag{4}$$

where, ρ is the density, α is the attenuation coefficient, and λ is the measured wave length in the nanofluid.

But the attenuation does not make a significant contribution to the acoustic impedance. Even for highly concentrated dense colloids at high frequency, the maximum possible contribution by attenuation to the acoustic impedance is only 2%. This leads us to approximate the acoustic impedance as a real number for nanofluids, by neglecting the complex part, and is equal to the product of the density and ultrasonic velocity. Thus, the acoustic impedance of nanofluid is calculated by the following equation

$$Z = \rho v \tag{5}$$

Figure 5 illustrates the variation of acoustic impedance with concentration at various temperatures, in the CuO—coconut oil nanofluid [57]. There is a slight increase in acoustic impedance at lower concentration, but as concentration increases beyond 1.5 wt%, one can observe a significant increase in acoustic impedance. Being the coefficient of proportionality between the pressure and velocity of the particles, acoustic impedance reveals the reason for a significant reduction in the particle velocity beyond a critical concentration.

The reduction observed in the acoustic impedance in CuO—coconut oil nanofluid, and the increasing acoustic impedance in cobalt ferrite magnetic nanofluid (as shown in Fig. 6), are attributed to the variations in density and compressibility with respect to concentration.

To get a better understanding regarding the temperature effect on acoustic impedance in CuO—coconut oil nanofluid [57], plots are made with acoustic impedance as a function of temperature, and are shown as Fig. 7. It depicts a sharp

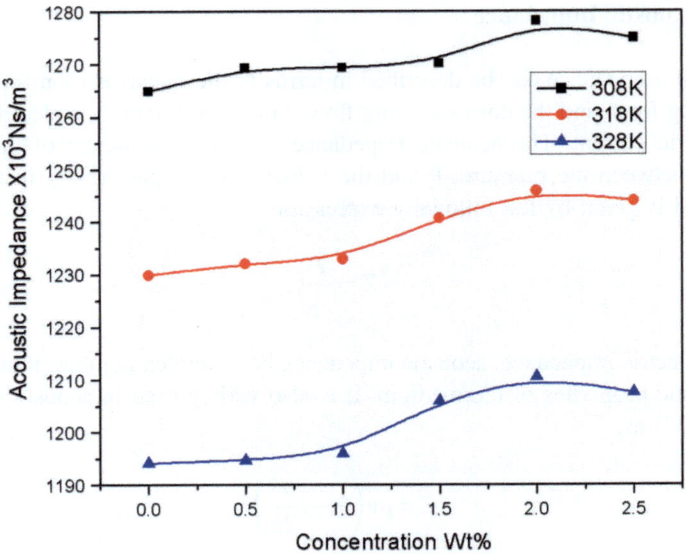

Fig. 5 Plots of acoustic impedance versus concentration in CuO—coconut oil nanofluid [57]

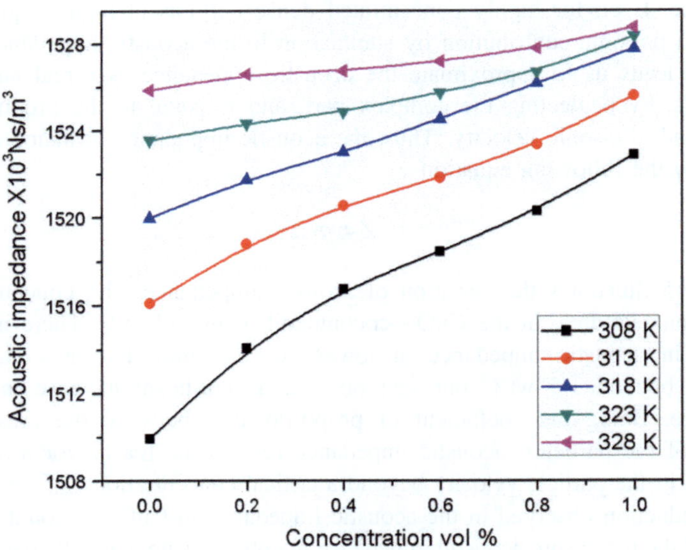

Fig. 6 Plots of acoustic impedance versus concentration in cobalt ferrite magnetic nanofluid [58]

reduction in acoustic impedance with temperature, which is attributed to the increase in compressibility and decrease in density. Such an interpretation can also be found in the case of cobalt ferrite [58] and magnetite [56] nanofluids.

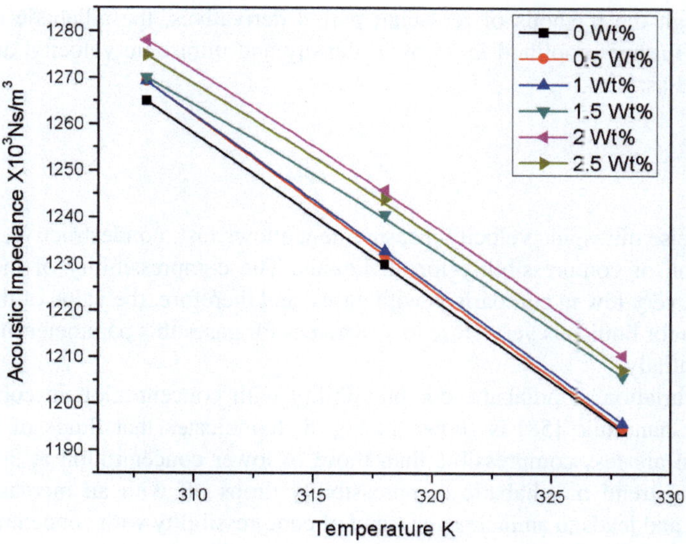

Fig. 7 Variation of acoustic impedance with temperature in CuO—coconut oil nanofluid [57]

2.2.3 Adiabatic Compressibility

Compressibility is a measure of the liquid elasticity. It is the change in the liquid volume with respect to the unit variation of pressure [33].

$$\beta = -\frac{1}{V}\frac{\partial V}{\partial P} \tag{6}$$

where, V is the volume. The definition above mentioned is incomplete, because for any system the magnitude of compressibility depends strongly on whether the process is adiabatic or isothermal. Thereby, compressibility is classified into two categories: Viz., isothermal and adiabatic.

Accordingly, adiabatic (isentropic) compressibility is defined as the following,

$$\beta_s = -\frac{1}{V}\left(\frac{\partial V}{\partial P}\right)_s \tag{7}$$

where, the subscript S indicates that the partial differential is to be taken at constant entropy.

The ultrasonic velocity is defined in classical mechanics as:

$$v^2 = \left(\frac{\partial P}{\partial \rho}\right)_s \tag{8}$$

Through the methods of replacing partial derivatives, the adiabatic compressibility is further simplified in terms of density and ultrasonic velocity, and can be expressed as:

$$\beta_s = \frac{1}{\rho v^2} \tag{9}$$

A precise ultrasonic velocity measurement allows fast, nondestructive, and exact calculation of compressibility for all liquids. The compressibility of most of the liquids is very low in comparison with gases, and therefore, the value of the specific heat ratio for liquids is very close to 1, whereas for gases this parameter may exceed 1 substantially.

The variation of adiabatic compressibility with concentration in cobalt ferrite magnetic nanofluid [58] is shown in Fig. 8. It indicates that fluids of high concentration are less compressible than those of lower concentration at 308 K. The decreasing trend of adiabatic compressibility drops off with an increase in temperature, and leads to an increasing trend of compressibility with concentration. The decreasing trend of adiabatic compressibility is attributed to the increase in density with respect to concentration. The increasing trend of compressibility with concentration at higher temperatures is in accordance with the rise in percentage drop in the magnitude of velocity [58].

The results lead to an interesting conclusion that the dependence of the sound speed with particle concentration cannot be explained as the sole product of the increase in density, resulting from the added solid phase. According to Eq. (9), the

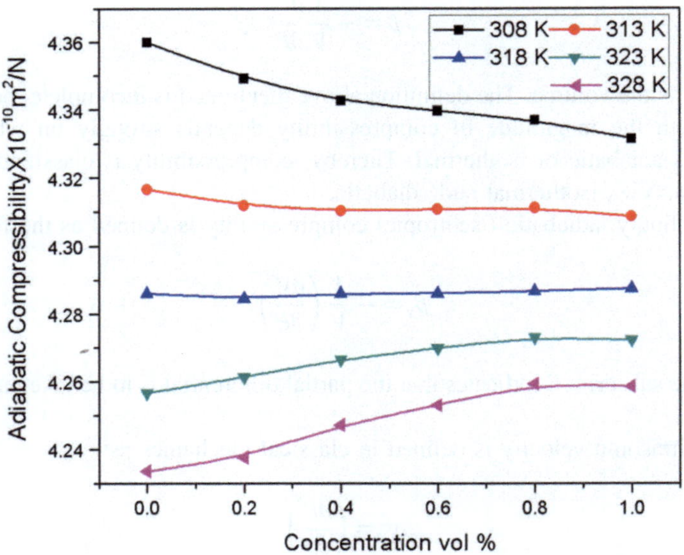

Fig. 8 Variation of adiabatic compressibility with respect to concentrations at various temperatures for cobalt ferrite magnetic nanofluid [58]

ultrasonic velocity should decrease with increasing density, with the assumption of compressibility being constant. However, the above plots also indicate a conflicting trend. The increase in sound speed with density suggests that the compressibility is not constant, but is also a function of the particle concentration. It follows that the compressibility of a liquid must be an outcome of intermolecular interactions.

In general, it is difficult to measure the elasticity of a nanoparticle in a colloidal system, especially, in its natural environment. But ultrasonics makes it possible to obtain some information about the elastic properties of individual nanoparticles, through the measurement of ultrasound velocity in the nanofluid [39]. Assuming that all molecules of the surfactant are adsorbed on the surface of the magnetic particles, the adiabatic compressibility of the nanoparticle, β_2 can be found from the following relation [39]

$$\beta_2 = \frac{\beta_s - (1-\phi)\beta_1}{\phi} \tag{10}$$

where, β_1 is the adiabatic compressibility of the base fluid.

The ultrasonic propagation in a water–based magnetic fluid with a double layered surfactant shell is analyzed [60] and the, adiabatic compressibility of the particle aggregates is measured using the ultrasonic velocity in the absence of a magnetic field.

2.2.4 Bulk Modulus

The elastic properties of the liquid are defined by a bulk modulus, which is a measure of the rigidity of the fluid medium. The bulk modulus is defined as the ratio of the infinitesimal pressure increase to the resultant relative decrease of the volume. It is generally represented by K, and is the reciprocal of the liquid compressibility.

$$K = \frac{1}{\beta_s} \tag{11}$$

Substituting Eq. (6) in Eq. (11)

$$K = \rho v^2 \tag{12}$$

The effect of concentration and temperature on the bulk modulus of the cobalt ferrite magnetic nanofluid, is depicted in Fig. 9. The data curves show an exact inverse trend to the adiabatic compressibility curves (Fig. 8).

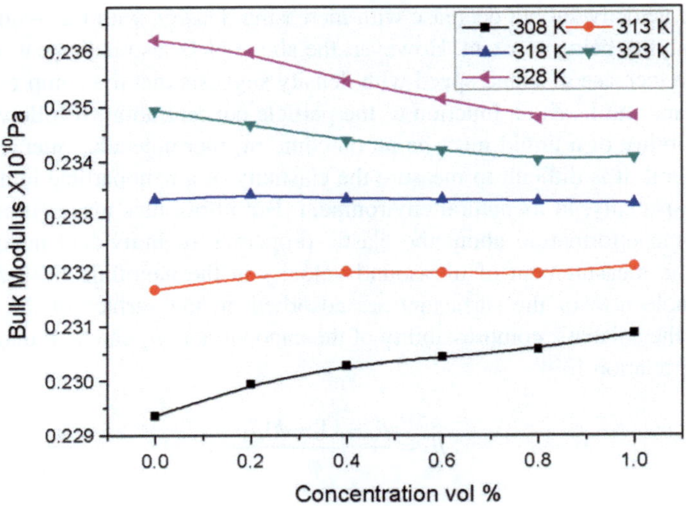

Fig. 9 Variation of Bulk modulus with respect to concentrations at various temperatures for cobalt ferrite magnetic nanofluid

2.2.5 Attenuation

In physics, attenuation is the gradual loss in intensity of any kind of flux through a medium. Attenuation in ultrasound is the reduction in amplitude of the ultrasound beam as a function of distance through the medium, or it is the measure of the energy loss of sound propagation in media.

The attenuation coefficient α determines how quickly the pressure amplitude decays with distance, caused by various dissipative effects. It has the dimension of [Np/m]. The attenuation expressed in decibels (dB) is correlated to attenuation expressed in nepers (Np) through the relation $[\mathrm{dB}/\mathrm{m} = -8.686\,\mathrm{np}/\mathrm{m}]$.

In contrast with ultrasonic velocity which is practically independent of frequency, attenuation is roughly proportional to frequency, at least over the frequency range of 1–100 MHz. Also, even in a transmission technique precise measurements of attenuation necessitate a reasonably precise knowledge of the ultrasonic velocity in the sample. An error in ultrasonic velocity data may lead to an apparent decrease in the signal level, because the received pulse is sampled at the wrong time [33]. At the same time, the attenuation data is used to improve the accuracy of the ultrasonic velocity measurements, by recording only the data at those gaps for which the attenuation is not so great as to exclude an adequate signal-to-noise ratio [33].

In comparison with other ultrasonic parameters and experimental methods, attenuation is the least sensitive to the chemical composition of the medium. This fact makes attenuation an excellent candidate for an accurate particle sizing technique.

Indeed, the viscous properties of the liquid are generally characterized by two dynamic viscosity coefficients; viz., shear dynamic viscosity and volume dynamic viscosity. It is hard to measure volume viscosity, and therefore, it is usually assumed to be zero. But, in fact all liquids exhibit a volume viscosity. For example, the volume viscosity of water is roughly three times larger than its shear viscosity. Ultrasonic attenuation is the only recognized way to measure the volume viscosity [33]. The structure of the liquid can be correlated to the volume viscosity. Hence, acoustic attenuation measurements provide potential information regarding nanofluid structure.

It is also possible to calculate the attenuation coefficient from ultrasonic velocity, by using some other raw data though it is substantially more complicated when compared to acoustic spectroscopy, where the attenuation coefficient is measured directly. The famous expression for ultrasound attenuation is derived by Stokes [33] from the Navier-Stokes [33] equation for an incompressible liquid, and is given as the following equation

$$\alpha = \frac{2\omega^2\eta}{3\rho v^3} \tag{13}$$

where, η is the dynamic viscosity and ω is the angular frequency. The substitution of $\omega = 2\pi f$ in Eq. (13) gives the following expression

$$\alpha = \frac{8\pi^2 f^2\eta}{3\rho v^3} \tag{14}$$

Stokes' expression neglects the liquid compressibility, and consequently, the bulk viscosity term η^b is missing, when comparing it with the general solution (15) for the Navier-Stokes equation. The attenuation predicted by Stokes' equation is generally known as *classical*.

$$\alpha = \frac{\omega^2}{2\rho v^3}\left[\frac{4}{3}\eta + \eta^b\right] \tag{15}$$

Instead of Stokes' equation, the more general Eq. (15) provides a better theoretical basis for attenuation, or for calculating the bulk viscosity from the measured attenuation and ultrasonic velocity. All other parameters in this equation, including the dynamic viscosity, are assumed to be known from independent experimental measurements [33].

Hemalatha et al. [41] have made a comparative study of the particle–fluid interactions in micro and nanofluids of aluminum oxide, based on the acoustic parameters. Investigations of ultrasonic velocity, adiabatic compressibility, attenuation and acoustic impedance are made at temperatures 303–323 K. From the analysis of the above parameters it is evident, that the particle–fluid interaction in microfluids decreases with increasing particle loading. But for the nanofluids, the particle–fluid interaction increases with an increase in the concentration up to a critical concentration of 0.6 wt%, above which the particle–fluid interaction weakens, due to the strong particle–particle interaction that leads to agglomeration.

The difference in the behavior of the nanofluid from that of the microfluid is attributed to the relatively high surface area to volume ratio of the nanoparticles, which is 1000 times larger than that of the microparticles.

In addition, Nabeel Rashin and Hemalatha have recently proposed [61], a new methodology to find the thermal conductivity of nanofluids, using ultrasonic velocity. The results, thus obtained through the ultrasonic method, are found to agree well with those obtained through the transient line heat source technique, which confirms the aptness of the Nabeel-Hemalatha (N-H) method.

$$k_{N-H} = 3\,\delta \left(\frac{\rho_{nf} N_A}{M_{nf}} \right)^{2/3} k_B\, v \tag{16}$$

where, k_{N-H}, stands for the thermal conductivity through the proposed N-H model. ρ_{nf} is the density of the nanofluid, and $M_{nf} = x_{bf}M_{bf} + x_pM_p$ is the molar mass of the nanofluid. x_{bf} and x_p are the molar fractions of the base fluid and nanoparticle respectively, whereas M_{bf} and M_p are the respective molar masses of the base fluid and nanoparticle.

Summarizing, the *ultrasonic velocity* studies of nanofluids ascertain the predominance of the nanoparticle–base fluid interactions over the particle–particle interactions or vice versa, at all temperatures and concentrations. The ultrasonic velocity data of the nanofluids, at low temperatures, identify the specific *critical concentrations*, below which, the particle–fluid interactions dominate the particle–particle interactions, and above the critical concentration, the effect of the particle–particle interactions surpasses that of the particle–fluid interactions. The investigation of the *adiabatic compressibility* of the nanofluids reveals the strong intermolecular interactions in each nanofluid system, which shift their elastic properties and make them more rigid with the loading of the nanoparticle. The opposition offered by the nanofluids to the passage of the ultrasonic wave is characterized by *acoustic impedance*, which signifies the decrease in the average molecular velocity caused by the increase in the particle concentration. The temperature effect on the aqueous magnetic nanofluids can be explained successfully, using the open- and close- packed structure of water. Indeed, the magneto-acoustic studies performed on the magnetic nanofluids, portray the clustering of the nanoparticles along the direction of the magnetic field, and the formation of chain-like structures at higher fields, which are the root causes of the observed magneto-viscous effects. In addition to the molecular interaction studies, a new ultrasonic methodology is proposed, to find the thermal conductivity of non-magnetic nanofluids through ultrasonic velocity. Thus, it is proved that the ultrasonic investigations can provide a potential and economical alternative to the precise determination of the thermal conductivity of non-magnetic nanofluids, which has been a formidable challenge, using the available conventional instruments.

In short, ultrasonics provides a versatile means to explore the nanofluids and ferrofluids. In other traditional optical characterization methods such as DLS, UV-Visible spectrometer, particle size analyzer etc., the characteristic light cannot propagate through concentrated and opaque liquid systems, which is the major drawback. Thus, they require super–dilution which can destroy aggregates or flocs. The electron microscopy, demands specific sample preparation, particularly drying of the nanofluid samples and therefore, the corresponding measured parameters would not be as exact as the original. But, the ultrasonic method works as an efficient approach to bring out the hidden information of nano–colloidal systems in their natural environment without demanding *super dilution* and thereby characterizes the systems as they are. The non-invasive nature, excellent efficiency, and relatively low cost, make this method a more admirable and novel tool to characterize nanofluids; this would play an important role in the emerging heat transfer applications.

References

1. Philip J, Shima PD (2012) Thermal properties of nanofluids. Adv Colloid Interface Sci 30:183–184
2. Das SK, Choi SUS, Yu W, Pradeep T (2008) Nanofluids: science and technology. John Wiley & Sons Inc, Hoboken, NJ
3. Choi SUS (1995) Enhancing thermal conductivity of fluids with nanoparticles. In: Singer A, Wang HP (eds) Developments and applications of non-newtonian flows, vol. 66. American Society of Mechanical Engineers, New York, pp 99–105
4. Eastman JA, Choi SUS, Li S, Yu W, Thompson LJ (2001) Anomalously increased effective thermal conductivities of ethylene glycol-based nanofluids containing copper nanoparticles. Appl Phys Lett 78:718–720
5. Buongiorno J (2009) A benchmark study on the thermal conductivity of nanofluids. J Appl Phys 106:094312–094312
6. Kim SJ, Bang IC, Buongiorno J, Hu LW (2007) Study of pool boiling and critical heat flux enhancement in nanofluids. Bull Polish Acad Sci 55:211–216
7. Gerardi C, Cory D, Buongiorno J, Hu LW, McKrell T (2009) Nuclear magnetic resonance-based study of ordered layering on the surface of alumina nanoparticles in water. Appl Phys Lett 95:253104-1–253104-3
8. Maxwell JC (1881) A treatise on electricity and magnetism. Clarendon, Oxford
9. Hamilton RL, Crosser OK (1962) Thermal conductivity of heterogeneous two-component systems. Ind Eng Chem Fundamen 1:187–191
10. Patel HE, Das SK, Sundararajan T, Nair AS, George B, Pradeep T (2003) Thermal conductivities of naked and monolayer protected metal nanoparticle based nanofluids: manifestation of anomalous enhancement and chemical effects. Appl Phys Lett 83:2931–2933
11. Hong TK, Yang HS, Choi CJ (2005) Study of the enhanced thermal conductivity of Fe nanofluids. J Appl Phys 97:064311–064311
12. Kang HU, Kim SH, Oh JM (2006) Estimation of thermal conductivity of nanofluid using experimental effective particle volume. Exp Heat Transf 19:181–191
13. Murshed SMS, Leong KC, Yang C (2005) Enhanced thermal conductivity of TiO_2—water based nanofluids. Int J Therm Sci 44:367–373
14. Zhang X, Gu H, Fujii M (2006) Experimental study on the effective thermal conductivity and thermal diffusivity of nanofluids. Int J Thermophys 27:569–580
15. Zhu HT, Zhang CY, Tang YM, Wang JX (2007) Novel Synthesis and thermal conductivity of CuO nanofluid. J Phys Chem C 111:1646–1650

16. Li Q, Xuan Y, Wang J (2005) Experimental investigations on transport properties of magnetic fluids. Exp Therm Fluid Sci 30:109–116
17. Chon CH, Kihm KD, Lee SP, Choi SUS (2005) Empirical correlation finding the role of temperature and particle size for nanofluid (A_2lO_3) thermal conductivity enhancement. Appl Phys Lett 87:153107-1–153107-4
18. Chopkar M, Das PK, Manna I (2006) Synthesis and characterization of nanofluid for advanced heat transfer applications. Scr Mater 55:549–552
19. Li CH, Peterson GP (2007) The effect of particle size on the effective thermal conductivity of Al_2O_3-water nanofluids. J Appl Phys 101:044312–044312
20. Hwang D, Hong KS, Yang HS (2007) Study of thermal conductivity of nanofluids for the application of heat transfer fluids. Thermochim Acta 455:66–69
21. Sinha K, Kavlicoglu B, Liu Y, Gordaninejad F, Graeve OA (2009) A comparative study of thermal behavior of iron and copper nanofluids. J Appl Phys 106:064307-1–064307-7
22. Gharagozloo PE, Eaton JK, Goodson KE (2008) Diffusion, aggregation, and the thermal conductivity of nanofluids. Appl Phys Lett 93:103110-1–103110-3
23. Shalkevich N, Escher W, Burgi T, Michel B, Ahmed LS, Poulikakos D (2010) On the thermal conductivity of gold nanoparticle colloids. Langmuir 26:663–670
24. Rusconi R, Rodari E, Piazza R (2006) Optical measurements of the thermal properties of nanofluids. Appl Phys Lett 89:261916-1–261916-3
25. Singh D, Timofeeva E, Yu W, Routbort J, France D, Smith D et al (2009) An investigation of silicon carbide-water nanofluid for heat transfer applications. J Appl Phys 105:064306
26. Venerus DC, Kabadi MS, Lee S, Luna VP (2006) Study of thermal transport in nanoparticle suspensions using forced Rayleigh scattering. J Appl Phys 100:094310-1–094310-5
27. Zhang X, Gu H, Fujii M (2006) Effective thermal conductivity and thermal diffusivity of nanofluids containing spherical and cylindrical nanoparticles. J Appl Phys 100:044325-1–044325-5
28. Ju YS, Kim J, Hung MT (2008) Experimental study of heat conduction in aqueous suspensions of aluminum oxide nanoparticles. ASME J Heat Transf 130:092403-1–092403-6
29. Putnam SA, Cahill DG, Braun PV, Ge Z, Shimmin RG (2006) Thermal conductivity of nanoparticle suspensions. J Appl Phys 99:084308-1–084308-6
30. Timofeeva EV, Gavrilov AN, McCloskey JM, Tolmachev YV, Sprunt S, Lopatina LM, Selinger JV (2007) Thermal conductivity and particle agglomeration in alumina nanofluids: experiment and theory. Phys Rev E 76:061203-1–061203-1-16
31. Pastoriza-Gallego MJ, Lugo L, Legido JL, Piñeiro MM (2011) Thermal conductivity and viscosity measurements of ethylene glycol-based Al_2O_3 nanofluids. Nanoscale Res Lett 6:221-1–221-11
32. Rayleigh L (1896) The theory of sound, 2nd edn. Macmillan, New York
33. Dukhin AS, Goetz PJ (2010) Characterization of liquids, nano and microparticulates and porous bodies using ultrasound, 2nd edn. Elsevier, New York
34. Nabeel Rashin M, Hemalatha J (2011) Ultrasonic studies and microchannel flow behavior of copper oxide nanofluid. AIP Conf Proc 1349:335–335
35. Singh DK, Pandey DK, Yadav RR (2009) An ultrasonic characterization of ferrofluid. ultrasonics 49:634–637
36. Nabeel Rashin M, Hemalatha J (2013) Acoustical studies on the interaction of copper oxide—ethylene glycol nanofluid. In: Giri PK, Goswami DK, Perumal A (eds) Advanced nanomaterials and nanotechnology. Springer, Berlin, Heidelberg, New York, pp 225–229
37. Hornowski T, Józefczak A, Łabowski M, Skumiel A (2008) Ultrasonic determination of the particle size distribution in water-based magnetic liquid. Ultrasonics 48:594–597
38. Sayan P, Ulrich J (2002) The effect of particle size and suspension density on the measurement of ultrasonic velocity in aqueous solutions. Chem Eng Process 41:281–287
39. Józefczak A, Skumiel A (2011) Ultrasonic investigation of magnetic nanoparticles suspension with PEG biocompatible coating. J Magn Magn Mater 323:1509–1516

40. Motozawa M, Iizuka Y, Sawada T (2008) Experimental measurements of ultrasonic propagation velocity and attenuation in a magnetic fluid. J Phys: Condens Matter 20:204117-1–204117-5
41. Hemalatha J, Prabhakaran T, Nalini RP (2011) A comparative study on particle–fluid interactions in micro and nanofluids of aluminium oxide. Microfluid Nanofluid 10:263–270
42. Shima PD, Philip J (2011) Tuning of thermal conductivity and rheology of nanofluids using an external stimulus. J Phys Chem C 115:20097–20104
43. Philip J, Shima PD, Raj B (2008) Nanofluid with tunable thermal properties. Appl Phys Lett 92:043108–1–043108-3
44. Philip J, Jaykumar T, Kalyanasundaram P, Raj B (2003) A tunable optical filter. Meas Sci Technol 14:1289–1294
45. Józefczak A (2003) The time dependence of the changes of ultrasonic wave velocity in ferrofluid under parallel magnetic field. J Magn Magn Mater 256:267–270
46. Skumiel A, Hornowski T, Józefczak A (2000) Investigation of magnetic fluids by ultrasonic and magnetic methods. Ultrasonics 38:864–867
47. Muller HW, Jiang Y, Liu M (2003) Sound damping in ferrofluids: magnetically enhanced compressional viscosity. Phys Rev E 67:031201–1–031201-5
48. Sinha DN (1998) Non-invasive identification of fluids by swept-frequency acoustic interferometry. US Patent 5, 767, 407, 1998
49. McClements JD, Powey MJW (1989) Scattering of ultrasound by emulsions. J Phys D Appl Phys 22:38–47
50. McClements JD (1998) Ultrasonic characterization of food emulsions. In Hackley VA, Texter J (eds) Ultrasonic and dielectric characterization techniques for suspended particulates, Am Ceramic Soc, pp 305–317
51. McClements JD (1996) Principles of ultrasonic droplet size determination in emulsions. Langmuir 12:3454–3461
52. Sette D (1968) Ultrasonic studies. In: Physics of simple liquids. Amsterdam, North-Holland
53. Pellam JR, Galt JK (1946) Ultrasonic propagation in liquids: application of pulse technique to velocity and absorption measurement at 15 megacycles. J Chem Phys 14:608–613
54. Pinkerton JMM (1947) A pulse method for measurement of ultrasonic absorption in liquids. Nature 160:128–129
55. Andreae J, Joyce P (1962) 30 to 230 Megacycle pulse technique for ultrasonic absorption measurements in liquids. Br J Appl Phys 13:462–467
56. Nabeel Rashin M, Hemalatha J (2012) Magnetic and ultrasonic investigations on magnetite nanofluids. Ultrasonics 52:1024–1029
57. Nabeel Rashin M, Hemalatha J (2012) Acoustic study on the interactions of coconut oil based copper oxide nanofluid. Int J Eng Appl Sci 6:216–220
58. Nabeel Rashin M, Hemalatha J (2014) Magnetic and ultrasonic studies on stable cobalt ferrite magnetic nanofluid. Ultrasonics 54:834–840
59. Nabeel Rashin M, Hemalatha J (2013) Synthesis and viscosity studies of novel ecofriendly ZnO–coconut oil nanofluid. Exp Therm Fluid Sci 51:312–318
60. Józefczak A, Hornowski T, Zavisova V, Skumiel A, Kubovcikova M, Timko M (2014) Acoustic wave in a suspension of magnetic nanoparticle with sodium oleate coating. J Nanopart Res 16:2271
61. Nabeel Rashin M, Hemalatha J (2014) A novel ultrasonic approach to determine thermal conductivity in CuO–ethylene glycol nanofluids. J Mol Liq 197:257–262

RF Nanostructured Security

Mohamed Kheir, Heinz Kreft, Iris Hölken and Reinhard Knöchel

Abstract This chapter gives an intensive overview of some recent micro- and nanostructured Radio Frequency (RF) security issues. It identifies the challenges of tomorrow's security problems and why this has been a big relevance not only to nano-communications but also to other applications. A short overview on the traditional Physical Unclonable Functions (PUFs) introduces the reader into the concept of applied electromagnetic waves interacting with nanomaterials. Major security and fingerprinting contributions, which are newly-proposed and implemented by the authors, are concluded in this chapter. These security techniques are based on artificially-synthesized disordered micro and nano materials. A potential on-chip realization and integration scenario of such approach is also discussed. Novel material synthesis technologies and functional prototype production processes are illustrated. Extraction process of RF fingerprints, based on near-field scattering measurements, is included as well. Finally, statistical analysis and distance measures of similarity, uniqueness and orthogonality of the extracted fingerprints are carefully investigated at the end of this chapter.

M. Kheir (✉) · H. Kreft · R. Knöchel
Institute of Electrical and Information Engineering, Microwave Techniques Group, Christian-Albrechts-University of Kiel, 24143 Kiel, Germany
e-mail: mkh@tf.uni-kiel.de

H. Kreft
e-mail: hk@tf.uni-kiel.de

R. Knöchel
e-mail: rk@tf.uni-kiel.de

I. Hölken
Institute of Material Science and Engineering, Functional Nanomaterials Group, Christian-Albrechts-University of Kiel, 24143 Kiel, Germany
e-mail: ih@tf.uni-kiel.de

J. Suzuki et al. (eds.), *Modeling, Methodologies and Tools for Molecular and Nano-scale Communications*, Modeling and Optimization in Science and Technologies 9, DOI 10.1007/978-3-319-50688-3_17

1 Introduction

Secure hardware systems are being demanded for numerous civil and military applications. Many contemporary emerging technologies require highly resilient and operational security systems. Several techniques that are capable of securing hardware and communication systems already exist in literature [1–7]. Among these security techniques, Physical Unclonable Functions (PUFs) have a great role in securing hardware and integrated circuits. They have been proposed in the last decade to create physical unique entities as a physical security anchor required for security applications [1].

Radio Frequency security systems that utilize PUFs are being considered as a potential solution to data security and counterfeiting problems [2, 5]. However, most of the known approaches are mainly designed for products and human identification and do not thoroughly deal with security matters related to integrated circuits and chips. On-chip security systems can be certainly cost-effective and simple in realization. Relying on 3D chip integration techniques, PUFs can be implemented within the same process of chip production.

Nanostructured hardware security has recently been proposed as a promising PUF security solution [8–13]. Throughout this chapter, the new concept of utilizing micro- and nanostructured security is going to be illustrated. The basic electromagnetic theory of mixed media is briefly summarized. Some early-stage experimental results are introduced and investigated from the perspective of material sciences and microwave technology.

2 Electromagnetic Properties of Mixed Materials

The proposed nanostructured security approach lies in using a dielectric host medium, or a fixing matrix, with mixed micro or nano-particles or so-called "granules" along with an applied electromagnetic wave. Both the fixing matrix and the granules have two different dielectric constants ε_e and ε_i, respectively. In this section, two different models are being dealt with, namely "single-phase" and "multi-phase" mixtures.

In the single-phase mixtures model shown in Fig. 1a, all granules are supposed to be in a spherical shape and have the same dielectric properties [12]. These spherical granules can be of different sizes. All granules are randomly distributed inside the homogenous fixing matrix. The overall effective medium permittivity ε_{eff} is best modeled by the well-known Maxwell-Garnett mixing formula [14, 15]

$$\varepsilon_{eff} = \varepsilon_e + 3f\varepsilon_e \frac{\varepsilon_i - \varepsilon_e}{\varepsilon_i + 2\varepsilon_e - f(\varepsilon_i - \varepsilon_e)} \tag{1}$$

where f represents the volume fraction of granules to the host medium.

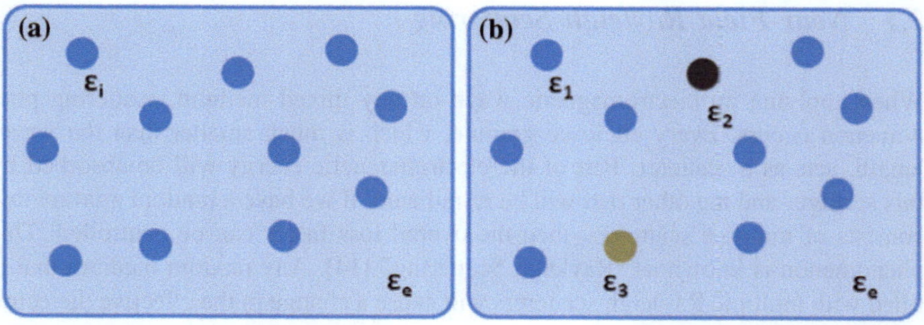

Fig. 1 Electromagnetic mixed media. **a** The single-phase model. **b** The multi-phase model

In case of multiple granules with different dielectric constants, the "multi-phase" model shown in Fig. 1b can be employed. It assumes different inclusions with k different dielectric constants. Formulation for the overall effective medium properties is given by [14]

$$\varepsilon_{eff} = \varepsilon_e + 3\varepsilon_e \frac{\sum\limits_{k=1}^{K} f_k \frac{\varepsilon_k - \varepsilon_e}{\varepsilon_k + 2\varepsilon_e}}{1 - \sum\limits_{k=1}^{K} f_k \frac{\varepsilon_k - \varepsilon_e}{\varepsilon_k + 2\varepsilon_e}} \tag{2}$$

Equation (2) gives a comprehensive insight into the overall effective medium properties, dielectric or magnetic, in the most generalized case.

It should be stressed on the fact that magnetic materials can be equally employed. In such case, ε can be directly substituted by the material permeability μ in (2).

In the proposed nanostructured technology, the latter model will be used. In such case the degree-of-freedom in choosing various parameter configurations consequently increases. Among those degrees-of-freedom, the following parameters are considered:

1. Dielectric material of the fixing matrix.
2. Materials of inclusions (permittivity, permeability or conductivity).
3. Fractional volume of inclusions.
4. Geometric shape and size of inclusions.
5. Randomization technique of inclusions.

All these parameters have a strong influence on the generated RF fingerprint which is going to be illustrated by the measured results in the next sections.

2.1 Near-Field Rayleigh Scattering

When applying an electromagnetic wave on any mixed medium, scattering phenomenon occurs. Every enclosed granule, which is much smaller than the wavelength, acts as a scatterer. Part of the electromagnetic energy will be absorbed by this scatterer and the other part will be re-radiated. If we have a random mixture that consists of multiple scatterers, then the overall loss factor can be controlled. This phenomenon is known as "Rayleigh Scattering" [14]. Any random medium that is filled with multiple Rayleigh scatterers will cause a change in the effective dielectric constant according to

$$\frac{\varepsilon''_{eff}}{\varepsilon_0} = \frac{n\sigma_s}{\omega\sqrt{\mu_0\varepsilon_0}} \quad (3)$$

where ε''_{eff} is the imaginary part of the effective permittivity, n is the density of scatterers, ω is the angular frequency of the applied wave and σ_s is the scattering cross-section. The previous formula is assuming free-space with a permittivity ε_0 and a permeability μ_0. However, the same concept is still applicable to dielectrics.

If we confine this wave inside a bounded medium (i.e. a chip package), then we can insure a fixed signature (or a fingerprint) and this is the main concept behind our RF security technique. As a proof-of-concept realization of the idea, our proposed Cocoon-PUF structure is investigated in the next sections.

3 RF Fingerprinting and Nanostructured Security

The concept of hardware fingerprinting is similar to that of human fingerprinting which is based on distinct attributes. Fingerprinting application using RF technology is the focus of this study. Examples to RF fingerprinting using PUFs are given elsewhere [2, 5, 7]. The newly-developed nanostructured security technique is hereby presented.

3.1 Cocoon-PUF

The Cocoon-PUF works from the perspective of an application chip designer in the way depicted in Fig. 2a [10]. The word "Cocoon" refers to the mixing-material that wraps around an inner protected electronic circuit. Examples are related to metallic and/or polymer-based particles/composites in dielectric and/or magnetic matrices. In Fig. 2a, the key information is a part of the distribution function for position and size of the particles used as a unique fingerprint. The RF hardware extracts the fingerprint out of the cocoon by applying the principle of near-field scattering explained above. This is achieved by the emitter/sensor phalanx of electrodes as depicted in Fig. 2b. The Fuzzy Extractor contains the digital signal processing crypto part to transform the previously mentioned fingerprint into our secret.

Figure 2b depicts the circuit carrier as the base for the mounted bare-die or chip connected in this example as flip-chip through the use of ball grids or bumps forming the protected nano-safe. The cocoon builds the surrounding wrapping consisting of a potting material including a plurality of different types of granulates from nano- up to micro-sized particles. All I/O terminals are going through the cocoon to connect the nano-safe chip with the Printed Circuit Board (PCB). There could be a shielding coat in combination with a finishing housing on the outer side of the cocoon.

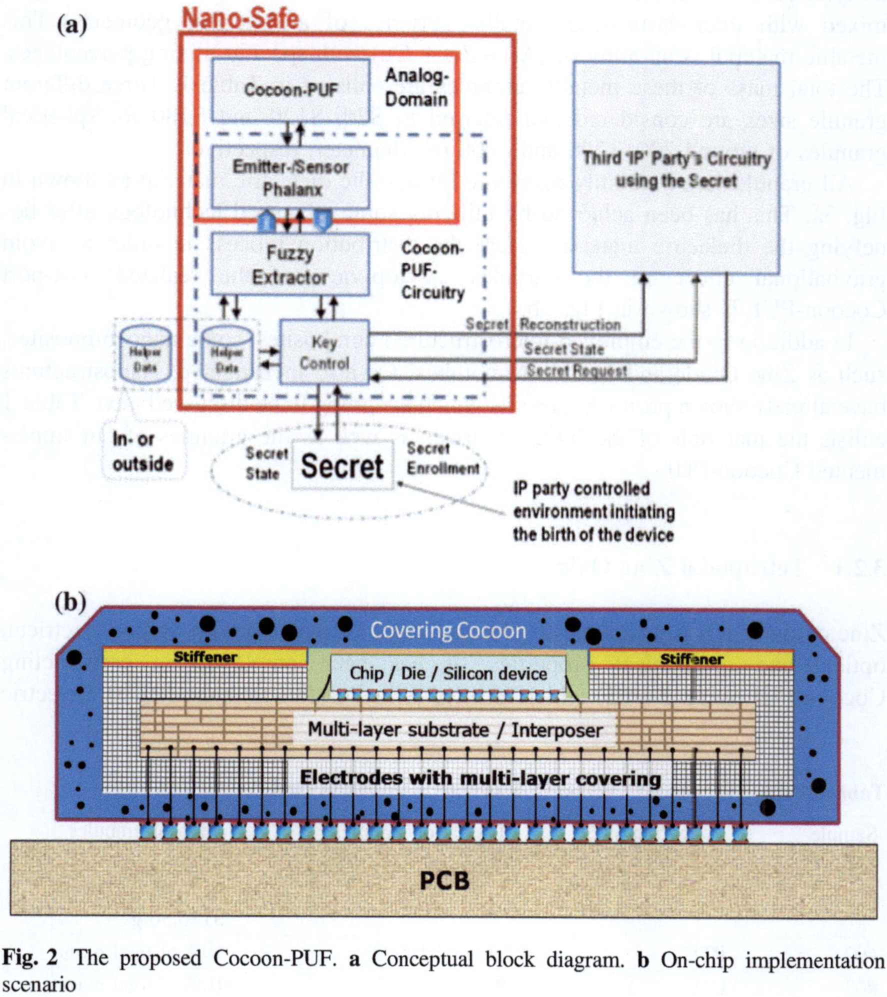

Fig. 2 The proposed Cocoon-PUF. **a** Conceptual block diagram. **b** On-chip implementation scenario

The intended effect of shielding comprises two principle strategies. The first one is to prevent undesirable emission to the outer world beyond the limits of the cocoon. The second reason is to prevent the mixture from any external electromagnetic interference that might influence the RF fingerprint.

3.2 *Synthesized Micro and Nano Composites*

Each Cocoon-PUF is a cylindrical cavity with four SMA-type connectors. The cavity is made of a Brass alloy with 60% Zinc and 40% Copper to maintain lower losses and better environmental immunity. It has a total diameter of 95 mm including the screw fixtures with 25 mm height. The first sample (#01) is filled with a Styrol dielectric material with a dielectric constant of 3.3. This dielectric filling is mixed with micro-structured metallic particles of a spherical geometry. This metallic material is an alloy of (Al + Zn + Cu) with specific mixing percentages. The total mass of these metallic granules are enlisted in Table 1. Three different granule sizes are considered and referred to S20, S120 and S180 for spherical granules of around 200, 1200 and 1800 μm diameter, respectively.

All granules are randomly distributed inside the dielectric material as shown in Fig. 3a. This has been achieved by utilizing some advanced technology after liquefying the dielectric material before the distribution process in order to avoid gravitational effect on the particles. A top-view of the realized four-port Cocoon-PUF is shown in Fig. 3b.

In addition to the employed microstructured composites, some nanocomposites, such as Zinc Oxide and Carbon Nanotubes, are also investigated. Nanostructures have already shown promising results which are going to be discussed next. Table 1 enlists the materials of the fixing matrices as well as the granules of our implemented Cocoon-PUFs.

3.2.1 Tetrapodal Zinc Oxide

Zinc oxide (ZnO) is a ceramic II-VI semiconductor material with unique electrical, optical and mechanical properties. It has been employed in constructing Cocoon-PUF sample (#02) which is filled with Polythiourethane (PTU) dielectric

Table 1 The implemented Cocoon-PUF samples used in this study

Sample	Fixing matrix	Material of granules	Mass of granules
#01	Styrol	Alloy (Al + Zn + Cu)	S20:30 g, S120:50 g S180:50 g
#02	PTU	Tetrapodal ZnO	10% of total mass
#03	PTU	CNTs	10% of total mass

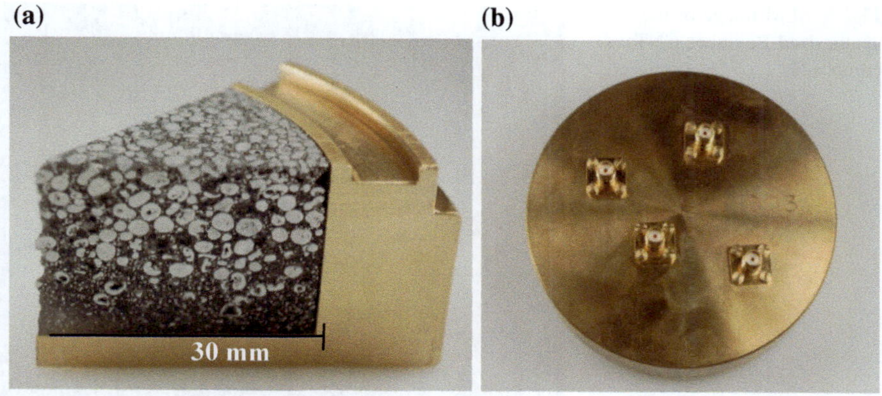

Fig. 3 **a** A section of the microstructured Cocoon-PUF #01. **b** *Top* view of the four-port Cocoon-PUF cavity

fixing matrix. The morphology of ZnO can differ in various ways depending on the manufacturing parameters. It can be of a spherical, comb-like, spring-like or tetrapodal shape [16, 17]. The tetrapod-shaped ZnO particles (T-ZnO) used in this work offer a special kind of morphology which acts as a strongly reinforcing filler if combined with polymeric materials. The tetrahedral coordination of wurtzite ZnO is shown in Fig. 4a.

A scanning electron microscopy (SEM) image of the T-ZnO particles is shown in Fig. 4b. The arms of the tetrapods have a mean length of around 20 μm and their shape is not uniform since multipods can be identified. The shown T-ZnO particles were produced by a patented flame transport synthesis technique at the Nano Laboratory of University of Kiel. During this synthesis, a mixture of polyvinyl bytural (PVB) and Zinc is heated up to 900 °C and held at this temperature for 30 min. The PVB is needed as a sacrificial polymer to allow unimpeded transition

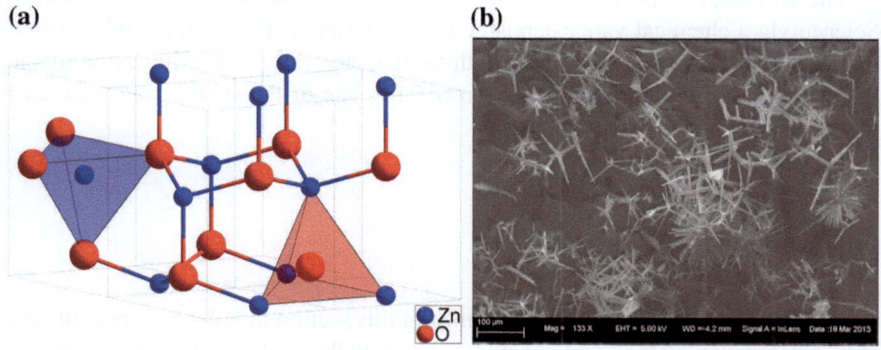

Fig. 4 Zinc oxide: **a** Wurtzite polyhedra with lattice parameters a = 0.3296 nm, c = 0.52065 nm. **b** SEM image of the T-ZnO particles

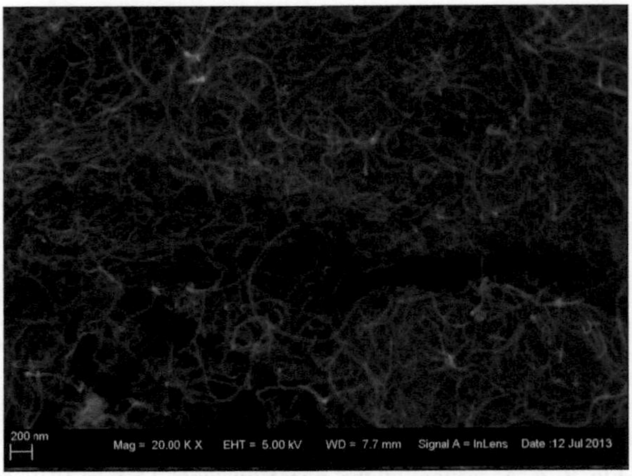

Fig. 5 SEM image of the synthesized Baytubes CNTs composites

to single T-ZnO particles and to avoid amalgamation. During the non-residual combustion of PVB, CO is generated at 500 °C which acts as a catalyst for the core-growth during the reaction to ZnO.

3.2.2 Carbon Nanotubes

The discovery of carbon nanotubes (CNTs) goes back to Sumio Iijima who in 1991 accidentally produced the quasi one-dimensional structures during the fullerene synthesis. Carbon nanotubes (CNTs) belong to the allotropes of carbon, the element with the richest family of chemical compounds [18]. CNTs can be described as multiple rolled-up sheets of Graphene and are categorized into single-walled (SWNT) and multi-walled (MWNT) nanotubes. The length of a single CNT covers a range from several micrometers to some millimeters while the diameter is in the nanometer range. Depending on the way the graphene-sheet is rolled up (chirality), SWNTs can either be metallic (armchair) or semiconducting (zigzag).

The Baytubes C150 P used in sample #03 was produced by Bayer® Material Science via a chemical vapor deposition based process. They are considered to be multi-walled with an ultra-high length-to-diameter ratio and with extraordinary thermal and electrical conductivity. An SEM-image of the synthesized material is shown in Fig. 5.

4 Experimental Results

Experimental measurements are performed in this section in order to verify different essential aspects of the implemented Cocoon-PUFs and to prove their physical robustness. Such measurements are intended to give an indication of the validity of the proposed technology with respect to reproducibility, uniqueness and aging robustness of the generated fingerprints. The measurements are performed using an Agilent E8361A vector network analyzer (VNA). The maximum frequency of the measured results is 15 GHz. The number of sweep points is chosen to be (20,000 points) in order to obtain sufficient accuracy. Only specific results are shown here due to the fact that there is no enough room to show all performed measurements.

The measured results are based on the complex scattering parameters. They are measured under extreme temperature points (−20 to 70 °C) to verify fingerprint robustness to different environmental variation. Moreover, the response to aging effects and external attack attempts are also investigated.

4.1 Scattering Parameters Measurements

The block diagram of Fig. 6 depicts the scattering parameters measurement principle that is typically performed by a VNA. It consists of a reference generator that excites the device under test (DUT) at a certain frequency, a power coupling network for power division and a relay switching matrix. The DUT in such case is the investigated four-port Cocoon-PUF. A microprocessor (μP) is used to control the operation of the instrument as well as transforming and storing the acquired signals into the digital domain for further processing.

The whole measurement system is planned to be integrated on-chip forming an embedded VNA. Such embedded design will be a part of existing electronic systems which will significantly help in reducing the overall cost of traditional laboratory experiments. Moreover, this system will also provide the necessary measurement precision required for extracting the RF fingerprints.

Figure 7 shows an example of the measured transmission scattering parameters magnitude (S_{ij}) of the CNT sample (#03) in dB from (1–15) GHz. A four-port

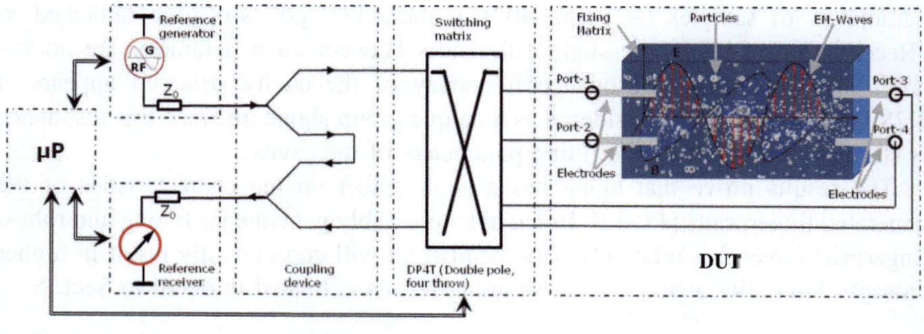

Fig. 6 Block diagram of the scattering parameters measurement setup

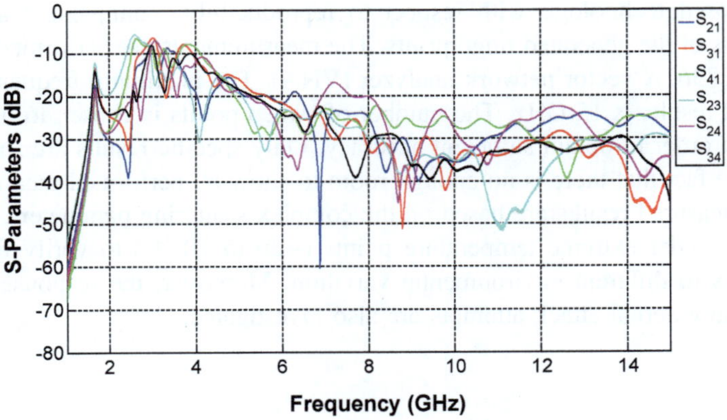

Fig. 7 Measured scattering parameters of the CNT sample #03

Cocoon-PUF device results six independent and different transmission combinations which are overlapped to each other in the same figure.

Our preliminary assumptions and calculations indicate that there are around 2^{31} different fingerprints resulting from each Cocoon-PUF. Due to the fact that PUFs are inherently noisy, error corrections mechanisms, such as fuzzy extraction, remove the noise effects and reduce the valid number of output bits of the uniformly random string entropy. How many bits are needed for fuzzy extraction, is still under further investigations by the authors.

4.2 Aging Effects

A gap of 16-weeks is considered for investigating the aging effects and how reliable the fingerprints should be due to this period. Figures 8 and 9 show the measured *S*-parameters of samples (#01 and #02) at these two points-of-time indicated as "Recent" and "Old" where slight differences between each parameter are noticeable. The fundamental resonance frequency of the cavity structure appears at 1.28 GHz. This can be considered as a unique group signature since this resonance is strongly controlled by the filling parameters of the cavity.

The results prove that aging has a slight effect on the characteristics of the generated fingerprints [11, 12]. Being able to reliably generate such long and robust fingerprints over this relatively wide bandwidth will consequently result in higher entropy. Similarity analysis of these aging results is treated in detail in Sect. 5.

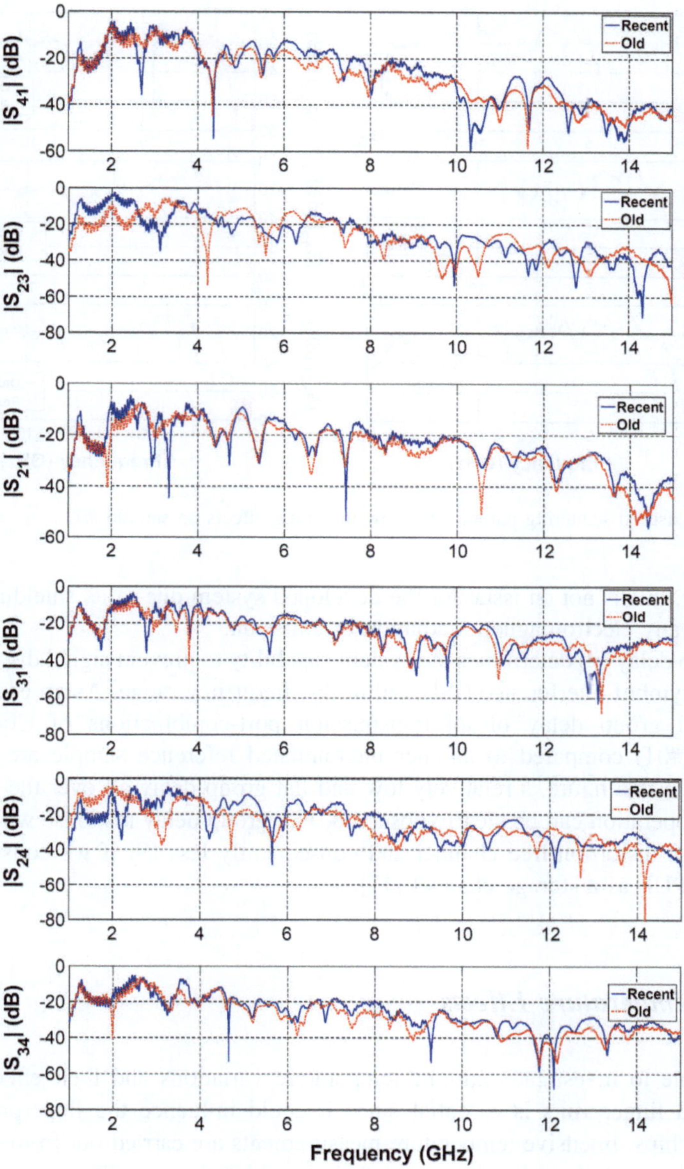

Fig. 8 Measured scattering parameters show the aging effects on sample #01

4.3 UWB Characteristics

Ultra Wideband (UWB) is the frequency band (3.1–10.6 GHz) assigned by the federal communications commission (FCC) for low spectral power unlicensed use.

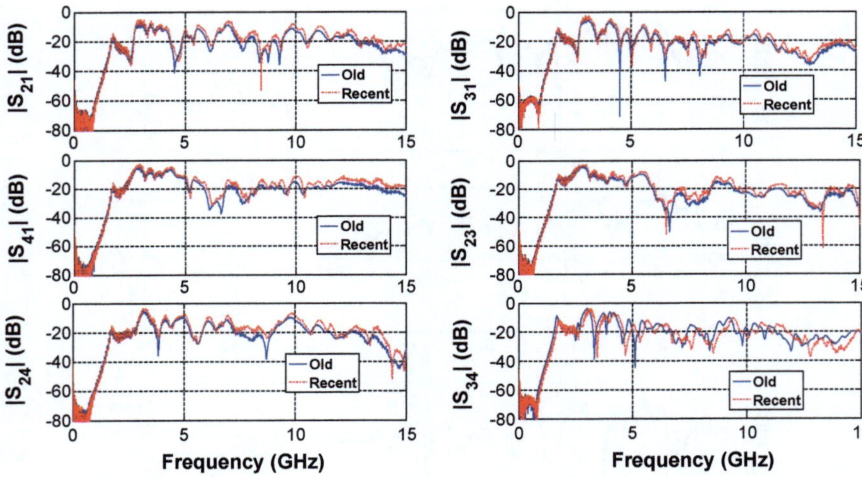

Fig. 9 Measured scattering parameters show the aging effects on sample #02

However, this is not an issue for the developed system due to its shielding since it prevents any electromagnetic radiation or emission.

Group delay (τ) determines the system capability to prevent signal distortion due to inter-symbol interference (ISI) within this frequency range. As an example, the measured group delay of all transmission port-combinations of Cocoon-PUF sample (#01) compared to another ungranulated reference sample are shown in Fig. 10. In this figure, a relatively low and flat group delay all over the frequency band of operation can be clearly observed. Flat group delay response will typically result in a distortion-free channel and consequently less ISI if we considered the Cocoon-PUF as a storage channel [11].

4.4 *Temperature Effects*

Being able to investigate extreme temperature variations and their effects on the generated fingerprints is essential since it could influence the fingerprint of the secured chips. Intensive temperature measurements are carried out from −20 up to 70 °C using a climate chamber controlled via a Matlab script. The measured S_{21} of samples #01 and #02 at the applied temperatures are depicted in Fig. 11a, b respectively [8, 12].

A slight deviation can be observed between the measured data at different temperature degrees. This change only occurs as a shift in the frequency

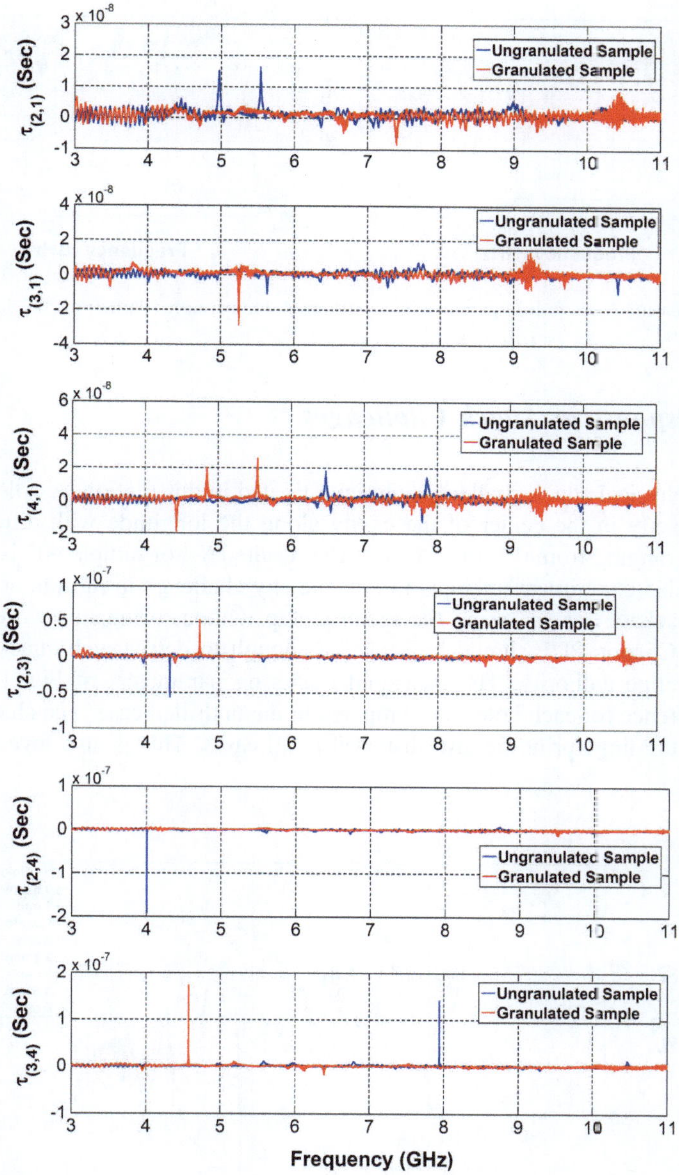

Fig. 10 Measured group delay responses between different independent port combinations of Cocoon-PUF sample #01

characteristics and does not distort the fingerprint characteristic points. Mode splitting phenomenon is still clear at the resonance frequency of sample #02 above 50 °C which can be due to the change in the dielectric constant of the fixing matrix.

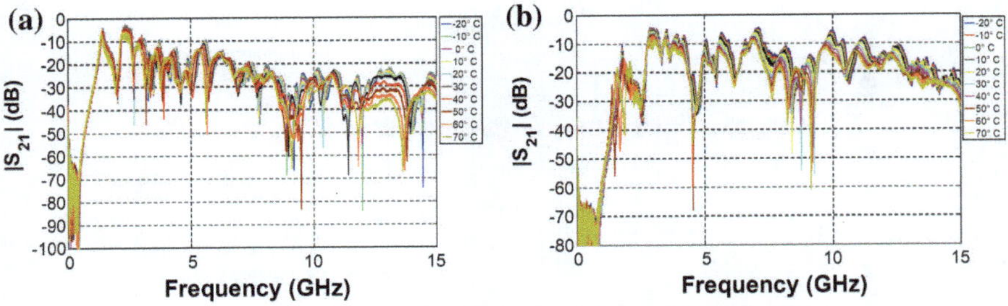

Fig. 11 Measured scattering parameters at different temperature variations. **a** Sample #01. **b** Sample #02

4.5 *Response to Attack Challenges*

External physical attack challenges are investigated in this section. A hole has been drilled exactly in the center of the cavity along the longitude with different drill diameters ranging from 0.5 up to 4 mm. The resulted S_{23} of sample #01 is shown in Fig. 12. This experiment attempts to emulate any challenge to intrude or penetrate the real package for accessing the secured chip without using its own embedded key. The Cocoon-PUF should be sensitive enough to detect such attack attempts and to decline authority. The measured scattering parameters of Fig. 12 show a clear difference for each hole size compared to the undrilled case. The characteristic points of the fingerprint are also distorted in all cases. This is also investigated in Sect. 5.2.

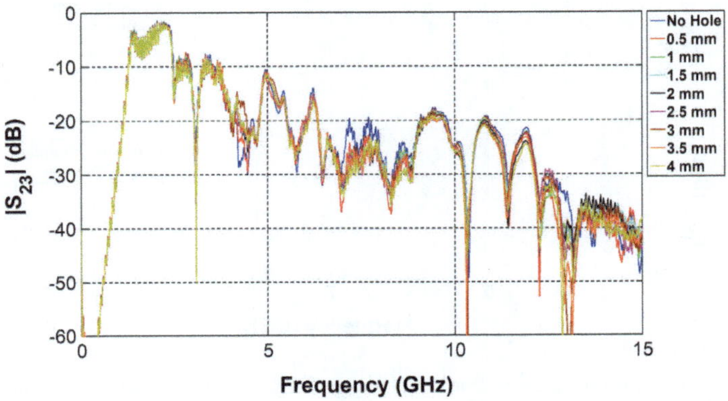

Fig. 12 Measured S_{23} at different attack challenges of sample #01

5 Similarity Analysis of Fingerprints

Similarity between any pair of the fabricated prototypes is an essential measure that is needed to compare and distinguish between different fingerprints [19–21]. Different similarity measures are employed and compared in this section. However, a complete fuzzy extraction process is still required for a better fingerprint identification.

5.1 *Euclidean Distance Metric*

The Euclidean distance metric (d_E) is a well-known and effective method for assessing the similarity/dissimilarity between a pair of data vectors. This method is adopted in this section in order to accurately judge how similar or dissimilar the measured fingerprints are in terms of distance metrics [8]. Meanwhile, it can be an initial indicator for the uniqueness of the produced and reproduced fingerprints. The Euclidean distance between two variables a_j and b_j with k-dimensions is typically calculated as [20]

$$d_E = \sqrt{\sum_{j=1}^{k} \left(a_j - b_j\right)^2} \quad (4)$$

The range of the linear magnitude values of S-parameters is from 0 to 1, thus a direct mapping of dissimilarity percentage between two different scattering parameters can be evaluated according to

$$Diss.(\%) = \frac{\left|\left|S_{ij}^{1}\right| - \left|S_{ij}^{2}\right|\right|}{r_i} \times 100 \quad (5)$$

where r_i represents the interval length that corresponds to the maximum value of the transmission parameter in linear scale which is '1' in such case [20].

Figure 13 depicts the evaluated dissimilarity between the old and recent scattering parameter measurements of Cocoon-PUF sample #01. The figure shows a maximum dissimilarity of 50% at around 3.2 and 6.2 GHz only for both S_{23} and S_{24}. However, it does not exceed 20% for all other parameters all the way after 6.2 GHz. Dissimilarity percentages between different temperature values are also evaluated for both samples as shown in Fig. 14. The applied extreme temperatures have a stronger effect on the ungranulated sample compared to the granulated one. It does not exceed 5% above 6 GHz for the granulated Cocoon-PUF while it is above 50% for the ungranulated one all over the frequency range. Such feature adds another benefit to the proposed system as it helps increasing the temperature-robustness of the chip.

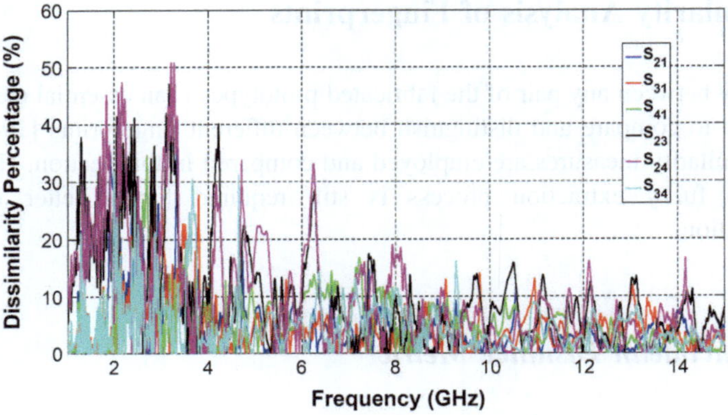

Fig. 13 Dissimilarity percentages representing the aging effects on the Cocoon-PUF sample #01

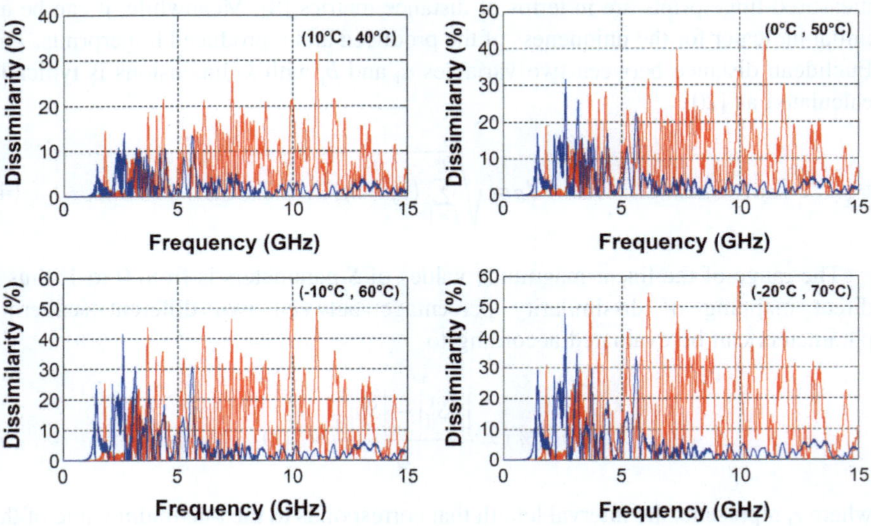

Fig. 14 Dissimilarity percentages of sample #01 at different temperatures compared to the ungranulated one. The *red* curve represents the ungranulated sample while the blue curve represents the granulated one

5.2 *Jaccard (Tanimoto) Distance Metric*

Another widely-used similarity metric is the so-called Jaccard (or Tanimoto) distance measure. It can be equally employed for both real- and discrete-valued vectors where it represents a special form of the normalized inner product value between two non-binary variables [19, 20]. It has been utilized in this section to investigate the effects of physical attacks on the realized samples. Jaccard distance metric is defined as

$$d_J = \frac{\sum_{i=1}^{k} a_i b_i}{\sum_{i=1}^{k} |a_i|^2 + \sum_{i=1}^{k} |b_i|^2 - \sum_{i=1}^{k} a_i b_i} \tag{6}$$

The calculated Jaccard similarity distances are evaluated and shown in Fig. 15 for all investigated hole diameters of Cocoon-PUF sample #01 as an example. The calculated metrics show a relatively low similarity for all drilled hole diameters. The similarity values do not exceed 40% only at few frequency points. This declares the fact that any attempt to attack the Cocoon-PUF can be easily detected.

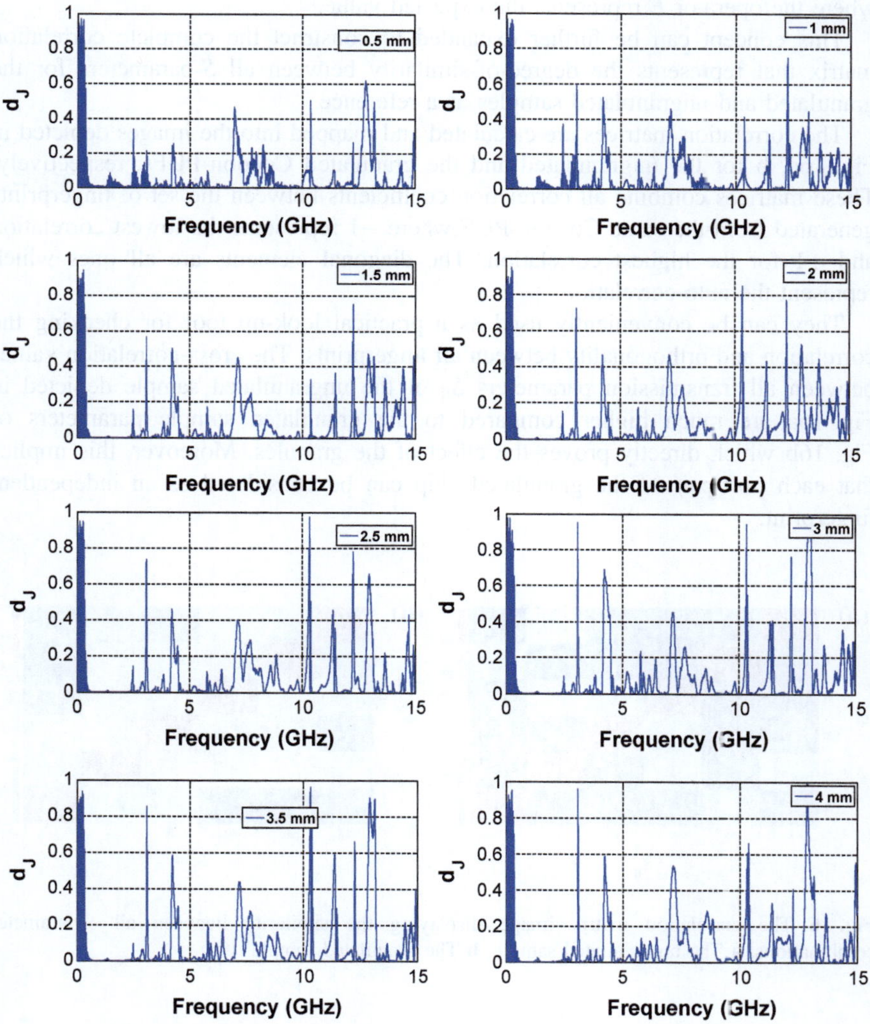

Fig. 15 Jaccard distance metrics at different hole diameters of sample #01

5.3 Correlation and Orthogonality

If each fingerprint, obtained by measurements, is considered as a sample of a stationary stochastic process then the correlation concept can be a meaningful measure of similarity. Pearson's correlation coefficient (ρ) between two real-valued random vectors $\mathbf{X}$ and $\mathbf{Y}$ can be evaluated as [20]

$$\rho_{\mathbf{XY}} = \frac{E(\mathbf{XY}) - E(\mathbf{X})E(\mathbf{Y})}{\sqrt{E(\mathbf{X}^2) - (E(\mathbf{X}))^2}\sqrt{E(\mathbf{Y}^2) - (E(\mathbf{Y}))^2}} \tag{7}$$

where the operator E represents the expected value.

This concept can be further expanded to construct the complete correlation matrix that represents the degree-of-similarity between all S-parameters for the granulated and ungranulated samples as a reference.

The correlation matrices are calculated and mapped into the images depicted in Fig. 16a, b for the ungranulated and the granulated Cocoon-PUFs, respectively. These matrices combine all correlation coefficients between the set of fingerprints generated from the same Cocoon-PUF where −1 represents the lowest correlation and +1 for the highest correlation. The diagonal elements are all ones which represent the auto-correlation.

They can be conveniently used as a practical look-up tool for checking the correlation and orthogonality between all fingerprints. The cross-correlation values between all transmission parameters S_{ij} of the ungranulated sample depicted in Fig. 16a are much higher compared to the granulated sample parameters of Fig. 16b which directly proves the effect of the granules. Moreover, this implies that each port-pair of the granulated chip can be considered as an independent fingerprint.

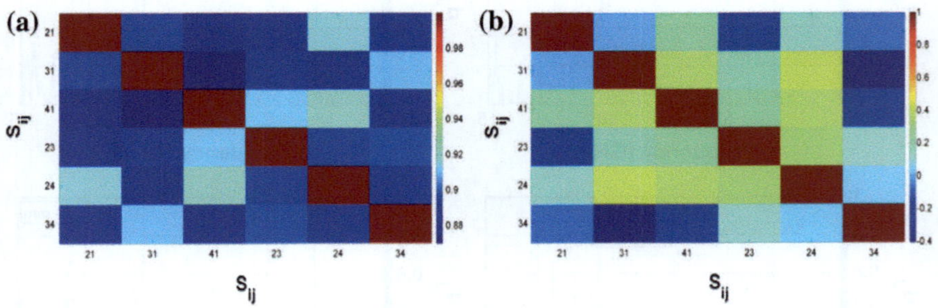

Fig. 16 The correlation matrix image displaying the similarity between all S-parameter combinations. **a** The ungranulated sample. **b** The granulated sample

6 Conclusions

In this chapter, the new concept of nanostructured RF security and fingerprinting of integrated circuits has been demonstrated. This concept has been proven by a number of implemented Cocoon-PUF prototypes fabricated of various micro and nanocomposites. Tetrapodal ZnO and CNTs have been thoroughly investigated where both have shown rather reliable results and have proven to be promising candidates for omnipresent hardware security purposes. Physical robustness investigations have also been carried out in order to insure the reliability and reproducibility of the fingerprints under extreme conditions. Moreover, different similarity and distance measures have been employed so as to have an initial insight into the characteristics of these fingerprints. It should be indicated that a fuzzy extraction technique is still required to be constructed to accurately evaluate the capacity of the digitized keys. It is also needed for the final process of identification. This part is still, however, under detailed investigations by the authors.

Acknowledgements The authors wish to acknowledge the support of Prof. Rainer Adelung, Head of Functional Nanomaterials Group, University of Kiel, Germany, for providing the nanocomposites used in our experiments and for the informative discussions.

References

1. Pappu R, Recht B, Taylor J, Gershenfeld N (2002) Physical one-way functions. Science 297:2026–2030
2. Lakafosis V, Traille A, Lee H, Gebara E, Tentzeris MM, De Jean GR, Kirovski D (2011) RF fingerprinting physical objects for anticounterfeiting applications. IEEE Trans Microw Theor Tech 59:504–514
3. Tuyls P, Schrijen G-J, Skoric B, Geloven J-V, Verhaegh N, Wolters R (2006) Read-proof hardware from protective coatings. In: 8th International Conference on Cryptographic Hardware and Embedded Systems, pp 369–383
4. Juels A (2006) RFID security and privacy: a research survey. IEEE J Sel Areas Commun 24:381–394
5. Bolotnyy L, Robins R (2007) Physically unclonable function-based security and privacy in RFID systems. In: 5th IEEE international conference on pervasive computing and communications, pp 211–220
6. Maiti A, Kim I, Schaumont P (2012) A robust physical unclonable function with enhanced challenge-response set. IEEE Trans Inf Forensics Secur 7:333–345
7. Cobb WE, Laspe ED, Baldwin RO, Temple Michael A, Kim YC (2012) Intrinsic physical-layer authentication of integrated circuits. IEEE Trans Inf Forensics Secur 7:14–24
8. Kheir M, Kreft H, Knöchel R (2013) Nanostructured RFID technology: attack-response and temperature-robustness investigations. In: 43rd European microwave conference (EuMC), Nuremberg—Germany, 2013
9. Kreft H. Tamper-protected hardware and methods for using same. International Patent PCT/EP2012/054248, European Patent Office Ref. no. WO123400
10. Kreft H, Adi W (2012) Cocoon-PUF, a novel mechatronic secure element technology. In: NASA/ESA conference on adaptive hardware and systems, Erlangen—Germany, 2012

11. Kheir M, Kreft H, Knöchel R (2013) RF identification and security technique for highly-secure UWB information systems. In: 24th IEEE personal, indoor and mobile radio communications (PIMRC), London—UK, 2013
12. Kheir M, Kreft H, Hölken I, Knöchel R (2014) On the physical robustness of RF on-chip nanostructured security. J Inf Secur Appl (Elsevier). doi:10.1016/j.jisa.2014.09.007
13. Kheir M, Kreft H, Knöchel R (2014) A novel RF fingerprinting approach for hardware integrated security. J Inf Secur Appl (Elsevier). ISSN 2214-2126. http://dx.doi.org/10.1016/j.jisa.2014.02.001
14. Shivola Ari (1999) Electromagnetic mixing formulas and applications. IET Press, London—UK
15. Merrill WM, Diaz RE, LoRe MM, Squires MC, Alexopoulos NG (1999) Effective medium theories for artificial materials composed of multiple sizes of spherical inclusions in a host continuum. IEEE Trans Antennas Propagation 47:142–111
16. Charinpanitkul T, Nartpochananon P, Satitpitakun T, Wilcox J, Seto T, Otani Y (2012) Facile synthesis of tetrapodal ZnO nanoparticles by modified french process and its photoluminescence. J Ind Eng Chem 18:469–473
17. Zhou RF, Xu XY, Feng HT, Yan D, Li HJ, Cheng S, Yan PX (2010) Morphology controlled syntheses, growth mechanisms, and optical properties of ZnO nanocombs, nanotetrapods. Adv Mater Res 97–101:960–964
18. Gogotsi Y (2006) Carbon nanomaterials. Taylor and Francis Group
19. Santini S, Jain R (1999) Similarity measures. IEEE Trans Pattern Anal Mach Intell 12:871–883
20. Theodoridis S, Koutroumbas K (2006) Pattern recognition, 3rd edn. Elsevier, USA
21. Cha S-H (2007) Comprehensive survey on distance/similarity measures between probability density functions. Int J Math Models Methods Appl Sci 1:300–307

Reliable Design for Crossbar Nano-architectures

Masoud Zamani and Mehdi B. Tahoori

Abstract The conventional CMOS technology faces various challenges in the continues down-scaling. Therefore, different emerging technologies based on bottom-up and self-assembly nanofabrications are being explored to overcome these challenges. These technologies exploit different nano-materials in the regular structures such as the crossbar nano-architecture, which is a two-dimensional grid with configurable switches at the crosspoints. Exploiting nano-materials in crossbar nano-architectures offers the possibility of significantly denser circuits at reduced fabrication costs compared to the existing lithography-based manufacturing. However, in these nano-architectures atomic device sizes and poor control on the fabrication processes impairs the reliability of these circuits. In this chapter, we investigate reliability issues in crossbar nano-architectures in terms of variation and defect tolerance. We study two approaches, namely logic mapping and self-timed architecture design, to provide variation and defect tolerance. In the logic mapping approach, different configurations, a.k.a mappings, of a logic function on a crossbar nano-architecture are explored to find the configuration with the required variation and defect tolerance. Simulation results, on a set of benchmark circuits, show that the proposed logic mapping approach achieves variation tolerance more than 98% of the cases, while in 100% of the cases all defects are tolerated. The efficiency of these algorithms is independent of crossbar size. At the architecture-level, a self-timed nano-architecture is introduced to reduce the circuit vulnerability to delay variations. Compared to the synchronous counterparts, with around 50% overhead in the number of activated switches, the proposed architecture provides 100% tolerance of delay variations.

M. Zamani (✉)
Qualcomm Incorporated, 5828 Pacific Center Blvd, San Diego, CA 92121, USA
e-mail: mzamani@qti.qualcomm.com

M.B. Tahoori
CDNC - Chair of Dependable Nano Computing, Department of Computer Science, Karlsruhe Institute of Technology, Haid-und-Neu-Str. 7 (Technologiefabrik Karlsruhe), 76131 Karlsruhe, Germany
e-mail: mehdi.tahoori@kit.edu

J. Suzuki et al. (eds.), *Modeling, Methodologies and Tools for Molecular and Nano-scale Communications*, Modeling and Optimization in Science and Technologies 9, DOI 10.1007/978-3-319-50688-3_18

1 Introduction

The *Complementary Metal Oxide Semiconductor* (CMOS) technology is facing major challenges in the continuation of the Moore's law as the device feature sizes scales down to nanometer ranges. Inherent physical issues such as ultra-thin gate oxides, short channel effects, doping fluctuations across the chip, and diffraction effects of sub-wavelength lithography cause significant complications in each successive device feature size reduction. Therefore, researchers are investigating alternative technologies, devices, and architectures to overcome such limitations. Nano materials such as *Carbon Nano-Tube* (CNT) and *Nano-Wires* (NW) are exploited in emerging fabrication processes. CNTs can be fabricated with the length in the range of micrometre while their diameter varies from a few nanometres to several tens of nanometers [1]. Fundamental electronic devices such as diode, Field Effect Transistors (FET), and memory elements have been assembled from these well-defined nanoscale building blocks [2–5]. However, the regular structures such as *crossbar nano-architectures* are naturally more amenable to fabricate in self-assembly nano fabrication. A crossbar nano-architecture is a two-dimensional grid structure with configurable switches and devices (FET or diode) at the crosspoints [6]. In such nano-crossbars, CNTs and NWs are aligned in one direction and then orthogonally superimposed with another set of CNTs or NWs to construct architectures similar to *Programmable Logic Array (PLA)* [2, 4].

Various fabrication methods are used to produce carbon nano-tubes. Some of these techniques include laser ablation, arc discharge, chemical vapour deposition (CVD) techniques, and bottom-up organic approach. CVD techniques are more applicable since they provide more control on the orientation, alignment, nanotube length, and diameter of carbon nano-tubes [1].

However, these emerging technologies have major reliability and robustness challenges. These reliability issues manifest themselves as extreme parametric variations, high defect rate at manufacturing, and high failure rate during lifetime operation. In the nanoscale regime, due to atomic-scale of device sizes, a small deviation in the device parameters can significantly change device characteristics [7]. Furthermore, a slight deviation in diameter of CNTs as well as process variation in fabrication of CNTs may result in extreme variation on resistance and capacitance of interconnects.

Furthermore, poor control in nanofabrication processes increases the process variation [8]. On the other hand, high defect rate, which detracts the manufacturing yield, is more likely and significant for circuit design in such emerging technologies [9]. Therefore, the reliability concerns must explicitly be addressed in these emerging technologies.

In this chapter, we investigate the reliability issues in crossbar nano-architectures in terms of variation and defect tolerance. We present two approaches, one at logic-level mapping and the other one based on self-timed architectures, to achieve variation and defect tolerance for crossbar nano-architectures. The mapping of a logic function into a crossbar array (aka configuration), i.e. which crosspoints are used (activated) and which are unused (deactivated), has a considerable impact on the delay as well as the correct functionality of the mapped circuit. In the logic mapping approach, different configurations of a logic function on a crossbar nano-architecture are explored to find a configuration which results in an optimal variation and defect tolerance. Furthermore, a set of logic transformations are introduced to increase the number of possible configurations for a logic function on a crossbar nano-architecture. Using these transformations, a larger number of configurations can be explored to increase the efficiency of the logic mapping approach in providing variation and defect tolerance. The proposed logic mapping algorithms provide configurations with more than 98% variation tolerance and 100% defect tolerance. Due to delay variations in crossbar nano-architectures a precise timing control is very difficult. Therefore, these architecture are highly vulnerable to variations on switching delays of devices. Therefore, in addition to mapping approaches, we also exploit asynchronous design methodologies to propose a self-time crossbar nano-architecture, which allows us to eliminate global clocking signals (by replacing with local handshake signals) to reduce the circuit vulnerability to delay variations. The proposed self-timed nano-PLA provides robust communication between crossbars by providing variation tolerant computation on the crossbar as well as tolerating delay variation on interconnects between the crossbars.

The rest of this chapter is organized as follows. Section 2 presents background regarding crossbar nano-architectures and related work. Our proposed logic mapping approach is introduced in Sect. 3. The proposed self-timed nano-architecture is presented in Sect. 4. Finally, Sect. 5 concludes the chapter.

2 Preliminaries

2.1 Crossbar Nano-architectures

Carbon NanoTubes (CNTs) and semiconductor *Nano Wires* (NWs) fabrication based on bottom-up self-assembly are some of the promising emerging nano-devices which provide the opportunity of significantly denser circuits compared to the existing lithography-based manufacturing [2, 6]. Selectively-doped semiconducting CNTs and NWs are orthogonally superimposed to construct fundamental electronic devices such as diode and FET [2–4, 6, 10]. Programmable interconnect, logic cores, and memory blocks can be implemented by means of configurable junctions in two-dimensional crossed horizontal and vertical arrays of CNTs or NWs [6, 11–13]. This two-dimensional grid structure with configurable switches at the crosspoints (junctions) is known as the *crossbar*. Cascaded crossbar blocks construct a *crossbar array*. Many proposed nano-architectures for nanoscale electronics have focused on this structure due to its simplicity for self-assembly based nanofabrication [6]. Figure 1 shows the general structure of a crossbar with activated and deactivated crosspoints.

Fig. 1 Crossbar array, as a possible structure for nanoelectronics

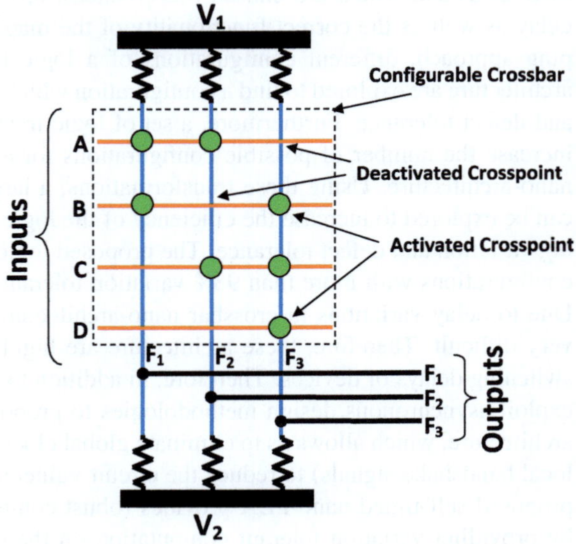

The activated crosspoints refer to crosspoints which are configured during the logic mapping phase. Depending on the type of the wire doping, a crosspoint shows different device behaviors (FET, diode, or resistor). As seen in this figure, the "*horizontal*" lines are the crossbar inputs and the outputs are taken from the "*vertical*" lines.

Based on the crossbar structure, various architectures have been proposed. The hybrid combination of CMOS and nanowire-based devices are used to propose CMOL (standing for CMOS/nanowire/molecular hybrids [14]), an architecture similar to *Field Programmable Gate Array* (FPGA) [15]. In this architecture, nano crossbars are used as storage elements, routing blocks, and wired-OR logic blocks. However, the other functions (e.g. inversion, signal restoration, and demultiplexing) are moved into the CMOS substrate. Field-programmable nanowire interconnect (FPNI) is an architecture which exploits the nano-crossbars as routing blocks, while the entire logic is implemented in the CMOS substrate [12].

Dehon and et al. have proposed a nano-architecture similar to Programmable Logic Array (PLA), called nano-PLA [13]. Cascaded logic blocks of nano-PLA provide NOR-OR logic using a selective inversion scheme at the inputs of each block. Nano-PLA uses a set of nano-wire based FET devices to provide the inversion and the restoration on the signals between the logic blocks. The FET-based crossbar can also be used for logic realization in PLA-like crossbar nano-architectures [2, 4, 16, 17]. In these nano-architectures, a PLA block consists of FET-based devices to implement logical NAND (NOR) functions. Cascaded blocks of such crossbars provide a set of NAND-NAND or NOR-NOR blocks, similar to the PLA structure.

2.2 *Related Work*

An application-independent defect tolerant design flow has been proposed in [18]. In this flow, the higher levels of the design stack are unaware of the exact defect locations, while defect consideration is done at logic mapping level, by using a set of recursive and greedy algorithms. Defect tolerant techniques for crossbar nano-architectures take the advantage of abundance of resources in order to introduce redundancy in the logic mapping [19]. The logic mapping problem and the problem of finding the largest subset of sub-crossbar with no defects have been shown to be NP-hard [20]. Therefore, the proposed defect tolerant logic mapping techniques include heuristic algorithms during the mapping step to bypass defective crosspoints.

The problem of finding a defect-free mapping is translated to *monomorphism* search in a graph [21]. In this approach, the Boolean function is converted to a graph. The nodes of this graph represent the inputs and outputs of the Boolean function. The edges of the graph represent input/output relations. On the other hand, the crossbar is represented by another graph. In this graph, the inputs and the outputs of the crossbar are represented by nodes. There is an edge between an input and an output node if and only if the crosspoint constructed by the corresponding horizontal and vertical wires is defect-free. Using these graphs the problem of defect tolerant mapping is converted to searching for a graph monomorphism in the crossbar graph. This problem is translated as finding a sub-graph in the crossbar graph such that it matches the Boolean function graph, which is a classical NP-complete problem [22].

A SAT-based defect-aware logic mapping framework has been introduced in [9]. In this method, the mapping problem is converted to a SAT instance. By solving the corresponding SAT instance, the method indicates whether there can be a defect-free mapping on the crossbar or not. Using ILP formulations, a logic mapping method is introduced in [23, 24]. These formulations provide a defect-free mapping for logic function on crossbar nano-architecture if such mapping exists.

The performance of the nanowire and nanotube based FET devices under process variation is analyzed in [25]. It is shown that nanowire and nanotube based FETs are less sensitive to process variations compared to CMOS and FinFET counterparts. However, in this comparison it is assumed that all these three technologies follow the same distribution in parameter variations, which neglects the inherent poor control in the self-assembly nano-fabrication. On the other hand, although it is shown that nanowire and nanotube based FETs are less sensitive to process variations, but still for the $3\sigma = 30\%$ of the normal distribution of the process parameters, the normal distribution of the overall circuit delay follows a normal distribution with the same variance.

To address the process variation, a method to reduce the net range of path delays is introduced in [26]. This method matches the fanout of logical nets with the transistor threshold voltages to tolerate the effect of variations on the threshold voltages of devices. This method, which is based on a greedy algorithm, tries to match a high-fanout product term with a low resistance NAND-term.

The *Simulated Annealing (SA)* algorithm is applied for variation and defect tolerant mapping on a crossbar in the approach presented in [27]. In this method, the mappings on the vertical and the horizontal lines are swapped to change the variation cost of a crossbar. An on-the-fly mapping technique without any pre-characterization has been introduced in [28]. This method enables the built-in self-mapping capability.

3 Proposed Logic Mapping Approaches

Different configurations of a logic function on a crossbar array nano-architecture can be explored to find a reliable configuration which results in the optimal variation and defect tolerance. These configurations are generated and evaluated by applying a set of post synthesis transformations. These transformations are used in a set of mapping algorithms to find such optimal configuration for the crossbar array to map the given function.

3.1 Post-synthesis Transformations

The output of a synthesis tool is the configuration of the crossbar array which provides the desired functionality. The configuration of a crossbar determines which crosspoints are used (activated) and which are deactivated to implement the required function. In other words, the outcome of the logic synthesis tools is the optimized sets of boolean functions to be mapped into each crossbar. It also determines how different crossbars are interconnected to implement the circuit (logic) functionality. However, in the generation of such configuration, the reliability concerns are not taken into account. Therefore, a set of *post-synthesis* transformations are required to be applied after the synthesis step to optimize the configuration in terms of reliability.

Here, we present a set of *logic transformations*, which preserves the logic functionality, while changing the way the crossbar resources are utilized for the implementation of that function [29–31]. Some of these transformations are *local* (intra block), meaning that they preserve the functionality of the portion of the logic function mapped into a crossbar by modifying the configuration of that crossbar. On the other hand, *global* transformations (inter block) may modify the portion of the logic function mapped into different crossbars while preserving the entire logic functionality of the circuit. For the sake of clarity, in order to distinguish between a single crossbar and the entire crossbar array, throughout this chapter we use the term "*block*" for a single crossbar of a crossbar array.

3.1.1 Intra Block Transformations

- **Swapping**: In the swapping transformation, the configurations of two rows or two columns of a crossbar are swapped. This transformation is used as the basic operation in variety of algorithms (e.g. [23, 27, 32–35]).
- **Input Duplication**: The crossbar array is a regular architecture, and the crossbar itself is a complete structure (there is a crosspoint at every intersection, in contrast to FPGAs). Therefore, during the logic mapping phase, there are always unused input rows. These unused input rows can be used to duplicate some of the inputs [29, 35].
- **Output Decomposition**: In the wired-OR logic, an output can be decomposed into two or more terms, each mapped into a separate vertical (output) line. However, all these lines must be wired to construct the output.

3.1.2 Inter Block Transformations

In an inter block transformation, the configuration of a single block is modified such that the (portion of the) logic function mapped by this block is distributed between two crossbars. In this transformation, the entire logic function mapped into a crossbar, or a portion of that is decomposed into two sets of functions each mapped into a separate crossbar.

- **Function Decomposition**: A function decomposition transformation decomposes some of the outputs into sub-functions, where the sub-functions together form the original function of the decomposed output. If $O_i = i_1 + i_2 + \cdots + i_m$ is an output of a block, then it can be decomposed into two sub-functions O_i^1 and O_i^2, where $O_i^1 = i_1 + i_2 + \cdots + i_t$ and $O_i^2 = O_i^1 + i_t + i_{t+1} + \cdots + i_m$ (for any $t < m$). Since the blocks of the crossbar array produce an OR function, the function decomposition does not affect the functionality of O_i ($O_i = O_i^2$). This transformation reduces the complexity of the logic mapping into a crossbar (the block which produces O_i^1), by reducing the complexity of O_i to O_i^1 by adding to the complexity of the block producing O_i^2, in terms of an extra output and a set of extra inputs. It also introduces more complexity to the routing crossbar which needs to route extra lines for the decomposed outputs.
- **Block Decomposition**: In the block decomposition transformation, a subset of the output functions generated by a block B_i is moved into a new block B_j. So if B_i originally implements n outputs, it will implement m ($m < n$) outputs and the remaining $n - m$ outputs are implemented by another block. Block decomposition introduces some area overhead in terms of an extra logic block and the corresponding routing block. However, since it reduces the number of outputs of that block, the mapping complexity of that block will be reduced.

We show these transformations by an example. Figure 2a shows the configurations of two logic blocks of a nano-PLA architecture. For the sake of simplicity,

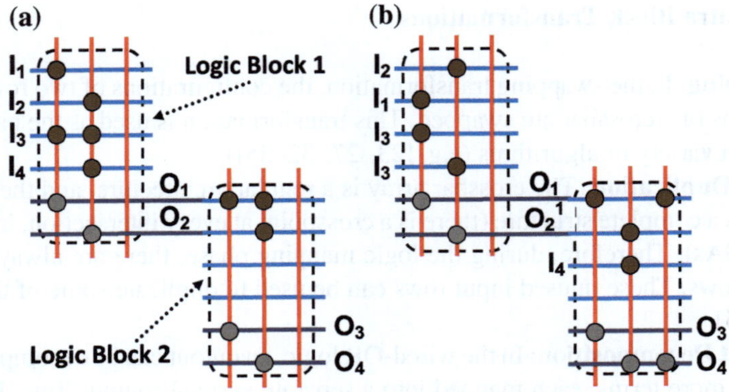

Fig. 2 Mapping of two logic functions (O_3 and O_4) into two logic blocks of a nano-PLA. **a** The configurations of the logic blocks before applying any logic transformation. **b** The configurations of the logic blocks after applying a set of intra and inter block logic transformations

the restoration units as well as routing crossbars are not shown in this figure. These configurations correspond to the mapping of two logic functions ($O_3 = I_1 + I_3$ and $O_4 = I_1 + I_2 + I_3 + I_4$). Using a set of intra and inter block logic transformations, the configurations are modified in Fig. 2b. The intra block transformations include the swapping of I_1 and I_2 (block 1), the duplication of I_3 over two rows (block 1), and the output decomposition of O_4 (block 2). Using the function decomposition transformation, O_2 is decomposed into two sub-functions, one of them ($O_2^1 = I_2 + I_3$) is implemented on block 1, and the other sub-function ($O_4 = O_2^1 + I_4$) is implemented on block 2.

3.2 Proposed Logic Mapping Algorithms

The proposed variation and defect tolerant mapping algorithms can be categorized into two groups: (1) *local mapping* (the intra block mapping algorithm) and (2) *global mapping* (the inter block mapping algorithm). The local mapping algorithm uses intra block transformations to provide a robust mapping for each nano-PLA block. If the local mapping fails to provide the required mapping for a block, then that block is decomposed into two nano-PLA blocks using inter block transformations (global mapping).

3.2.1 Definitions

Before we explain the mapping algorithms, some definitions are provided which are used in the rest of this chapter.

- *Output Delay*; the delay between a transition of an input and the corresponding transition on the output of a crossbar is defined as the output delay.
- *Threshold Delay*; The maximum output delay of a crossbar which does not falsify the timing requirement for the correct functionality of the circuit is defined as the threshold delay.
- *Defect-Free Configuration*; A configuration which avoids using any defective crosspoint is defined as the defect-free configuration (defect tolerant configuration).
- *Variation Tolerant Configuration*; A configuration is variation tolerance if it satisfies the delay requirements of the circuit.
- *Reliable Mapping*; reliable mapping is a mapping of a logic function into a crossbar array with a defect-free and variation tolerant configuration.
- A logic function and a crossbar can be represented by a set of matrices which are defined in the following [27, 28].
  - *Mapping Matrix (MM)*; the configuration of a crossbar is determined by a matrix called *Mapping Matrix (MM)*. MM[i][j] is 1 if the crosspoint at row *i* and column *j* is activated, otherwise it is 0.
  - *Variation Matrix (VM)*; deviation on the switching delay of each crosspoint from the expected nominal delay is determined by a matrix called *Variation Matrix (VM)*. Each entry of this matrix denotes the delay variation of corresponding crosspoint which can be an arbitrary value. The entries of this matrix can be extracted by using delay test techniques [27, 28]. Defective crosspoints are denoted by infinity in the corresponding elements of VM.

MM is determined by the logic mapping of the function. The delay variation on outputs of a crossbar as well as the correctness of the logic function mapped into the crossbar (with respect to defective crosspoints) depend on MM, which defines the configuration of the crossbar.

3.2.2 Intra/Inter Block Mapping Algorithm

The proposed logic mapping algorithm consists of two phases: 1- local mapping, 2- global mapping. The local mapping uses intra block transformations, and in case it fails on a block, the global mapping step is executed on the failed blocks, which uses inter block transformations.

- **Local Mapping**
  The local mapping can be done using various algorithms (e.g. ILP formulations [24], cardinality-based and greedy algorithms [30, 31]). Here, we introduce a heuristic algorithm [29] to apply the intra block transformations on each block. The proposed intra block mapping algorithm converts a configuration (MM_1) to a reliable mapping (MM_2). In the beginning, it applies a set of random swapping transformations on the MM_1 to avoid local optima. In each step, the algorithm maps a non-zero entry of MM_1 ($MM_1[i][j]$) into MM_2 ($MM_2[t][k]$), then it

changes $MM_1[i][j]$ to zero. Therefore, the algorithm terminates successfully, when all entries of the MM_1 are zero. After mapping $MM_1[i][j]$ into $MM_2[t][k]$, the row t and the column k of MM_2 are assigned to the input i and the output j, respectively. Therefore, the algorithm starts with the swapping transformation, then it tries to map $MM_1[i][j]$ into $MM_2[t][k]$, where the row t (column k) has been assigned to the input i (the output j). If such row (column) cannot be used (due to reliability constraints), the algorithm assigns a new row (column) for that input (output), using the input duplication (output decomposition) transformation. If the local mapping algorithm fails, then the last updated MM_1 and MM_2 are saved to be used in the global mapping (the inter block mapping algorithm); otherwise, only MM_2 is saved.

In summary, the algorithm repeats the following steps until all non-zero entries of MM_1 are set to zero.

1. Apply a set of random swapping transformations on the MM_1. In this step two rows (columns) are randomly selected and swapped.
2. Assign a weight to each entry of MM_1; If the entry $MM_1[i][j]$ is zero, then its weight is zero. If an entry $MM_1[i][j] = 1$, then its weight is the sum of the weights of the row i and the column j. The number of ones in each row (column) determines its weight.
3. Select an entry of MM_1; A non-zero entry of MM_1 is randomly selected. The probability of selecting an entry is proportional to its weight.
4. Map the selected entry into MM_2; All defect-free crosspoints of the MM_2, which have delay less than the threshold delay, can be used in the mapping of $MM_1[i][j]$. The rows (columns) of MM_2, which have been assigned to the input i (output j), have the highest priority in the mapping of $MM_1[i][j]$. If the entries on such rows (columns) are defective or they have the delay more than the threshold delay, then a new row (column) is assigned to the input i (output j), randomly.
5. Remove $MM_1[i][j]$; After mapping an entry, it will be set to zero in MM_1.

If the number of rows and column of an MM is m and n, respectively, then runtime complexity of step 1 is $O(nm)$. This is because the weight assignment runs on each entry of the matrix. On the other hand, in step 3 the entries on each row are searched. Therefore the runtime complexity of step 3 is $O(nm)$. However, since all steps run for every non-zero entry of MM, therefore, in the worst case there are nm non-zero entries. As a result, the runtime complexity of the local mapping for the worst case is $O(n^2m^2)$.

If the local mapping (intra block mapping algorithm) fails in providing reliable mapping for a block, then the global mapping is applied to that block. Two global mapping algorithms are presented here. The first algorithm (a function decomposition algorithm) uses the function decomposition transformation to map the remaining entries of the failed block into one of the existing blocks (without adding area overhead in terms of using an extra block). However, if in this way a reliable mapping is not found, the second algorithm (the block decomposition algorithm) is executed to decompose the block by exploiting an extra block.

- **Global Mapping**

  The function decomposition algorithm is executed on all of the MM_1 matrices which have at least one non-zero entry. The algorithm tries to map the unmapped entries of each MM_1 into the existing MM_2 matrices. For example, if the expected function of a vertical line of a crossbar (an output port) is $i_1 + i_2 + i_3 + i_4$, but a portion of it (e.g. $i_1 + i_2$) is already mapped into an output (O_1), the algorithm tries to map $O_1 + i_3 + i_4$ into one of the existing MM_2 matrices.

  At first, all of the unmapped or partially mapped outputs are inserted to a list (called *unmapped list*). An output is added to this list if it has at least one non-zero entry in an MM_1. Furthermore, the input set of each MM_2 matrix is determined. This set includes all the inputs which are used in the configuration of the corresponding block. Finally, for each output in the unmapped list a set (called *input set*) is determined. The "input set" of an output includes all the required inputs in the mapping of that output. An input i is added to the input set of an output j if $MM_1[i][j] = 1$.

  The algorithm selects the outputs from the unmapped list, one by one. It tries to map the selected output into one of the existing MM_2 matrices. However, an MM_2 matrix can be used as a candidate for mapping an output if: (1) the "input set" of that output is a subset of the input set of the MM_2 matrix, and (2) the matrix has at least one unused row and one unused column (if needed). The first condition eliminates the extra complexity in the local mapping of the candidate matrix, in terms of adding extra inputs. The unused column is required to map the output (if needed), while the unused row is used to include the already mapped part of the function. If using the local mapping algorithm the output can be mapped into the candidate MM_2, the output is removed from the unmapped list. The algorithm repeats this procedure until all entries of the unmapped list are removed.

  If the crossbar array consists of k cascaded crossbars, the runtime complexity of finding unmapped list will be $O(knm)$. On the other hand, the runtime complexity of finding the input set is $O(knm)$. Since the algorithm repeats for each entry of unmapped list, and in each iteration it searches among the crossbars, the complexity of this algorithm is $O(k^2 nm)$.

  If the function decomposition algorithm cannot successfully find a reliable mapping for a block, the block decomposition transformation is used. The block decomposition partitions the block into two separate blocks, each to be mapped by the local mapping step. However, the efficiency of the block decomposition transformation depends on the partitioning algorithm. Since for a larger number of unused rows and columns in a block the flexibility of the local mapping increases proportionally, the objective of the block decomposition transformation is to maximize the number of unused rows and columns in both blocks.

  In order to maximize the number of unused columns, half of the outputs are mapped into one block and the rest are mapped into another block. Therefore, in each block at least half of the columns will be unused. On the other hand, each shared input between the blocks reduces the number of unused rows by two (one from each block), while each unshared input reduces the number of unused rows only by one. Therefore, the total number of unused rows in the blocks increases by reducing the number of the shared inputs between the blocks. An input is shared between two blocks, if at least one of the outputs in each block is a function of that input. In order to reduce the number of the shared inputs between the blocks, the partitioning algorithm selects the outputs with the maximum number of shared inputs to be mapped into the same block. These outputs must also have the maximum number of unshared inputs with the outputs of another block. Therefore, the partitioning is done based on the *joint differential input dependency* of the outputs. The joint differential input dependency of two outputs is defined as the difference between the number of shared and unshared inputs of those outputs. Furthermore, the joint differential input dependency of an output and a set of outputs is defined as the difference between the number of shared and unshared inputs of that output with all the outputs in that set. The outputs with highest joint differential input dependency are mapped into the same block.

The joint differential input dependency of the outputs can be identified by a bipartite graph. The nodes in each side of the graph correspond to the outputs. For each common input there is an edge between the corresponding output nodes. Therefore, the problem is finding a partition of the bipartite graph which results in two sub-graphs with minimum connectivity.

3.2.3 Simulation Results

We have synthesized a set of MCNC benchmarks by RASP PLA synthesis tool [36] for different sizes of the PLA. The synthesized circuits are mapped into the crossbar array. Table 1 shows the average number of unused rows, columns and crosspoints with respect to different sizes of crossbars. As can be seen in this table, more than 50% of rows and columns as well as more than 85% of crosspoints are unused for typical crossbar mappings.

In order to evaluate the efficiency of the proposed method, the proposed algorithm has been compared to the *Simulated Annealing* (SA) [27] which exploits the swapping operation in the logic mapping. The size of the crossbar arrays were considered

Table 1 Average unused rows, columns, and crosspoints with respect to crossbar size over all benchmarks

Crossbar size	Unused rows (%)	Unused columns (%)	Unused crosspoints (%)
4×4	57	61	87
6×6	60	62	91
8×8	59	59	92
10×10	59	57	92
12×12	60	53	92
14×14	63	55	93
16×16	63	50	94

Table 2 Percentage of successful reliable mapping of blocks (block yield) and the circuits (circuit yield) achieved by Simulated Annealing (SA) and the proposed method

Circuit	SA [27]		Proposed method		
	Block yield (%)	Circuit yield (%)	Block yield (%)	Circuit yield (%)	Overhead (%)
alu4	58.6	0.0	99.9	99	12.0
apex4	95.2	0.0	99.9	99	6.9
b9	97.9	75.7	100	100	0.0
C880	75.31	0.0	100	100	0.0
C1355	46.42	0.0	100	100	0.7
C3540	69.6	0.0	100	100	2.61
duke2	88.2	0.0	100	100	3.2
ex5p	15.1	0.0	100	100	41.3
k2	91.8	0.0	100	100	4.6
rd84	25.8	0.0	100	100	66.3
t481	89.4	0.0	100	100	0.4
Average	68.9	6.9	99.9	99.9	12.5

as 16×16 (16 input 16 and 16 outputs). The size of crossbar affects the runtime of the algorithms. As mentioned in Sect. 3.2.2, the runtime complexity of the proposed local mapping algorithm is $O(m^2n^2)$. Where, m and n are the number of rows and columns of a crossbar, respectively. On the other hand, the runtime complexity of the proposed global mapping algorithm is $O(k^2\, nm)$, where k is the number of crossbars in a crossbar array. Since the timing complexity of these algorithms is polonominal, they can be applied for crossbars with larger dimensions. Each circuit is mapped into 1000 crossbar arrays, and the average results are reported Table 2. The VMs for the crossbars were generated using Gaussian random distribution ($\mu = 50$, $\sigma = 15$), while defects were inserted by uniform random function (defect rate = 20%).

We have reported two yields: (1) *Block Yield*, and (2) *Circuit Yield*. The block yield is defined as the number of reliable mapping of blocks of each circuit to the total number of blocks of that circuit in 1000 mappings. The circuit yield for the mapping of an entire circuit is defined as the number of cases were all the blocks of that circuit are reliably mapped. The timing constraint (threshold delay) of nano-PLA is set to 65 ($\mu + \sigma$).

As seen in Table 2, the proposed algorithm can achieve more than 99.9% block and circuit yields, while SA achieves 68.9% block yield and only 6.9% circuit yield. It must be noted although for some cases the block yield for SA is high (more than 97%) but the yield for reliable mapping of circuit is very low. This is due to the fact that if at least one of the blocks of a circuit cannot be mapped reliably the mapping of the circuit will not be reliable. The area overhead of the proposed method compared to SA is 12.5%, in average. In fact the area overhead of SA is zero, because it does not require any additional crossbar more than what is determined by the synthesis tool.

4 Architecture-Level Approach: Self-Timed Nano-PLA

The use of the nano-PLA for the logic implementation in nano-architectures introduces a set of challenges. One issue is non-deterministic values on the buffered signals during the pre-charge phase. In addition, this architecture is highly vulnerable to variations on the switching delay of diodes at the logic block. It is due to the fact that all the valid outputs of a logic block must be ready before the evaluation phase of the successor blocks starts. On the other hand, the logic mappings on this architecture assumes the triggers on the evaluation and pre-charge signals occur simultaneously for all crossbars. Also, the pre-charge and the evaluation signals must be routed to the entire fabric, which requires wiring at the micro-level. Therefore, a self-timed method, with a localized timing control, is a promising approach to deal with these challenges in this emerging technology.

We propose a self-timed nano-PLA architecture which consists of multiple cascaded blocks [37, 38]. In each block the logic computations on the data signals are done in diode-based PLA-like logic units. Figure 3 shows the structure of a block in this architecture. Each block consists of three major parts:

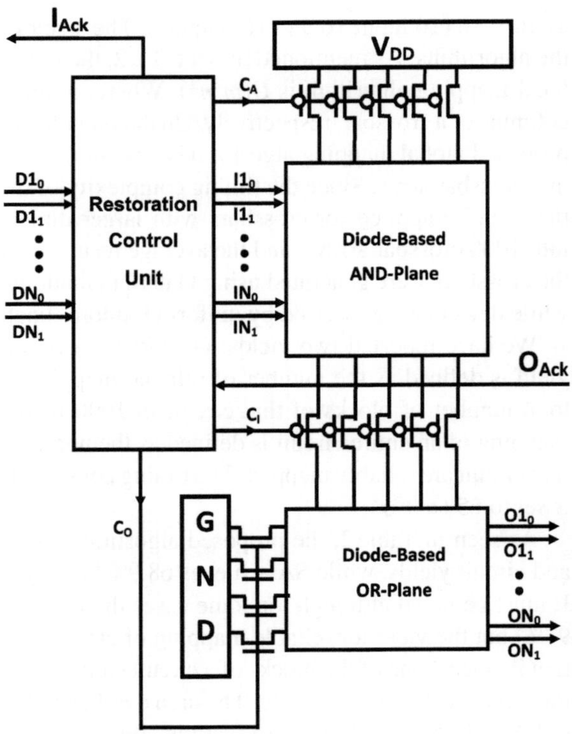

Fig. 3 The structure of a block in the self-timed nano-PLA

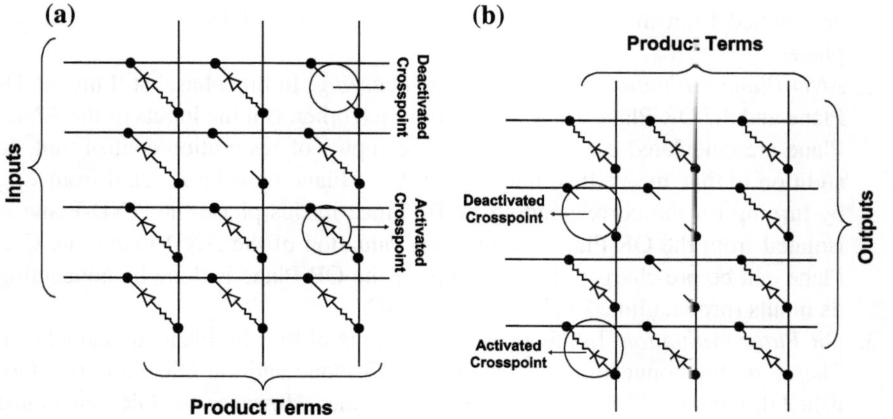

Fig. 4 The structure of logic computation unit: **a** crossbar structure to implement AND logic. **b** crossbar structure to implement OR logic

1. *Restoration/Control unit*. This unit is constructed from non-programmable FET-based devices. Pre-charging and evaluation phases are controlled by this unit. It also restores the signals generated by the diode-based block at the previous stage.
2. *AND-Plane*. A programmable diode-based crossbar is used to implement logical AND operation on the inputs (i.e. to generate the product terms).
3. *OR-Plane*. Another programmable diode-based crossbar, which is fed by the outputs of AND-Plane, produces the OR function on the product terms (sum of products).

Figure 4 shows the general structure of the AND-Plane and the OR-Plane.

4.1 Sequence of Operations

Since diode devices are passive elements, before each logic computation step the diodes in the logic block must be initialized to the appropriate states. In the AND-Plane before each computation, the horizontal and vertical lines (inputs and product terms) must be initialized to V_{DD}. On the other hand, all lines at the OR-Plane must be initialized to *GND*. After this initialization, the computation can be done by the logic block. These initializations and computations (pre-charges and evaluations) are done in three major steps, which determine the sequence of operations in a logic block. These three steps are the followings:

1. *AND-Plane pre-charging*; In this phase the inputs of the AND-Plane must be connected to V_{DD}. In addition to this, the vertical lines of the AND-Plane must be connected to V_{DD} through the upper FETs. Furthermore, the AND-Plane must

be isolated from the subsequent OR-Plane using the FETs between these two planes.

2. *AND-Plane evaluation & OR-Plane pre-charging*; In this phase, still the AND-Plane and the OR-Plane are isolated from each other. But the inputs of the AND-Plane are calculated corresponding to the inputs of restoration/control unit. In addition to this, the vertical lines of the AND-Plane must be isolated from V_{DD} by turning off the corresponding FETs. Since in this phase the AND-Plane is isolated from the OR-Plane, during the evaluation of the AND-Plane, the OR-Plane can be pre-charged. Pre-charging of the OR-Plane is done by connecting its inputs (product lines) and outputs to *GND*.
3. *OR-Plane evaluation*; In this phase, the outputs of the OR-Plane are calculated. Therefore, the connection between the AND-Plane and the OR-Plane is established through the FETs between these two planes. However, the OR-Plane must be disconnected from *GND* through the gate control of the corresponding FETs.

In order to distinguish valid and invalid values on the data signals, we have used dual-rail coding to represent data signals. In this coding scheme, each data bit is coded with two bits (D_0D_1). A valid data is represented by $D_0D_1 = 10$ or $D_0D_1 = 01$. If both are 0 or both are 1, then the data is empty or invalid, respectively. Invalid and empty data are separated to distinguish the pre-charge value from the faulty ones.

Furthermore, there is a signal (acknowledgment signal) between each pair of subsequent blocks. This signal acknowledges the completion of the data computation for the successor block. Using the acknowledgment signal as well as dual-rail coding, the three major operation steps of a self-timed nano-PLA logic block are divided to 6 sub-phases. The order of these sub-phases is as the followings:

1. *Reset the AND-Plane*; The control unit waits for the empty inputs (00 on dual-rail inputs) to reset the inputs of the AND-Plane. However, to reset these signals the acknowledge signal between this block and the successor block (called O_{Ack}) must be 0. It means that the block must ensure that the successor block has consumed its previous input data. Therefore, this block waits for the completion of the computation in the successor block. In this phase, $C_A = 0$ and all inputs of the AND-Plane are connected to V_{DD} through the restoration unit. In addition, $C_i = 1$ to isolate the AND-Plane from the OR-Plane.
2. *Reset the OR-Plane*; Since $O_{Ack} = 0$, the block resets the OR-Plane by forcing C_O to 1.
3. *Acknowledge the previous block*; After resetting the AND-Plane, the control unit acknowledges the previous block by placing 1 on the acknowledge signal (called I_{Ack}) which is between these two blocks. This signal informs the previous block that this block can accept new valid data.
4. *Wait for valid inputs*; After capturing 1 on I_{Ack} by the previous block, it can generate the valid data. The control unit waits for such valid data to generate valid inputs for the AND-Plane. After all inputs become valid, the control unit changes C_A to 1 and calculates the inputs of the AND-Plane.

5. *OR-Plane computations*; The block waits for 1 on O_{Ack} as well as the computation completion of the AND-Plane to start the computation of the OR-Plane. In this phase, $C_A = 1$, $C_I = 0$ and $C_O = 0$.
6. *Acknowledge the previous block*; After completion of the computation, the control unit acknowledges the previous block by changing I_{Ack} to 0. Now, the previous block can generate empty data to start a new round of computation.

4.2 Implementation

The implementation of a logic function on the self-timed nano-PLA can be divided to the implementation of two units: (1) the logic unit, and (2) the control and restoration unit. The control and restoration unit has a fixed (i.e. non-configurable) application-independent implementation, which must be done during the fabrication. However, the implementation of the logic unit is flexible and depends on the logic function, which is configurable.

4.2.1 Logic Unit Implementation

The dual-rail mapping on the logic blocks can be achieved by using conventional PLA synthesis tools. However, these tools are developed for single-rail logic. Therefore, the following steps are performed for the dual-rail conversion.

- Logic decomposition; Logic functions which are described in *Berkeley Logic Interchange Format* (blif) format [39], must be converted to custom-sized PLA blocks. PLAMAP function of RASP synthesis tool [36] can be used to convert the logic function to custom-sized cascaded PLA blocks.
- Truth table generation; Each PLA block produced by PLAMAP is in blif format for the single-rail logic. Therefore, these files must be converted to the equivalent truth table in order to be modified and simplified by the logic optimization tool.
- Dual-rail conversion; In the truth tables, a dual output is added for each output. For each entry in which the output is 1, its dual output is 0. For the remaining entries, the dual output is 1.
- Logic minimization; The truth tables for dual-rail functions are not optimized. Therefore, a conventional logic optimization tool is used to optimize the truth tables. We use ESPRESSO [40] logic optimization tool to optimize the dual-rail truth tables.
- Dual-rail mapping matrices; The configuration of a logical block is determined by the *Mapping Matrix (MM)*. Each logic block is determined by two MMs: the AND-Plane Mapping Matrix (AMM) and the OR-Plane Mapping Matrix (OMM). AMM determines the configuration of the AND-Plane and OMM determines the configuration of the OR-Plane. The rows of an AMM are the inputs of logic block,

while the rows of an OMM are the outputs of that logic block. For both of these matrices, the columns are the product terms.
The outputs of the optimized truth tables are dual-rail outputs, while the inputs are still single-rail. These optimized tables must be converted to AMMs and OMMs. In this conversion, a dual input is added for each input. An input may have three values in the product terms: 1, 0, and x(don't care). If an input is 1 for a product term, there will be a 1 at the row of that input and the column of that product term in the AMM entry, while the entry at the row of the dual of that input and the column of that product term is 0. The situation is similar for 0 entries. If the input is x, both of the entries will be 0. Since, the outputs are dual-rail, the OMMs is directly generated from the truth tables without any modifications.

4.2.2 Restoration/Control Unit Implementation

The implementation of each component of the restoration and control unit is done as follows:

- *AND-Plane input reset and calculation unit*: An input is reset to 1, if the both corresponding data signals are 0 (empty), and O_{Ack} is 0. The pull-up network of Fig. 5a shows the implementation of this condition. The pull-down networks of these circuits implement the input calculation of the AND-Plane. During the reset phase, both signals which are the inputs of the AND-Plane (I_0 and I_1) are set to 1. If the inputs to the block (D_0 and D_1) are changed from empty to valid, the inputs of the AND-Plane are calculated. Assuming D_0D_1 changes from empty (00) to 10, then the pull-down network of I_1 will be connected and I_1 changes to 0. However, since D_1 is 0, therefore I_0 remains at 1.
- *AND-Plane reset*; The AND-Plane is reset if the inputs are empty and $O_{Ack} = 0$, otherwise it will be in the computation mode. Figure 5b shows the implementation of the AND-Plane reset unit.
- *OR-Plane reset*; The OR-Plane is reset if $O_{Ack} = 0$, otherwise it must be at the computation mode ($C_O = \neg O_{Ack}$). In this state, the AND-plane can not accept new data.
- *AND-Plane/OR-Plane isolation*; The AND-Plane and the OR-Plane must be isolated when at least one of them is in the reset mode (i.e. if $C_A = 0$ or $C_O = 1$), otherwise these two planes must not be isolated. Since, $C_O = \neg O_{Ack}$, therefore, the isolation control is $C_I = \neg(C_A.O_{Ack})$.
- *Input acknowledgment signal*; I_{Ack} is 1 if the AND-Plane is reset ($C_A = 0$). When the outputs of the OR-Plane are valid (the computation is completed) and the AND-Plane is not in reset mode, then I_{Ack} returns to 0. The implementation of I_{Ack} control unit has been shown in Fig. 5c.

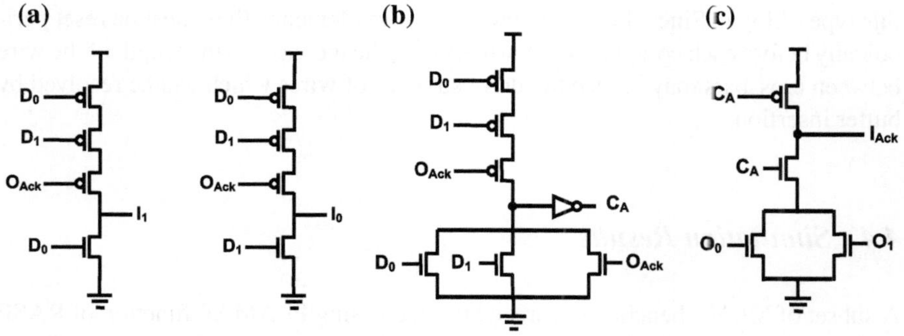

Fig. 5 A part of control implementation: **a** implementation of AND-Plane inputs, **b** implementation of AND-Plane reset, **c** implementation of I_{Ack} generator

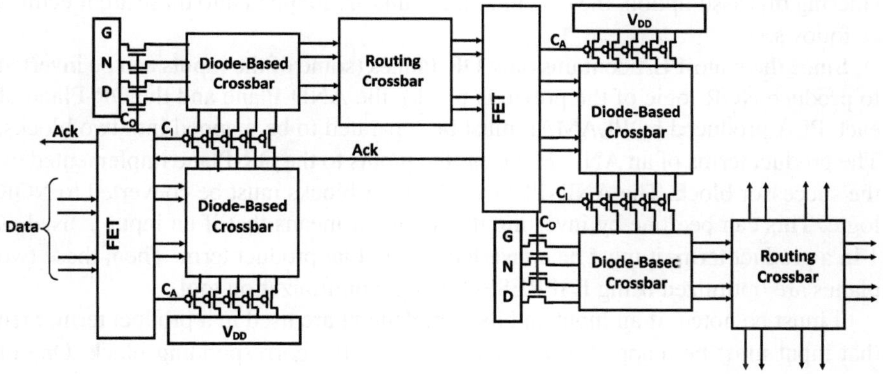

Fig. 6 Two cascaded blocks of self-timed nano-PLA on 2D crossbars for logic and routing units, and lithography based fabricated FET devices

4.3 Discussion

While the lithography-based fabrication is utilized to manufacture FET-based devices, the logic and routing units can fit in high density two dimensional (2D) regular crossbar structure, fabricated using self-assembly nano-fabrication. Two cascaded blocks of the self-timed nano-PLA in a 2D structure are shown in Fig. 6. As seen in the figure, a set of signals controlling the FET devices are placed between the logic units and the pull-up and the pull-down networks (C_A, C_I, and C_O). Since these signals are carried by nano-wires, they can drive a limited number of FET devices. To drive more FET devices, these nano-wires can be replaced by micro-wires or alternatively, each signal can be carried by multiple nano-wires (each nano-wire controls a subset of FET devices).

It must be noted, the self-timed nano-PLA is designed based on the dynamic logic, similar to [2, 13]. Although the dynamic logic has some drawbacks in terms of noise immunity and robustness, however, the diode structure in this technology dictates this type of logic. Since diode devices are passive elements, they must be reset periodically before each computation. Furthermore, the increase on the length of the wire between crossbars may reduce the drive strength of wires which can be resolved by buffer insertion.

4.4 Simulation Results

A subset of MCNC benchmarks are synthesized using PLAMAP function of RASP logic synthesis tool [36]. It converts a logic circuit to custom-sized multi-stage PLAs. In defining the size of PLAs, we assumed the crossbar array of 16×16 crossbars (16 inputs and 16 outputs) for both the nano-PLA and the self-timed architecture. Considering this assumption, the benchmark circuits are mapped into these architectures as follows.

Since the nano-PLA contains only OR-Planes (some of the inputs can be inverted to produce NOR logic of the previous block), the AND-Plane and the OR-Plane of each PLA produced by PLAMAP must be separated to be mapped into two blocks. The product terms of an AND-Plane are the inputs to the OR-Plane implemented on the successor block. The AND-Plane of the PLA blocks must be converted to NOR logic. This can be done by inverting the inputs. It means that if an input is used as 1 in a product term, it must be converted to 0 in that product term. Then, these two planes are optimized using ESPRESSO (a logic minimization tool).

It must be noted, if an input and its complement are used in a product term, then that input must be mapped in two input lines of the corresponding block. One of these lines generates the buffered input for the block, and the other line generates the inverted input. This is due to the fact that nano-PLA supports one of the buffered or inverted inputs at each line. Since, we have assumed all logic blocks have 16 rows (16 inputs), therefore 8 rows of the input lines of a block are reserved for the buffered inputs and the other 8 rows are reserved for the inverted inputs. Therefore, in PLAMAP the number of inputs for each PLA is set to 8. However, the number of product terms and outputs is set to 16.

The mapping matrices of the logic blocks are generated based on the dual rail mapping method presented in Sect. 4.2. Since in the self-timed nano-PLA, the inputs and the outputs are dual rail, in PLAMAP the input and output sizes are set to 8. But, the number of product terms is set to 16.

After mapping the PLA blocks on these architectures, only a subset of lines are used to map inputs or outputs. In Table 3 (I, O, P) shows the average number of activated inputs, product terms, and the outputs for each benchmark mapping into the self-timed nano-PLA. In this table, (I, O) shows the average number of inputs and outputs utilized for mapping into the nano-PLA.

The simulation results (shown in Table 3) compares the proposed architecture with the original nano-PLA [2] in terms of area and variation tolerance.

We have considered same-sized crossbars (the same number of vertical and horizontal lines per crossbar) for both architectures. Therefore, the number of stages

Table 3 Simulation results on a set of benchmarks, Total number of FET devices (*FET*), total number of diode devices (*D*), average number of activated inputs (*I*), products (*P*), and outputs (*O*) per block, number of stages (*S*), variation tolerance (*VT*), the improvements, and overheads

Circuit	Nano-PLA [2]					Self-Timed Nano-PLA				Improvements		Overhead
	FET	D	(I, O)	S	VT (%)	FET	D	(I, P, O)	S	FET (%)	S (%)	D (%)
alu4	18432	8837	(8, 6)	156	38	18720	11659	(14, 10, 4)	78	−1	50	31
b9	3328	1388	(7, 5)	26	46	3120	1745	(14, 8, 5)	13	6	50	25
C1355	6912	3543	(8, 8)	54	46	7200	4714	(12, 12, 6)	30	−4	44	33
duke2	15872	4798	(7, 4)	128	25	15360	6917	(15, 7, 5)	64	3	50	44
ex5p	7680	5298	(9, 8)	60	45	9840	7915	(15, 13, 4)	41	−28	31	49
k2	32256	10210	(9, 5)	252	37	28080	15580	(15, 7, 7)	117	13	53	52
rd84	7424	849	(3, 2)	58	26	6960	4154	(15, 12, 3)	29	6	50	389
t481	20992	5808	(7, 4)	164	18	18480	10482	(15, 8, 4)	77	12	53	80
term1	5120	2148	(8, 5)	40	41	5040	2850	(14, 9, 5)	21	1	47	32
termm	5888	2355	(8, 5)	46	43	5520	3228	(14, 9, 5)	23	6	50	37
vda	12544	5376	(9, 6)	98	37	12000	7669	(15, 9, 6)	50	4	48	42
x4	13824	4100	(8, 4)	108	39	10080	5268	(15, 7, 5)	42	27	61	28
Average	14421	4559	(8, 6)	99	37	11700	6848	(15, 10, 6)	49	18	50	50

and the number of *FET* devices indicates the difference of these two architectures in terms of area. As can be seen in Table 3, the number of stages used in the self-timed nano-PLA is considerably lower than that in the original nano-PLA. This is because in the nano-PLA each of the AND-Plane and the OR-Plane must be mapped in two separate nano-PLA blocks, while in the self-timed nano-PLA, both planes are mapped into the same block. However, the number of stages for the nano-PLAs is not exactly twice since the settings for their synthesis are not the same. The improvement on the number of stages achieved by the self-timed nano-PLA has been shown in the table.

The number of FET devices depends on the size of logic blocks (the number of inputs and outputs), the number of FET devices per block, and the number of stages. The number of FET devices per block in the self-timed nano-PLA is more than that in the nano-PLA. However, due to the reduction in the number of stages in the self-time nano-PLA, the total number of required FET devices for the entire circuit mapped into the self-timed nano-PLA is 18% (in average) less than that for the nano-PLA.

This table also shows the total number of activated diodes in each architecture. Since, in the self-timed nano-PLA the dual of the outputs must be computed, the number of activated diodes is more than that for the nano-PLA architecture. However, since the size of crossbars are same for both architectures, this does not translate to an actual area overhead. This means a better crosspoint utilization in self-timed nano-PLA. Nevertheless, it must be noted that more crosspoint utilization (more activated diodes) results in higher power consumption.

Extensive Monte Carlo simulations for delay variations have been done to estimate the variation immunity of the benchmark circuits implemented on these architectures. For these simulations, the average delay of the blocks of the nano-PLA in 1000 simulation runs is used to determine the periods of pre-charge and evaluation signals. Then, if the delay of a block exceeds the period of evaluation signal, that block fails the timing constraint of the nano-PLA. The average number of the cases where the nano-PLA can tolerate variation over the total number of cases, *variation tolerance*, is shown in *VT* column. For all of these simulations, our propose self-timed architecture can achieve 100% variation tolerance compared to only 37% in the original nano-PLA architecture.

5 Summary and Conclusions

Various emerging technologies such as crossbar nano-architectures have been investigated as the potential solutions to overcome the challenges faced by the conventional CMOS technology at nanoscale. However, the atomic scale of devices as well as poor control in the self-assembly nanofabrication raises reliability issues for these emerging nanotechnologies in terms of manufacturing yield, predictability in the presence of extreme process variation, testing, and runtime reliability.

In this chapter, we proposed a set of methods to address the reliability concerns in crossbar nano-architectures. The proposed methods include an architecture-level approach by introducing the self-timed nano-architecture as well as a set of logic mapping techniques to provide defect and variation tolerance. The proposed self-time crossbar nano-architecture allows us to eliminate global clocking signals, by replacing them with local handshake signals, to reduce the circuit vulnerability to delay variations. While, in the logic mapping approaches, different configurations of a logic function on a crossbar nano-architecture are explored to provide variation and defect tolerant mapping. The simulation results indicate that these techniques improve the reliability of the circuits significantly, which enables the realization of high yield circuits in this emerging nanotechnology.

References

1. Prasek J et al (2011) Methods for carbon nanotubes synthesis review. J Mater Chem 21(40):15872–15884
2. Dehon A (2005) Nanowire-based programmable architectures. ACM J Emerg Technol Comput Syst 1:109–162
3. Rueckes T, Kim K, Joselevich E, Tseng GY, Cheung C, Lieber CM (2000) Carbon nanotube-based nonvolatile random access memory for molecular computing. Science 289(5476):94–97
4. Liu B (2010) Advancements on crossbar-based nanoscale reconfigurable computing platforms. In: IEEE international midwest symposium on circuits and systems, pp 17–20
5. Sverdlov VA, Walls TJ, Likharev KK (2003) Nanoscale silicon mosfets: a theoretical study. IEEE Trans Electron Dev 50(9):1926–1933
6. Lu W, Lieber CM (2007) Nanoelectronics from the bottom up. Nat Mater 6:841–850
7. Yang C, Lu W, Lieber MC, Wu Y, Xiang J (2003) Single-crystal metallic nanowires and metal/semiconductor nanowire heterostructures. Nature, 430(6995):61–65
8. Fujita S, Okajima M, Lee TH, Wong H, Nishi Y, Paul BC (2007) Impact of a process variation on nanowire and nanotube device performance. IEEE Trans Electron Dev 54(9):2369–2376
9. Zheng Y, Huang C (2009) Defect-aware logic mapping for nanowire-based programmable logic arrays via satisfiability. In: Design, automation test in Europe, pp 1279–1283
10. Cui Y, Lieber CM (2001) Functional nanoscale electronic devices assembled using silicon nanowire building blocks. Science 291:851–853
11. Stan MR, Franzon PD, Goldstein SC, Lach JC, Ziegler MM (2003) Molecular electronics: from devices and interconnect to circuits and architecture. Proc IEEE:1940–1957
12. Snider GS, Williams RS (2007) Nano/cmos architectures using a field-programmable nanowire interconnect. Nanotechnology 18(3):035204
13. Manem H, Rose GS, DeHon A, Gojman B (2009) Inversion schemes for sublithographic programmable logic arrays. Comput Digit Tech IET
14. Likharev KK, Strukov DB (2005) CMOL: devices, circuits, and architectures. Introducing molecular electronics. Springer, pp 447–477
15. Strukov DB, Likharev KK (2005) CMOL FPGA: a reconfigurable architecture for hybrid digital circuits with two-terminal nanodevices. Nanotechnology 16:888–900
16. DeHon A (2003) Array-based architecture for fet-based, nanoscale electronics. IEEE Trans Nanotechnol 2:23–32
17. Joshi MV, Al-Assadi, WK (2007) Nanofabric PLA architecture with redundancy enhancement, pp 427–438
18. Tahoori MB (2006) Application-independent defect tolerance of reconfigurable nanoarchitectures. J Emerg Technol Comput Syst 2(3):197–218 July
19. Rao W, Orailoglu A, Karri R (2007) Logic level fault tolerance approaches targeting nanoelectronics plas. In: Design, automation test in Europe, pp 1–5

20. Shrestha ST, Ueno S. Orthogonal ray graphs and nano-*PLA* design. In: IEEE international symposium on circuits and systems, pp 2930–2933
21. Hogg T, Snider G (2007) Defect-tolerant logic with nanoscale crossbar circuits. J. Electron. Test. 23:117–129
22. Garey MR, Johnson DS (1979) Computers and intractability: a guide to the theory of NP-completeness. Freeman & Co, New York
23. Yang JS, Datta R (2011) Efficient function mapping in nanoscale crossbar architecture. In: 2011 IEEE international symposium on defect and fault tolerance in VLSI and nanotechnology systems (DFT), pp 190–196
24. Zamani M, Mirzaei H, Tahoori MB (2013) ILP formulations for variation/defect tolerant logic mapping on crossbar nano-architectures. ACM J Emerg Technol Comput Syst
25. Paul BC, Fujita S, Okajima M, Lee T, Wong HSP, Nishi Y (2007) Impact of process variation on nanowire and nanotube device performance. In: Device research conference, 2007 65th annual, pp 269–270
26. Gojman B, DeHon A (2009) *VMATCH*: using logical variation to counteract physical variation in bottom-up, nanoscale systems. In: International conference on field-programmable technology, pp 78–87
27. Tunc C, Tahoori MB (2010) Variation tolerant logic mapping for crossbar array nano architectures. In: Design automation conference Asia and South Pacific, pp 855–860
28. Tunc C, Tahoori MB (2010) On-the-fly variation tolerant mapping in crossbar nano-architectures. In: VLSI test symposium, pp 105–110
29. Zamani M, Tahoori MB (2012) Reliable logic mapping on Nano-PLA architectures. In: Proceedings of the great lakes symposium on VLSI (GLSVLSI), pp 107–110
30. Zamani M, Tahoori MB (2011) Variation-aware logic mapping for crossbar nano-architectures. In: Asia and South Pacific design automation conference (ASP-DAC), pp 317–322
31. Zamani M, Tahoori MB (2011) Variation tolerance for Nano-PLA architectures. In: 20th IEEE North Atlantic test workshop (NATW)
32. Zheng Y, Huang C (2009) Defect-aware logic mapping for nanowire-based programmable logic arrays via satisfiability. In: DATE, pp 1279–1283
33. Orailoglu A, Rao W, Karri R (2009) Logic mapping in crossbar-based nanoarchitectures. IEEE Des Test Comput 26(1):68–77
34. Ugurdag H, Goren S, Palaz O (2011) Defect-aware nanocrossbar logic mapping through matrix canonization using two-dimensional radix sort. ACM J Emerg Technol Comput Syst, 7(3):12:1–12:16
35. Choi M, Yellambalase Y, Kim Y (2006) Inherited redundancy and configurability utilization for repairing nanowire crossbars with clustered defects. In: DFT, pp 98–106
36. Cong J, Chen D, Ercegovac M, Huang Z (2001) Performance-driven mapping for CPLA architecture. In: Proceedings of the ACM international symposium on FPGA, pp 39–47
37. Zamani M, Tahoori MB (2011) Self-timed nano-PLA. In: International symposium on nanoscale architectures (NANOARCH), pp 78–85
38. Zamani M, Tahoori MB (2010) A transient error tolerant self-timed asynchronous architecture. In: IEEE European test symposium (ETS), pp 88–93
39. http://embedded.eecs.berkeley.edu/research/vis//usrdoc.html
40. Brayton R et al (1984) Logic minimization algorithms for VLSI synthesis. Kluwer Academic Publishers, Boston, MA

Part V
Medical Applications of Nanoscale Communication

Effect of Aging, Disease Versus Health Conditions in the Design of Nano-communications in Blood Vessels

Luca Felicetti, Mauro Femminella, Pietro Liò and Gianluca Reali

Abstract This chapter illustrates the analysis of a nano-communication system, implemented in blood vessels, designed for detecting tumor cells This system may be used for diagnostic purposes in the early stage of a disease or to check any relapse of a previous disease already treated.The tumor detection happens through revealing tumor biomarkers, such as the CD47 protein, on the cell surface. Once a biomarker has been revealed, a molecular communication system distributes the information to a number of nano-machines having a size allowing them to flow through the vessel at the maximum speed of the bloodstream. The final information detection is extra-body, and based on a smart probe, which triggers a decision tree computing which aims to find if any tumor is present and the most likely location. Effects of aging and serious disease, such as the diabetes, are highlighted.

1 Introduction

Nanoscale communications is a rapidly emerging research area, promising innovative applications in many fields [1, 2], including medicine [3, 4]. Even if the research on the design of biocompatible nanomachines has been active for many years [5], organizing effective and reliable communications among them is still

L. Felicetti · M. Femminella (✉) · G. Reali
Department of Engineering, University of Perugia Perugia, 06125 Perugia, Italy
e-mail: mauro.femminella@unipg.it

L. Felicetti
e-mail: ing.luca.felicetti@gmail.com

G. Reali
e-mail: gianluca.reali@unipg.it

P. Liò
Computer Laboratory, University of Cambridge, 15 JJ Thomson Avenue, Cambridge CB3 0FD, UK
e-mail: pl219@cam.ac.uk

J. Suzuki et al. (eds.), *Modeling, Methodologies and Tools for Molecular and Nano-scale Communications*, Modeling and Optimization in Science and Technologies 9, DOI 10.1007/978-3-319-50688-3_19

challenging, especially when they have to be established within a human body. Hence, in order to design nanomachines able to exploit the communications potentials of nanoscale environments, it is necessary to identify and analyze the basic communications mechanisms available and how they are affected by common biological phenomena such as aging and diseases, since they can significantly alter the basic parameters affecting the information exchange.

A special environment for implementing nanoscale communications is the cardiocirculatory system. In fact, the continuous monitoring of the concentrations of specific parameters in the blood could greatly help doctors detecting and analyzing potentially critical heath conditions. For example, through the information exchanged both within vessels and with the external, it is possible to both initiate the early therapies in standard ways, and release of specific medicines by nano-actuators deployed in the human body. Since the blood requires about one minute to complete the large and small circulation, it is an extraordinary medium to quickly transport information to different parts of the body. Thus, understanding the available communications mechanisms in the blood, along with their dependence on other biological phenomena, is of great interest.

In this chapter we illustrate a proposal for detecting tumor cells through the interception of biomarkers by mobile sensors circulating in the blood flow. The captured information is delivered to the intended recipients through a molecular communication system. This system involves mobile transmitters and mobile receivers.

Consider that the cardiocirculatory system is intrinsically a multiscale system, due to the different size of particles and vessels involved. Our analysis approach is coherently multiscale, involving tissues, cells, molecules, within a vessel length up to 4 mm. As for the section of the vessel, our analysis will consider also microcirculation, which is the blood delivery to the smallest blood vessels. This environment is particularly important since it is mostly affected by the diseases, such as diabetes, that restrict the actual size of the vessels and modify the elastic properties of blood vessels.

The analysis illustrated provides a model showing the response of the nano-communication system to a pulse of carriers released by a transmitter, measured at both different receivers, located along the vessel wall and by other mobile receivers. This analysis has been performed by using the BiNS simulator [6], which is a simulator of biological nano-communication systems, the operation of which has been tuned by using the results of in-vitro experiments. The latest release of BiNS allows simulating bounded spaces, through the use of parallel simulation techniques [7, 8] to effectively handle the very large number of events happening within a portion of a blood vessel. On the basis of the achieved results, we show a simple but effective scheme for implementing a biocompatible digital receiver, able to auto-synchronize and decode the information received by means of repeated pulses of information molecules [9, 10].

A further objective of this work is to analyze the dependence of the information propagation features in blood vessels as a result of biological process due to aging and diabetes. In fact, as mentioned above, these processes may significantly alter some fundamental parameters such as the elasticity of the endothelium. Thus the analysis will also consider the link between mechanical and diffusion features with communication features.

Section 2 illustrates the medical objective that has stimulated our research. Section 3 reports the state of the art for what concerns molecular communications and simulated biological environment. Section 4 reports some details about the simulation platforms, the models used therein, together with the simulation results. Finally, our conclusions are reported in Sect. 5.

2 The Medical Scenario and Research Objectives

It is well known that in healthy conditions the immune system can eliminate occasional circulating tumor cells (CTCs) in blood vessels, thus preventing the development of a disease. Differently, in pathologic conditions, the early detection of tumor cells is essential and regarded as fundamental component of the so-called P4 medicine (proactive, personalized, predictive, and participatory) [11]. Monitoring CTC concentration is also a method for detecting any relapse of a disease after a treatment considered successful.

CTCs originate from a primary tumor site. Though the blood stream they can metastasize and reach any part of the body, thus spreading the disease. Thus, the knowledge of the number of CTCs in the bloodstream provides an effective diagnostic and prognostic estimate of the location and progression of an existing cancer.

Recently, some researchers have implemented a micro-fluidic chip capable of capturing CTCs with 90% success rate [12]. This achievement, and similar ones in the field [13–15], have stimulated further the research activities.

Typically, the detection of CTCs happens through revealing the presence of biomarkers on the cell surface. In addition, in order to gain a deep understanding about the nature and the possible location of a tumor, future cancer medicine will benefit from evaluating more than one marker at a time, which may possibly create a more precise index for diagnosis and prognosis, based on a computer-based decision tree.

The most important biomarkers are illustrated in what follows.

A very important class of biomarkers, which has driven a significant research effort, is generated by some basic mechanisms through which tumor cells try to avoid macrophages, which are highly specialized in removal of dying or dead cells and cellular debris. They have also a specific role in the immune system for eliminating pathogens.

For example, tumors try to evade the immune system by exploiting the regulatory mechanisms that protect healthy cells from immune-mediated attacks [16]. In particular, the CD47 protein, an immunoglobulin (Ig)-like receptor, is exposed by many cell types on their surface in order to indicate to macrophages that they should not be destroyed. Most or all cancers over-express CD47. The binding of CD47 expressed on the cell surface with the signal regulatory protein-α (SIRP-α) protein carried by macrophages allows tumor cells to escape the macrophage's mediated innate immune response. Tackling this mechanism through modifying the mutual relationship between CD47 and SIRP-α is an active research area [17]. Current nanomedicine approaches consider RNA interference (RNAi) technology by means of liposomes made with protamine-hyaluronic acid and loaded with anti-CD47 siRNA. This has resulted in an efficient silencing of CD47 and tumor regression [14].

It is worth mentioning other important biomarkers, such as circulating microRNA-101 for hepatocellular carcinoma [18], plectin, for pancreatic cancer [19], Apolipoprotein C-II for cervical cancer [20], CD164 protein for ovarian cancer [21], plasma osteopontin, for non-small cell lung cancer [22], and Carcinoembryonic antigen (CEA) for different types of lung cancer [23].

The medical research areas mentioned above can be successfully fertilized by biological nano-communication technologies. This is a quite novel research area [24–28], the development of which requires the combined expertise of ICT and biological researchers.

Our proposal contributes to the research objectives mentioned above and consists of a CTC detection and communication system including sensors detecting tumor cells, actuators transmitting the collected information, and receivers, as illustrated in what follows.

The typical concentration of tumor cells in the early stages of a disease, or relapse of it, makes their detection from a fixed receiving point extremely difficult. For this reason, as mentioned above, a fundamental component of our proposal, inspired by the state of the art in the relevant technologies, consists of making use of mobile nano-sensors/detectors free to move within the blood vessels.

The mobile CTC sensors are assumed to be made of biodegradable components. Their role is to increase the system's sensitivity of identifying CTCs at very low density, i.e. when their density is still small enough to allow defining a successful treatment. The biodegradable composition of CTC sensors may be suitably used for avoiding any possible accumulation of sensors in lymph nodes or elsewhere. In particular, after the introduction of a number of sensors/detectors in the blood stream and their usage as CTC detectors, we assume their elimination by injecting the enzyme able to degrade sensors. The subsequent actions of antibodies and macrophages will eliminate them from the blood circulatory system.

The detected information has to be delivered to a specific probe. According to the biomarkers detected, the subsequent processing has the aim of driving further diagnostic analysis. The simplest approach consists of using an extra-body probe. Nevertheless, the capacity of this smart probe of capturing the circulating CTC detectors depends on their concentration. Since the additional number of CTC in the early stage of a disease could be small, even the number of captures could be not enough to be clearly distinguished from the normal concentration in healthy conditions. Thus, it is necessary to use them to stimulate a more powerful process having the aim of transferring the available information to the maximum number of particles at the highest speed.

For this purpose, a molecular communication system is needed in order to distribute the point-wise information collected from the sensors to larger particles. The communication system, illustrated in what follows in more detail, consists of the emission of a burst of carrier molecules by the mobile sensors upon a CTC detection. These carriers propagate within the blood vessel and are hopefully received by all compliant receivers passing through the cloud of the propagating carriers. The objective is to maximize the number of receivers able to receive these propagating carriers. This way, receivers are made aware of the detected CTCs and can carry this information to the smart probe. Clearly, the higher the number of CTCs detected, the higher the alarm level that they can generate.

Since the detection of CTCs is assumed to happen by a mechanism that can be assimilated to a cell particle absorption, the size of the freely moving sensors cannot cause obstruction of vessels. In addition, they can be implemented by using biodegradable material. In our system, we assume that the smaller sensors, pushed through the blood flow, can detect the cells having their surface populated by the CD47 protein, or other biomarkers, since it can be put in relation with a number of tumors. In fact, it was shown that blocking CD47 can reduce tumor growth by enabling macrophages to eliminate cancer cells [29]. In simply words, expression of the CD47 by a cell has the effect of communicating to macrophages a "don't eat me" signal [16, 29]. In fact, macrophages can recognize CD47 molecules by a signal regulatory protein-α (SIRPα), also known as SHPS1. It acts as a receptor for CD47, thus blocking the phagocytosis of macrophages, as schematically shown in Fig. 1.

Hence, the antibodies able to block CD47–SIRPα interactions can allow phagocytosis of tumour cells by macrophages [17, 16].

In order to allow such treatments, detection of cells exposing the CD47 is of paramount importance. This detection happens through a physical contact of the tumor cells and sensors.

Detection of sensors can be implemented by exploiting the RNA interference (RNAi) technology [14]. We can assume the usage of RNAi to implement the detection mechanisms in mobile sensors of the CD47 proteins, by associating it to the antibody of CD47.

Fig. 1 Tumor cell blocking phagocytosis though CD47 exposure

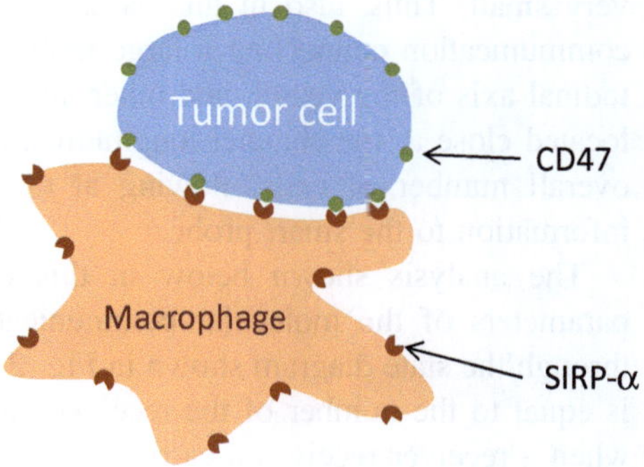

Each time a sensor detects a tumor cell, it updates the stored information by comparing the number and the frequency of detections against a number of thresholds and a decision tree. Note that the implementation of the decision tree could be biological. For example, two detections of the same type of stimulus in a given time window can trigger the production of specific protein expressed in the biosensor surface. This way, the current status of the sensor reflects the likelihood of the tumor growth. The key aspect is that this information could regulate/trigger the further release of liposomes-loaded siRNA anti CD47.

In regard to the capture of CTCs and similar information in the tumor site, some considerations about the size of the sensors are necessary. It is known that larger particles, such as white blood cells, tend to occupy the central portions of blood vessels, in proximity to the longitudinal axis. Differently, smaller particles, such as platelets, are pushed towards the endothelium. CTC are typically large cells, having a size similar to that of white blood cells. For this reason, small sensors/detectors are more suitable to get in touch with the tumor site, whilst larger sensors can get in touch with CTCs more frequently.

The use of different sensor sizes is thus envisaged. In addition, the search for different biomarkers requires the use of different types of sensors, each able to detect a specific biomarker.

Given the small size of the sensors needed for getting in touch with the tumor site, when they flow through the blood circulatory system, they are pushed towards the endothelium. Since the flow speed at the endothelium is very low, and the number of detections should be (hopefully) small in the early stage of a disease, it follows that the possibility of being detected in a reasonable time (few minutes or even one hour) by a smart probe is very low. For this reason, we propose to establish a molecular communication between a small, low-speed sensor, which is the transmitter, and many other larger particles, flowing at a significantly larger speed, located close to the longitudinal axis of the blood vessels. This way, the relatively large number of bigger particles, which have a size similar to the white blood cells, can spread the information through the body with a concentration significantly larger than that of CTCs, thus making their detection faster and more reliable.

It is also worth noting that the number of larger sensors that have collected the desired information, designed to interact with the large circulating CTCs, could be very small. Thus, also in this case, it is necessary to make use of a molecular communication connecting a large mobile transmitter, flowing close to the longitudinal axis of the vessel, and other large nearby cells, having a similar size, also located close to the channel longitudinal axis. This way, we aim at increasing the overall number of cells, flowing at the maximum speed, carrying the desired information to the smart probe.

The analysis shown below in this chapter aims to determine the suitable parameters of the molecular communication system. We will model a receiver through the state diagram shown in Fig. 2. We assume that the state of the receiver is equal to the number of the received carriers in a given time window W. Thus, when a receiver receives a carrier, the state value of the receiver increases, while

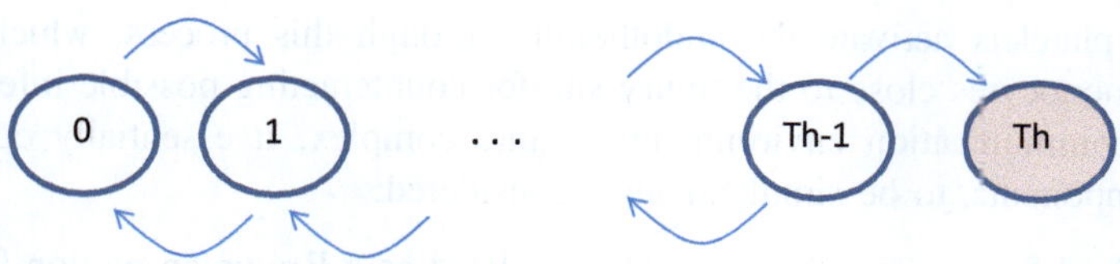

Fig. 2 Model of the receiver behavior

after *W* time units since the reception of a carrier the state value of the receiver decreases. The state *Th* is an absorbing state, and we call the receiver *activated*. An activated receiver is a receiver that carries the information "*tumor detected*", which should be delivered to the smart probe.

The use of a threshold *Th* values significantly higher than 1 is due to the need of avoiding false detections, which may happen when the CTCs are detected in healthy people. They can be regarded as a noise component affecting the detection process. The numerical results shown in what follows will consider different *Th* values and will determine the number of receivers that can be activated for different burst sizes of the transmitted carriers.

We first show the numerical results of a communication system implemented within a healthy, sufficiently young person so as not to include aging effects. Then, we will repeat the numerical experiments with the aim of analyzing the achievable performance of the molecular communication system when it is implemented in elderly people, affected by stiffness of blood vessels and in increased blood pressure, and people affected by a serious disease such as the diabetes. In the latter case, some different effects may appear due to an increased glucose concentrations in the cardio-circulatory system. In particular, lipids and glucose may combine and cause soft deposits over the endothelium and, in very serious cases, on particle surface. This way, the interactions in blood vessels are significantly altered, with a direct effect on the carrier propagation. These effects need a specific analysis for gaining a deep understanding, since could potentially modify the detection capabilities of the system illustrated in this chapter.

3 State of the Art

3.1 The Biological Environment

Our research activities on communication systems within the bloodstream were inspired by the signaling system existing between platelets and endothelial cells. Platelets can release specific proteins, known as CD40L [30], which can bind to the relevant receptors (CD40) present on the endothelial surface. In normal conditions, this process regulates the response to injuries in vessels. In order to repair vessel damage, platelets activate the endothelium through this process, which in turn recruits monocytes close to the injury site for counteracting possible infections.

This communication environment is quite complex. It essentially consists of three components, to be simultaneously considered:

- particle diffusion, which is typically modeled as a Brownian motion [31];
- positive drift caused by the bloodstream, the extent of which depends on the distance of the considered particle from the longitudinal axis of the vessels. Clearly, the drift is maximum on the axis and minimum at the vessel walls. Two well established drift models are the Poiseuille or Casson laws [32]);
- interactions (collisions) between particles in blood vessels and between particles and vessel walls.

Whilst different theoretical diffusion models have been proposed [31] also including the effects of drift (see [32] and references therein), a comprehensive analysis of all these effects has been published only in recent times [33]. It was made evident that the red blood cells are those that most influence the particle motion, due to their number and size. The latter is slightly inferior to only that of the white blood cells. The resulting effect is that smaller particles, such as platelets, are pushed close to the vessel walls.

3.2 Molecular Communications

In molecular communications, the typical signal propagation occurs by way of particle diffusion through a fluid medium, which is usually modeled as a Brownian motion. Mathematical models of transmitter, receiver, and channel, are proposed in [28], where nodes are assumed to be fixed. Information is associated with the concentration of the emitted particles. Noise components affecting the diffusion-based molecular communications are analyzed in [25]. The number of the so-called bound receptors, which are receptor-ligand pairs, or complexes, is considered the input signal to the receiver [34]. The first example of a complete protocol for molecular communications is presented in [35], which extends [36]. An example of intra-cell communications is described in [37]. All these papers consider diffusion-based propagation only, which is a model that cannot be used in our reference environment.

A slightly different model is illustrated in [24], where the transmitter is assumed to sit within a fluid and emits a burst of molecules, which diffuse in the fluid through a Brownian motion, until a receiver capable of measuring their arrival times absorbs them. The particle release time is associated with the transmitted information as in a pulse position modulation. The scenario shown in [27] is slightly different. Two communicating nanomachines and emitted molecules propagate through a fluid medium. Their motion includes both a drift velocity and Brownian component. The communication model consists of the release of one or two molecules within the medium. The same authors show in [26] that the additive inverse Gaussian noise model is appropriate for molecular communication channels in fluid media with drift. They derived an upper bound and a lower bound of channel capacity and proposed a maximum likelihood receiver model. Nevertheless, the propagation model does not include the collisions, which are of paramount importance in our scenario, as observed also in [33].

Finally, a recent work describes a complete view of a layered network architecture of molecular communications [38]. Following the layered architecture of traditional communication networks such as the Open Systems Interconnection model (OSI) and the TCP/IP reference model, it develops a formal model for each layer, explains how each layer behaves, and identifies potential research directions for each layer.

4 Communication Model and Numerical Analysis

4.1 Model of Particle Motion in Blood Vessels

Blood vessels under consideration are assumed to be at a distance from the heart so as to allow modeling the bloodstream without turbulence. This way, the flow properties, such as velocity and pressure, can be considered constant over time, and the resulting motion laminar. In particular, in the longitudinal direction the vessel can be viewed as a set of concentric cylinders. The space between concentric cylinders is a lamina, and a laminar flow consists of fluid particles moving in straight lines in each lamina.

The velocity profile v of this laminar flow was shown to be parabolic, and is expressed by the well-known Hagen–Poiseuille equation, which can be derived from the Navier-Stokes equations [39–41]:

$$v(r) = \frac{1}{4\mu}\frac{\Delta P}{L}\left(R^2 - r^2\right) \tag{1}$$

where R is the vessel radius, μ is the dynamic fluid viscosity, ΔP is the pressure decrease happening through a vessel section of length L, and r is the transversal distance from the longitudinal axis of the vessel. Table 1 reports some known parameters for a sample blood vessel.

It is worth noting that this velocity profile needs some adaptation when large vessels are considered. In particular, some authors consider the Casson profile more accurate [33].

This model generalizes the motion model of particles within a fluid, typical of blood vessels. In particular, the existing flow pushes particles, thus creating a drag effect. The drag force F_d is given by the Stokes Law [39] and can be applied on a particle with a low Reynolds number (such as red blood cells) in a continuous viscous fluid [42]:

Table 1 Simulation parameters

General parameters		Elasticity	
Vessel length	4.00 mm	Restitution coefficient of vessel wall (normal)	0.6
Transversal diameter	60 μm	Restitution coefficient of vessel wall (healthy aging)	0.9
Mean flow velocity	0.5 mm/s	Restitution coefficient of vessel wall (diabetes)	0.2
Viscosity	1.3 mPa × s	Longitudinal attenuation (diabetes)	0.7
		Target cells/white blood cells	
Temperature	310° K	Concentration	4 × 10^3 U/mm^3
Transmitter nodes		# CD40 receptors	5000
Radius	2.8 μm	Radius	5 μm
Burst size (thousands of carriers)	3000, 5000, or 10000	*Red blood cells (RBC)*	
		Concentration	4 × 10^6 U/mm^3
Carrier radius	1.75 nm	Radius	3.5 μm

$$F_d = 6\pi\mu R v_p \tag{2}$$

where v_p is the relative velocity of the particle with respect to the flow, given by (1). By using this value of the drag force, it is possible to estimate the acceleration and thus the velocity of particles. In addition, a diffusion component must be considered. It is modeled by the Brownian diffusion coefficient D_m of a quiescent fluid [43]:

$$D_m = \frac{k_B T}{6\pi\mu a} \tag{3}$$

where k_B is the Boltzmann constant, a is the particle's radius, and T is the temperature in Kelvin.

Since the Reynolds number for nano particles is high, Eq. (2) cannot be used as it is. It is necessary to consider both convection and diffusion effects in addition to the velocity component (1), governed by the following equation [32]:

$$\frac{\partial C}{\partial t} + v\nabla C = D_m \nabla^2 C \tag{4}$$

where C is the particle concentration, v is the fluid velocity vector, and D_m is the diffusion coefficient. A similar model, alternative to solving Eq. (4), has been proposed in 1950s and further investigated in other papers, such as [32] and [44]. It consists of the using an effective longitudinal diffusion coefficient D_{eff}, applied only along the propagation direction. A recent study, [33], considers the presence of the blood cells by a numerical analysis validated by experimental results. It shows that these cells can significantly influence nano-particle propagation and assimilations, so it is proposed to include them explicitly in the model [33].

In fact, what happens is that larger and heavier cells move essentially along the longitudinal vessel axis, whilst the smaller elements, such as platelets, when collide with red blood cells are pushed towards the vessel walls. They remain confined close to walls since they find a less obstructed propagation path. The collisions between particles or between particles and endothelium can be modeled as inelastic, by using the values of the coefficient of restitution reported in [45]. Thus, the simulation analysis presented in what follows integrates the results shown in [33] by modeling collisions between nano particles and white and red blood cells. This is clearly a significant contribution for suitably addressing the propagation of the information carriers.

4.2 The BiNS2 Simulator

This section, illustrates the main functions implemented in the BiNS2 simulator, particularly those important for the model; the interested reader can find more details in [6–8].

BiNS2 allows modeling biological entities, which are regarded as nodes, either transmitting or receiving carriers, or affecting the signal propagation in a molecular-based communication. It is possible to model key properties of the communication channel, which is the blood stream in this chapter, in some detail.

The simulator is implemented in Java. The program classes, such as nodes and carriers, implement and integrate a general abstract class, called *Nano Object*. Although they share general features, they also expose different functions. This way, for any scenario, it is possible to differentiate the node object type at any time, thus obtaining a multitude of different node objects. The same properties are available for modeling carriers as *Nano Objects*.

Simulations proceed in discrete time steps, each including different phases, in which objects execute the operations relevant to their specific behavior. The most significant phases are:

- transmission, which executes the carrier emission;
- reception, which implements carrier assimilation;
- information processing, the behavior of which, depending on the received signal intensity, is evaluated as the number of received carriers in a time step;
- motion, which implements the mobility models of nodes and carriers;
- object destruction, introduced for removing objects due to either lifetime expiration or their exit from the region of interest;
- collision management, which includes a collision check phase and a relocation phase, used to locate nodes and carriers after an (in)elastic collision.

A collision can result in either a bounce or an assimilation. The latter happens when a carrier comes in contact with a compliant receptor located on a node surface.

The simulated spatial region can be either unbounded or bounded by a configurable surface. Since in this chapter the simulated environment is a blood vessel, we have bounded the simulated region by a cylindrical surface.

The most computationally intensive phases, such as the collision management and motion phases, are implemented by parallel threads. The list of simulated nano objects is split into smaller lists, which are handled in parallel by a suitably synchronized thread pool. This approach is clearly effective in multicore platforms, since they can efficiently execute different simultaneous threads. The larger the number of available cores, the larger is the number of sub-lists that can be generated, and the smaller is the number of objects in each of them. It is fundamental to determine the suitable number of threads and their workload distribution, in order that a single thread does not hamper the execution of the other threads in a time step. It is also clearly necessary to take into account the management overhead of each thread. A rule of thumb determined experimentally suggests using a number of threads larger by 4 to 8 times than the number of available CPU cores.

The simulated region is handled hierarchically, by splitting it into smaller domain volumes. Each domain is aware of the nano objects it embraces and manages their mobility and lifetime independently of what happens in the other domains. This approach is useful for simulating different regions of the simulated environment showing different features.

In regard to the CPU time assigned to each domain, it is necessary to take into account of the need to implement collision management, which is the most computational intensive phase. In fact, the use of domains requests the differentiation of the collision types. Just to give an idea, Fig. 3 shows two different collision types, a collision between two nano objects (Fig. 3a) and collisions between a nano object and a domain boundary. The latter differs between the inner domain surface

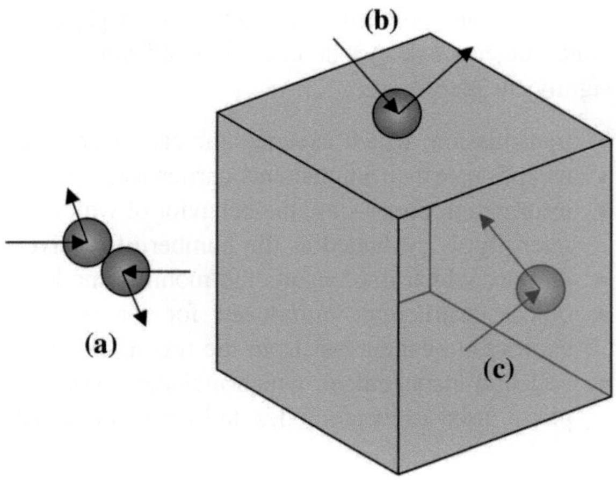

Fig. 3 Different types of collisions. **a** Between nano objects, **b** external collision, **c** internal collision

(Fig. 3c) and the outer domain surface (Fig. 3b). The domain shape is also a further discrimination. More details can be found in [7], which shows that the asymptotic average complexity of the collision management in each domain is $O(n \cdot log(n))$, being n the number of managed nano-objects.

The volume within each domain is split into 8 cubes, each one managed in isolation by a (set of) thread(s), and this splitting process is applied recursively, according to the well-known octree algorithm. More details about the implementation of the octree in the BiNS2 simulator can be found in [8].

If the total number of nano objects is denoted by N, and it is split into D domains, each sub domain will manage $M = N/D$ nano objects. Then, the complexity of the collision checks will be $O(M \cdot D \cdot log(M)) < O(N \cdot log(N))$.

In terms of simulation time, by deploying D parallel threads, it is possible to get the theoretical limit of $O(M \cdot log(M))$, if the number of available CPU cores is at least D. Clearly, the management overhead of multiple threads, concurrency, and synchronization issues, can have a significant impact on the total computational effort required.

4.3 Numerical Results

The block scheme of the end-to-end model is depicted in Fig. 4. It highlights some details of the transmitter and the receiver. The transmitter encodes the received stimulus in a train of pulses, each one composed of a burst of B molecules. These pulses are spaced in time each other by a time τ. A bit 0 is encoded with no release of molecules, thus we are using an on-off keying. The first pulse is used to allow the receiver to synchronize with the transmitter. It is followed by a fixed number of N pulses to encode the stimulus, i.e. the information (see also [9]). In this regard, we consider different choices to encode the information to be transmitted. One of them is to map each bit with a symbol b_k. This means that a train of N pulses b_k will encode a sequence of N bits. Another choice is to encode a single group of M

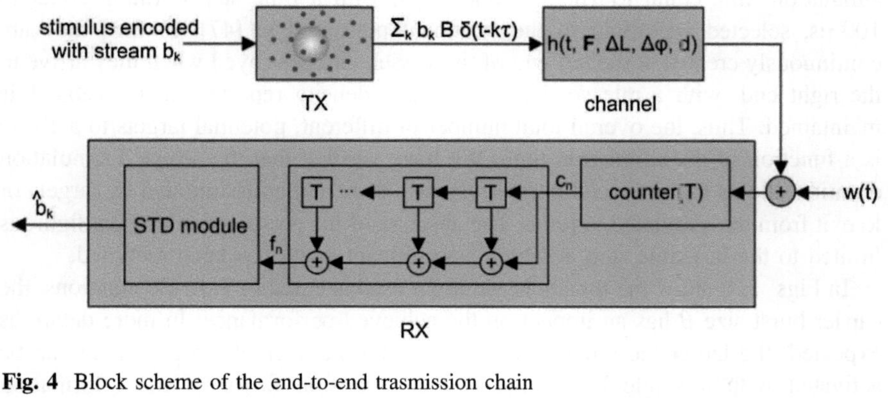

Fig. 4 Block scheme of the end-to-end trasmission chain

information bits (symbols) with *K* pulses b_k, with $K > M$. As for the receiver, it is composed by a number of modules. The received signal consists of the number of nanoparticles (ligands) that get in contact with the surface of the receiver, which is populated by a number of receptors, compliant with the specific ligand. The module counter (*T*) counts the number of nanoparticles assimilated during the last time window of duration *T*, which represent the complexes able to transmit a signal towards the nucleus of the receiving cells/nanomachine. In this chapter, we assume to limit the information transmission to a single bit.

We will exploit in future works the possibility to transmit different information by releasing a train of bursts of molecules. The possibility to set up a reliable communication strongly depends on the relative position of the transmitter and receiver in the blood flow. In fact, if the transmitter is mobile and the receiver is fixed, or vice versa, it is highly unlikely that a communication beyond a single burst (i.e. 1 bit) can be delivered. Instead, if both of them are mobile or fixed, it needs to be deeply analyzed.

The subsequent set of figures shows the achieved numerical results. A significant parameter is the distance of the center of the transmitting particle from the endothelium. We have used the following values: $d_1 = 2.828$ μm, $d_2 = 8.263$ μm, and $d_3 = 30$ μm, corresponding the a transmitter adjacent to the endothelium, a transmitter very close to the endothelium, and a transmitter located over the longitudinal axis of the vessel, respectively. The first two transmitters are small and slow transmitters, i.e. they suffer (and exploit) the effect of particle margination [46], which pushes small particles toward vessel walls. Please note that this effect does not necessarily apply to all nanoparticles, but only to those with a size similar to platelets, due to the interaction with red blood cells. In this way, they can capture the information at the tumor site. The third case is relevant to a large sensor, which can get in contact with a CTC. In fact, a large sensor, just like a natural large cell such as a white blood cell or a CTC, will not suffer of the process of margination, and will be able to travel in the blood flow close to the axis of the vessels. In this position, it is more likely to get in contact with and thus to detect CTCs.

Simulation parameters are reported in Table 1. The number of potential targets which are present in the simulation space are in the order of few hundreds for a simulation time equal to 180000 time steps, with a time step duration equal to 100 μs, selected according to the treatment presented in [47]. In fact, they are continuously created at the left end of the vessel, and destroyed when they arrive to the right end, with a rate so that the average density reported in the Table 1 is maintained. Thus, the overall total number of different, potential targets to activate is a function of the simulation time. We have verified that the selected simulation duration allows the produced burst of carriers to be either assimilated by targets or to exit from the simulated volume. The abscissa of the presented numerical figure is limited to the last time step at which a significant event has been recorded.

In Figs. 5, 6 and 7 the threshold value *Th* used is equal to 3. In all situations, the carrier burst size *B* has an impact on the achieved performance. In more detail, as expected, the larger the burst size, the larger the number of targets which can be activated with a single burst, that is the expansion factor of the information

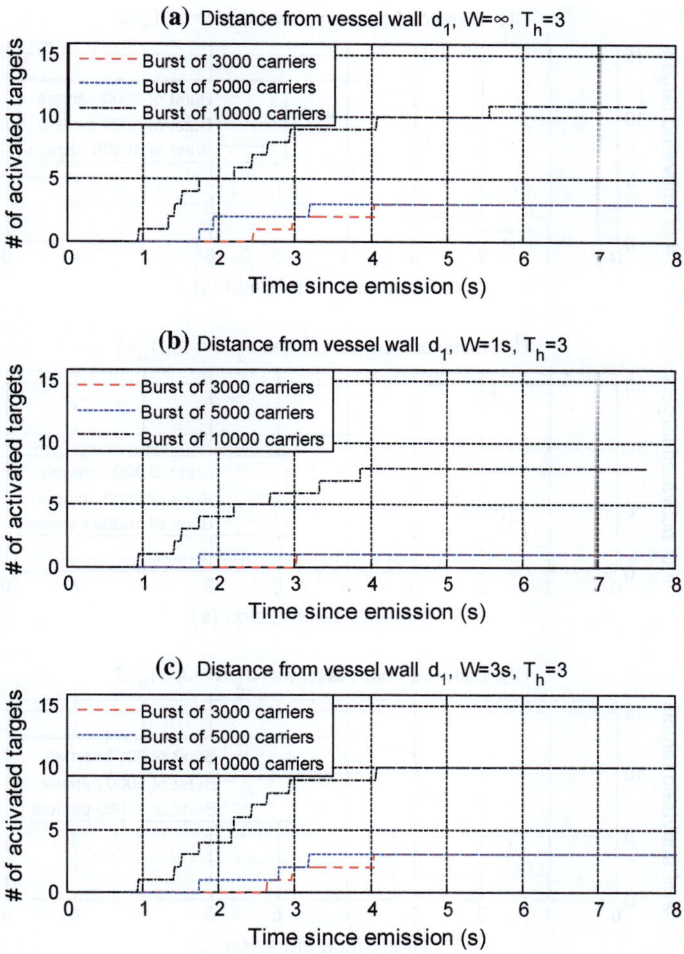

Fig. 5 Number of activated receivers for different carrier burst size. Threshold value Th = 3. Distance from the endothelium d_1

transmission. In addition, it is quite evident that a burst of 3000 carriers is often too small, since it is able to activate very few targets, and a significant benefit can be reached by using a burst of 10000 carriers. Please note that a value of 3000 carriers is typical for stimulated cells (see experimental data in [34]).

It can be observed that the window size *W*, introduced to avoid the effect of possible spurious CTC capture, has an impact on the achieved performance if its value is small. In addition, the combined usage of a threshold and a time window provides a basic model for the behavior of biological entities. In fact, usually an internal reaction is triggered when an external stimulus is strong enough and it is sustained for a limited time, and not spread over very large times. As it is possible to see from the above figures, values of *W* larger than 1 s are preferable, since in

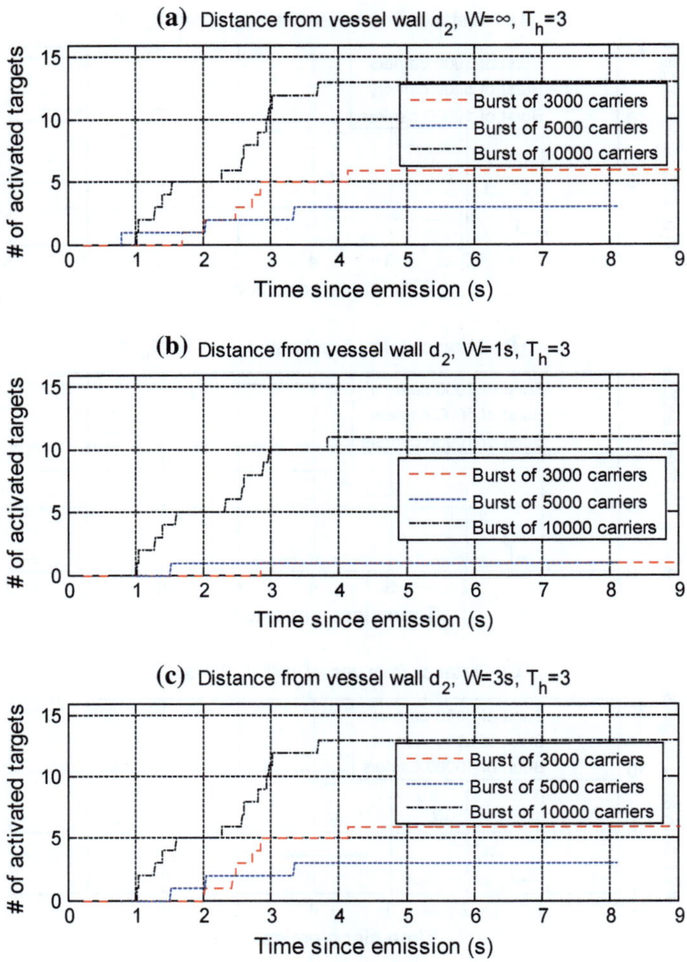

Fig. 6 Number of activated receivers for different carrier burst size. Threshold value Th = 3. Distance from the endothelium d_2

this way the performance penalty associated to a limited observation of the received stimulus can be neglected, especially when the burst size is large. A value of W equal to 3 s seems to be a good choice, as shown in the figures.

Figure 8 compares the effect of the position of the transmitter with respect to the vessel wall, for a burst size B = 10000 carriers, a threshold $Th = 3$, and a window size $W = 3$ s. As expected, the closer is the position to the vessel axis, the larger the number of targets (large mobile sensors) which can be activated. This behavior can be explained easily, since large mobile sensors usually occupy the central region of the vessel, whereas both small transmitter (margination) and ligands [33] are

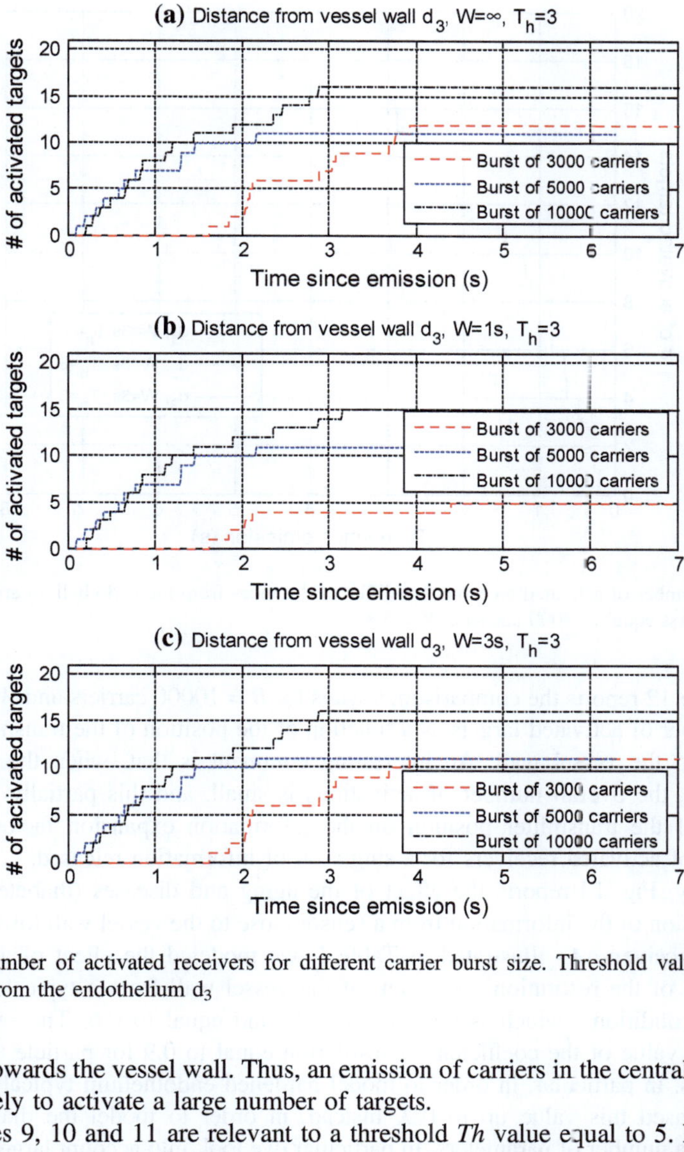

Fig. 7 Number of activated receivers for different carrier burst size. Threshold value Th = 3. Distance from the endothelium d_3

pushed towards the vessel wall. Thus, an emission of carriers in the central region is more likely to activate a large number of targets.

Figures 9, 10 and 11 are relevant to a threshold *Th* value equal to 5. Basically, the same comments made for the smaller value of the threshold apply. The general comment is that the total number of activated receiver clearly decreases, and the gap between the case with a very larger burst of carriers (i.e. $B = 10000$ carriers) and all the other cases becomes more evident. We stress that, increasing the threshold, the unsuitability of the smaller bursts (3000 or 5000 carriers) is even more evident than in previous performance figures.

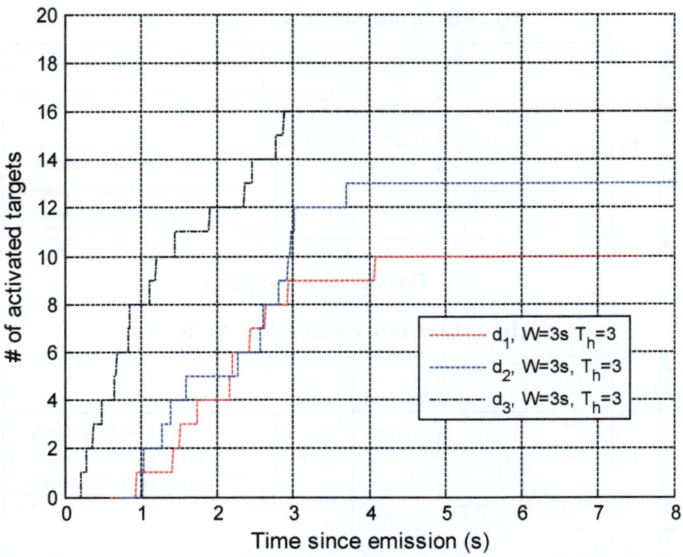

Fig. 8 Number of activated receivers for different distances from the endothelium and threshold values. Burst equal to 3000 carriers, $W = 3$ s

Figure 12 reports the comparison, always for $B = 10000$ carriers and $W = 3$ s, of the number of activated targets as a function of the position of the transmitter with respect to the vessel wall. An interesting comment is that, when the threshold increases, the overall number of activations is small, and this partially limits the impact of the transmitter position on the information expansion factor, i.e. the number of activated receivers for a single bit of information released.

Finally, Fig. 13 reports the effect of the aging and diseases (diabetes) on the transmission of the information from a sensor close to the vessel wall towards large mobile receivers. As illustrated in Table 1, we modeled the effect of aging as a deviation of the restitution coefficient of the vessel wall from a typical value for healthy conditions, which is reported in [45] and equal to 0.6. The same work reports a value of the coefficient of restitution equal to 0.9 for particle to particle collisions. In particular, in order to model hardened endothelium typical of aging, we increased this value up to 0.9. Instead, in order to model the diabetes, we changed a number of parameters. In particular, we took into account larger value of viscosity, as reported in [48]. Then, in order to model the compounds of lipids and glucose over the endothelium, we lowered to the restitution coefficient from 0.6 (normal) to 0.2. In addition, we also attenuated the longitudinal displacement of particles when these get in contact with endothelium, always to model the presence of the compounds of lipids and glucose.

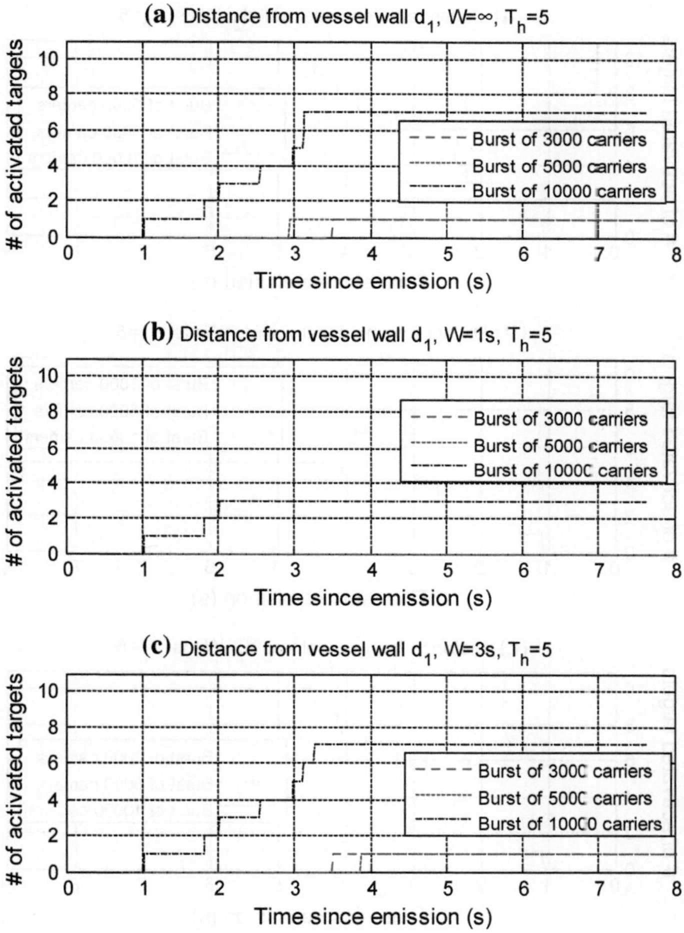

Fig. 9 Number of activated receivers for different carrier burst size. Threshold value Th = 5. Distance from the endothelium d_1

The results, plotted in Fig. 13, are quite interesting. In particular, the stiffness of endothelium has the effect of scattering nanoparticles away at high speed (almost an elastic collision with the endothelium happens), which favors their exiting from the simulated volume. Thus, molecular communications in blood vessels seems to be impeded by aging. Also the presence of diabetes seems to impede these communications. The small advantage given by diabetes over aging can be explained as follows. When diabetes is present, not only carriers remain closer to the blood vessels, but also a possible larger cells, if pushed towards the vessel, is hampered

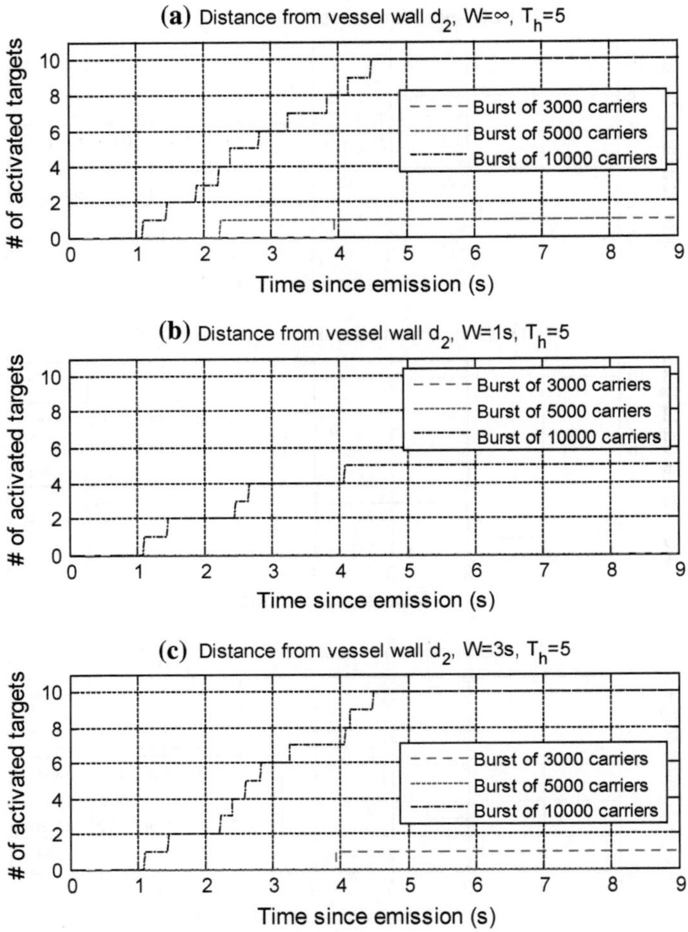

Fig. 10 Number of activated receivers for different carrier burst size. Threshold value Th = 5. Distance from the endothelium d_2

and, if it is close to the cloud of released carriers, this facilitates the (occasional) assimilations. From the simulations results presented in Fig. 13, the effect of aging and/or diabetes is not dramatic with respect to normal conditions, but marked, and can surely have an impact on the molecular communications happening inside vessels. Please note that the difference becomes evident in the long run (steady conditions for travelling carriers), since, in the first 2 s of simulations, the total number of mobile receivers activated is the same for normal, diabetes, and aging.

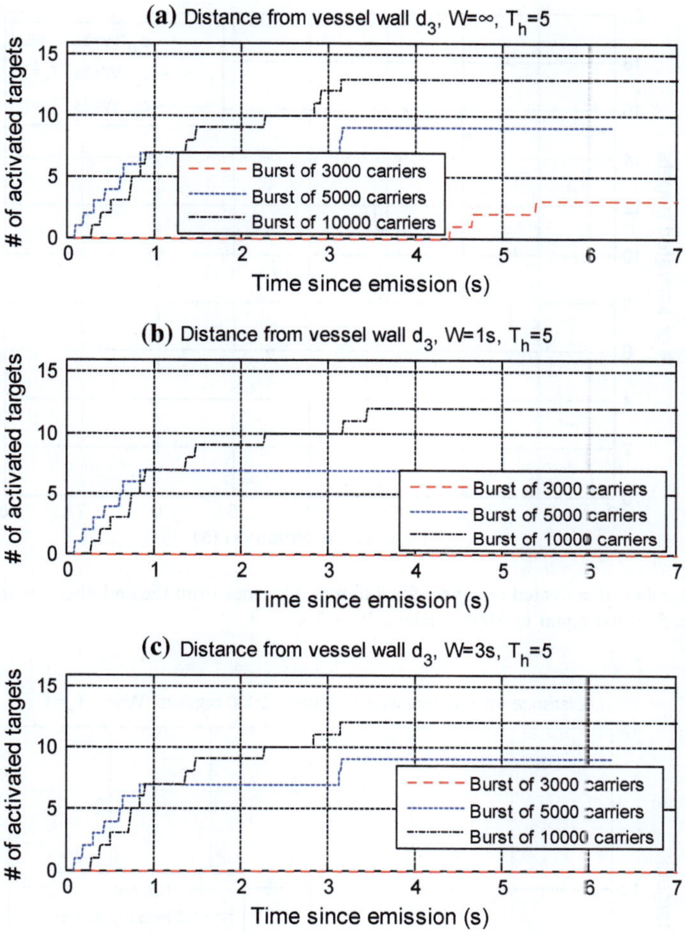

Fig. 11 Number of activated receivers for different carrier burst size. Threshold value Th = 5. Distance from the endothelium d_3

Please note that, applying a threshold value and a window value similar to the previous figures, diabetes and aging case studies would result in no assimilations, thus we have not shown this case. To sum up, when we move away from the trivial case in which we consider just a single carrier assimilation enough to trigger signal reception, the effect of aging and/or diabetes is really important. However, by analyzing Figs. 5 and 9, it is evident that, also in normal conditions, the impact of *Th* and *W* is significant, especially for large threshold values.

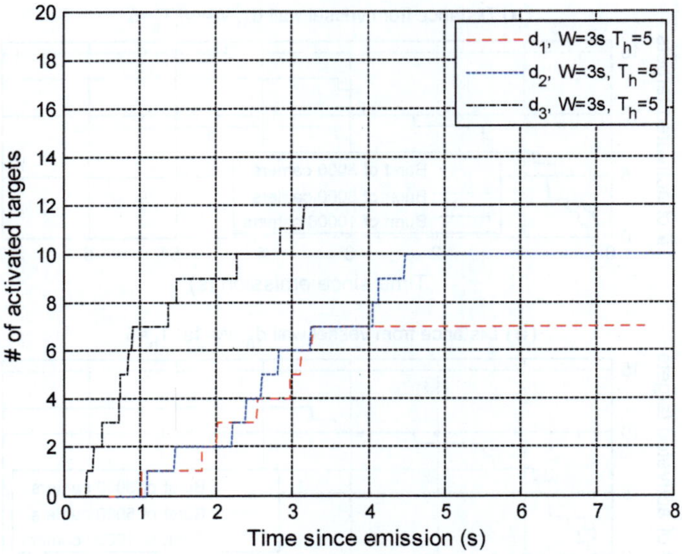

Fig. 12 Number of activated receivers for different distances from the endothelium and threshold value Th = 5. Burst equal to 3000 carriers, $W = 3$ s

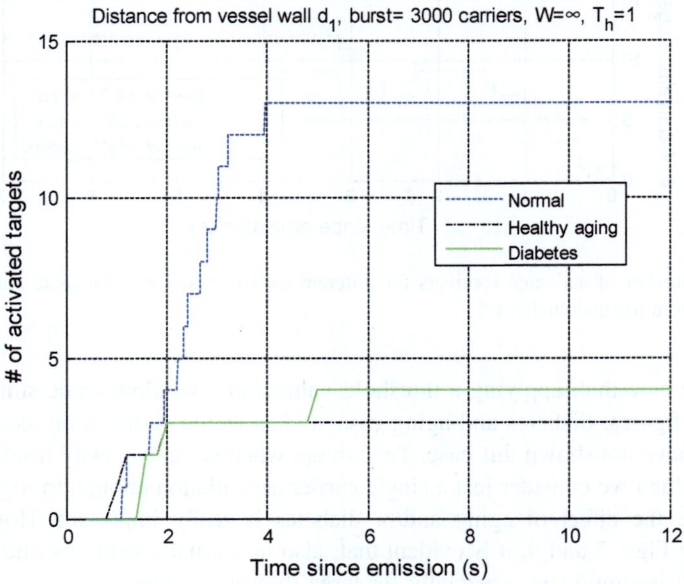

Fig. 13 Number of activated receivers for a distance from the endothelium equal to d_1. Burst equal to 3000 carriers, threshold set to 1 and time window unlimited. The curvers are associated to normal conditions, healthy aging, and diabetes

5 Conclusion

This chapter shows a proposal for implementing a molecular communication system in blood vessels. This system has the task of delivering any detected information about the presence of tumors to an extra-body smart probe. This communication system consists of the emission of a burst of carriers that propagate through the vessels, which are captured by fast moving particles which, in turn, deliver the information to the smart probe.

The results of the numerical analysis, obtained by using the BiNS2 simulator, shows remarkable effects of aging and unhealthy conditions. In particular, a healthy aging favorites the particle propagation, due to the more elastic bounces of carriers over a hardened endothelium. Conversely, serious unhealthy conditions, as in the case of diabetic patients, make the propagation difficult due to the compounds of lipids and glucose over the endothelium.

The results obtained allow sizing the parameters that characterize the communication model in order to obtain a good chance of detecting a disease, even in its initial stage, under varying conditions of health and age. In addition, the results obtained provide useful information for the introduction of molecular communications in blood vessels, with any purpose.

This work is the basis of further research activities aiming at the final result of implementing a nanoscale communication system for medical purposes. These activities, already in place, consists of the integration of transmission techniques with protocol architectures necessary for actually distribute information in the considered challenging environment, and the implementation of experimental testbeds.

Acknowledgements This work is supported by the EU project H2020 FET Open CIRCLE (project No. 665564) and by the MolML project funded by University of Perugia.

References

1. Akyildiz IF, Jornet JM (2010) The internet of nano-things. IEEE Wirel Commun Mag 17(6)
2. Nakano T et al (2012) Molecular communication and networking: opportunities and challenges. IEEE Trans Nanobiosci 11(2):135–148
3. Atakan B, Akan OB, Balasubramaniam S (2012) Body area nanonetworks with molecular communications in nanomedicine. IEEE Commun Mag
4. Felicetti L, Femminella M, Liò P, Reali G (2016) Applications of molecular communications to medicine: a survey. Nano Commun Netw 7(1). doi:10.1016/j.nancom.2015.08.004
5. Freitas RA (1999) Nanomedicine, volume I: basic capabilities. Landes Bioscience, Georgetown, TX
6. Felicetti L, Femminella M, Reali G (2012) A simulation tool for nanoscale biological networks. Nano Commun Netw 3(1)
7. Felicetti L, Femminella M, Reali G (2013) Simulation of molecular signaling in blood vessels: software design and application to atherogenesis. Nano Commun Netw 4(3)

8. Felicetti L, Femminella M, Reali G, Gresele P, Malvestiti M (2013) Simulating an in vitro experiment on nanoscale communications by using BiNS2. Nano Commun Netw 4(4): 172–180
9. Felicetti L, Femminella M, Reali G, Liò P (2014) A molecular communication system in blood vessels for tumor detection. ACM Nanocom, Atlanta, US
10. Felicetti L, Femminella M, Reali G (2013) Establishing digital molecular communications in blood vessels. In: BlackSeaCom 2013—first international black sea conference on communications and networking, Batumi, Georgia, 3–5 Jul 2013
11. Lv P, Tang Z, Liang X, Guo M, Han RPS (2013) Spatially gradated segregation and recovery of circulating tumor cells from peripheral blood of cancer patients. AIP Biomicrofluidics 7:034109
12. Alunni-Fabbroni M, Sandri MT (2010) Circulating tumour cells in clinical practice: methods of detection and possible characterization. Methods 50:289–297
13. Ben Hsieh H, Marrinucci D, Bethel K, Curry DN, Humphrey M, Krivacic RT, Kroener J, Kroener L, Ladanyi A (2006) High speed detection of circulating tumor cells. Biosens Bioelectron 21:1893–1899
14. Wang Y, Xu Z, Guo S, Zhang L, Sharma A, Robertson GP, Huang L (2013) Intravenous delivery of siRNA Targeting CD47 effectively inhibits melanoma tumor growth and lung metastasis. Mol Ther 21(10):1919–1929
15. Yu M, Stott S, Toner M, Maheswaran S, Haber DA (2011) Circulating tumor cells: approaches to isolation and characterization. J Cell Biol 192(3)
16. Bordon Y (2013) Cracking the combination. Nat Rev 12 (Macmillan Publishers Limited)
17. Weiskopf K, Ring AM, Ho CCM, Volkmer J-P, Levin AM, Volkmer AK, Özkan E, Fernhoff NB, Van de Rijn M, Weissman IL, Garcia KC (2013) Engineered SIRPα variants as immunotherapeutic adjuvants to anticancer antibodies. Sci Am Assoc Adv Sci 341
18. Fu Y, Wei X, Tang C, Li J, Liu R, Shen A, Wu Z (2013) Circulating microRNA-101 as a potential biomarker for hepatitis B virus-related hepatocellular carcinoma. Oncol Lett 6(6):1811–1815
19. Shin SJ, Smith JA, Rezniczek GA, Pan S, Chen R, Brentnall TA, Wiche G, Kelly KA (2013) Unexpected gain of function for the scaffolding protein plectin due to mislocalization in pancreatic cancer. Proc Natl Acad Sci USA 110(48):19414–19419
20. Harima Y, Ikeda K, Utsunomiya K, Komemushi A, Kanno S, Shiga T, Tanigawa N (2013) Apolipoprotein C-II Is a potential serum biomarker as a prognostic factor of locally advanced cervical cancer after chemoradiation therapy. Int J Radiat Oncol Biol Phys 1; 87(5):1155–61
21. Huang AF, Chen MW, Huang SM, Kao CL, Lai HC, Chan JY (2013) CD164 regulates the tumorigenesis of ovarian surface epithelial cells through the SDF-1α/CXCR4 axis. Mol Cancer 12(1):115
22. Han SS, Lee SJ, Kim WJ, Ryu DR, Won JY, Park S, Cheon MJ (2013) Plasma osteopontin is a useful diagnostic biomarker for advanced non-small cell lung cancer. Tuberc Respir Dis (Seoul) 75(3):104–110
23. Grunnet M, Sorensen JB (2012) Carcinoembryonic antigen (CEA) as tumor marker in lung cancer. Lung Cancer 76(2):138–143
24. Farsad N et al (2012) On-chip molecular communication: analysis and design. IEEE Trans Nanobiosci 11(3):304–314
25. Pierobon M, Akyildiz I (2011) Diffusion-based noise analysis for molecular communication in nanonetworks. IEEE Trans Signal Process 59(6):2532–2547
26. Srinivas KV, Eckford AW, Adve RS (2012) Molecular communication in fluid media: the additive inverse Gaussian noise channel. IEEE Trans Inf Theor 58(7):4678–4692
27. Kadloor S, Adve R, Eckford A (2012) Molecular communication using brownian motion with drift. IEEE Trans NanoBiosci 11(2):89–99
28. Pierobon M, Akyildiz IF (2010) A physical end-to-end model for molecular communication in nanonetworks. IEEE J Sel Areas Commun 28(4):602–611

29. Unanue ER (2013) Perspectives on anti-CD47 antibody treatment for experimental cancer. In: Proceedings of the national academy of sciences of the United States of America (PNAS), vol 110, no 27, 2 Jul 2013
30. Schonbeck U, Libby P (2001) The CD40/CD154 receptor/ligand dyad. Cell Mol Life Sci 58(1):4–43
31. Philibert J (2006) One and a half century of diffusion: Fick, Einstein, before and beyond. Diff Fundam 4:6.1–6.19
32. Gentile F, Ferrari M, Decuzzi P (2008) The transport of nanoparticles in blood vessels: the effect of vessel permeability and blood rheology. Ann Biomed Eng 36(2):254–261
33. Tan J, Thomas A, Liu Y (2012) Influence of red blood cells on nanoparticle targeted delivery in microcirculation. Soft Matter 8:1934–1946
34. Felicetti L, Femminella M, Reali G, Daigle JN, Gresele P, Malvestiti M (2014) Modeling CD40-based molecular communications in blood vessels. IEEE Trans NanoBioSci
35. Felicetti L, Femminella M, Reali G, Nakano T, Vasilakos A (2014) TCP-like molecular communications. IEEE J Sel Areas Commun
36. Nakano T, Okaie Y, Vasilakos AV (2013) Transmission rate control for molecular communication among biological nanomachines. IEEE J Sel Areas Commun 31(12)
37. Demiray D, Cabellos-Aparicio A, Alarcón E, Altilar DT, Llatser I, Felicetti L, Femminella M, Reali G DIRECT: a model for molecular communication nanonetworks based on discrete entities. Nano Commun Netw 4(4):181–188
38. Nakano T, Suda T, Okaie Y, Moore M, Vasilakos A (2014) Molecular communication among biological nanomachines: a layered architecture and research issues. IEEE Trans NanoBiosci. http://dx.doi.org/10.1109/TNB.2014.2316674
39. Batchelor GK (2000) An introduction to fluid dynamics. Cambridge University Press
40. White FM (2008) Fluid mechanics. McGraw-Hill
41. Hubert C (2009) Applied hydrodynamics: an introduction to ideal and real fluid flows. CRC Press, Taylor & Francis Group
42. Rhodes MJ (1998) Introduction to particle technology. Wiley, Chichester, New York
43. Dobkin D, Zuraw K (2010) Principles of chemical vapor deposition. Springer
44. Decuzzi P, Causa F, Ferrari M, Netti P (2006) The effective dispersion of nanovectors within the tumor microvasculature. Ann Biomed Eng 34(4):633–641
45. Gidaspow D, Huang J (2009) Kinetic theory based model for blood flow and its viscosity. Ann Biomed Eng 37(8):1534–1545
46. Reasor DA Jr, Mehrabadi M, Ku DN, Aidun CK (2012) Determination of critical parameters in platelet margination. Ann Biomed Eng. doi:10.1007/s10439-012-0648-7
47. Li T, Kheifets S, Medellin D, Raizen MG Measurement of the Instantaneous velocity of a brownian particle. Science 328(5986):1673–1675
48. Brun JF, Aloulou I, Varlet-Marie E (2004) Type 2 diabetics with higher plasma viscosity exhibit a higher blood pressure. Clin Hemorheol Microcirc 30(3–4):365–372

Electromagnetic Nanonetworks for Sensing and Drug Delivery

Renato Iovine, Valeria Loscrì, Sara Pizzi, Richard Tarparelli and Anna Maria Vegni

Abstract The use of nanodevices for biomedical applications has recently been object of study by researchers. Novel prospectives can be envisaged in the field of nanomedicine, also supported by innovative nanodevices with specific properties. In this chapter, we present the electromagnetic properties of different metal nanoparticles (i.e., nanocube, nanocylinder, nanorod, bow-tie, biconical nanoparticle, etc.), opportunely functionalized for sensing applications, as well as drugged with medicament to be released to specific locations, for innovative therapeutic treatments. After modeling the design of such nanoparticles, we investigate the *channel model* adopted in electromagnetic nanonetworks. Basically, we focus on the nanoparticle transmission, diffusion and reception processes, both for extra- and in-vivo applications i.e., for the detection of target cells in a biological tissue sample, and for drug delivery via nanoparticle adsorption, respectively. Numerical results obtained through full-wave simulations have shown the effectiveness of electromagnetic nanoparticles for specific biomedical applications (e.g., DNA alteration detection). Finally, we highlight that in this chapter the electromagnetic properties that are described are used for sensing and drug delivery, and not for communication among nanoparticles.

1 Introduction

The concept of *nanomedicine* arises from the visionary idea that miniaturized devices at nanoscale level (i.e., nanodevices or nanorobots) could be designed, manufactured, and introduced into the human body, for therapeutic aims (e.g.,

R. Iovine · R. Tarparelli · A.M. Vegni
Department of Engineering, Roma Tre University, Rome, Italy

V. Loscrì (✉)
INRIA, Lille-Nord Europe, Lille, France
e-mail: valeria.loscri@inria.fr

S. Pizzi
DIIES, Mediterranea University of Reggio Calabria, Reggio Calabria, Italy

J. Suzuki et al. (eds.), *Modeling, Methodologies and Tools for Molecular and Nano-scale Communications*, Modeling and Optimization in Science and Technologies 9, DOI 10.1007/978-3-319-50688-3_20

cellular repairs at the molecular level) [1]. The possibility of applying *nanotechnology* to medicine has been object under study for the last decades, and it represents a new approach based on the comprehension and deep knowledge of the properties of the matter at the nanoscale level [2].

The intrinsic behavior and characteristics of nanodevices distinguish them from traditional devices working at the macroscale level, and particular features at the nanoscale level should be addressed [3]. Indeed, the properties of the matter drastically change, making it necessary a synergy among several different disciplines, in order to define novel communications techniques and design efficient nanodevices [4].

Generally, nanodevices represent the most basic functional unit with passive features, which allow performing very easy tasks, like sensing or actuation. A set of nanodevices, sharing the same medium (e.g., the biological tissue or the blood flow) and performing multiple tasks, form a *nanonetwork* [5]. Nanonetworks allow to expand the number and range of applications envisioned for single nanodevices, since collaborative tasks can be done by different nanodevices. Nowadays, applications are foreseen in four main fields, namely (*i*) biomedical, (*ii*) environmental, (*iii*) industrial and consumer goods, and (*iv*) military and defense [3, 6].

Communication and signal transmission techniques to be utilized/used in nanonetworks are one of the most challenging topics, due to the limited computation skills of single nanodevices. Classical communication and network paradigms cannot be directly utilized in nanonetworks, since the poor capabilities of nanodevices pose novel challenges, establish new requirements and show novel properties that need to be opportunely addressed. As an instance, current encoding and decoding techniques are not feasible due to very limited processing capability of nanodevices, as well as traditional transceiver circuitries cannot be mounted into them due to the limitation of the nanoscale. Also novel mobility models should be addressed accordingly to this particular field, due to specific physical rules in this regime [7].

The biomedical field is one of the most challenging area of application of nanonetworks, as well as the most intriguing due to a variety of biomedical scenarios. Indeed, in the biomedical field, nanonetworks are expected to provide a perfect interface to interact with single molecules, proteins, DNA sequences and the major components of cells. Both in-vivo and extra-vivo applications of biocompatible nanodevices are largely investigated. As an instance, the use of nanosensors to detect chemical compounds in concentrations, or the presence of different infectious agents, such as virus or bacteria is an objective of several research studies [8, 9].

In the biomedical field, we highlight three main applications i.e., (*i*) *health monitoring systems,* (*ii*) *Drug Delivery Systems (DDS)*, and (*iii*) *bio-hybrid implants*. In health monitoring systems, the use of nanosensors allows detecting and monitoring different levels of molecule concentration in the blood (e.g., sodium, glucose and other ions), as well as the presence of infectious intra-body agents. The DDS use nanoactuators capable of releasing nanoparticles, drugs or biomolecules in specific locations of the body; this means that drug molecules are released locally and are adsorbed only by the diseased cellular membranes, so that patients benefit from a less invasive and much more efficient treatment. Finally, bio-hybrid implants rely on nanodevices able to cooperate not only with each others, but also with biological components (e.g., to restore the central nervous tracks or to support the immune system).

As the design and manufacturing of devices at the nanoscale advance, new possibilities are given for the interconnection among nanodevices and new challenges rise in the development of protocols and channel models for nanonetworks. Based on the different types of nanodevices (i.e., biological, and electromagnetic), nanonetworks are mainly classified as (*i*) *molecular* [10], and (*ii*) *electromagnetic* [3].

In this chapter, we investigate electromagnetic (EM) nanonetworks properties with different nanoparticles, working in the THz band, used for sensing and drug delivery, and foreseeing the transmission and reception of electromagnetic radiation from components based on nanomaterials. In [3], Akyildiz and Jornet present the architecture of a nanosensor device as comprised of nanosensors, nanoactuator, nano-memory, nano-antenna, nano-EM transceiver, nano-processor and nano-power unit. All these components allow the integrated device to sense, compute or even perform local actuation. Furthermore, the authors foresee that nanosensor devices will potentially communicate among them in the terahertz band (i.e., 0.1–10.0 THz).

The main challenges of electromagnetic-based nanonetworks are expressed in terms of THz channel modeling, information encoding and protocols for nanosensor networks. A physical channel model for wireless communication in the THz band has been developed by Jornet and Akyildiz in [11]. The presented model computes the signal path loss, the molecular absorption noise and, ultimately, the channel capacity of EM nanonetworks. In [12], a modulation and channel sharing mechanism based on the asynchronous exchange of femtosecond-long pulses transmitted through an on-off keying modulation is proposed for the transmission of binary streams among nanodevices of an EM nanonetwork. A medium access control protocol for EM nanonetworks built on the top of the pulse-based communication scheme in [12] for the coordination of multiple simultaneous transmissions is presented in [13]. The proposed protocol is tailored to the peculiarities of the terahertz band and is constituted by two main stages i.e., (*i*) the handshaking process, and (*ii*) the transmission process. Finally, in [14] Jornet and Akyildiz have developed an energy model for self-powered nanosensor motes, which successfully captures the correlation between the energy harvesting and the energy consumption processes. The energy harvesting process is realized by means of a piezoelectric nanogenerator, for which a new circuital model is developed that can accurately reproduce existing experimental data. The energy consumption process is due to the communication among nanosensor motes in the THz band.

The application of nanoparticles for combined targeting and delivery of diagnostic and therapeutic agents has received significant attention in the last years [15]. Colloidal metallic nanoparticles have been recently investigated in the field of nanomedicine, especially for drug delivery systems [16]. Noble metals, like silver and gold, are largely used for the design of nanoparticles, in the field of sensing applications, as well as acting as carriers of medicament molecules. Particularly, colloidal silver has been used as an antibacterial agent by weakening DNA replication and inactivating proteins, while gold has low toxicity to biological systems, and so results inefficient for antibiotic therapy [17]. In addition, nanoparticles are also engineered to provide sustained drug release [18, 19], especially beneficial for chronic therapies. In [20], the authors use gold nanoparticles for diagnostic and drug delivery applications by exploiting chitosan. The use of chitosan serves dual purpose by acting as a reducing agent in the synthesis of gold nanoparticles, and also promotes the penetration and uptake of peptide hormone insulin across the mucosa. As a first step towards real implementable solutions, in [21] nanomachines have been designed for medical applications to colonize and autonomously work inside the human body.

Models for particulate DDS, based on the injection of drug molecules, have been proposed in [22, 23]. In [22], the authors consider the human blood vessels to model the medium where molecules diffuse, subjected to the cardiac input. Drug molecules are then allowed to move in every location of the cardiovascular system. On the other side, the possibility of using a swarm of bio-nanomachines (i.e., nanoscale devices composed of natural or synthetic biological materials) with collective behaviour, in order to perform collaborative tasks like target detection and molecule guiding, has been investigated by Nakano et al. in [23]. The same concept of swarm of nanomachines has been utilized in [24], where endogenous diseases of the brain are treated by means of nanodevices that communicate through acoustic signals. Finally, a multi-source nanonetwork model for extra-vivo biomedical diagnosis applications, specifically the detection of DNA alterations, is presented in [25].

In this chapter we focus on the use of electromagnetic nanodevices for biomedical applications, specifically sensing of DNA alterations and drug delivery through oral ingestion. After describing the main features of electromagnetic nanoparticles used in biomedical applications, we consider a physical end-to-end model, and investigate how nanoparticles are transmitted, diffuse, and finally are captured. Furthermore, we also present simulation results that show the capability of nanosensors to operate in biomedical applications.

This book chapter is organized as follows. In Sect. 2, we present different types of electromagnetic nanodevices (i.e., nanoparticles), and describe their specific features and properties. All these nanoparticles represent simple nanonodes, comprising an electromagnetic nanonetwork, and are then subjected to specific processes (i.e., transmission, diffusion, and reception). These are then investigated in Sect. 3. The sensing and drug delivery applications of nanoparticles are finally presented in Sect. 4; we show specific use cases, by means of simulation results. Finally, conclusions and considerations on future investigations are drawn at the end of the chapter.

2 Electromagnetic Nanoparticles for Biomedical Applications

In the last few years the fabrication of nanostructures received too much attention. Optical properties of metallic nanoparticles make them suitable for biomedical applications [26]. In particular, gold nanoparticles have inner electromagnetic properties, depending on the size, shape, geometrical parameters, and the surrounding dielectric environment Refractive Index (i.e., *RI*). To be more precise, the strongly enhanced Localized Surface Plasmon Resonance (LSPR) of this metal, at optical frequencies, makes it good light scatters and absorbers [27]. In addition to this, gold nanoparticles offer good bio-compatibility, optimal synthesis and conjugation properties [28], and are useful tools as contrast agents in cellular and biological imaging [29].

Nanoparticles are of great interest in biomedical applications such as light scattering microscopy-based imaging, sensing applications, and photothermal therapy for superficial and deep tumors treatment [30]. New sensing techniques to reveal malignant tissues are needed. For many optical sensors, proposed in literature, the presence of the biological sample is detected by measuring its refractive index in several ways [31].

One of the most used techniques for biosensing is based on the LSPR phenomenon [32], which occurs when an electromagnetic plane wave impinges on metallic nanoparticles that are electrically small. In this condition the free electrons of the nanoparticle follow collectively the electromagnetic oscillations. This phenomenon derives from the peculiar metallic nanoparticles optical properties, and leads to a strong local electromagnetic field enhancement. Therefore, the resonant frequency of the electron motion strongly depends on the nanoparticle size, shape, composition, and surrounding dielectric environment [33].

In order to explain the nanoparticle electromagnetic properties, in terms of scattering and absorption cross-section, the following assumptions need to be established:

- The particle size is much smaller than the wavelength in the surrounding medium. In this case, under the limit of electrically small particles, the electromagnetic field is approximately constant over the particle volume, and then the resonant behavior of the structure can be studied in terms of a quasi-static approximation;
- The considered particle is homogeneous and isotropic. In addition, the surrounding material is a homogeneous, isotropic and non-absorbing medium.

Under previous assumptions, it is possible to relate the geometrical properties of the structure to its electromagnetic ones (i.e., scattering and absorption), by developing its polarizability analytical expression:

$$\underline{\underline{\alpha}} = V\varepsilon_e \sum_{n=1}^{3} \frac{\varepsilon_i - \varepsilon_e}{\varepsilon_e + L_n(\varepsilon_i - \varepsilon_e)} \underline{u}_n \underline{u}_n, \tag{1}$$

where V is the nanoparticle volume, ε_e is the dielectric permittivity of the surrounding environment, ε_i is the complex dielectric permittivity of the metallic nanoparticle, $\underline{u}_n$ are unit vectors in the direction of the principal axes of the nanoparticle, and L_n (with $n = [1\text{–}3]$) are the three components of the corresponding depolarization dyadic, that is

$$\underline{\underline{L}} = L_1 \underline{u}_1 \underline{u}_1 + L_2 \underline{u}_2 \underline{u}_2 + L_3 \underline{u}_3 \underline{u}_3. \tag{2}$$

The depolarization dyadic L is an essential concept in the evaluation of the electric field in the source region, in order to establish the electromagnetic response of an arbitrary shape of nanoparticle through the polarizability expression. From a physical point of view, it is a dimensionless matrix that allows taking into account anisotropic nanoparticles.

Following the same procedure in [34], it is possible to develop new depolarization factors for specific nanoparticles, and then new analytical closed-form formulas for scattering and absorption cross-section can be derived.

The general expression describing the *extinction cross-section* properties for different nanoparticles shapes can be written as the sum of absorption and scattering phenomenon as follows:

$$C_{ext} = k.\mathrm{Im}[\alpha] + \frac{k^4}{6\pi} |\alpha|^2, \tag{3}$$

where $k = \frac{2\pi n}{\lambda}$ is the wavenumber, λ is the wavelength, and $n = \sqrt{\varepsilon_e}$ is the refractive index of the surrounding dielectric environment. The extinction cross-section represents the effective area that governs the probability of scattering and absorption event by a nanoparticle. In general, the extinction cross-section is different from the geometrical cross-section of a particle, and it depends upon the wavelength of light and the permittivity, shape and size of the particle. In terms of area, the extinction cross-section [nm^2] is the sum of the cross-sections due to absorption and scattering.

Following this way, it is possible to predict in accurate manner the electromagnetic response of each nanoparticle shape. This aspect is crucial in project phase, and allows optimizing the nanoparticles sensibility for specific applications.

As shown in [35], considering cube, rod and elliptical cylinder nanoparticles, for each of the considered structures the analytical models have been calculated. By assuming that the structures are excited by a plane wave, having the electric field **E** and the vector propagation **k** directed as depicted in Fig. 1, the polarizability of cube, rod and elliptical cylinder nanoparticles follow, respectively:

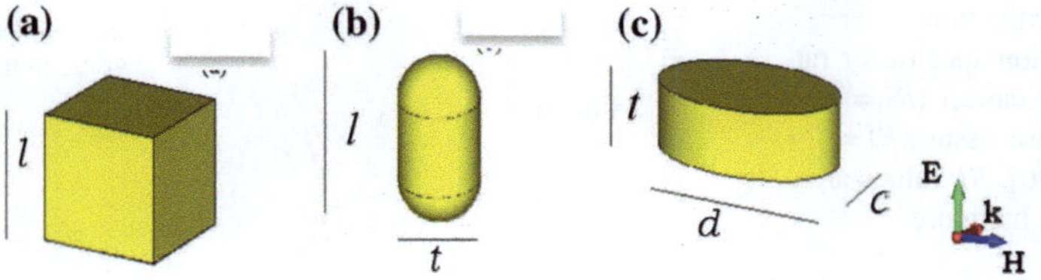

Fig. 1 Geometrical sketch of the nanostructures: **a** cube, **b** rod, and **c** elliptical cylinder

$$L_{\text{cube}} = 10\sqrt{2}\left(1 - \frac{1}{\sqrt{2}}\right)\pi \cdot \frac{1}{l}, \tag{4}$$

$$L_{\text{nanorod}} = 1 - \frac{1}{\sqrt{1 + \left(\frac{t}{2}\right)^2 \cdot \frac{1}{\left(\frac{t}{2} + \frac{l}{4}\right)^2}}}, \tag{5}$$

$$L_{\text{elliptical cylinder}} = \frac{1}{\pi}\left(1 - \frac{t}{4\sqrt{\frac{t^2}{2} + \frac{t^2}{16}}}\right) \cdot E\left[\frac{1}{1 - \left[1 - \left(\frac{c}{2}\right) \cdot \left(\frac{2}{t}\right)\right]^2} - 1\right], \tag{6}$$

where E [ ] is the complete Elliptic Integral of the second kind, l [nm] is the cube side length, as well as the height of the nanorod and of the elliptical cylinder, d [nm] and c [nm] are the elliptical cylinder base axes lengths, t [nm] is the rod thickness, as well as the length of elliptical cylinder.

Replacing (4), (5) and (6) in (1), and later in (3), the extinction cross-section spectrum for cube, rod and elliptical cylinder nanoparticles can be obtained [35]. In this way it is possible to predict the optimal geometrical parameters, in order to optimize the nanostructure for specific applications (i.e., biosensing applications). For example, in [36], a sensor revealing tumor and adipose tissue is presented. It consists of a gold linear chain of four nanocubes deposited on a silica substrate, and allows revealing the rat mammary cancer and adipose tissue by LSPR shift, as shown in Fig. 2.

By using very small inter-particle distance among the nanoparticles it is possible to obtain high-scattering and low-absorption efficiencies. These properties are very important for biosensing applications. In fact, high-absorption efficiency could heat the biological sample invalidating medical diagnosis. The biological sample used to test this device is an in-silico replica with values of *RI* taken from the literature. In particular the *RI* values of rat mammary adipose and tumor tissue have been considered.

Finally, by using the same physical principle, in [37], a label-free immunosensor was designed and fabricated for sensitive of alpha-fetoprotein (AFP) of gold nanorods.

Fig. 2 Extinction cross-section spectra for rat mammary cancer (*RI* = 1.39), and adipose tissue (*RI* = 1.467), [36]. *RI* values are known in literature

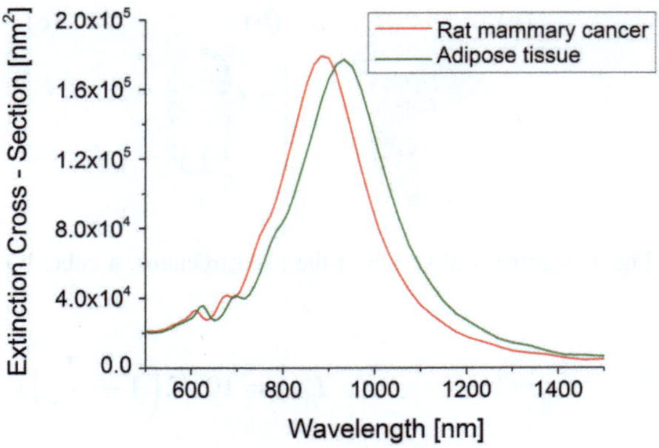

In literature various shapes of metallic nanoparticles are used for biosensing applications. For example, in [38] the ellipsoidal gold nanoparticles have been analyzed, and a new analytical study of metallic nanoparticles working in the infrared and visible frequency range has been presented. The approach proposed in [38] is a useful tool to design nanostructures for sensing applications.

Recently, another important property of metallic nanoparticles has been analyzed and exploited. In [39], an analytical and numerical investigation for modified gold nanorods, operating in the visible and in the infrared regime, is proposed. The modified particle consists in a core/shell structure (i.e., silica core, and gold shell) embedded in a dielectric environment, as shown in Fig. 3. In order to study and to tune the electromagnetic nanostructure, a new analytical model has been developed. The electric field distribution at the resonant wavelength demonstrates the intensity of the electromagnetic field in the neighborhood nanoparticle, as shown in Fig. 4. This result derives from the silica core that allows the field intensification.

Exploiting the obtained model, the nanoparticle sensitivity was studied and verified by full-wave simulations. In particular, being an asymmetric structure, the electromagnetic properties, in terms of extinction cross section (i.e., absorption and scattering) for both longitudinal and transverse modes excitation, have been evaluated.

Fig. 3 Core/shell nanorod particles, and two mode excitations

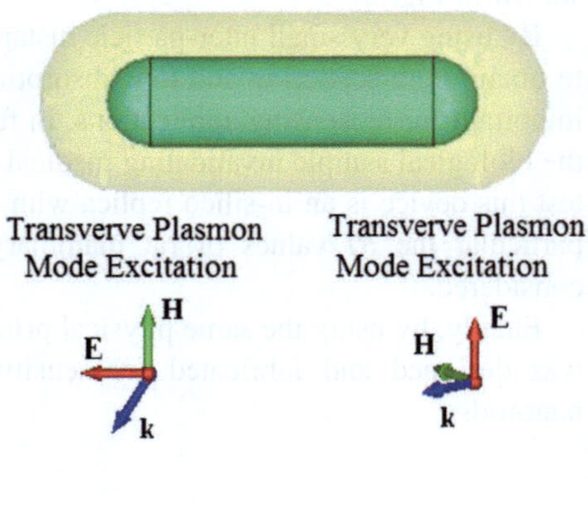

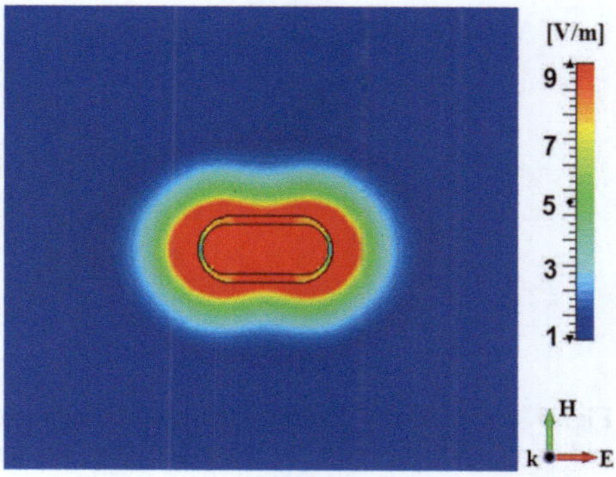

Fig. 4 Near electric field distribution at the resonant wavelength (i.e., 696 nm) of nanorod

As explained in [39], the difference between the longitudinal and transverse sensitivity can be explained in the following manner: it is known that nanoshells possess two resonances arising from the interaction between the classical plasmon resonance of the solid particle, and the resonance of the dielectric core inside the metal. The particle plasmon resonance holds increased sensitivity to the dielectric environment variation, especially for the longitudinal polarization. Instead, the dielectric cavity resonance is much more sensitive to changes in dielectric properties within the nanoparticle core and shell dimensions for transverse polarization.

Modified nanorods represent useful tools for sensing, since they combine two main optical properties of both nanorods (i.e., the high Aspect Ratio), and nanoshells (i.e., core/shell thickness), in order to reach the higher sensitivity. Thus, it allows additional degrees of freedom for the optical tenability of such particles.

In the last few years several research have focused the attention to metallic nanoparticles in a coupled configuration. In fact, to arrange gold nanoparticles in this way allows obtaining a greater intensification of the electromagnetic field, with respect to the single-element configuration. This aspect provides a nanostructure with major sensitivity in terms of LSPR shift [40]. Because the resonance of this coupled mode is sensitive to the gap distance change in the order of few tens of nanometers or less, it is inversely possible to measure such distances by monitoring the scattering of the particle pair, the so-called *plasmon ruler*, as well described in [40].

Another property of metallic nanoparticles consists of multi-resonant approach. In [41], a multi-resonant bow-tie structure is presented. The classical bow-tie nanoparticle consists of two opposing truncated gold prisms, as depicted in Fig. 5a. In the classical bow-tie particle, the reason for the employment of the dielectric hole, as shown in Fig. 5, is the possibility to excite a new resonant frequency on the same structure in order to achieve a multi-band behavior.

In this configuration the particles exhibit an additional resonant frequency, as reported in the following Fig. 6. In order to tune and control the physical phenomenon, and to design it with specific requirements, in [41] the analytical model has been developed. A good agreement among numerical and analytical results was

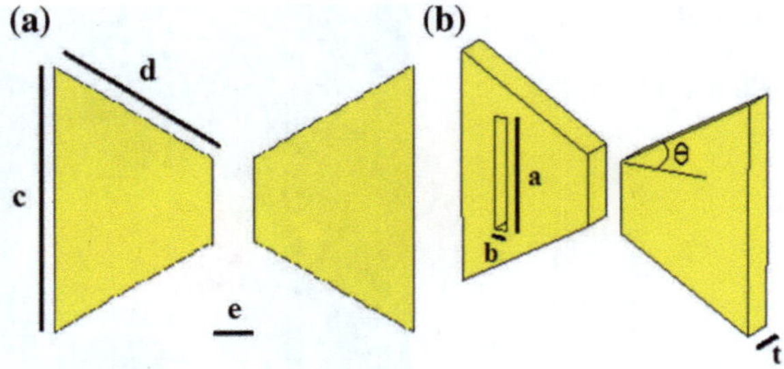

Fig. 5 Geometrical sketch of the bow-tie nanoparticle. **a** *Top* view of classical bow-tie particle, and **b** perspective view of modified bow-tie nanoparticle by dielectric incision. Geometrical parameters: $a = 80$ nm, $b = 10$ nm, $c = 160$ nm, $d = 155$ nm, $e = 20$ nm $\theta = 30°$, and $t = 25$ nm

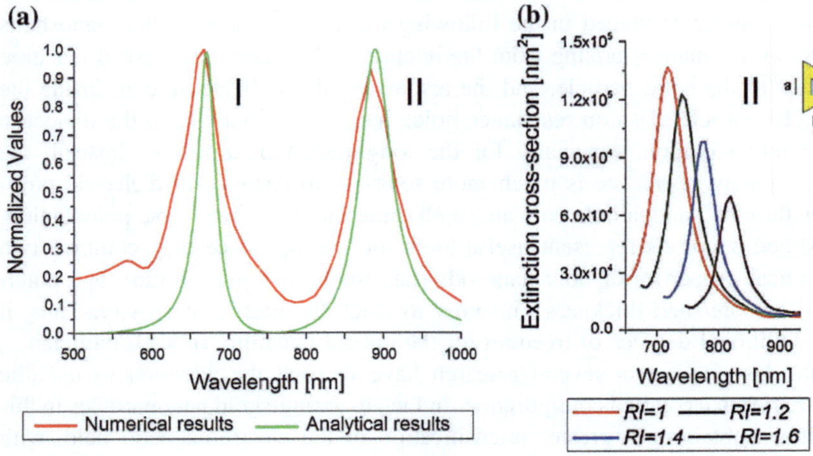

Fig. 6 Extinction cross-section spectra for modified gold bow-tie nanoparticles, in the case of (**a**) analytical-numerical model comparison (normalized values) [41], and (**b**) local *RI* variation, inside the teal *blue* area

achieved, as shown in Fig. 6a. We notice that the nanostructures have been analyzed in terms of sensitivity properties, and results reveal that the modified bow-tie structure can be applied for biomedical applications.

For example, it is well known that in the VIS-NIR tissues have a specific *RI* value, representing a unique physical property. Local *RI* changes in the tissue are related to different pathological conditions. In disease states, such as neoplasm or inflammation, color of tissue changes due to the change in *RI*, which in turn is related to the relative permittivity of the tissue. Therefore, the dielectric constant changes as a result of different re-distributions in electron density of the tissue and it can be associated to different pathological conditions.

The obtained results are shown in Fig. 6b, where the local *RI* variation causes a selective shift in the extinction cross-section spectra (Peak II). In particular, Peak II associated to the local *RI* variation inside the teal blue area shifts from 717 to 820 nm for $1 < RI < 1.6$. The mean sensitivity value is 171 nm/*RIU*.

3 Transmission, Diffusion and Reception Process in Electromagnetic Nanonetworks

After describing the main features of electromagnetic nanoparticles used in biomedical applications, like sensing and drug delivery, in this section we investigate how nanoparticles are *transmitted*, *diffuse*, and finally are *captured* in a physical end-to-end model.

The concept of electromagnetic nanonetworks holds classical basics of both information theory, and electromagnetic propagation. The specific feature is the use of engineered metallic nanoparticles, working in the THz frequency range, which are exploited by means of the use of LSPR phenomenon.

From the communication and channel modelling point of view, the electromagnetic nanonetworks represent a nanonetwork where the nanonodes are metallic nanoparticles, as small as molecules, and then depending on diffusion-based nanocommunications. Following this consideration, we can compare a flow of nanoparticles as a flow of molecules, and the electromagnetic theory laws are applied for sensing applications, by exploiting the interaction of an impinging electromagnetic wave with the metallic surface of the nanostructures.

In a nanonetwork utilized for sensing or drug delivery purposes, from the point of view of the information theory, the *transmitter* is represented by the source (i.e., nanomachine) that emits nanoparticles, while the *receiver* is the set of target cells laying in the area where a phenomenon needs to be sensed or a drug needs to be delivered. Due to the particular nature of nanoparticles (i.e., very small devices at nanoscale level), the propagation process allows nanoparticles to move along the space linking the transmitter to the receiver, according to a diffusion model.

Communications via Diffusions (CvD) arise from molecular nanonetworks [42, 43], where specific molecules, called *messenger molecules*, act as the information carriers between two nanomachines residing in close-to-medium proximity to each other in a fluid environment.

In our vision, CvD systems can be also applied to electromagnetic nanonetworks, due to common features between molecules and metallic nanoparticles. Indeed, a single nanoparticle is an indivisible object, like a molecule, which is released (during the emission process), or collected by means of chemical reactions (during the reception process). At the same time, a nanoparticle can act as *messenger nanoparticle*, since it can carry information, such as drug concentration. Furthermore, metallic nanoparticles constituting the electromagnetic nanonetwork are passive nanodevices that is, they cannot transmit data information (e.g., drug molecules) by themselves, but need to be impinged by an electromagnetic wave in order to release information. Due to all these features, electromagnetic nanoparticles are assumed to be very similar to molecules.

Several works have thought of the transmitter as a nanomachine or a bio-engineered cell capable of emitting nanoparticles and releasing them into the medium to change their concentration [44]. Similarly, the receiver is capable of capturing the nanoparticles, by using *ligand-receptor* bindings [45, 46].

If a nanoparticle collides with a receiver, it means that the nanoparticle hits the receiver, and the nanoparticle is then removed from the system since the couple ligand-receptor at a receiver forms a chemical bond with the messenger nanoparticle [47]. Indeed, it is assumed that the whole surface of the receiver is composed of receptors, which are able to bind with the messenger nanoparticles. It follows that each received nanoparticle constitutes the signal just once. This process is named *first hitting process* [48]. On the other side, if a nanoparticle hits a transmitter, it bounces back from the transmitter since a transmitter does not have the same ligand receptors on their outer shell.

The receiver is often a biological sample (i.e., a tumor tissue), and has a large number of binding places, so that it can estimate the concentration by averaging over all the created bonds.

From all previous considerations, a nanonetwork can be represented by the following main processes [6, 42]:

1. *Emission*: this process investigates how nanoparticles are transmitted from a nanomachine (or a set of nanomachines);
2. *Diffusion*: illustrates how nanoparticles diffuse along the gap that separates the transmitter from the receiver lying in the common space *S*;
3. *Reception*: describes how nanoparticles are captured by the receiver, by means of ligand-receptor bindings.

The transmitter releases a number of nanoparticles in a time slotted fashion. These messenger nanoparticles scatter in the medium following the probabilistic diffusion dynamics in the environment. Some of these released nanoparticles are received via receptors in the cell membrane.

All the above-mentioned processes take place inside a space *S*, which is strictly related to the application for which the nanonetwork is designed, and it is initially filled with a homogeneous concentration of particles equal to zero. The physical end-to-end model is depicted through the scheme in Fig. 7, where the main modules of *emission*, *diffusion* and *reception* of nanoparticles are represented.

In the recent years, many studies have focused on the channel capacity and propagation dynamics of the CvD medium [47, 49–51]. Some of these propagation process studies consider the probabilistic behaviour of the channel as the transfer function of the system, while others model it as a unique noise source inherent to a diffusion medium. According to the aforementioned studies on channel capacity, it has been shown that the reliability of the transmission diminishes exponentially

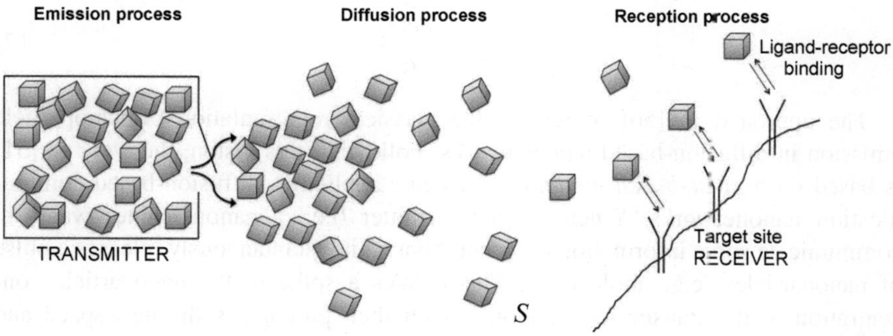

Fig. 7 End-to-end physical model of an electromagnetic nanonetwork for sensing or drug delivery purposes. The transmitter is represented by a nanomachine, filled with a nanoparticle concentration, exiting and diffusing into the medium (i.e., the space *S*). The nanoparticles act either as drug carriers for drug delivery applications, or as bio-functionalized nanodevices for sensing applications. The receiver is often represented by a target site (e.g., a group of tumor cells), with receptors for capturing the nanoparticles. The sensing and drug delivery applications occur only when the nanoparticles have been bound to the receiver's receptors

with increasing transmission range, while the average end-to-end delay increases exponentially. These results limit the effective communication range of the CvD systems to a few tens of micrometers.

Several studies on CvD systems focus on a single transmitter single receiver systems. However, when there are more communicating couples in the environment, additional issues arise (i.e., interference and noise). An important issue is the interference between closely placed transmitting couples in the same medium. When two or more transmitting pairs try to communicate simultaneously using the same technique and the same type of messenger nanoparticles, their signals affect each other and reduce/increase the Signal-to-Noise and Interference Ratio (SINR) of all nearby transmissions.

3.1 Emission Process

The emission process aims to modulate the nanoparticle concentration rate (i.e., $r_T(t)$) at the transmitter. This is the first process occurring in the end-to-end physical model, as depicted in Fig. 7.

In [50], Pierobon and Akyildiz developed a mathematical framework to interpret the diffusion-based nanoparticle communications, in the simple case of one transmitter and one receiver. The emission process has been modelled by means of the electrical circuit theory, considering a RC circuit, where the capacitor charging/discharging current is related to the net flux of nanoparticles. Then, the particle concentration rate $r_T(t)$ can be expressed as:

$$r_T(t) = \frac{dc(t)}{dt} \tag{7}$$

The approach in [50] represents the classical representation of nanoparticle emission in diffusion-based nanonetworks. Following this vision, the work in [51] is based on a *pulse-based modulation scheme* applied to diffusion-based communication nanonetworks. Whenever a transmitter (i.e., a nanomachine) wants to communicate some information to its neighbours, it instantaneously releases a pulse of nanoparticles (e.g., molecules). This creates a spike in the nanoparticle concentration at the transmitter location, which then propagates through space and time. Notice that the nanoparticle concentration does not only depend on time, but it also varies along the space. The propagation of this pulse can be analytically modelled by solving Fick's Laws of Diffusion.

If the transmitter releases Q molecules at the instant $t = 0$, the molecular concentration at any position x [nm] in space, from the transmitter location is given by:

$$c(x,t) = \frac{Q}{(4\pi Dt)^{3/2}} e^{-x^2/4Dt}, \tag{8}$$

where t [s] is the time, and D [cm^2/s] is the diffusion coefficient, assumed as a constant value for a given fluidic medium, and depending on the size and shape of the nanoparticles, as well as the interaction with the solvent and viscosity of the solvent. Equation (8) allows obtaining the concentration measured by a receiver located at a distance $x = r$ [nm] from the transmitter as a function of time. We can observe that the concentration initially measured by the receiver is zero, but it sharply increases until reaching its maximum. The time instant at which this maximum occurs can be interpreted as the pulse delay. After the concentration peak is reached, the impulse response slowly decreases, forming a long tail due to the effect of diffusion. In 1-D environments, it is easy to obtain the closed form solution for the first hitting probability function, since the nanoparticles diffusing along the fluid hit the receiver with probability 1, thus representing a recurrent process. The expression of the first hitting probability F_h (r_0, t) for a point source in 1-D environment is

$$F_h(r_0,t) = \frac{r_0}{\sqrt{4\pi Dt^3}} e^{-r_0^2/4Dt}, \tag{9}$$

where r_0 [nm] is the distance to the receiver. On the other side, in 3-D environments solving the first hitting probability function is a hard surface integration or differential equation problem, since there is a nonzero probability for a diffusing nanoparticle to miss the receiver [52]. As a solution for the 3-D case, Yilmaz et al. [48] simulate the first hitting process following a Gaussian distribution for each movement at each dimension in one time step made by messenger molecules, i.e.

$$\Delta x_i \simeq \mathcal{N}(0, 2D\Delta t), \tag{10}$$

where Δx_i [nm] is the displacement at the i-th dimension (with $i = \{1, 2, \ldots, n\}$), Δt [ns] is the time step, and assuming a reflection or removal of those particles that hit to transmitter or receiver, respectively.

In some biomedical applications, like extra-vivo sensing, a *multi-source nanoparticle emission* process can be required. As an instance, *target cell detecting* can be tailored to the detection of DNA alterations of the BReast CAncer susceptibility gene 1 (BRCA1). As known, BRCA1 is a human gene belonging to a class of genes known as tumor suppressors. Mutation of these gene causes a genetic susceptibility to breast cancer, and changes in its alternative splicing profile have been associated with malignant transformations that greatly increase woman's risk of developing breast cancer [53, 54].

In this scenario, the emission process should be designed to allow the capture of target BRCA1 DNA sequences through a set of unit cells, each of them composed of square gold patches (*receptors*) deposited on a silica substrate. Each receptor is functionalized with the structure of the BRCA1 splice with the corresponding sandwich assays. The chemical receptors are BRCA1 alternative splice variants i.e., $\Delta(5q, 6)$, $\Delta(9, 10)$, and $\Delta(11q)$ [55]. Figure 8a depicts the end-to-end physical model for the detection of BRCA1 DNA alterations [25].

In this case, each nanomachine can emit a particular type of nanoparticles (i.e., with a given shape and size), and the chemical receptors are accordingly

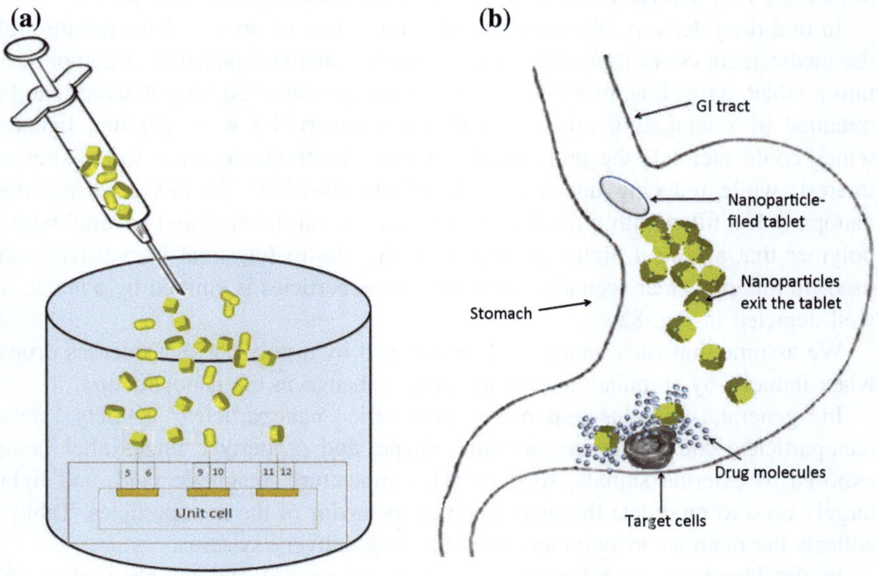

Fig. 8 End-to-end physical model in (**a**) sensing (i.e., multiple-detection of DNA alterations) [25], and (**b**) drug delivery (i.e., stomach disease therapy) applications

functionalized for the capture of one type of nanoparticle. Then, the transmitter (i.e., represented by the syringe) is comprised of three nanoparticles flows, injected all together in a sink with a biological tissue, and each constituted of different shapes nanoparticles (see Fig. 8a). The receiver is represented by the unit cell of a sensing platform, as described in the next section.

A multi-source nanonetwork can suffer from synchronous and asynchronous transmissions, which can degrade network performance with an increase of interference. The same consideration exists at the receiver side, where a selective reception of nanoparticles occurs. As expected, a nanoparticle (a) synchronous transmission corresponds to a (a) synchronous reception. In the case of *asynchronous emission*, just one nanomachine has transmitted a flow of nanoparticles, and then just one kind of ligand-receptor bind can occur (e.g., the nanomachine 2 has transmitted a flow of nanocubes), while in the *synchronous* case, all the nanomachines have transmitted the own nanoparticles.

In the case of synchronous emission, nanoparticles detection issues are limited and not likely to occur since each nanoparticle is allowed forming a ligand-receptor binding with a given receptor. As a result, no ambiguity issue regarding the nanoparticle detection can occur. However, during the diffusion process, the flows of nanoparticles can be affected by interferences (i.e., other nanoparticles can be recognized as belonging to the same flow), and this provides a change in the emission rate of the single nanomachine.

Finally, in oral drug delivery systems, the drug delivery process can occur via different modalities, such as oral ingestion, and injection.

In [22], the diffusion of nanoparticles within the human body is modelled as a diffusion-based nanonetwork, under the specific features of the blood flow.

In oral drug delivery, the recommended total dose of drug is delivered through the medicament concentration filled into a sachet, and encapsulated or compressed into a tablet, as well as in a liquid fluid. Spatially controlled drug delivery can be obtained by conjugating drug-encapsulated nanoparticles with targeting ligands, which could facilitate the preferential delivery of nanotherapeutics to the sites of interest, while reducing undesired side effects elsewhere. In this case, metallic nanoparticles, filled with typical drugs for antitumoral therapy, and covered with a polymer that allows a higher resistance to the Gastro-Intestinal tract barrier, are used. In this particular scenario, the flow of nanoparticles is emitted by a tablet, as well depicted in Fig. 8b.

We assume that each nanoparticle is covered by a polymer that releases drugs, when induced by stimulation, able to cause a change in the nanostructure.

In general, stimulus-responsive polymeric nanoparticles, namely smart nanoparticles, could undergo structure, shape, and property changes after being exposed to external signals, such as pH, temperature, magnetic field, and light, largely used to modulate the macroscopical behavior of the nanoparticles. Table 1 collects the main smart polymers used for drug delivery systems.

In the literature several works have focused on pH and temperature as the predominant stimulus signals, so that pH- and thermo-responsive nanoparticles

Table 1 Examples of smart polymers used for drug delivery systems [74]

Stimulus	Polymer	Drug released
pH	Poly(methacrylic-g-ethylene glicol)—p (MMA-g-EG)	Insulin
Electric field	Poly(methacrylic acid)—PMA	Pilocarpine and rafinose
Glucose concentration	Poly(methacrylic acid-co-butyl methacrylate)	Insulin
Temperature	Layer of Chitosan Pluronic on PLGA microparticles	Indomethacin
Morphine concentration	Methyl vinyl ether-Co-anhydride maleic copolymer	Naltrexone
Urea concentration	Methyl vinyl ether-Co-anhydride maleic copolymer	Hydrocortisone

have been extensively studied [56–59]. As an instance, the most commonly used thermo- and pH-sensitive segments are poly(N-isopropy- lacrylamide) (PNIPP Am) and poly(acrylic acid) (PAA), respectively [59–62].

3.2 *Diffusion Process*

The characterization of the diffusion process derives from the capability to learn from biology and has mainly inspired molecular nanocommunication systems.

In Fig. 9, we can see an example of how the diffusion process works. Independently from the type of communication (i.e., molecular or electromagnetic), it is worth to discuss about the main physical principles of the fluid dynamics that are the milestones of the diffusion process. In fact, the main mechanism that drives the communication among nanoparticles is the free diffusion of nanoparticles in the space.

First of all it is useful to start from the beginning, the Fick's First Law of Diffusion. This law sums up the diffusion process through the following mathematical expression [63]:

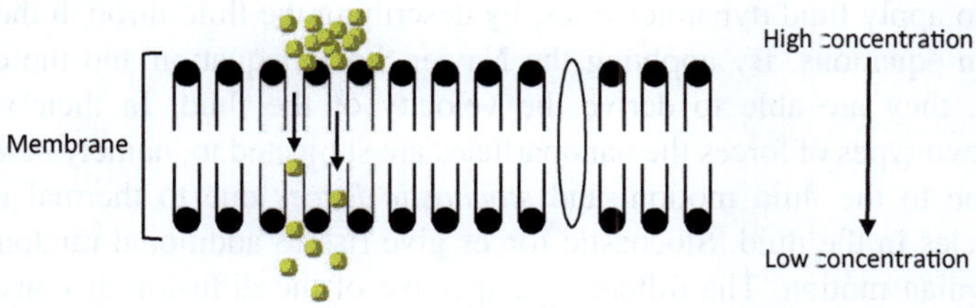

Fig. 9 Schematic of diffusion process, where nanoparticles move from high to low concentration levels

$$J_T(x,t) = -D\nabla_c(x,t) \tag{11}$$

where $\nabla_c(x,t)$ represents the concentration gradient per unit length. It follows that $J_T(x, t)$ is the flow of solute, and then Eq. (9) can be identified with the nanoparticle concentration flux at the output of the transmitter, dependent on the nanoparticle concentration gradient at time t and position x.

In practice, the Fick's First Law of Diffusion describes the diffusion as the process where a solute moves from a region of high concentration to a lower concentration area. Usually, there is a predominant direction of the flow, from the highest to the smaller concentration region, but diffusion is a complex phenomenon and also occurs in the opposite direction.

This more complex phenomenon can be described as a function of time by describing the net change in this way:

$$\frac{\partial c}{\partial t} = \frac{1}{dx}[J_T(x) - J_T(x+dx)] = -\frac{\partial J_T}{\partial x}, \tag{12}$$

and so, we can derive the Fick's Second Law of Diffusion as:

$$\frac{\partial c}{\partial t} = D\frac{\partial^2 c}{\partial x^2}. \tag{13}$$

Some studies present the propagation of *messenger nanoparticles* by means of free diffusion. In [64], the authors discuss about two simple models of pheromones diffusion in still air that are different in terms of the emitting scheme. The first one treats instantaneous emission whereas the second considers the continuous emission of pheromones. Communications via Diffusion (CvD) in a fluid are investigated in [65], where the authors show how the most of the studies about CvD systems is based on a very simple assumption, namely the systems are with a single transmitter and a single receiver. They analyzed a more realistic situation, that is, there are more communicating couples that make the systems much more complicated. In fact, the authors focus on the interference between closely located transmitting couples that share the same medium.

In [21] authors consider the diffusion process from a very interesting perspective, and for a very important application i.e., the development of nanorobots with sensors for nanomedicine. They show how, for an intra-body application, it is possible to apply fluid dynamics rules, by describing the fluid through the classical continuum equations. By applying the Navier-Stokes equation and the continuity condition, they are able to derive the velocity of the fluid. In their work they consider two types of forces the nanoparticles are subjected to, namely *deterministic forces*, due to the fluid motion, and *stochastic forces* due to thermal motion of nanoparticles in the fluid. Stochastic forces give rise to additional random motion, i.e., Brownian motion. The different perspective of the diffusion as considered by Cavalcanti et al. consists on the fact that a nanorobot is considered at rest with respect to the rest to the fluid, but it will be able to collect biomolecules (that is the main task of the nanorobot), due to the diffusion of the biomolecules.

Based on the main concept and the same principles of the CvD, authors in [25] show how the diffusion process works in the case of electromagnetic nanoparticles. More specifically, the authors show that gold nanoparticles can behave similarly to the molecules. In particular, different geometries are assumed, but the different nanomachines are similar in terms of volume occupancy. This characteristic simplifies the analysis of the diffusion process, even if the authors consider a three-dimensional space i.e., a lattice, to describe the diffusion phenomenon.

3.3 Reception Process

In sensing applications designed for the detection of chemical and biological agents, nanoparticles (specifically, *nanosensors*) are usually composed of (*i*) a *recognition system* or receptor, and (*ii*) a *transduction mechanism*.

Nanosensors can be based on the use of noble metal nanoparticles [66] (e.g., gold nanoparticles) and are functionalized with a biological receptor, such as an antibody, that is bound to the specific antigen of a given disease. The ligand-receptor binding [67] is, therefore, the mechanism that starts the reception process. In several works [68, 69], a particle receiver model is developed by taking the ligand-receptor binding mechanism into account.

The binding event between the recognition element and the target can alter physicochemical properties of the transducer that, in turn, can generate a detectable response signal. In particular, the binding creation changes the resonant frequency of the surface plasmons resulting from light irradiation. Through the LSPR technique, it is then possible to perform chemical or biological sensing.

The two main operations (i.e., binding creation, and signalling transduction) that occur in the reception process are depicted in Fig. 10, for the case of sensing application through metallic nanoparticles.

From the above considerations, it follows that the reception process has the task of (*i*) sensing the particle concentration at the receiver, and (*ii*) producing,

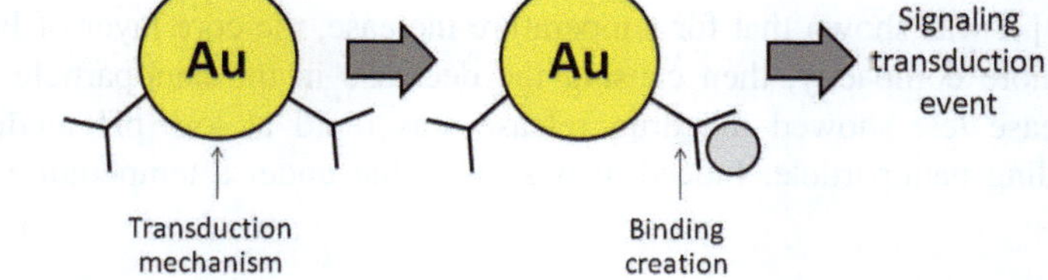

Fig. 10 Schematization of the reception process in electromagnetic nanonetworks

accordingly, the output signal. As an example of sensing application, in the case of detection of BRCA1 DNA alterations [25], receptors take place only in correspondence of BRCA1 alternative splice variants, i.e., $\Delta(5q, 6)$, $\Delta(9, 10)$, and $\Delta(11q)$ [55].

The binding reaction occurs when the receptor was not previously bound to a particle. A chemical receptor, depending whether it is involved in a complex or not, triggers an output signal accordingly. The output signal of the end-to-end model results proportional to the rate of change in the ratio τ of the number of bound chemical receptors over the total number of chemical receptors. The trend is to reach a ratio between the number of bound chemical receptors over the total number of chemical receptors. The variation of the number of bound chemical receptors n_c (t) inside the reception space at time t is related to the number of receptor N_R according to the following equation [50]:

$$\frac{dn_c(t)}{dt} = N_R \frac{d\tau}{dt}. \tag{14}$$

According to the ligand-receptor binding reaction kinetic, when a set of nanoparticles (i.e., *NP*), accordingly functionalized with a given antigen, are emitted by the transmitter nanomachine, encounters with receptors (i.e., *R*) lying on the receiver, nanoparticles bind the receptors. These bound nanoparticle-receptors (i.e., *NP* + *R*) constitute complexes (i.e., bound receptors), as well as it is possible to release molecules *NP* from receptors *R*, respectively according to the following chemical reactions [68],

$$(NP + R) \underset{k_{-1}}{\overset{k_1}{\rightleftarrows}} C, \tag{15}$$

where k_1 [μmol/L/s] and k_{-1} [s $^{-1}$] is the *rate of binding reaction*, and the *rate of release reaction*, respectively.

In the case of drug delivery systems that make use of hollow metallic nanoparticles opportunely filled with drug concentration to be released locally, as the oral administration of a tablet for stomach disease therapy, depicted in Fig. 8b, the further step of *drug release* needs to be accomplished.

The breakage of the polymer that covers the nanoparticle is also realized by means of an electromagnetic impinging plane wave that causes the polymer chains to collapse, exposing the holes on the metallic nanoparticle, and thereby releasing the pre-loaded effector.

In [70] it was shown that for temperature increase, the core layer of PNIPAAm packed more compactly, then causing the decrease in the nanoparticle size. The drug release test showed the drug release was rapid at low pH medium from drug-loading nanoparticle. Indeed, it is known that under a temperature stimulus, the LSPR phenomenon is established, and the free electrons of the nanoparticles follow collectively the electromagnetic oscillations. As a result, the absorbed incident electromagnetic field at the resonant wavelength is converted into heat through the photothermal effect [71]. The increase of temperature causes the polymer chains to collapse, thus exposing the holes on the nanoparticle, and thereby releasing the drug concentration. Therefore, the drug is released at the appropriate time and place.

Also, Li et al. synthesized a Y-shaped pH-/thermo-responsive copolymer P (UA-Y-NIPAAm) composed of pH sensitive poly(undecylenic acid) (PUA) segment and temperature-sensitive poly(N- isopropylacrylamide) (PNIPAAm) segment [72]. Soppimath et al. prepared novel temperature responsive nanoparticles with the transition at physiological temperature [73], whose structure could be deformed in acid environments to induce the release of encapsulated drug.

To summarize, the drug release mechanism based on polymers aims to [74]:

- improve the pharmaceutical profile and stability of a drug,
- ensure its correct concentration,
- achieve maximum biocompatibility,
- minimize side effects,
- stabilize the drug in vivo and in vitro,
- facilitate the accumulation of the drug at a specific site of action,
- increase exposure time in the target cell.

The above-mentioned nanoscale sensing technologies require the use of external excitation and measurement equipment to operate. In the vision of future nanosensors devices [3] equipped with nanosensors, nanoactuator, nano-memory, nano-antenna, nano-EM transceiver, nano-processor and nano-power unit, nanodevices will be able to exchange information through *nano-electromagnetic communications*. By means of communication, the nanomachines will be able to accomplish more complex missions in a cooperative manner. As an example, nanosensors will be able to transmit the sensed information in a multi-hop fashion to a sink or a command centre.

For electromagnetic nanonetworks, the use of modulation and channel sharing mechanism based on the asynchronous exchange of femtosecond-long pulses, which are transmitted following an on-off keying modulation spread in time, has been proposed [12]. In [75], a receiver architecture for electromagnetic nanonetworks that make use of pulse-based modulation has been proposed. The receiver is designed in order to be simple and robust, and it is based on a Continuous-Time Moving Average (CTMA) symbol detection scheme. This scheme bases its decision in the received signal power maximum peak after the CTMA, which is implemented with a single low-pass filter. Afterwards, to decode the symbol, this maximum is compared with a previously defined threshold.

4 Nanoparticulate Sensing and Drug Delivery

Recently, by functionalizing the nanoparticles with biological agents such as antibodies or single stranded DNA chains, nanoparticles are forced to bind preferably to specific target cells. This aspect is the focus of extra-vivo sensing application, such as the multi-detection of DNA alterations of the BRCA1.

Several works addressing the issue of DNA detection exploit metallic nanoparticles, in order to enhance the Raman signal of fluorescent target absorbed on the metallic surface. The predominant mechanism responsible for this enhancement arises from the local electromagnetic field intensification at the surface of noble metal nanoparticles. This method is called Surface-Enhanced Raman Scattering (SERS), [76]. Among main SERS-based studies, the majority uses fluorophores as Raman labels, but it has been shown that they reduces the Raman scattering efficiency, due to the displacement of fluorophores by biological media [76].

On the other hand, techniques based on the use of metallic nanoparticles illuminated by an electromagnetic wave (visible and near infrared region) are largely exploited, as alternative approaches for gene alterations detection [77, 78]. More specifically in [25], the LSPR phenomenon for the multi-detection of BRCA1 DNA alteration, when biological gold nanoparticles are captured at the receiver, has been presented.

In this approach, it is crucial to obtain different and independent electromagnetic responses from each nanoparticle in terms of resonant wavelength, amplitude and magnitude width. Cube, rod and elliptical cylinder nanoparticles have been functionalized with the corresponding *probe sequences* of alternative splicing junctions of BRCA1, as shown in Fig. 11. The three corresponding DNA *capture sequences* of alternative splicing junctions of BRCA1 are allocated on three square gold patches of silica substrate, as depicted in Fig. 11.

Furthermore, Fig. 11 shows the unit cell of the sensing platform, composed of three square gold patches, functionalized with three different Capture Sequences (specifically, depicted in blue, red and green). The sensing platform is excited by an impinging plane wave, the excitation is employed to analyze the electromagnetic properties, in terms of extinction cross-section spectra, as reported in Fig. 12 for the case of cube, rod, and elliptical cylinder nanoparticles, in binding and no-binding.

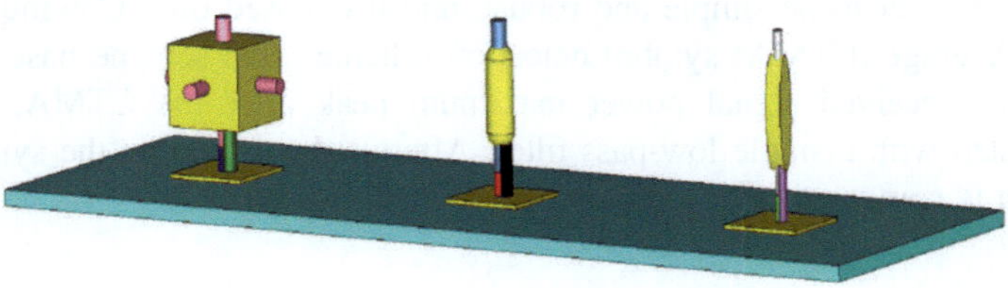

Fig. 11 Ligand-receptor binding of three nanoparticles. *Green*, *black* and *violet* sections represent the target sequences for each nanoparticle [25]

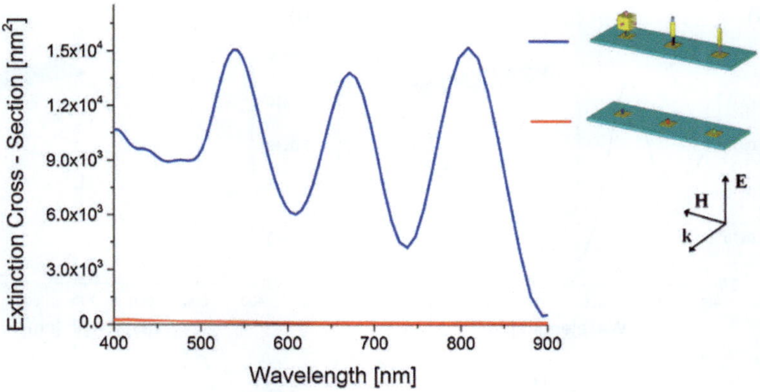

Fig. 12 Synchronous nanoparticle detection [25]. Extinction cross-section spectra obtained when all the target sequences are bound (*blue line*), and are not bound (*red line*)

The meaning of this result is the lack of overlapping of the extinction cross-section spectra generated by the nanoparticles. In other words, it is crucial that the resonant peaks do not overlap, in order to obtain a multi-sensing approach. This result has been reached by using the aforementioned analytical model; in fact, they have allowed tuning the electromagnetic response for each nanoparticle. In addition to this, two cases of nanoparticle detection have been considered: the *asynchronous* and *synchronous*. In the first case, just one nanomachine has transmitted a flow of nanoparticles, and so we have just one kind of ligand-receptor bind, while in the synchronous case, all the nanomachines have transmitted the own nanoparticles, as shown in Fig. 12. In this way, it is still possible to detect the DNA alteration as in the asynchronous case. The synchronous case is very important for multi-detection, since it allows the detection of different DNA alterations.

To demonstrate this ability, the extinction cross-section spectra for cube, rod and elliptical cylinder binding in the asynchronous case are shown in Fig. 13a, b and c, respectively. Figure 13 shows the capability of the nanosensors to distinguish the different BRCA1 alterations. Furthermore, these results prove the structure capability to be able to reveal BRCA1 alterations in the case of double binding e.g. (*i*) nanocube, and nanorod particle binding, (*ii*) nanocube and elliptical cylinder binding, and (*iii*) nanorod and elliptical cylinder binding.

In the last few years several researches have focused the attention on the drug release by using functionalized gold nanoparticles [79]. The goal of this technique is to use specifics nanoparticles as drug carriers. Among the different used approaches, the photothermal effect appears as a great promise in the drug delivery field. When a metallic nanoparticle is illuminated by electromagnetic field in the visible and near infrared frequencies regime, the LSPR-induced local heating causes the release of loaded drug; this leads to enhanced drug efficacy with high spatio-temporal resolution, and very few side effects. In fact, the high electric field concentration on the nanoparticles surface ensures the heat of particle and,

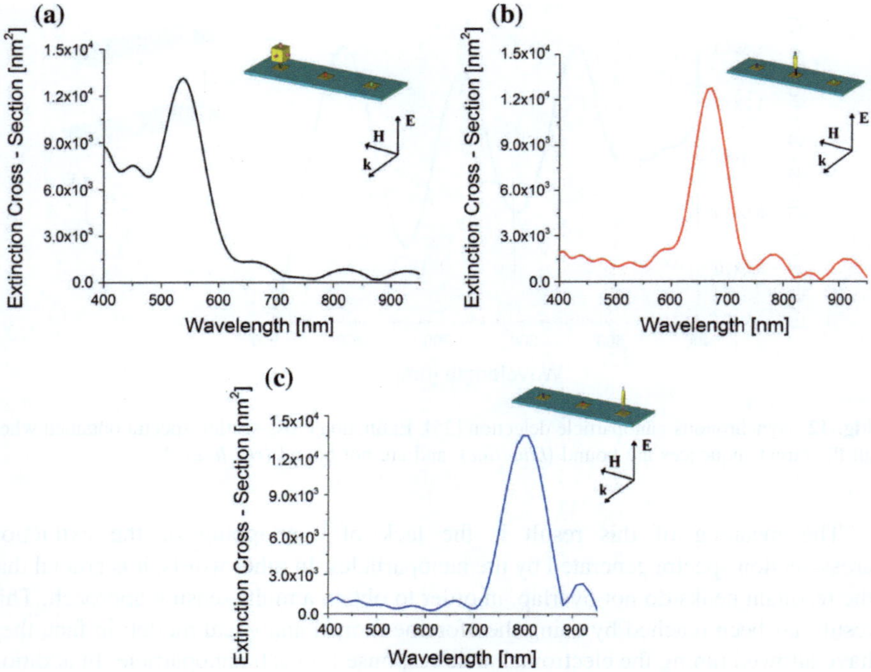

Fig. 13 Asynchronous detection [25]. Extinction cross-section spectra obtained for binding of **a** nanocube, **b** nanorod, and **c** elliptical cylinder nanoparticle

therefore, it causes the polymer chains to collapse, exposing the holes on the nanoparticles and thereby releasing the pre-loaded effector.

The same principle used in [25] could be efficiently applied to *in-vivo* drug delivery system. More specifically, it is possible to use different shapes of nanoparticles to release more types of drug. As shown in [80], there are cases that need of a multiple drug release in different times. Selective releases were induced by selective hot-spots of gold nanorods by electromagnetic field irradiation at the resonance frequency of the nanorod. In this way, at specific wavelengths it is possible to excite one type of gold nanorods, and selectively release one type of DNA strand [80].

By using the aforementioned nanoparticles, it is possible to load different drugs into distinct different nanoparticles shapes and by tuning the electromagnetic field frequency will be possible to excite specific nanoparticles. Therefore, by exploiting the photothermal effect, only selective drugs will be released, in the case of drug delivery applications and only selective DNA chains will be detected, in the case of sensitive applications. As results, the use of different nanoparticles shapes enables independent control over the release of each drug by tuning the nanoparticles, leading to a programmed release of multiple drugs.

Finally, another work dealing on the use of nanoparticles for sensing applications (i.e., detection of DNA alterations) is given in [81]. Specifically, two-layers nanostructures are considered, comprising of a SiO2 core, and a gold shell with different thicknesses. By varying the ratio of the core/shell radius, it is possible to tune the electromagnetic response. Indeed, the optical cross-section propriety obtained for nanoshells with different radii is higher than that one obtained for the nanoshells with the same volumes.

Simulation results in [81] show the extinction cross-section spectra in the case of (*i*) synchronous nanoparticle transmission, (*ii*) asynchronous nanoparticle transmission, and (*iii*) no nanoparticle reception. The extinction cross-section spectra show a peak at 550 nm for both synchronous and asynchronous transmission, but the magnitude of the spectral signal is higher for the synchronous transmission, with respect to the asynchronous case. In both two cases, no shift of the resonant wavelength occurs. From a physical point of view, this means that there are no coupling effects among nanoparticles. Finally, as expected, in the case of no nanoparticle reception, the extinction cross-section spectrum is approximately zero.

5 Conclusions

In this chapter we have addressed recent advances in electromagnetic nanonetworks for biomedical applications. Starting from the description of most used nanoparticles for sensing and drug delivery systems (i.e., nanorods, nanocubes, bow-tie nanoparticles, etc.), we have presented the analytical models behind their specific properties. By exploiting the well-known LSPR phenomenon, electromagnetic nanoparticles are shown to be very suitable for sensing (i.e., analysis of concentration of specific substances), and drug delivery applications (i.e., nanoparticles are intended as drug carriers for localized drug release).

Furthermore, under the assumption of similarity of electromagnetic nanoparticles with biological molecules (i.e., small size of the structure), we rely on specific laws in the nanoscale. The channel model linking one transmitter to one receiver is generally represented by a liquid environment (i.e., the human blood flood or the biological tissue), and is subjected to specific processes (i.e., nanoparticle transmission, diffusion, and reception). Once nanoparticles are captured by the receptors, the detection process occurs by means of LSPR phenomenon.

Specific sensing and drug delivery applications of nanoparticles are also presented, regarding the detection of DNA alterations of BRCA1 gene, as well as for fighting stomach diseases. The case of different shapes of nanoparticles has been addressed, both for sensing and drug delivery applications, by showing how different geometric forms can be exploited for multi-detection sensing and to effectively regulate programmed drug delivery, respectively.

Simulation results assess the validity of the use of electromagnetic nanoparticles for these biomedical applications.

As a future investigation, novel nanomaterials are promising for the design of innovative nanodevices. As an instance, graphene-based nanoantennas can provide outstanding sensing capabilities, as well as graphene-based transistors are not only smaller, but predictably faster too.

References

1. Freitas RA Jr (2005) "What is nanomedicine?", nanomedicine: nanotechnology. Biol Med 1 (1):2–9
2. Freitas RA (1999) Nanomedicine, Volume I: basic capabilities, landes bioscience, Georgetown, TX
3. Akyildiz IF, Jornet JM (2010) Electromagnetic wireless nanosensor networks. Nano Commun Netw 1:3–19
4. Freitas RA Jr (2005) Current status of nanomedicine and medical nanorobotics. J Comput Theor Nanosci 2:1–25
5. Akyildiz IF, Jornet JM, Pierobon M (2011) Nanonetworks: a new frontier in communications. Commun ACM 54(11):84–89
6. Akyildiz IF, Brunetti F, Blzquez C (2008) Nanonetworks: a new communication paradigm. Comput Netw 52(12):2260–2279
7. Jornet JM, Akyildiz IF (2010) Channel capacity of electromagnetic nanonetworks in the terahertz band. In: Proceedings of the IEEE international conference on communications, ICC 2010, Cape Town, South Africa, May 2010
8. Wang X, Lou X, Wang Y, Guo Q, Fang Z, Zhong X, Mao H, Jin Q, Wu L, Zhao H, Zhao J (2010) QDs-DNA nanosensor for the detection of hepatitis B virus DNA and the single-base mutants. Biosens Bioelectron 25(8):1934–1940
9. Abraham A, Kannangai R, Sridharan G (2008) Nanotechnology: a new frontier in virus detection in clinical practice. Indian J Med Microbiol 26(4):297–301
10. Akyildiz IF, Fekri F, Sivakumar R, Forest CR, Hammer BR (2012) Monaco: fundamentals of molecular nano-communication networks. IEEE Wirel Commun 19(5):12–18
11. Jornet JM, Akyildiz IF (2011) Channel modeling and capacity analysis for electromagnetic wireless nanonetworks in the terahertz band. IEEE Trans Wirel Commun 10(10):3211–3221
12. Jornet JM, Akyildiz IF (2011) Low-weight channel coding for interference mitigation in electromagnetic nanonetworks in the terahertz band. In: Proceedings of the IEEE international conference on communications (ICC 2011), June 5–9, Kyoto, Japan
13. Jornet JM, Akyildiz IF (2011) PHALME: a physical layer aware mac protocol for electromagnetic nanonetworks. In: Proceedings of the IEEE international conference on communications (ICC 2011), June 5–9, Kyoto, Japan
14. Jornet JM, Akyildiz IF (2012) Joint energy harvesting and communication analysis for perpetual wireless nanosensor networks in the terahertz band. IEEE Trans Wirel Commun 11 (3):570–580
15. Svenson S, Prud'homme RK (2012) Multifunctional nanoparticles for drug delivery applications: imaging, targeting, and delivery. Springer
16. Swami A, Shi J, Gadde S, Votruba A, Kolishetti N, Farokhzad O (2012) Nanoparticles for targeted and temporally controlled drug delivery. In: Svenson S, Prud'homme R (eds) Multifunctional nanoparticles for drug delivery applications: imaging, targeting, and deliver. Springer
17. Zhou Y, Kong Y, Kundu S, Cirillo J, Liang H (2012) Antibacterial activities of gold and silver nanoparticles against escherichia coli and bacillus calmette-gurin. J Nanobiotechnol 10

18. Hossen M, Kajimoto K, Akita H, Hyodo M, Harashima H (2012) Vascular-targeted nanotherapy for obesity: unexpected passive targeting mechanism to obese fat for the enhancement of active drug delivery. J Control Release 163:101–110
19. Vllasaliu D, Alexander C, Garnett M, Eaton M, Stolnik S (2012) Fc-mediated transport of nanoparticles across airway epithelial cell layers. J Control Release 158:479–486
20. Bhumkar D, Joshi H, Sastry M, Pokharkar V (2007) Chitosan reduced gold nanoparticles as novel carriers for transmucosal delivery of insulin. Pharm Res 24
21. Cavalcanti A, Shirinzade B, Freitas RA, Hogg T (2008) Nanorobot architecture for medical target identification. Nanotechnology 19(1)
22. Chahibi Y, Pierobon M, Song S, Akyildiz I (2013) A molecular communication system model for particulate drug delivery systems. IEEE Trans Biomed Eng
23. Nakano T, Moore MJ, Okaie Y, Enomoto A, Suda T (2012) Swarming biological nanomachines through molecular communication for targeted drug delivery. In: Proceedings of IEEE conference on soft computing and intelligent systems and symposium on advanced intelligent systems, November 2012
24. Loscrì V, Natalizio E, Mannara V, Aloi G (2012) A novel communication technique for nanobots based on acoustic signals. In: Proceedings of the 7th international conference on bio-inspired models of network, information, and computing systems, ser. Bionetics'12, Lugano, Switzerland
25. Iovine R, Loscrì V, Pizzi S, Tarparelli R, Vegni AM (2013) Model of multi-source nanonetworks for the detection of BRCA1 DNA alterations based on LSPR phenomenon. Adv Nanoparticles 2(4):301–312
26. Dykman L, Khlebtsov N (2012) Gold nanoparticles in biomedical applications: recent advances and perspectives. Chem Soc Rev 41(6):2256–2282
27. Kumar A, Boruah BM, Ling X-J (2011) Gold nanoparticles: promising nanomaterials for the diagnosis of cancer and HIV/AIDS. J Nanomaterials 2011:1–17
28. Patra CR, Bhattacharya R, Mukhopadhyay D, Mukherjee P (2010) Fabrication of gold nanoparticles for targeted therapy in pancreatic cancer. Adv Drug Deliv Rev 62(3):346–361
29. Cho EC, Glaus C, Chen J, Welch MJ, Xia X (2010) Inorganic nanoparticle-based contrast agents for molecular imaging. Trends Mol Med 16(12):561–573
30. Cai W, Gao T, Hong H, Sun J (2008) Applications of gold nanoparticles in cancer nanotechnology. Nanotechonology 2008(1):17–32
31. Salamon Z, Macleod HA, Tollin G (1997) Surface plasmon resonance spectroscopy as a tool for investigating the biochemical and biophysical properties of membrane protein systems. I: theoretical principles. Biochimica and Biophysica Acta 1331(2):117–129
32. Sagle LB, Ruvuna LK, Ruemmele JA, Van Duyne RP (2011) Advances in localized surface plasmon resonance spectroscopy biosensing. Nanomedicine 6(8):1447–1462
33. Moores A, Goettmann F (2006) The plasmon band in noble metal nanoparticles: an introduction to theory and applications. New J Chem 30:1121–1132
34. Van Bladel JG (2007) Electromagnetic fields. Wiley, Hoboken
35. La Spada L, Iovine R, Vegni L (2012) Nanoparticle electromagnetic properties for sensing applications. Adv Nanoparticles 1:9–14
36. Iovine R, La Spada L, Vegni L (2013) Nanoparticle device for biomedical and optoelectronics applications. COMPEL 32(5):1596–1608
37. Xu X, Ying Y, Li Y (2011) Gold nanorods based lspr biosensor for label-free detection of alpha-fetoprotein. Procedia Eng 25:67–70
38. La Spada L, Iovine R, Vegni L (2013) Electromagnetic modeling of ellipsoidal nanoparticles for sensing applications. Opt Eng 52(5):1–5
39. Iovine R, La Spada L, Vegni L (2014) Optical properties of modified nanorod particles for biomedical sensing. IEEE Trans Magn 50(2) (to appear)
40. Tanaka A, Nakamura B (2012) Optical imaging: technology, methods and applications. Nova Science Publisher
41. Iovine R, La Spada L, Vegni L (2013) Modified bow-tie nanoparticles operating in the visible and near infrared frequency regime. Adv Nanoparticles 2(1):21–27

42. Suda T, Moore M, Nakano T, Egashira R, Enomoto A (2005) Exploratory research on molecular communication between nanomachines. In: Proceedings of genetic and evolutionary computation conference, (GECCO'05). ACM
43. Atakan B, Akan OB (2008) On molecular multiple-access, broadcast, and relay channels in nanonetworks. In: Proceedings of the ICST/ACM Conference BIONETICS 2008, Japan, Nov 25–28, 2008
44. Einolghozati A, Sardari M, Beirami A, Fekri F (2011) Capacity of discrete molecular diffusion channels. In: Proceedings of international symposium on information theory (ISIT 2011), Saint Petersburg, Russia, July 2011
45. Keramidas A, Moorhouse AJ, Schofield PR, Barry PH (2004) Ligand-gated ion channels: mechanisms underlying ion selectivity. Prog Biophys Mol Biol 86(2):161–204
46. Model MA, Omann GM (1995) Ligand-receptor interaction rates in the presence of convective mass transport. Biophys J 69(5):1712–1720
47. Atakan B, Akan OB (2008) On channel capacity and error compensation in molecular communication. Trans Comput Syst Biol X 59–80
48. Yilmaz HB, Kim N-R, Chae C-B (2014) Effect of ISI mitigation on modulation techniques in communication via diffusion. In: Proceedings of 1st ACM international conference on nanoscale computing and communication, Atlanta, May 13–14, 2014
49. Atakan B, Akan OB (2010) Deterministic capacity of information flow in molecular nanonetworks. Nano Commun Netw (Elsevier) 1(1):31–42
50. Pierobon M, Akyildiz I (2010) A physical end-to-end model for molecular communication in nanonetworks. IEEE J Sel Areas Commun 28(4):602–611
51. Nakano T, Okaie Y, Liu J-Q (2012) Channel model and capacity analysis of molecular communication with brownian motion. IEEE Commun Lett 16(6)
52. Redner S (2001) A guide to first-passage processes. Cambridge University Press
53. Llatser I, Alarcòn E, Pierobon M (2011) Diffusion-based channel characterization in molecular nanonetworks. In: Proceedings of IEEE conference on computer communications workshops (INFOCOM WKSHPS), pp 467–472, 10–15 April 2011
54. Kadouri L, Hubert A, Rotenberg Y, Hamburger T, Sagi M, Nechushtan C, Abeliovich D, Peretz T (2007) Cancer risks in carriers of the BRCA1/2 ashkenazi founder mutations. J Med Genet 44(7):467–471
55. Thompson D, Easton D, Consortium BCL (2002) Cancer incidence in BRCA1 mutation carriers. J Natl Cancer Inst 94(18):1358–1365
56. Li YY, Dong HQ, Wang K, Shi DL, Zhang XZ, Zhuo RX (2010) Stimulus-responsive polymeric nanoparticles for biomedical applications. Sci China Chem 53(3):447–457
57. Schmaljohann D (2006) Thermo- and pH-responsive polymers in drug delivery. Adv. Drug Deliver Rev 58(15):1655–1670
58. Suchaoin N, Chirachanchai S, Perrier S (2009) PH- and thermo-multi-responsive fluorescent micelles from block copolymers via reversible addition fragmentation chain transfer (RAFT) polymerization. Polymer 50(17):4151–4158
59. Zhang QS, Zha LS, Ma JH, Liang BR (2009) A novel route to prepare pH– and temperature-sensitive nanogels via a semibatch process. J. Colloid Interf Sci 330(2):330–336
60. Ganta S, Devalapally H, Shahiwala A, Amiji M (2008) A review of stimuli-responsive nanocarriers for drug and gene delivery. J. Control Release 126(3):187–204
61. Alexander C, Shakesheff KM (2006) Responsive polymers at the biology/materials science interface. Adv Mater 18(24):3321–3328
62. Rapoport N (2007) Physical stimuli-responsive polymeric micelles for anti-cancer drug delivery. Prog Polym Sci 32(8–9):962–990
63. Sun L, Yu C, Irudayaraj J (2008) Raman multiplexers for alternative gene splicing. Anal Chem 80(9):3342–3349
64. Cussler EL (1997) Diffusion: mass transfer in fluid systems, 2nd edn. Cambridge University Press
65. Parcerisa L, Akyildiz IF (2009) Molecular communication options for long range nanonetworks. Comput Netw J 53(16): 2753–2766 (Elsevier)

66. Kuran MS, Yilmaz HB, Tugcu T, Akyildiz IF (2012) Interference effects on modulation techniques in diffusion based nanonetworks. Nano Commun Netw 3(1):65–73 (Elsevier)
67. Saha K, Agasti SS, Kim C, Li X, Rotello VM (2012) Gold nanoparticles in chemical and biological sensing. Chem Rev 112(5):2739–2779
68. Atakan B, Akan OB (2007) An information theoretical approach for molecular communication. In: Proceedings of 2nd bio-inspired models of network, information and computing systems, bionetics 2007, pp 33–40, 10–12 Dec 2007
69. Rospars J-P, Krivan V, Lansky P (2000) Perireceptor and receptor events in olfaction. comparison of concentration and flux detectors: a modeling study. Chem Senses 25:293–311
70. Hong SW, Kim DY, Lee JU, Jo WH (2009) Synthesis of polymeric temperature sensor based on photophysical property of fullerene and thermal sensitivity of poly(N-isopropylacrylamide). Macromolecules 42:2756–2761
71. Lee J-H, Jang J-T, Jang J-T, Choi J-S, Moon SH, Noh S-H, Kim J-W, Kim J-G, Park KI, Cheon J (2011) Exchange-coupled magnetic nanoparticles for efficient heat induction. Nat Nanotechnol Lett 6:418–422
72. Li YY, Zhang XZ, Cheng H, Kim GC, Cheng SX, Zhuo RX (2006) Novel stimuli-responsive micelle self-assembled from Y—shaped P(UA-Y—NIPAAm) copolymer for drug delivery. Biomacromolecules 7(11):2956–2960
73. Soppimath KS, Tan DCW, Yang YY (2005) PH-triggered thermally responsive polymer core-shell nanoparticles for drug delivery. Adv Mater 17(3):318–323
74. Vilar G, Tulla-Puche J, Albericio F (2012) Polymers and drug delivery systems. Curr Drug Deliv 9(4):367–394
75. Cid-Fuentes RG, Jornet JM, Akyildiz IF, Alarcon E (2012) Receiver architecture for pulse-based electromagnetic nanonetworks in the terahertz band. In: Proceedings of international conference on communications, Ottawa, Canada, June 10–15
76. Docherty FT, Clark M, McNay G, Graham D, Smith WE (2003) Multiple labelled nanoparticles for bio detection. Faraday Discuss 126:281–288
77. Zhou W, Ma Y, Yang H, Ding Y, Luo X (2011) A label-free biosensor based on silver nanoparticles array for clinical detection of serum p53 in head and neck squamous cell carcinoma. Int J Nanomed 2011(6):381–386
78. Tan YN, Su X, Zhu Y, Lee JY (2010) Sensing of transcription factor through controlled-assembly of metal nanoparticles modified with segmented DNA elements. ACS Nano 4(9):5101–5110
79. Rana S, Bajaj A, Mout R, Rotello VM (2012) Monolayer coated gold nanoparticles for delivery applications. Adv Drug Deliv Rev 64(2):200–216
80. Wijaya A, Schaffer SB, Pallares IG, Hamad-Schifferli K (2009) Selective release of multiple DNA oligonucleotides from gold nanorods. ACS Nano 3(1):80–86
81. Iovine R, Tarparelli R, Vegni AM (2013) Detection of DNA alterations using gold nanoparticles exploiting the LSP phenomenon. In: Proceedings of 21st IEEE international conference on applied electromagnetics and communications (ICECOM), Dubrovnick, Croatia, Oct 14–16, 2013

Communication of Drug Loaded Nanogels with Cancer Cell Receptors for Targeted Delivery

Govind Soni and Khushwant S. Yadav

Abstract The human body is a massive nanoscale molecular communications network composed of billions of interacting cells Drug delivery is an important application of molecular communication. Nanogels are aqueous dispersions of nanoscale size formed by cross-linking of hydrophilic polymers, capable of retaining large amounts of water yet remaining insoluble and maintaining a three-dimensional structure. Nanogel structure enables easy attachment of vector groups for effective communication with cells to reach the desired targeted site. This chapter highlights communication of drug loaded nanogels for targeting cancer through receptors. The chapter critically discusses receptors like- integrin αvβ3, EphA2, folate, Hyaluronan and monoclonal antibody for communication with nanogels.

Keywords Nanogels · Integrin αvβ3–targeting · Epha2 · Folate receptor hyaluronan and monoclonal antibody

1 Introduction

The human body is a massive nanoscale molecular communications network composed of billions of interacting cells [17]. These cells primarily function by nanoscale molecular communications. In molecular communication inside the human body, molecules are used to encode, transmit and receive information [3]. Human biological systems are connected to each other and communicate primarily through molecular transactions. Engineered biological nanomachines communicate with biological systems at the molecular level to enable future applications in human body. Drug delivery through such nanomachines is an important application of molecular communication. These biological nanomachines encode information

G. Soni · K.S. Yadav (✉)
Department of Pharmaceutics, Rajeev Gandhi College of Pharmacy,
Bhopal-42, Madhya Pradesh, India
e-mail: khush.yadav@gmail.com

J. Suzuki et al. (eds.), *Modeling, Methodologies and Tools for Molecular and Nano-scale Communications*, Modeling and Optimization in Science and Technologies 9, DOI 10.1007/978-3-319-50688-3_21

on molecules and release the molecules in the environment, the molecules then propagate in the environment to receiver biological nanomachines, and the receiver biological nanomachines biochemically react with the molecules to decode information [22]. Molecular communication provides mechanisms for nanomachines to communicate by propagating such molecules. Nanogels are one such example of nanomachine used for communicating with cancer cells inside the body at cellular level.

2 Nanomachines, Nanogels and Molecular Communication

Nanomachines and communications come together to propose a fast and unified solution to combat diseases. Biological nanomachines communicate to fulfill the needs of human body. A biological nanomachine may be defined as a device that performs a useful function using components of nanometer-scale and defined molecular structure.

Nanogels are aqueous dispersions of nanoscale size formed by cross-linking of hydrophilic polymers, capable of retaining large amounts of water yet remaining insoluble and maintaining a three-dimensional structure. The phenomenon of cross-linking is closely related to self-assembly principal and leads to formation of a swollen network having nanoscale size [31]. Such cross linking may be physical or chemical. Nanogels are promising drug delivery carriers due to their high loading capacity, high stability, and responsiveness to environmental factors, such as ionic strength, pH, and temperature that are unprecedented for common pharmaceutical nanocarriers [32]. Loading efficiency in nanogels is high and is achieved spontaneously through electrostatic, vander Waals and/or hydrophobic interactions between the agent and the polymer matrix. As a result, nanogels collapse forming stable nanoparticles, in which biological agent becomes entrapped.

Molecular communication enables nanomachines to exchange information with each other by emitting molecules to their surrounding environment (Nakano and Moore, [21]. For example, the molecular communication paradigms of nervous system is the nanoscale neuro-spike communication channel, for cardiovascular system it is the action potential based cardiomyocyte molecular communication channel, and hormonal molecular communication channel is for endocrine nanonetworks. A network of communicating nanomachines can be programmed to share nanoscale information over a network so as to fulfill more complex tasks such as drug delivery. The receptors located on the cell membrane; act as molecular transceivers for inter-cell communication.

Nanogels are three dimensional networks capable of detection and localization of cells. They respond to malicious agents and cells, such as cancer cells, resulting in a less aggressive and invasive treatments compared to the existing ones.

A typical communication process includes the following phases [2]: The encoding phase (in which the transmitter forms the information molecule), the transmission phase (release of these molecules to the environment), the propagation phase (information molecules trough the medium), the reception phase (information molecules receiving) and finally the decoding phase (received information of molecules).

3 How Nanogels Effectively Communicate with Cancer Cells for Targeting?

It is well known fact that the drugs fight cancer either by attacking the tumor cell itself or by disrupting the physical interaction between the tumor and the body, which dislodges the cell. Nanogels communicate within the body to reach the desired target cell by targeting. Targeted therapies enhance cancer cell sensitivity to chemotherapeutic agents. Targeted drug delivery systems increases drug efficacy and reduces toxicity to healthy cells. Including targeting ligands on the surface of nanogels enables active targeting of the tumor which improves therapeutic index of the anticancer drug. Nanogels are unique as a drug delivery system for targeting cancer.

Below are the important applications of nanogels which make them to communicate effectively with the cancer cells for targeting:

i. Nanogels can be engineered with a definite control over their stability, size, biodegradability, and functionality for bioconjugation which guides them to a targeting receptor.
ii. Nanogels can be designed to present proteins or antibodies on their surface to precisely target specific cells.
iii. Apart from subcutaneous or intramuscular injection, nanogels may also be applied for intravenous administration and easily reach the intended target.
iv. Nanogels can be engineered with suitable functional groups for bioconjugation which guides them to a targeting receptor [36].
v. Nanogel structure enables easy attachment of vector groups for targeted delivery.
vi. There is a leaky vasculature observed in tumours which allows easy extravasation of particles having size up to 400 nm in diameter [6]. Nanogels extravasate through this capillary endothelium to reach the target tissue.
vii. Nanogels evade the reticuloendothelial system due to their small size and hydrophilicity [29].
viii. Nanogels can overcome the impermeable cellular barrier and allow intracellular release of the encapsulated therapeutics.
ix. Nanogels decorated with monoclonal antibody may be promising for tumor targeting towards improved cancer diagnosis and therapy.

x. Virus mimetic (VM) nanogels entrapped with anticancer drug allows the transfer of drug from the endosomes to the cytosol then diffuses into the nucleus, and finally to the pharmacological target site [14].
xi. Nanogels enhance transport of oligonucleotides across the blood–brain barrier [36].
xii. Thermosensitive pluronic nanogels cross-linked with poly(ethylenimine) are useful for cytosolic delivery of therapeutic siRNAs.

Nanogels due to their small size offer intracellular delivery of therapeutic molecules with respect to cellular uptake via endocytosis and the enhanced permeation and retention (EPR) effect [4]. Intracellular drug delivery refers to the delivery of therapeutic agents to specific compartments or organelles within the cell. This targeted intracellular drug delivery results in higher bioavailability of a therapeutic agent at its site of action, increases the pharmacologic effect and reduces the side effects of the drug [26]. Choi et al. [7] al developed a novel self-assembled heparin-Pluronic nanogel incorporating RNase A for intracellular delivery of proteins. Their investigation showed that nanogels were more efficiently internalized into HeLa cells and even localized to the nucleus. The uptake mechanism was via caveolae/lipid-raft-mediated endocytosis.

To reach its targeted site a drug delivery device has to be present in the circulation for a suitable time before getting recognized by the opsonins (plasma proteins) and remove them from the circulation through the reticulo-endothelial system (RES) (Yadav et al. [40]). Surface modification of nanogels allows them to retain stealth properties and accumulate at the tumor site by passive targeting. Active targeting on the other hand helps in cell-specific recognition and internalization of the therapeutic carrier.

To effectively communicate with the cancer cells, the nanogels require the targets. In the next sections we discuss how the nanogels communicate with receptors to reach their target sites.

4 Integrin αvβ3 Targeted Nanogels

Integrins are a family of transmembrane receptors that contain two physically associated subunits, termed α and β. Integrins bind to extracellular matrix proteins such as vitronectin and fibronectin and also posses an intracellular binding domain, which allows cells to communicate to the extracellular environment. This outside-in and inside-out signaling pathway allows the internal cells to communicate and adapt to the external environment [33]. The integrin αvβ3 receptors are overexpressed during the formation of new blood vessels (neovascularization) by the hypoxic conditions in a growing tumor, and thus is an anti-angiogenic target candidate.

Murphy et al. designed an integrin αvβ3–targeted lipid coated nanogel system that was made up of a lipid bilayer enclosing a cross-linked protein/polymeric core

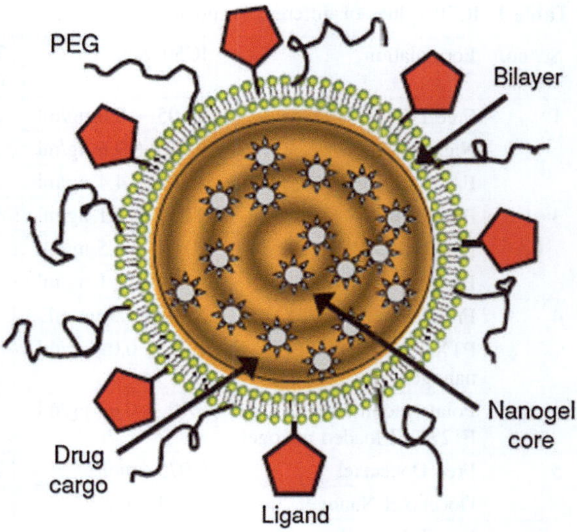

Fig. 1 Schematic representation of the final nanogel product with a lipid bilayer, presenting targeting ligands and polymeric coatings surrounding the gel core containing drug cargoes

[20]. They used M21 human melanoma cells which overexpress integrin αvβ3 and examined cyclic peptide cRGDfK (RGD) as a targeting moiety for the tumor neovasculature. The lipid bilayer incorporated targeting ligands and polymeric coatings such as polyethylene glycol (PEG), whereas the cross-linked core consists of proteins such as human serum albumin (HSA) which serve as carriers for the drugs (Fig. 1). Drug-loaded nanogels showed enhanced potency when compared to free drug after exposure to M21 cells (Table 1).

There was an important finding that the IC50 values of the cells exposed for 20 min with the nanogels were comparable with the cells exposed to the free drug for 72 h. Then they investigated the contribution of targeting integrin αvβ3 by comparing the cell viability of cells exposed to targeted, docetaxel-loaded nanogels (RGD-Doc-NG); untargeted, docetaxel-loaded nanogels (RAD-Doc-NG). The αvβ3 targeting resulted in a 13-fold enhancement in inhibiting cell viability with EC50 values of 0.018 and 0.238 μmol/L for RGD-Doc-NG versus RAD-Doc-NG. These results ascertained that targeted nanogels show better antitumor activity. After establishing these promising results of targeting in vitro, the RGD-Doc-NGs were tested for targeting the tumor vasculature in vivo. Fluorescent RGD-Doc-NGs (vascular targeted), RAD-Doc-NGs (nontargeted) nanogels were i.v. administered at a 1 mg/kg dose of docetaxel for 5 h, and the tumors were removed and imaged with confocal microscopy to view accumulation of the nanogels (green) in the tumor vasculature (red; Fig. 2). The vascular targeted RGD-Doc-NGs targeted the vascular beds within the tumor and bright, punctate nanogel fluorescence was colocalized with the tumor vessels throughout the tumor area. The nontargeted nanogels did not show any significant vascular targeting in the breast tumors. The authors also proved that benefits of active targeting over passive tumor uptake by the improved efficacy results of targeted nanogel (RGD-Doc-NG) compared to untargeted nanogel (RAD-Doc-NG).

Table 1 IC50 values of different formulations

Sr. no.	Formulation	IC50 value	Test duration	Cell line	References
1	Free DOX	54.05 ± 8.0 ng/ml	24 h	A2780	Duan et al. 2011
	Nanogel/DOX	42.75 ± 9.6 ng/ml			
	FA-Nanogel/DOX	30.09 ± 4.4 ng/ml			
3	Free DOX	27.05 ± 3.1 ng/ml	48 h	A2780	Duan et al. 2011
	Nanogel/DOX	20.32 ± 1.5 ng/ml			
	FA-Nanogel/DOX	18.29 ± 1.1 ng/ml			
4	Free Paclitaxel (PTX)	29.0 ± 3.3 μg/mL	–	HEPG2	Low and Kularatne [16]
	PTX-loaded F127/PEI nanogel	1.27 ± 0.09 μg/mL			
	Folate-modified PTX-F127/PEI loaded nanogel	0.40 ± 0.03 μg/mL			
5	Free Docetaxel	0.075 μmol/L	20 min	M21	Morrisse et al. 2005
	Docetaxel Nanogel	0.0045 μmol/L			
6	Free Paclitaxel	0.2354 μmol/L			
	Paclitaxel Nanogel	0.0050 μmol/L			
7	FATP alone	27 μM	4 h	MCF-7	Vinogradov et al. [37]
	Nanogel/FATP complex	3 μM			

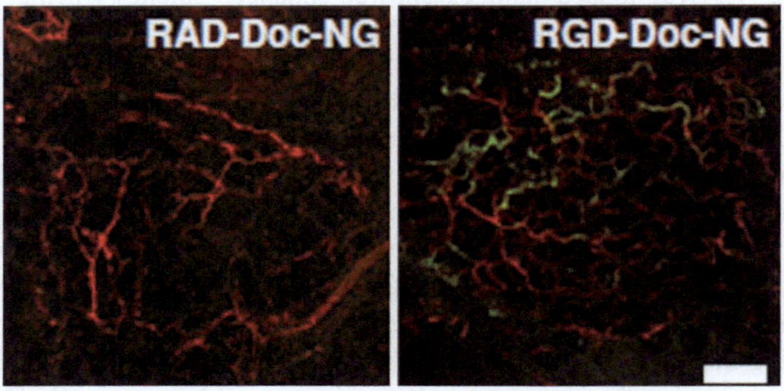

Fig. 2 Comparison of the RGD-Doc-NG and RAD-Doc-NG vascular targeting in MDA-MB-231/LM2-4 tumors on day 21. RGD-Doc-NGs or RAD-Doc-NGs labeled with 1% BODIPY 630/650–conjugated DSPE in the lipid formulation were i.v. injected (1 mg/kg docetaxel concentration) and the tumors were imaged with confocal microscopy at 5 h postinjection. *Red*, rhodamine-labeled G. simplicifolia lectin for staining the endothelium and *green* represents nanogel binding. Scale bar, 100 mm. (Reprinted with permission from American Association for Cancer Research) Ref–Murphy et al. [20]

5 Targeting the EphA2 Receptor

Ephrin type-A receptor 2 (EphA2) is a receptor of tyrosine kinases family [1]. They bind membrane-anchored ligands, ephrins, at sites of cell-cell contact and regulate the cellular organization. The EphA2 receptor tyrosine kinase is over expressed in a variety of human cancers including ovarian, cervical, breast, stomach, colon, prostate, kidney, skin and lung cancer. Expression correlates with degree of angiogenesis, metastasis and xenograft tumor growth. EphA2 receptors are potential therapeutic target and targeting it could inhibit several aspects of tumor progression [9, 13].

Dickerson et al. reported the use of nanogels functionalized with YSA peptides conjugated to the nanogels via maleimide coupling that specially target the EphA2 receptor to deliver small interfering RNAs (siRNAs) targeting epidermal growth factor receptor (EGFR) [8]. The role of the peptide-targeted receptor, EphA2, in nanogel uptake, and the level of nonspecific nanogels incorporation into cells was explored by using an EphA2 negative cell line, SK-OV-3 (these cells lack EphA2 expression). The authors showed that activation of EphA2 by the YSA peptide and subsequent EphA2 degradation may lead to a reduction in EGFR expression levels and significantly increase the sensitivity of this cell line to docetaxel.

6 Folate Receptor Targeted Nanogels

Folic acid (FA) is a prominent targeting moiety capable of specific interaction with cells expressing the folate receptor (FR) [16]. Tumor cells have increased amounts of the FA essential for DNA synthesis and the differential expression of FR in cancers makes it an attractive marker and target. Folate-targeting has attracted considerable attention for delivery of drugs to cancer cells [11].

Folate receptor medicated endocytosis of the nano carrier plays an important role in transporting the loaded drug within the cells [41]. Nayak et al. [23] described the design of fluorescent, thermoresponsive microgels surface-functionalized with folic acid. They incubated these particles with KB cells grown in folate-free medium which resulted in efficient endocytosis of the particles via a receptor-mediated pathway. Figure 3 gives illustration of the endocytosed particles, which after internalization in the cancer cells causes cell death.

Blanco et al. evaluated folate-conjugate poly [(p-nitrophenyl acrylate)-co-(N-isopropylacrylamide)] submicrogels (F-SubMGs) for delivery of 5-Fluorouracil (5-FU). The submicrogels showed enhanced internalization in HeLa cells (which are positive for the folate receptor). The microgels when injected subcutaneously to rats showed increase in the mean residence time and did not show any acute inflammatory response or rejection signs [5].

Nukolova et al. [24, 25] demonstrated possibility of delivery of FR-targeted nanogels to the cancer cells in vivo. They used diblock copolymer poly (ethylene

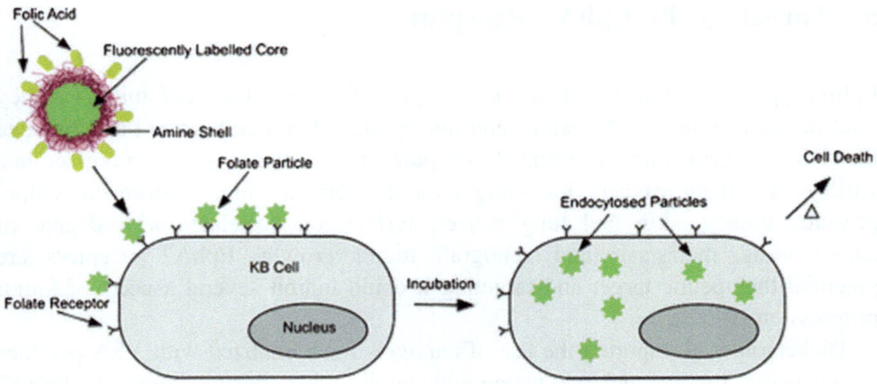

Fig. 3 Folate targeting

oxide)-b-poly (methacrylic acid) (PEO-b-PMA) for controlled template synthesis of nanogels by polyion complexation and cross-linking of doubly hydrophilic block ionomer. An optimal number of folate molecules were conjugated to nanogels to maintain good anti-cancer drug loading, stability and cellular uptake. Such optimized FA-nanogels targeted cell populations expressing FR and had superior anti-tumor efficacy of the anti-cancer drug, cis-dichlorodiamminoplatinum (II) (cisplatin, CDDP) delivered to a xenograft tumor along with decreased renal toxicity.

In another study folate was used to further improve the targeting capability of a nanogel loaded with PTX [15]. Folate was conjugated to the surface of Pluronic F127 micelles with polyethylenimine (F127/PEI) nanogel to further improve the targeting capability. Folic acid was mixed with the solution of F127/PEI nanogel in PBS to synthesis folate-modified Pluronic/PEI nanogel. In this case folate could be recognized and bound by the receptor expressed on the target cell surface, thus trigger receptor-mediated endocytosis to increase intracellular delivery of the drug. Moreover, the folate-modified F127/PEI nanogel demonstrated a significantly superior cytotoxicity compared with the non-modified F127/PEI nanogel. The cytotoxicity of the nanogel was seen against HEPG2 cells and the IC50 values (Table 1) suggested that PTX-loaded nanogel displayed higher cytotoxicity compared with that of the free drug. The folate-modified PTX-loaded nanogel demonstrated a significantly superior cytotoxicity as it had a much lower IC50 value.

Folate-modified PTX-loaded nanogel were uptaken efficiently into HEPG- 2 cells than the non-modified F127/PEI nanogel due to the interaction between the folate on the nanogel surface and the folate receptors on the HEPG-2 cell surface. This interaction ensured that more drugs were pumped into the tumor cells to give a better anti-cancer effect. Huang et al. used PF127 to produce amphiphilic nanocarriers for doxorubicin (DOX) [12]. In order to stabilize the nanocarriers, the hydroxyl groups on both termini of PF127were acrylated and reacted with methacrylated chondroitin sulfate (CSMA) to form CS-PF127 nanogels.

The better cellular internalization of FA-CS-PF127 into the FR overexpressing KB cells was evidenced by CLSM and flow cytometry. Flow cytometry analysis was applied to check the targeting efficiency of the FA-modified nanogel (FA-CS-PF127) and to investigate the cellular uptake in FR-positive KB cells. It was shown that FA-CS-PF127 nanogels entered into KB cells efficiently and this FA targeting moiety was responsible for the better internalization into KB cells.

7 Hyaluronan-CD44 Targeted Nanogel

Hyaluronan (also known as Hyaluronic acid or hyaluronate, HA) is found on the cell surface and in the extracellular matrix of most human tissues. HA binding receptors are overexpressed in cancer cells [34]. The principal cell surface receptor with which HA interacts with cells is cluster determinant 44 (CD44) [35]. When the CD44 of the tumor cell surface is followed by ligand binding, it provokes CD44 internalization which also helps in intracellular drug release and allows more selective access to the tumor cells [10]. Both CD44 on cancer cells and HA in the extra cellular matrix have been targets for anti-cancer therapy. The macromolecular carriers predominantly extravasate into the tumor and not normal tissue; thus CD44-HA targeted carriers localize preferentially into tumors [19, 28]. Although this receptor is present at lower levels on normal cells, its level is considerably elevated in various carcinoma, lymphoma and tumor cells. The overexpression of HA-binding receptors on cancer cells leads to enhanced cancer selectivity making it useful material for delivery of anti-cancer agents using nanogels.

To overcome the problem of hydrophilicity of HA, it was chemically modified by acetylation [27] and was employed to prepare self-organized nanogels for cancer cell selectivity laoded with DOX. The hydrophobic core allowed control of pharmacokinetic properties such as drug loading and release and DOX loaded acetylated HA nanogels showed cytotoxicity with cancer cells with HA-binding receptors. Such results show promise for use of acetylated HA for Cancer cell specific targeting of nanogels.

Wei et al. reported synthesis of nanogel—drug conjugates based on membranotropic cholesteryl-HA (CHA) [38]. The nanogels had a diameter 20–40 nm with a hydrophobic core demonstrated a sustained drug release. CHA–drug nanogels were efficiently internalized via CD44 receptor-mediated endocytosis and simultaneous interaction with the cancer cell membrane. Consequently, these nanogels had 2–7 times higher cytotoxicity in CD44-expressing drug-resistant human breast and pancreatic adenocarcinoma cells compared to that of free drugs and non modified HA–drug conjugates. Drug-resistant tumors have elevated levels of CD44 receptor to which HA binds; hence nanogel–drug conjugates based on CHA could be efficiently used for targeting and suppression of drug-resistant tumors.

Wu et al. developed a core shell structured multifunctional hybrid nanogels having Ag-Au bimetallic core, PEG based hydrogel as shell and hyaluronic acid (HA) chains on the surface [39]. These different layers of the hybrid nanogel served different functions such as optical sensing and cellular imaging was due to the bimetallic core which emitted fluorescent light; the PEG-based gel shell served as

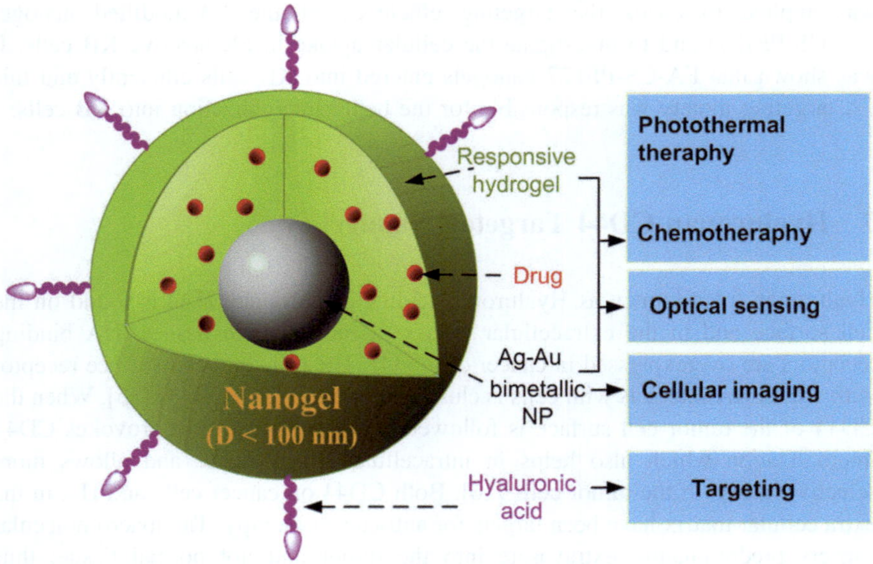

Fig. 4 Schematic illustration of multifunctional coreshell hybrid nanogels

intelligent drug carriers high drug loading capacity (of anticancer drug temozolomide) having responsiveness and the surface HA chains were useful for CD44 overexpressed on various tumors for targeting function. The schematic illustration of multifunctional core shell hybrid nanogels is represented in Fig. 4. Such a combination of the functions is useful for enhancing the therapeutic efficacy.

8 Nanogels Decorated with Monoclonal Antibody

According to Scott et al. "Monoclonal antibodies (mAb) are defined as a unique immunoglobulin of known specificity that is produced by a B-lymphocyte clone usually immortalized by cell fusion with a non-secreting myeloma cell line" [30]. The mAb bind to cancer cell-specific antigens and induce an immunological response against the target cancer cell [18]. Moreover, mAbs have specificity for their target molecules and the ability to either activate or suppress the molecular function of the target. A surface-functionalized cross-linked nanogels would act as a platform to allow conjugation of mAb for targeted drug delivery for cancer [24, 25]. A diblock copolymers of poly (ethylene glycol)-b-poly (methacrylic acid) (PEG-b-PMA) with PEG terminal aldehyde functionality was synthesized by atom transfer radical polymerization (ATRP) method for controlled attachment of mAb CC49. This approach shows promise for tumor targeting of nanogels towards improved cancer diagnosis and therapy.

9 Conclusion

Recent years have witnessed an extraordinary expansion in drug delivery research in the area of cancer. There is an increasing assurance that nanotechnology applied to medicine will bring significant advances in the diagnosis and treatment of cancer. When most of the chemotherapeutics fail to show effect clinically in the treatment of cancer due to their toxic side effects. Nanogels as nanomachines yields more effective therapies. Molecular communication is biologically inspired communication technique which uses nanoscale properties of materials. Nanogels have demonstrated promise for delivery of various anticancer drugs. Molecular communication seems to provide efficient mechanisms for networking of nanomachines. Nanogels in cancer chemotherapy not only improves the therapeutic efficacy of the anticancer drug but also reduces its side effects. Nanogels effectively communicate with several receptors for targeted drug delivery. These targeted therapies increase cancer cell sensitivity to chemotherapeutics by avoiding healthy cells and destroying resistant cells.

Acknowledgements Dr. K.S. Yadav thanks AICTE, New Delhi, for the award of Research Promotion Scheme project (Ref No. 8023/RID/RPS-71/POLICY III(PVT)/2011-12).

References

1. Aasheim HC, Delabie J, Finne EF (2005) Ephrin-A1 binding to CD4 + T lymphocytes stimulates migration and induces tyrosine phosphorylation of PYK2. Blood 105:2869–2870
2. Akyildiz IF, Brunetti F, Blazquez C (2008) Nanonetworks: a new communication paradigm. Comput Netw 52:2260–2279
3. Atakan B, Akan OB (2010) Deterministic capacity of information flow in molecular nanonetworks. Nano Commun Netw J 1:31–42
4. Ayame H, Morimoto N, Akioshi K (2008) Self-assembled cationic nanogels for intracellular protein delivery. Bioconjugate Chem 19:882–890
5. Blanco MD, Guerrero S, Benito M, Fernández A, Teijón C, Olmo R, Katime I, Teijón JM (2011) In vitro and in vivo evaluation of a folate-targeted copolymeric submicrohydrogel based on n-isopropylacrylamide as 5-Fluorouracil delivery. Polymers 3:1107–1125
6. Brown JM, Giaccia AJ (1998) The unique physiology of solid tumors: opportunities (and problems) for cancer therapy. Cancer Res 58(7):1408–1416
7. Choi JH, Jang JY, Joung YK, Kwon MH, Park KD (2010) Intracellular delivery and anti-cancer effect of self-assembled heparin-Pluronic nanogels with RNase A. J Control Release 147:420–427
8. Dickerson EB, Blackburn WH, Smith MH, Kapa LB, Lyon LA, McDonald JF (2010) Chemosensitization of cancer cells by siRNA using targeted nanogel delivery. BMC Cancer 10:10
9. Duxbury MS, Ito H, Zinner MJ, Ashley SW, Whang EE (2004) EphA2: a determinant of malignant cellular behavior and a potential therapeutic target in pancreatic adenocarcinoma. Oncogene 23:1448–1456
10. Ghosh SC, Neslihan Alpay S, Klostergaard J (2012) CD44: a validated target for improved delivery of cancer therapeutics. Expert Opin Ther Targets 16(7):635–650

11. Hilgenbrink AR, Low PS (2005) Folate receptor-mediated drug targeting: from therapeutics to diagnostics. J Pharm Sci 94:2135–2146
12. Huang SJ, Sun SL, Feng TH, Sung KH, Lui WL, Wang LF (2009) Folate-mediated chondroitin sulfate-Pluronic® 127 nanogels as a drug carrier. Eur J Pharm Sci 38:64–73
13. Ireton RC, Chen J (2005) EphA2 Receptor Tyrosine Kinase as a Promising Target for Cancer Therapeutics. Curr Cancer Drug Targets 5:149–157
14. Lee ES, Kim D, Youn YS, Oh KT, Bae YH (2008) A virus mimetic nanogel vehicle. Angew Chem 2008; 120(13): 2452–2455
15. Li N, Wang J, Yang X, Li L (2011) Novel nanogels as drug delivery systems for poorly soluble anticancer drugs. Colloids Surf B 83:237–244
16. Low PS, Kularatne SA (2009) Folate-targeted therapeutic and imaging agents for cancer. Curr Opin Chem Biol 13:256–262
17. Malak D, Akan OB (2012) Molecular communication nanonetworks inside human body. Nano Commun Netw 3:19–35
18. Melero I, Hervas-Stubbs S, Glennie M, Pardoll DM, Chen L (2007) Immunostimulatory monoclonal antibodies for cancer therapy. Nat Rev Cancer 7(2):95–106
19. Misra S, Heldin P, Hascall VC, Karamanos NK, Skandalis SS, Markwald RR, Ghatak S (2011) Hyaluronan–CD44 interactions as potential targets for cancer therapy. FEBS J 278(9):1429–1443
20. Murphy EA, Majeti BK, Mukthavaram R, Acevedo LM, Barnes LA, Cheresh DA (2011) Targeted nanogels: a versatile platform for drug delivery to tumors. Mol Cancer Ther 10(6):972–982
21. Nakano T, Moore M (2011) Molecular communication paradigm overview. J Next Gener Inf Technolo 2(1):9–16
22. Nakano T, Moore MJ, Wei Fang, Vasilakos AV, Shuai Jianwei (2012) Molecular communication and networking: opportunities and challenges. IEEE Trans Nanobiosci 11(2):135–148
23. Nayak S, Lee H, Chmielewski J, Lyon LA (2004) Folate-mediated cell targeting and cytotoxicity using thermoresponsive microgels. J Am Chem Soc 126(33):10258–10259
24. Nukolova NV, Oberoi HS, Cohen SM, Kabanov AV, Bronich TK (2011) Folate-decorated nanogels for targeted therapy of ovarian cancer. Biomaterials 32:5417–5426
25. Nukolova NV, Yang Z, Kim JO, Kabanov AV, Bronich TK (2011) Polyelectrolyte nanogels decorated with monoclonal antibody for targeted drug delivery. React Funct Polym 71:315–323
26. Panyam J, Williams D, Dash A, Leslie-Pelecky D, Labhasetwar V (2004) Solid-state solubility influences encapsulation and release of hydrophobic drugs from PLGA/PLA nanoparticles. J Pharm Sci 93(7):1804–1814
27. Park W, Kim K, Bae B, Kim Y, Na K (2010) Cancer cell specific targeting of nanogels from acetylated hyaluronic acid with low molecular weight. Eur J Pharm Sci 40:367–375
28. Platt VM, Szoka FC Jr (2008) Anticancer therapeutics: targeting macromolecules and nanocarriers to hyaluronan or CD44, a hyaluronan receptor. Mol Pharm 5(4):474–486
29. Raemdonck K, Demeester J, De Smedt S (2009) Advanced nanogel engineering for drug delivery. Soft Matter 5:707–715
30. Scott AM, Wolchok JD, Old LJ (2012) Antibody therapy of cancer. Nat Rev Cancer 12(4):278–287
31. Soni G, Yadav KS (2014) High encapsulation efficiency of poloxamer based injectable thermoresponsive hydrogels of etoposide. Pharm Dev Technol 19(6):651–661
32. Soni G, Yadav KS (2014) Nanogels as potential nanomedicine carrier for treatment of cancer: A mini review of the state of the art. Saudi Pharm J. doi:10.1016/j.jsps.2014.04.001
33. Sunderland CJ, Steiert M, Talmadge JE, Derfus AM, Barry SE (2006) Targeted nanoparticles for detecting and treating cancer. Drug Dev Res 67(1):70–93
34. Toole BP (2004) Hyaluronan: from extracellular glue to pericellular cue. Nat Rev Cancer 4:528–539

35. Toole BP (2009) Hyaluronan–CD44 interactions in cancer: paradoxes and possibilities. Clin Cancer Res 15:7462–7468
36. Vinogradov SV, Batrakova EV, Kabanov AV (2004) Nanogels for oligonucleotide delivery to the brain. Bioconjug. Chem. 15:50–60
37. Vinogradov SV, Zeman AD, Batrakova EV, Kabanov AV (2005) Polyplex nanogel formulations for drug delivery of cytotoxic nucleoside analogs. J Control Release 107:143–157
38. Wei X, Senanayake TH, Warren G, Vinogradov SV (2013) Hyaluronic acid-based nanogel-drug conjugates with enhanced anticancer activity designed for the targeting of CD44-positive and drug-resistant tumors. Bioconjugate Chem. 24(4):658–668
39. Wu W, Shen J, Banerjee P, Zhou S (2010) Core shell hybrid nanogels for integration of optical temperature-sensing, targeted tumor cell imaging, and combined chemo-photothermal treatment. Biomaterials 31:7555–7566
40. Yadav KS, Jacob S, Sachdeva G, Chuttani K, Mishra AK, Sawant KK (2011) Long circulating PEGylated PLGA nanoparticles of cytarabine for targeting leukemia. J Microencapsul 28(8):729–742
41. Yoo HS, Park TG (2004) Folate receptor targeted biodegradable polymeric doxorubicin micelles. J Control Release 96:273–283

Modeling, Analysis and Design of Bio-hybrid Micro-robotic Swarms for Medical Applications

Guopeng Wei, Paul Bogdan and Radu Marculescu

Abstract In this chapter, we present a cyber-physical approach towards the design of bio-hybrid micro-robotic swarms that can achieve various complex tasks at micro-level in a minimal invasive manner, such as abnormal tissue detection and drug delivery in hardly accessible regions of the human body. To this end, we cease to view micro-robots as passive point-like particles, but rather as interactive Turing machines performing complex biochemical processing, as well as physical interactions in a realistic 3D environment. Our theoretical framework is based on a non-equilibrium statistical physics approach capable of accounting for attraction and repulsion interactions among micro-robots, as well as the volume exclusion effects. To account for biological sensing, interacting, actuation dynamics, and the 3D complex tumor microenvironment, we also use an open-source 3D multiscale simulator specifically developed for this research. Taken together, the theoretical framework and computational platform we develop can enable various design trade-offs of interacting bio-hybrid micro-robotic swarms for future medical applications.

1 Introduction

Reductionism has enabled medical and pharmacological research to achieve significant progress in diagnosis and treatment techniques for life-threatening diseases ranging from pathogen infections to cancer. In most cases, the medical techniques depend on the macroscopic signatures of the disease and rely on a single or a

G. Wei · R. Marculescu
Carnegie Mellon University, 5000 Forbes Avenue, Pittsburgh, PA 15213, USA
e-mail: guopengw@ece.cmu.edu

R. Marculescu
e-mail: radum@ece.cmu.edu

P. Bogdan (✉)
University of Southern California,
3740 McClintock Ave., Los Angeles, CA 90089-2564, USA
e-mail: pbogdan@usc.edu

J. Suzuki et al. (eds.), *Modeling, Methodologies and Tools for Molecular and Nano-scale Communications*, Modeling and Optimization in Science and Technologies 9, DOI 10.1007/978-3-319-50688-3_22

combination of drugs, or even surgeries, in order to suppress the effects of a medical condition (e.g., hypertension, abdominal pain).

Despite these achievements, many challenges still remain to be solved. One important challenge is related to the early detection of complex medical conditions. For instance, in some situations, malfunctions happening at the molecular level can lead to an unfortunate cascade of protein misfoldings that give rise to diseases like cancer which spread silently through the human body [1]. These events cannot be easily detected by current standard tests or scanning (e.g., blood analysis or medical imaging) techniques and so the cancer cells can corrupt both neighboring and remote organs up to the point where surgery cannot help with a cure. Another important challenge is related to the drug dosing problem as a function of patient biological characteristics and disease progression [2]. There are numerous situations when inappropriate drug dosing can lead to multidrug resistance or the death of the patient rather than curing the disease. Consequently, there is a stringent need for developing personalized therapies that do not only maximize the efficacy of a drug, but also minimize its harmful impact on the body homeostasis [3, 4].

Systems biology aims to address these challenges by understanding the genomics, molecular, and cell biology while helping with new strategies that embrace a holistic perspective on disease progression and treatment. For instance, a large body of research in the analysis of tumor angiogenesis suggests that the silent progression of tumors is most of the time accompanied by an increased demand for nutrients and oxygen [1]. Starting from these premises, system biology can exploit the sensing capabilities of bio-hybrid micro-robots for detecting various disease biomarkers and harness the dense networks of bio-hybrid micro-robots to deliver drugs in hard to access regions of the human body [5–7] (see Fig. 1).

Towards this end, in this paper, we use bacteria-propelled micro-robots as an example of interacting bio-hybrid micro-robots [8, 9]. In this case, the detection of disease cues and the coordination of dense network of bio-hybrid micro-robots is realized by chemotaxis [9].[1] Simply speaking, the trans-membrane chemo-receptors sense the presence of certain ligands in the environment; in turn, the signaling pathway of bacteria serves a dual role: (i) First, it converts the receptor activation into a regulatory cytoplasmic response that controls the rotation of the flagella motor; and (ii) Second, it contributes to the production of chemo-attractant and chemo-repellent molecules which serve as information messengers for other bacteria. Consequently, chemotaxis not only allows bacteria to detect changes in the environment (e.g., abnormal cellular growth, higher chemical activity), but also helps the coordination of bacteria population to perform complex tasks (e.g., tumor invasion [5, 7]) at microscale.

---

[1] Chemotaxis is a physico-chemical process through which single- or multi-cellular organisms sense the concentration of specific chemicals in the environment via chemo-receptors distributed on the cell surface and direct their movement towards/away from these chemical gradients [10]. More precisely, if chemo-attractant gradient is detected in the environment, then bacteria swim up that gradient for a longer period of time (also called run motion) before tumbling and choosing a different direction of motion. In the run motion, the flagella rotate in a counter-clockwise fashion and align as a single rotating bundle. In contrast, in the tumble motion, the flagella rotate in a clockwise fashion which makes bacteria reorient and choose a different swimming direction [11].

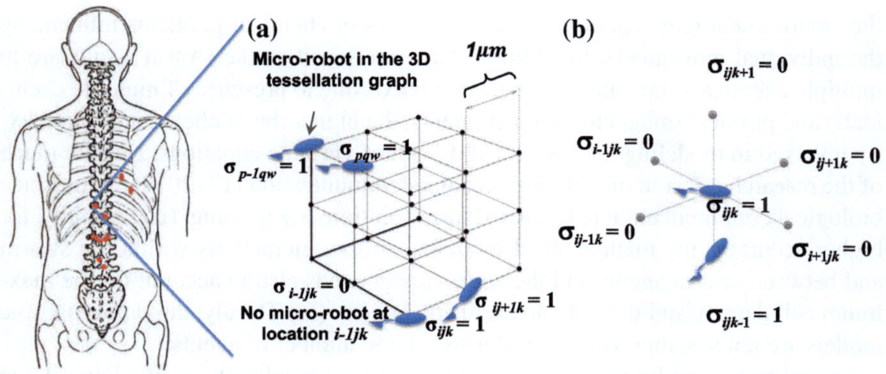

Fig. 1 Dense network of micro-robotic swarms swimming in hard to access regions of the body. The micro-robots dynamics is modeled as a collective behavior of multiple interacting random walks in a 3D graph tessellation of space. Each graph node has associated a binary random variable $\sigma_{i,j,k}$ denoting whether or not it is occupied by one micro-robot. For detection and monitoring purposes, the goal is to find the time-dependent coverage of the swarm. The goal of the drug delivery problem is to find the probability for the nodes in the disease area to be occupied by micro-robots

Although the use of dense networks of bio-hybrid for medical applications seems promising, there is a need for developing mathematical methodologies for modeling their dynamics as a function of the interactions between bacteria-propelled micro-robots and between micro-robots and the environment. This is the main focus of this chapter.

2 Related Work

Developing mathematical models for either maximizing the detection probability of abnormal processes at cellular-level or minimizing the drug delivery time for micro-robots require a comprehensive understanding of interactions between individuals of the micro-robotic swarm and those between the swarm and the biological environment. From this perspective, the mathematical modeling of swarms (including modeling of motile micro-robots, swimming fish, flying birds, or migratory herds of animals) can be classified into individual- (micro-scopic) and population-based (macro-scopic) models.

The *microscopic models* describe the behavior of each individual in a swarm through a continuous- or a discrete-time dynamical equation [12–16]. Many of such models assume swarm agents as point-like particles; they may encapsulate via mathematical terms and varying parameters the fact that the position and state of each individual at any point in time is a function of external influences and internal processing features (e.g., averaging of movement direction). Despite their simplicity (and accuracy) in capturing local or individual behavior, these models have several disadvantages: (i) Firstly, these models require precise local information about the environment (e.g., spatio-temporal dynamics of chemical gradients influencing the individual movement). In addition, the more sensitive the swarm agents are to multiple chemicals (i.e., multi-modal chemotaxis due to presence of multiple chemicals) and physical (magneto-taxis) gradients, the higher the mathematical complexity involved in modeling the individual dynamics via state equations. Despite much of the research effort in modeling molecular communication [17–20] and multiscale biological communication [21], closed form solutions are missing. (ii) Secondly, for higher accuracy, the mathematical modeling of the interactions within the swarm and between swarm agents and the environment needs also to account for the maximum cell density and the volume exclusion effects. (iii) Thirdly, these microscopic models are not scalable when considering a large number of agents.

In contrast, in order to avoid the computational complexity of simulating large swarms via micro-scopic equations, the *macro-scopic models* represent the biological swarm dynamics via advection-diffusion-reaction (ADR) partial differential equations (PDEs) [14, 22, 23]. These PDEs describe the evolution of a probability density function (PDF) in space and time. For instance, to model the collective behavior of animal herds or flocks of fish, the PDEs describe the probability distribution of group sizes or fraction of individuals with specific motion orientation [24–26]. Although theoretical and numerical strategies for solving PDE-based models have received significant attention, we still lack rigorous mathematical strategies for incorporating multiscale phenomena (e.g., molecular based interactions between swarm agents, volume exclusion effects, agent density dependent interactions) into the macroscopic models that can be further incorporated into population-level control methodologies. For instance, accounting for chemical- (via molecular communication processes [27–29]) or physical-based (via small magnetic fields) interactions is of crucial importance not only for constructing more accurate macroscopic models, but also for identifying more efficient and robust control strategies. In addition, this mesoscopic type of characterization can also contribute to a better understanding of the various physico-chemical processes that take place at cellular level and can affect the intra-cellular structure and function. For example, the importance of volume exclusion and interactions has been recognized for modeling the kinetics of bi-polymerization [30] via asymmetric simplex exclusion processes developed within the framework of non-equilibrium statistical physics and interacting particle systems [31–35].

In order to avoid all these above mentioned limitations, in this chapter, we discuss a non-equilibrium statistical physics inspired model which describes the dynamics of bacteria-propelled micro-robotics as a *collection of interacting random walks* in the three dimensional (3D) space while obeying the following rules: (i) attraction and repulsion rules are dictated by the molecular concentrations and internal chemical pathway processing and (ii) volume exclusion rules in a 3D space which forbids the possibility of having two micro-robots occupying the same space at the same time. Unlike the microscopic models which provide a local representation of the dynamics of each agent, the proposed mathematical formalism allows a *collective* description of the micro-robotic swarm via a multivariate probability distribution function. In addition, by relying on micro- and mesoscopic interactions, the proposed master equation governing the dynamics of a multivariate probability density function allows the derivation of more accurate macroscopic models.

3 Modeling Bacteria-Propelled Micro-robotic Swarms

In order to study the dynamics of dense networks of bacteria propelled micro-robotic swarms, we present an asymmetric simple exclusion process (ASEP)-inspired mathematical model [36, 37] having the following features: (i) The model describes the dynamics of bacteria propelled micro-robots as interacting random walks in a 3D space. (ii) The localized 3D directional movement of a single micro-robot in the absence of neighboring micro-robots is dictated by the environmental sensing of chemical cues and internal chemical processing of bacteria. This internal chemical processing contributes to self-propulsion of micro-robots towards targeted space locations. (iii) By accounting for the chemical gradients that are constantly diffusing as a result of thermal and hydrodynamics effects, the model captures the 3D attraction interactions among bacteria attached to different micro-robots. (iv) When two or more micro-robots are close to each other, the model captures the 3D repulsion interactions and prevents one space location to be occupied by multiple micro-robots at the same time.

In order to take into account all these swarm characteristics, the mathematical model considers a tessellation of the 3D space into a lattice of size $M \times N \times Q$ and associates a binary random variable $\sigma_{i,j,k}$ to each node (i, j, k). This binary random variable $\sigma_{i,j,k}$ represents whether or not there exists a micro-object (e.g., bacteria, macromolecules, blood vessels) at location (i, j, k), i.e., $\sigma_{i,j,k} = 1$ if a micro-robot is present and $\sigma_{i,j,k} = 0$ if it is not (as shown in Fig. 1). In order to capture the reality of the biological environment, we also assume that the graph tessellation of space is fine-grained (i.e., each cubicle has the same order of magnitude as a bacterium and other environmental micro-objects) and then impose the volume exclusion rules.

Within this mathematical framework, the dynamics of multiple interacting random walks is encapsulated by $P(\sigma_{1,1,1}, \dots, \sigma_{i,j,k}, \dots, \sigma_{M,N,Q}; t)$ the multivariate probability density function which represents whether or not a micro-robot is present at each space location i, j, k, $i = 1 : M$, $j = 1 : N$, $k = 1 : Q$. Unlike previous population-based models that postulate a particular diffusion equation, in what follows, we describe how the microscopic interactions contribute to defining the transition probabilities and the master equation characterizing the dynamics of the entire swarm. More precisely, we assume[2] that during an infinitesimal time interval δt, a random walker (e.g., bacterium or micro-robot) located at the (i, j, k) location moves to the $(i-1, j, k)$ location. From a mathematical point of view, this implies that, at time t, $\sigma_{i,j,k}(t) = 1$ and $\sigma_{i-1,j,k}(t) = 0$, while at time $t + \delta t$, $\sigma_{i-1\,j,k}(t + \delta t)$ changes

[2]For ease of the mathematical description, we only present the mathematical terms describing the movement of micro-robots backwards along the X-axis. The transition probabilities for other directions can be similarly derived.

from 0 to 1 (i.e., $\sigma_{i-1,j,k}(t+\delta t)=1$), and $\sigma_{i,j,k}(t+\delta t)$ changes from 1 to 0 (i.e., $\sigma_{i,j,k}(t+\delta t)=0$). Under this assumption, the transition probability satisfies the following relation:

$$Pr\left\{\sigma_{i-1,j,k}(t+\delta t)+1,\sigma_{i,j,k}(t+\delta t)-1|\sigma_{i,j,k}(t)=1\right\}=\sigma_{i,j,k}(t)(1-\sigma_{i-1,j,k}(t)) \cdot\left[\alpha_{i,j,k\rightarrow i-1,j,k}+\sum\nolimits_{(p,q,w)\in\Omega_{i\_j,k}}\beta_{i,j,k\rightarrow p,q,w}+\chi\sigma_{i+1,j,k}(t)\right]\delta t+O(\delta t) \quad (1)$$

where $\sigma_{i,j,k}(t)$, $\sigma_{i-1,j,k}(t)$, and $\sigma_{i+1,j,k}(t)$ are binary random variables indicating whether or not locations (i,j,k), $(i-1,j,k)$, and $(i+1,j,k)$ are occupied by a random walker at time t. In order to satisfy the basic rules of probability theory, the mathematical term $[\alpha_{i,j,k\rightarrow i-1,j,k}+\sum_{(p,q,w)\in\Omega_{i\_j,k}}\beta_{i,j,k\rightarrow p,q,w}+\chi\sigma_{i+1,j,k}(t)]\cdot\delta t$ must be less than or at most equal to one. The $O(\delta t)$ shows that all terms smaller than δt can be neglected.

The so-called directional random walk coefficient $\alpha_{i,j,k\rightarrow i-1,j,k}$ denotes the transition rate of a micro-robot towards a certain direction due to a chemotactic response to external stimuli. Regarding the chemical stimuli, we assume that there are no chemical interactions between the molecules diffusing from the target cancer cells or neighboring micro-robots. This allows us to model the dynamics of molecular communication affecting the transition rates $\alpha_{i,j,k\rightarrow i-1,j,k}$ via a classical diffusion equation [37]. Consequently, in our formalism, this directional random walk coefficient is not constant, but rather changes as a function of the bacteria sensing the environment.

The $\beta_{i,j,k\rightarrow p,q,w}$ coefficients are introduced to model the short-range repulsion and long-range attraction interactions. These interactions among micro-robots at various distances in space are mitigated by the inter-cellular molecular communication [38]. In other words, the release of the attractant and repellent molecules and their time varying diffusion dictates a wide range of values for the $\beta_{i,j,k\rightarrow p,q,w}$ coefficients and represent the strength of attraction or repulsion forces acting on a micro-robot at location (i,j,k) [37].

Without loss of generality, the interactions among micro-robots can be generalized to an exponentially decaying Morse potential as a function of distance R, which is a typical choice to reproduce collective motion observed in multi-agent systems; therefore, the definition of β is as follows:

$$\beta_{i,j,k\rightarrow p,q,w}=\sum cos(\theta)cos(\gamma)\sigma_{p,q,w}\times(C_a exp(\frac{-R^2}{e_a})-C_r exp(\frac{-R^2}{e_r})) \quad (2)$$

where $R=\sqrt{(p-i)^2+(q-j)^2+(w-k)^2}$ is the distance between micro-robots at the p,q,w and i,j,k locations, e_a and e_r are the effective range of attraction and repulsion interactions (see Fig. 2).

As shown in Fig. 2, β is first projected to calculate its impact on micro-robot probability to move along the X, Y, and Z axes. A positive value of β represents an attraction interaction, while a negative value represents a repulsion interaction. Hence, this pairwise potential induces repulsion at short ranges and attraction at longer ranges.

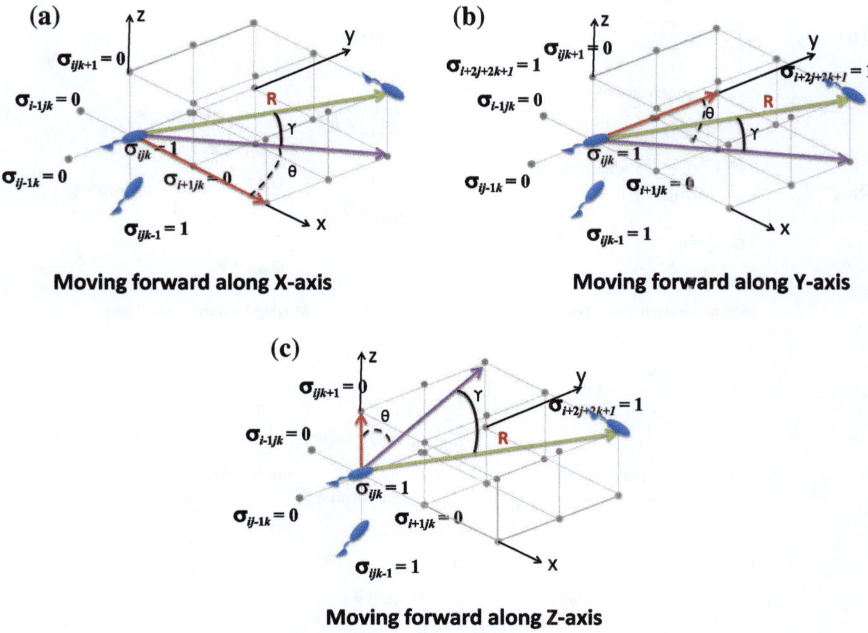

Fig. 2 Interaction potential calculation between micro-robots when moving forward along X, Y, Z axes. R is the distance between micro-robots located at p, q, w and i, j, k. $\theta; \gamma$ are used to decompose the attraction and repulsion interaction forces and project them to positive X, Y, Z directions. The decomposed forces are reflected in the transition probabilities of micro-robot$_{ijk}$ moving forward along the X, Y, Z directions

The χ parameter captures the possible "collisions"[3] among neighboring micro-robots. More precisely, if a neighboring location is already occupied, a micro-robot at location (i, j, k) may have an increased probability to move in the opposite direction. The magnitude of the χ parameter depends on the geometrical shape and biological features of the attached bacteria (e.g., length and number of flagella). For instance, the collision forces among engineered bacteria without flagella would be small since they are influenced by liquid viscosity and hydrodynamic effects and so the contribution of the χ parameter to the directional movement of micro-robots would be mainly driven by the molecular communication based interaction. In contrast, for flagellated bacteria propelled micro-robots swimming in highly viscous liquids the χ parameter plays a much more important role.

To better understand the intuition behind Eq. (1), let us consider that $\sigma_{i-1,j,k} = 1$ at time t. In this case, because the location $(i-1, j, k)$ is occupied by a random walker, the transition probability in Eq. (1) is zero and so the random walker at (i, j, k) cannot move to the $(i-1, j, k)$ location. This way, the transition probability in Eq. (1)

[3]Note that we assume a homogeneous micro-robots population here, hence χ is the same for every micro-robot. In general, if a heterogeneous population is used, χ can have multiple values.

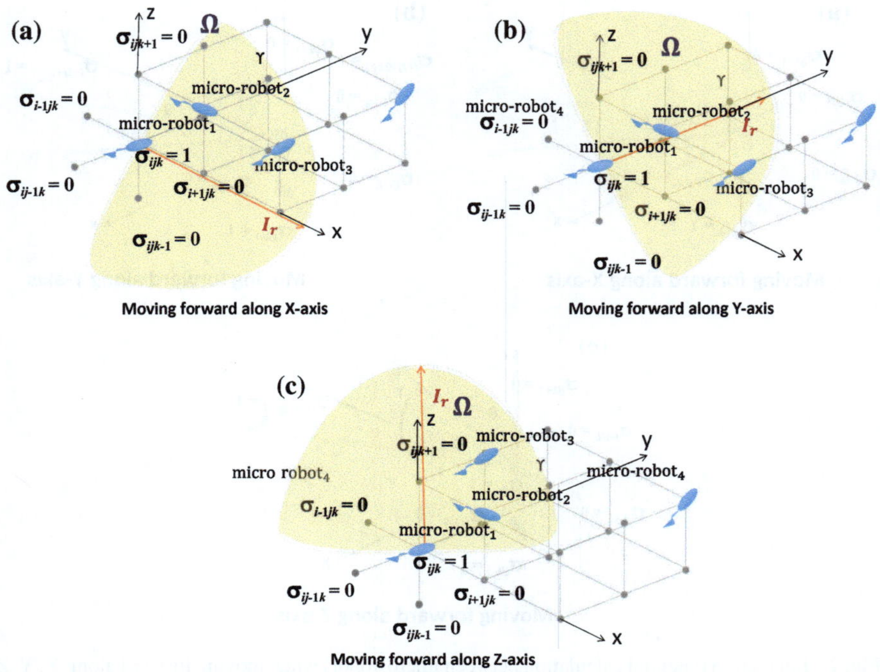

Fig. 3 Interaction potential calculation for all micro-robots located in a hemisphere $\Omega_{i_+,j,k}$, $\Omega_{i,j_+,k}$, Ω_{i,j,k_+} of radius I_r. From *bacterium*$_1$ perspective, in all three cases, micro-robot$_2$ and micro-robot$_3$ are within its interaction radius, while micro-robot$_4$ is not

accounts only for not the molecular communication based interactions and hydrodynamic interactions, but also for the volume exclusion effects. To summarize, the first term in Eq. (1) follows the volume exclusion principle, while the second term captures the attraction and repulsion interactions of micro-robots with a certain interaction range I_r, as shown in Fig. 3.

Defining the transition probabilities that encapsulate the interactions among the neighboring micro-robots along the X, Y, and Z directions allows us to write a master equation governing the evolution of the multivariate probability density function $P(\sigma_{1,1,1}, \dots, \sigma_{i,j,k}, \dots, \sigma_{M,N,Q}; t)$ as follows:

$$\frac{dP(\sigma_{1,1,1}, \dots, \sigma_{i,j,k}, \dots, \sigma_{M,N,Q}; t)}{dt} =$$

$$\sum_{i,j,k=1}^{M,N,Q} \{(\alpha_{i-1,j,k\to i,j,k} + \sum_{\substack{(p,q,w)\in \\ \Omega_{i_+-1,j,k}}} \beta_{i-1,j,k\to p,q,w} + \chi\sigma_{i-2,j,k})\sigma_{i-1,j,k}(t)$$

$$+ (\alpha_{i+1,j,k\to i,j,k} + \sum_{\substack{(p,q,w)\in \\ \Omega_{i_-+1,j,k}}} \beta_{i+1,j,k\to p,q,w} + \chi\sigma_{i+2,j,k})\sigma_{i+1,j,k}(t) +$$

$$
\begin{aligned}
&+(\alpha_{i,j-1,k\to i,j,k} + \sum_{\substack{(p,q,w)\in\\ \Omega_{i,j_+-1,k}}} \beta_{i,j-1,k\to p,q,w} + \chi\sigma_{i,j-2,k})\sigma_{i,j-1,k}(t) + \\
&+(\alpha_{i,j+1,k\to i,j,k} + \sum_{\substack{(p,q,w)\in\\ \Omega_{i,j_-+1,k}}} \beta_{i,j+1,k\to p,q,w} + \chi\sigma_{i,j+2,k})\sigma_{i,j+1,k}(t) + \\
&+(\alpha_{i,j,k-1\to i,j,k} + \sum_{\substack{(p,q,w)\in\\ \Omega_{i,j,k_+-1}}} \beta_{i,j,k-1\to p,q,w} + \chi\sigma_{i,j,k-2})\sigma_{i,j,k-1}(t) + \\
&+(\alpha_{i,j,k+1\to i,j,k} + \sum_{\substack{(p,q,w)\in\\ \Omega_{i,j,k_-+1}}} \beta_{i,j,k+1\to p,q,w} + \chi\sigma_{i,j,k+2})\sigma_{i,j,k+1}(t)\} \times \\
&\quad \times(1-\sigma_{i,j,k}(t))P(\sigma_{1,1,1},\dots,\sigma_{i,j,k},\dots,\sigma_{M,N,Q};t) - \qquad (3)\\
&\{\alpha_{i,j,k\to i-1,j,k} + \sum_{\substack{(p,q,w)\in\\ \Omega_{i_-j,k}}} \beta_{i,j,k\to p,q,w} + \chi\sigma_{i+1,j,k})(1-\sigma_{i-1,j,k}(t)) \times \\
&+(\alpha_{i,j,k\to i+1,j,k} + \sum_{\substack{(p,q,w)\in\\ \Omega_{i_+j,k}}} \beta_{i,j,k\to p,q,w} + \chi\sigma_{i-1,j,k})(1-\sigma_{i+1,j,k}(t)) + \\
&+(\alpha_{i,j,k\to i,j-1,k} + \sum_{\substack{(p,q,w)\in\\ \Omega_{i,j_-,k}}} \beta_{i,j,k\to p,q,w} + \chi\sigma_{i,j+1,k})(1-\sigma_{i,j-1,k}(t)) + \\
&+(\alpha_{i,j,k\to i,j+1,k} + \sum_{\substack{(p,q,w)\in\\ \Omega_{i,j_+,k}}} \beta_{i,j,k\to p,q,w} + \chi\sigma_{i,j-1,k})(1-\sigma_{i,j+1,k}(t)) \\
&+(\alpha_{i,j,k\to i,j,k-1} + \sum_{\substack{(p,q,w)\in\\ \Omega_{i,j,k_-}}} \beta_{i,j,k\to p,q,w} + \chi\sigma_{i,j,k+1})(1-\sigma_{i,j,k-1}(t)) \\
&+(\alpha_{i,j,k\to i,j,k+1} + \sum_{\substack{(p,q,w)\in\\ \Omega_{i,j,k_+}}} \beta_{i,j,k\to p,q,w} + \chi\sigma_{i,j,k-1})(1-\sigma_{i,j,k+1}(t))\} \times \\
&\quad \times\sigma_{i,j,k}(t)P(\sigma_{1,1,1},\dots,\sigma_{i,j,k},\dots,\sigma_{M,N,Q};t)
\end{aligned}
$$

For a complete mathematical description of the multiple interacting random walkers, we also need to consider the initial and the normalization conditions imposed on the evolution of the multivariate PDF $P(\sigma_{1,1,1},\dots,\sigma_{i,j,k},\dots,\sigma_{M,N,Q};t)$ (Fig. 4):

$$\sum_{i=1}^{M}\sum_{j=1}^{N}\sum_{k=1}^{Q}\sum_{\sigma_{ijk}=0,1} P(\sigma_{111},\dots,\sigma_{ijk},\dots,\sigma_{MNQ};t) = 1 \qquad (4)$$

$$\sum_{i=1}^{M}\sum_{j=1}^{N}\sum_{k=1}^{Q}\sum_{\sigma_{ijk}=0,1} \sigma_{ijk}P(\sigma_{111},\dots,\sigma_{ijk},\dots,\sigma_{MNQ};t) = N_W \qquad (5)$$

where N_W represents the total number of random walkers (e.g., bacteria propelled micro-robots) under investigation.

Simply speaking, the master Eq. (3) describes the dynamics of dense networks of micro-robots which can be used, for instance, to tag the cancerous cells for diagnos-

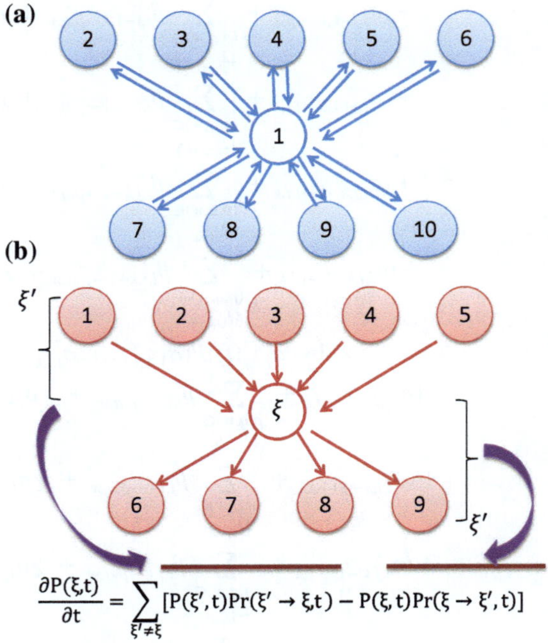

Fig. 4 Comparison of equilibrium and non-equilibrium systems. $P(\xi, t)$ is the probability that a system is in state ξ at time t, while $Pr(\xi' \rightarrow \xi, t)$ is the transition probability for a system in ξ' at time t moving to state ξ in an infinitesimal time dt. **a** Equilibrium systems satisfy the detailed balance condition, so models built on the equilibrium assumption can only describe steady-state behavior. **b** The master equation can capture the evolution of a system that is far from equilibrium

tic and monitoring purposes. Therefore, our analysis focuses on finding the spatial coverage of micro-robots as a function of time. In contrast, solving the drug delivery problem requires finding the probability of having a critical number of micro-robots hit the target region within a certain time interval.

Note that classical diffusion theory [39–41] cannot be used to describe such a scenario since the dynamics of such multiple random walks is affected by interactions (captured via the potential function β in Eq. (2)). As we show later in this chapter, the accurate modeling of trajectories and distances traveled by micro-robots is of crucial importance for solving the diagnostic and drug delivery problems.

4 Implications of Molecular and Volume Exclusion Interactions

4.1 Spatial Extension for Gillespie's Algorithm

Finding the analytical solution of the master Eq. (3) in a 3D space and under the constraints in Eqs. (4) and (5) is a challenging task; this is because it requires determining the state probabilities in a high dimensional space. Consequently, in order to gain some insight into the dynamics of the bacteria propelled micro-robotic swarm and quantify its performance in terms of the space covered by micro-robots (or the hitting times of the micro-robots to reach targeted locations) we develop a numerical strategy that uses concepts from kinetic Monte Carlo (KMC) simulation and Gillespie's stochastic simulation algorithm [42]. This allows us to obtain the realizations of the master Eq. (3) efficiently and quantify the spatio-temporal evolution of the swarm.

Getting now into more details, we label by $\mathbf{X}(t) \equiv (X_1(t), \ldots, X_i(t), \ldots, X_{N_W}(t))$ the state vector whose i-th element $X_i(t)$ represents the state (e.g., position) of micro-robot (bacterium) i at time t in a 3D space, where N_W is the number of micro-robots (bacteria) within the swarm. The state vector $\mathbf{X}(t) \equiv (X_1(t), \ldots, X_i(t), \ldots, X_{N_W}(t))$ obeys all the transition probabilities described in previous section (i.e., the directional transition probabilities of moving North, East, South, West, Up or Down within a confined 3D space), but has a reduced dimensionality compared with the state of the multivariate PDF $P(\sigma_{1,1,1}, \ldots, \sigma_{i,j,k}, \ldots, \sigma_{M,N,Q}; t)$ in the master Eq. (3). In this representation, each movement u of an i-th micro-robot represents an "action" R_u in the system.

Having defined the state vector $\mathbf{X}(t) \equiv (X_1(t), \ldots, X_i(t), \ldots, X_{N_W}(t))$ and the actions R_i that obey the above mentioned transition probabilities, we can define the *propensity function* similar to the Gillespie's algorithm. The propensity function we introduce $a_u(\mathbf{x})dt$ aims at encoding the physical actions (i.e., the probability of physical action R_u taking place in the infinitesimal time interval $[t, t + d\tau]$). For instance, if u is the action of a micro-robot (bacterium) located at site (i, j, k) moving forward along X-axis, then $a_u(\mathbf{x})dt$ is calculated as follows:

$$\begin{aligned} a_u(\mathbf{x})dt &= Pr(\sigma_{i+1,j,k}(t + \delta t) + 1, \sigma_{i,j,k}(t + \delta t) - 1; t) = \\ &= \sigma_{i,j,k}(t)((1 - \sigma_{i+1,j,k}(t))(\alpha_{i,j,k\to i+1,j,k} + \quad \& \\ &\textstyle\sum_{(p,q,w)\in\Omega_{i+,j,k}} \beta_{i,j,k\to p,q,w} + \chi\sigma_{i-1,j,k}(t))\delta t + O(\delta t) \end{aligned} \tag{6}$$

As shown in Eq. (6), if there exists another micro-robot already occupying the space ($\sigma_{i+1,j,k} = 1$), then the propensity function $a_u(\mathbf{x})dt = 0$, meaning that action R_u has zero probability to take place in the infinitesimal time interval $[t, t + dt]$. The rest of the kinetic Monte Carlo method that we are using is the same as the Gillespie's algorithm. Consequently, numerically simulating and determining the PDF of the state vector $\mathbf{X}(t) \equiv (X_1(t), \ldots, X_i(t), \ldots, X_{N_W}(t))$ allows us to retrieve the trajectories of $\xi(t) = \{\sigma_{111}(t), \ldots, \sigma_{MNQ}(t)\}$.

4.2 Simulation Setup

In order to investigate the impact of micro-robots volume and the attraction/repulsion interactions among micro-robots in diagnostic and drug delivery problems, we employ the above mentioned KMC method and record the trajectories of N_W interacting random walkers in 3D space. To determine the minimum number of required simulations that guarantee reliable estimates regarding both average space cover-

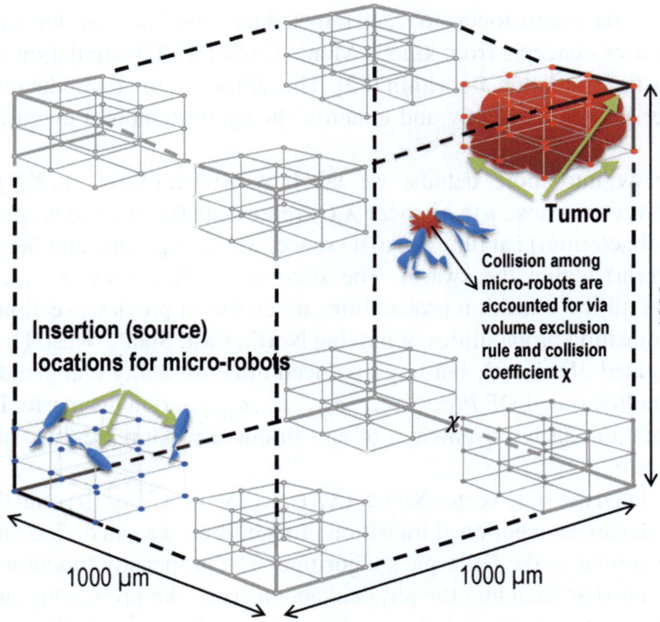

Fig. 5 Experimental setup where micro-robots are injected one millimeter away from the targeted region. Such micro-robots can sense the target (i.e., tumor) using chemotaxis and then propel themselves towards the target region

age and hitting time metrics, we use stochastic simulation rules outlined in [43]. More precisely, in order to guarantee a specific maximum deviation magnitude in the average hitting time obtained after k simulations, we use the Markov and Chebyshev inequalities to obtain the minimum number of required stochastic simulation k and then perform the averaging over the hitting time realizations. Relying on this statistical methodology, we collected 50 independent simulations of micro-robots trajectories; this took about two days of computational runtime on the CMU Condor cluster consisting of 300 Intel Xeon servers and CISCO Unified Computing System.

Following the KMC methodology, we simulate and record the trajectories of $N_W = 1000$ micro-robots starting from non-overlapping random locations on the lower left hand side and moving along the axes towards the targets located in the upper right hand side of a 3D lattice (as shown in Fig. 5). The preferential movement of micro-robots along certain direction is a result of various chemical interactions. For example, in a nutrient-free environment, any micro-robot has an equal probability of moving in any of the six possible directions if no micro-robot is in the immediate neighborhood. To model the affinity of the chemotactic micro-robots to move towards the target regions that exhibit abnormal behavior (e.g., due to the chemical gradients produced by tumors), we use the linear chemotactic response function and assume a constant chemical gradient; the coefficients α are therefore biased with a small value of 0.003 toward the target region. Obtaining the time-dependent trajectories of the N_W micro-robots allows us to estimate the probability $P(\sigma_{111}, \dots, \sigma_{MNQ}; t)$ and compute two metrics:

- **Space coverage**: the number of volume units (percentage of space) visited by at least one micro-robot during the simulation time.
- **Hitting time**: the minimum amount of time required for a micro-robot or (a group of micro-robots) to reach a set of target locations.

Throughout this section, we will use these two metrics, namely, node coverage and hitting time[4] to estimate the collective performance of the micro-robotic swarm. A high percentage of space coverage means that micro-robots can explore a significant portion of space in a short time interval and so increase the probability of detecting the abnormal behavior of cancerous cells. In contrast, a hitting time distribution peaking at small values and having short tail implies that micro-robots reach their targeted locations fast and deliver the drug before cancer cells population grows or migrates to other regions. Moreover, the efficacy of a drug is maximized if a certain level of drug concentration is maintained throughout the target region. Consequently, as more and more micro-robots hit the target region, the drug concentration can increase gradually. Therefore, it is also very important to study the distribution of hitting times, as for some scenarios, all micro-robots may hit the target at the same time and then increase the drug concentration abruptly up to a very high level which may induce undesirable side-effects. On the other hand, the hitting times may follow a uniform distribution and spread over a long time interval, in which case the drug concentration never reaches the required level. Consequently these metrics are both relevant for assessing the performance of the swarm on rapid tumor detection, environmental monitoring, or tumor suppression applications.

4.3 *Effects of Volume Exclusion*

Traditional point-like models do not take into account the physical volume occupied by micro-robots. Under this formalism, micro-robots are represented as points or particles moving freely in the environment without any physical interactions with other micro-robots or obstacles. In this subsection, we consider 1000 micro-robots which interact with each other in a space of $1000 \times 1000 \times 1000\,\mu m^3$ according to the relations given in Sect. 3 and investigate the impact of taking into account the volume occupied by micro-robots. More precisely, we consider that the micro-robots have a spherical shape. We also assume that the radius of the sphere volume that any micro-robot occupies effectively is r [μm], therefore its volume is $\frac{4}{3}\pi r^3$ [μm^3]. In our simulations, we consider r to be 0 μm (when $r = 0$ micro-robots are modeled as "points"), 1, 25 and 45 μm. We estimate the space coverage and hitting time distri-

[4]A mathematical definition of the hitting time for a collection of random walks moving along the edges of a graph can be found in [36].

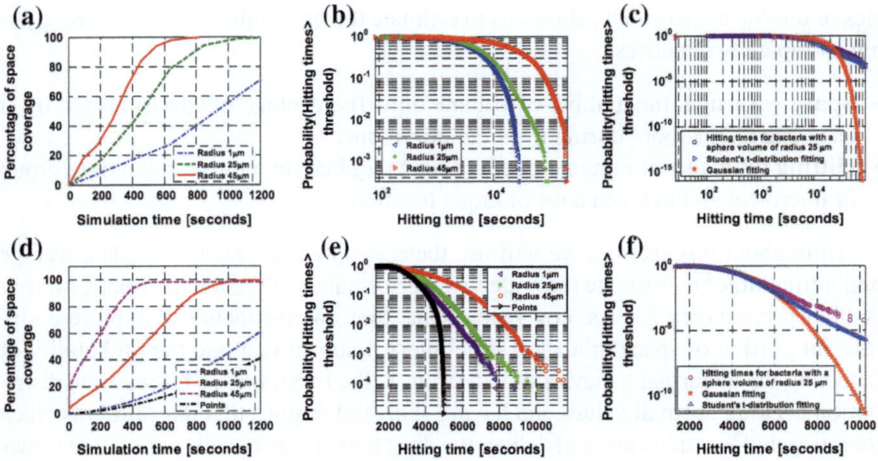

Fig. 6 Effects of volume exclusion. **a–c** are obtained from BNSim, and **d–f** are generated from the master equation. For **a** and **d**, the percentage of space covered by micro-robots. As shown, the increase of the radius (volume) of micro-robots can contribute to a significant increase in coverage over a short time interval. Note that the space coverage for micro-robots modeled as "points" are counted as the number of spheres that those "points" visit multiplied by the volume of one sphere. For **b** and **e**, the probability of hitting time to exceed a certain threshold as a function of the micro-robot radius. One can observe that by considering a larger radius, the probability of hitting time to exceed a given threshold tends to diverge from the Gaussian law and is better approximated by long tail type distributions. For **c** and **f**, the probability of hitting time to exceed a certain threshold for the micro-robot with radius 25 μm. The probability of hitting time to exceed a given threshold is better approximated by Student's t-distribution instead of the Gaussian law

bution via the proposed mathematical model and an open-source bacteria network simulator (i.e., BNSim) [44].

We measure the efficiency of the dense networks of micro-robots to detect abnormal behavior via the percentage of space covered by micro-robots. Figure 6a shows that using the power of multiple random walkers, while taking into account the volume occupied by micro-robots, leads to a significant spatial coverage within a short time interval.

Besides coverage, it is also important to quantify the time needed by a group of micro-robots to reach a certain target region; this represents the hitting time or first passage time and can be regarded as a performance metric for a successful (targeted) drug delivery. Figure 6b shows the probability of hitting time to exceed a certain threshold as a function of different radii. By increasing the radius of the exclusion sphere from 0 to 45 μm, one can clearly see that the probability of hitting time to exceed a certain threshold increases for larger volumes. As already explained, this may have detrimental effects on the treatment.

As shown in Fig. 6c, the probability of hitting time for larger micro-robots to exceed a given threshold tends to diverge from the classical Gaussian distribution; this can be better approximated by long-tail type of distributions (e.g., Burr or

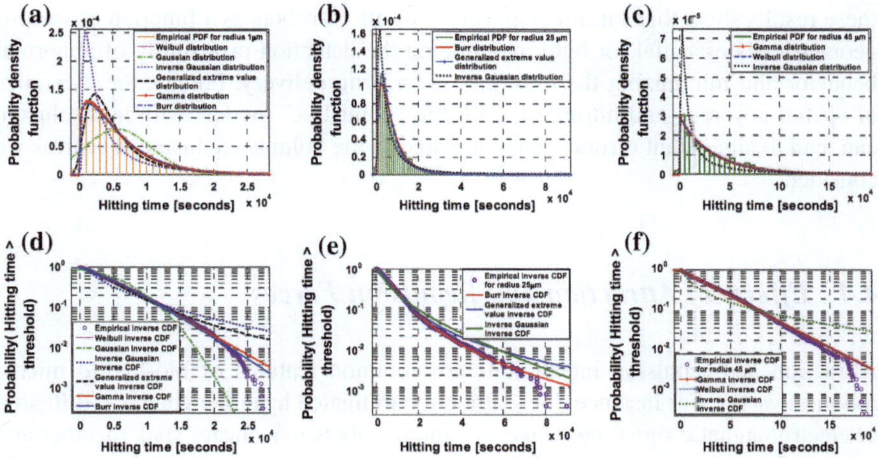

Fig. 7 Figure 7a–c show the empirical hitting time PDFs and a few fittings with well known distributions (Burr, Gaussian, Gamma, Weibull) as a function of the size of the micro-robot (i.e., 1 µm, 25 µm, and 45 µm). Figure 7d–f show the empirical hitting time exceedance probabilities and a few fittings with well known distributions (Burr, Gaussian, Gamma, Weibull) as a function of the size of the micro-robot (i.e., 1, 25, and 45 µm)

Weibull distributions). For completeness, Fig. 7 shows the probability density function and the exceedance hitting probability (i.e., the probability of hitting time to exceed a certain threshold) as a function of the side of the micro-robot (i.e., 1, 25 and 45 µm) obtained via BNSim simulation.

In addition to the empirical estimates of the hitting time PDFs and the exceedance probabilities, we also report the best graphical fittings obtain for these results. The statistical analysis reveals that Gaussian distribution does not fit well the hitting time realizations. In contrast, long-tail like distributions such as Burr and Weibull seem to offer a better fitting than Gaussian distribution.

We also investigate how well the inverse Gaussian distribution fits the empirical estimates of hitting time PDFs and exceedance probabilities. The rationale for considering the inverse Gaussian distribution as a possible candidate lies in the fact that this distribution characterized the times it takes for a Brownian motion with positive drift to reach a fixed position. From Fig. 7a–c we can clearly see that the inverse Gaussian distribution overestimates the peak of the hitting time distribution. Moreover, from Fig. 7d–f, we can conclude that the inverse Gaussian distribution does not offer a good fit for the exceedance probability. The main danger in relying on a crude upper bound given by the inverse Gaussian distribution is that if the simulation results did not capture well enough large hitting times, the prediction may underestimate the likelihood of rare events which has important implications on the performance of drug delivery.

The fact that the distribution of hitting times exhibits a long tail behavior implies that the optimization models that neglect the volume exclusion rules can lead to very optimistic predictions concerning the success of drug delivery tasks. In addition, these results show that careful engineering of micro-robots as a function of various geometries is essential for both maximizing the detection probability of abnormal behavior and minimizing the hitting time for drug delivery. Finally, the dynamics of spatial coverage and hitting time for the "point-like" models with no exclusion can lead to significant errors when compared to the volume exclusion scenarios we consider.

4.4 Effects of Attraction and Repulsion Forces

Attraction and repulsion interactions are common features of bio-hybrid micro-robotic swarms. For instance, local attraction mitigated by either chemical diffusion or electromagnetic signaling, makes the micro-robots in a cluster stick together and move in formation as shown in Fig. 8. In contrast, repulsion forces determine micro-robots not only move away from each other but also, in some cases, make them visit

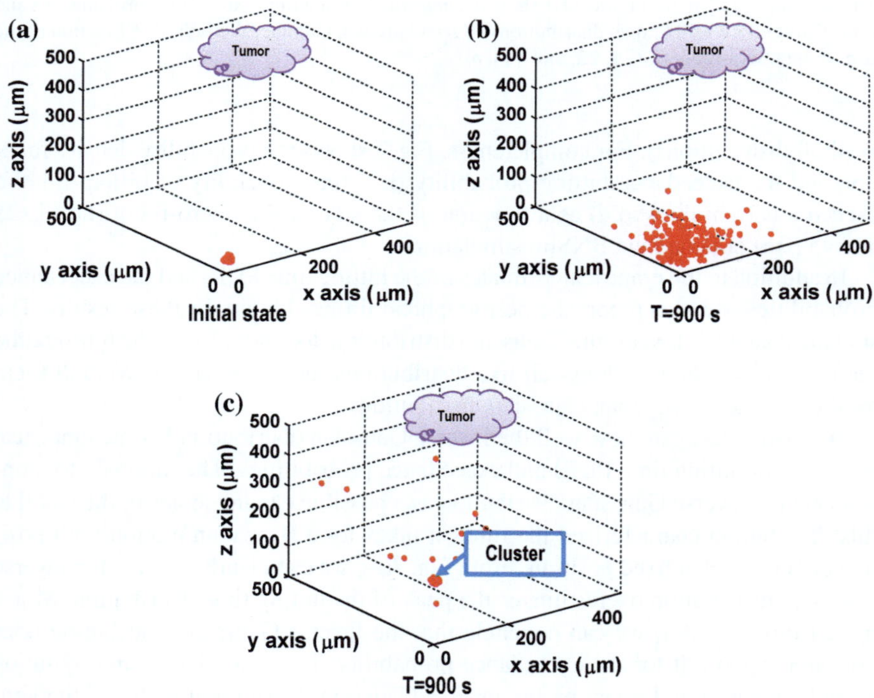

Fig. 8 Dense networks of micro-robots with and without attraction interaction. Figure **a** shows the initial locations of the injected micro-robots. Both interacting and non-interacting micro-robots populations are injected into the same initial locations. **b** shows the position of the swarms without interaction at time = 900 s. **c** shows the position of the population with interaction at time = 900 s. As shown, the local attraction interaction can cause population to form clusters and move in formation

new space and so increase the chances of detecting abnormal behavior. Aiming at quantifying the effects of attraction and repulsion forces on the tumor detection and drug delivery problem, we consider 200 identical micro-robots with radius $r = 1\,\mu\text{m}$ and volume $4.187\,\mu\text{m}^3$.

4.4.1 Local Attraction

Local attraction contributes to the creation of a self-organizing behavior or aggregation within the swarm. In our particular setup, local attractions not only helps micro-robots form coherent aggregates, but also, by sharing information about the environment, can enable collective decisions at the swarm level and increase the probability of success for delivering the drug to a specific location. To investigate the effects of local attraction, we set the attraction range to 30 μm, while we consider that the radius of the spherical model of any micro-robot is set to 2 μm. In addition, the parameters in Eq. (2) are as follows: $C_r = 0.04$, $C_a = 350$, $e_r = 15$ which correspond to a weak repulsion force induced by chemical repellents and hydrodynamics; the attraction interaction has a much larger effective range than the repulsion interaction. We show that repulsion helps micro-robots cover more space within the same time interval, while attraction helps micro-robots move in formation; this is beneficial for information exchange [38] and collective defense [45].

Figure 9a shows the space coverage of the micro-robotic swarm for various values of the attractant strength. One can notice that increasing the attraction strength makes the micro-robotic swarm cover less space (because the micro-robots spread less in space) than in the case without attraction rules. This has both beneficial and detrimental effects. On the beneficial side, the micro-robotic swarm is moving as a more cohesive cluster (i.e., less randomness in micro-robots trajectories); if the swarm is taking wise decisions regarding the movement direction toward the targeted region, then the hitting time is minimized. On the detrimental side, the micro-robotic swarm may only cover a limited space in the vicinity of its trajectory from the starting locations all the way to the target region. If multiple targeted regions exist and these are far apart from each other, the local attraction can cause the swarm reach one targeted region, but miss others which can be equally important for drug delivery efficiency. To avoid such situations, it is important to tune the attraction rules such that they do not completely overwrite the intrinsic stochastic characteristics of bacteria processing and allow the micro-robots once in a while, take random decisions. This would allow micro-robots swim away or join the swarm aggregate dynamically depending on the information exchanges between them, as well as their own (stochastic) decision-making process according to current conditions and recent memory. Another alternative is to consider a heterogeneous population of micro-robots that exhibits both collective (join the cluster) and competitive (swim away) behavior depending on the size of the cluster and the variability sensed in the environment (some micro-robots more sensitive to chemical cues than others and prone to swim towards regions). However, the topic of heterogeneity and cluster stability is left for future work.

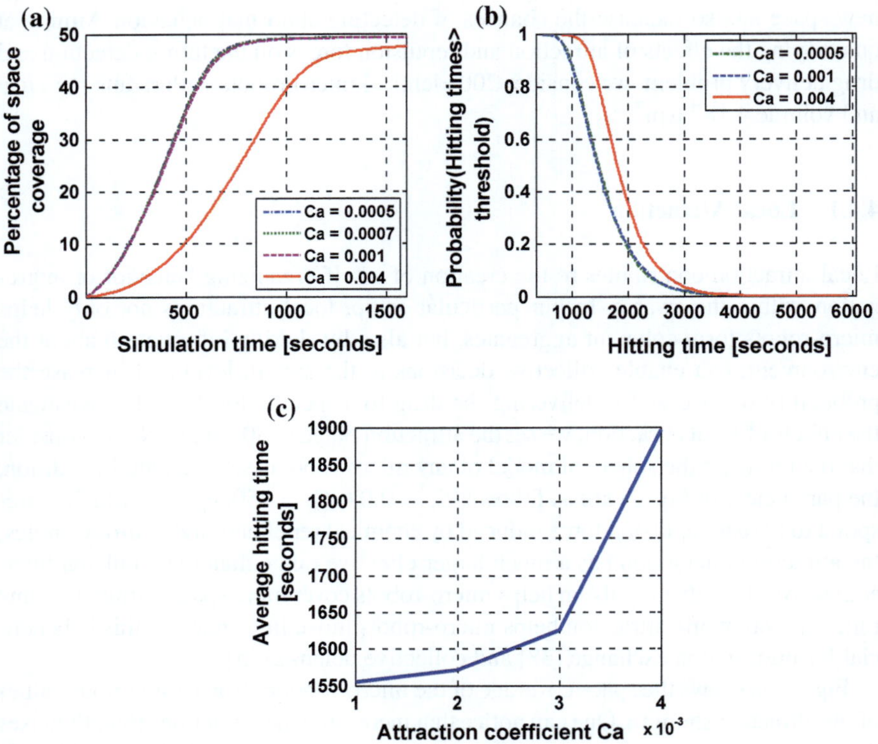

Fig. 9 **a** Space coverage of micro-robots population for different strengths of attractant forces. As shown, the percentage of space coverage does not significantly change for the following attractant parameters: attractant strength parameter $C_a = 0.0005$, $C_a = 0.0007$, $C_a = 0.001$. However, for $C_a = 0.004$, the difference is obvious and the micro-robots cover much less space; this is because they form a cluster and move in formation as shown in Fig. 8c. The results also suggest the existence of a phase-transition in cluster formation. The cluster of micro-robots allows for better information exchange, and so micro-robots follow a more straight trajectory to the target region. The cluster can also minimize the probability of inducing reactions from the human immune system. **b** Probability of hitting times being larger than a certain threshold for different strengths of attractant forces. The figure shows that clusters tend to converge to the target region more slowly because the micro-robots in the back of the clusters can drag the micro-robots in the front by attraction, therefore delaying the convergence of the entire population to the target region. The results also suggest that in practice, we may want to induce some heterogeneity in micro-robot population in order to better control them for drug delivery purposes. **c** Average hitting time of micro-robots as a function of the attraction coefficient strength C_a. The results show a phase transition around $C_a = 3 \times 10^{-3}$ indicating that when the attractant strength reaches a critical point, the micro-robots form a strongly connected network (cluster). (Figure reproduced with permission from [37])

Figure 9b shows the probability of the hitting time at which the micro-robots reach the targeted locations to exceed a specific threshold. As one can notice, without the true bias towards the targeted region the micro-robots moving in a more clustered (compact) formation converge to the targeted region more slowly. In addition, this

Table 1 Statistical moments of hitting times and the communication loss rate for various attraction strength coefficients

Attraction strength C_a	0.0005	0.001	0.004
Average of hitting times (s)	1,564	1,558	1,897
Variance of hitting times	305,690	304,280	274,110
Skewness of hitting times	1.3465	1.3405	1.3180
Kurtosis of hitting times	6.3340	6.7112	6.4550
Communication loss rate (%)	18	19	35

situation is also caused by the fact that micro-robots that are further away from the targeted region can slow down the cluster and divert it for some periods of time.

To provide some insights on the statistical nature of hitting time distribution, Table 1 reports the average, variance, skewness, and kurtosis of the hitting times as a function of the magnitude of attraction coefficient. One can notice that for attraction strengths of 0.0005 and 0.001, the hitting time statistics are approximately close to each other because no aggregation happens and micro-robots swim individually towards the target region without too much interactions with each other. In contrast, for a higher attraction strength of 0.004, the micro-robots aggregate into clusters and this reduces randomness (i.e., variance). In addition, the aggregation induces a significant delay as can be observed from the average hitting time and the communication loss rate.

In summary, these results suggest that if each micro-robot has the same impact on its peers in a dense network of micro-robots, then, the entire population will converge to the target region more slowly in the absence of the right target cues. Consequently, in order to increase the drug delivery efficiency by minimizing the hitting time, it is desirable to embrace swarm heterogeneity and consider micro-robots with varying swimming, sensing and interaction capabilities (e.g., use only a few "smart" micro-robots which have a greater impact on others as it can be often observed in biological swarms [46]). While here we focused on the impact of local attraction interactions within micro-robotic swarms, our mathematical framework allows the study of attraction forces in more general setups by encoding the various interaction cues existing among eukariotic cells [47].

Figure 9c shows the average hitting time of micro-robots to reach their target locations as a function of the attraction strength coefficient C_a. One can notice that there exists a critical point at which the micro-robots aggregate into larger clusters and this corresponds to higher average hitting times. The micro-robot aggregation is mitigated by various dynamical links that are established where micro-robots are close to each other and lost when they get further apart. The strength of these links is influenced by the chemical attraction interactions and the environment in which the micro-robots swim. Over consecutive time intervals, nodes get connected and disconnected as a function of physicochemical interactions among micro-robots; therefore they are not necessarily forming a strongly connected component. Consequently, the design of dense networks of interacting micro-robots for a particular task has to fine tune the attraction rules such that the swarm is more self-controllable and achieves its desired task in a timely manner.

Aggregation and moving in formation or exploring the space independently for random attacks are two very different approaches; it is hard to say which one may be better in the battle against growing and migrating cancer cells. Under the assumption that micro-robots are injected into hostile environments, it may be better for the micro-robots to aggregate. In contrast, if the targets are randomly located far apart from each other and the environment is less hostile for micro-robots, then the aggregation may prove to become detrimental to the drug delivery task. To make the dense networks of micro-robotics swarms a viable alternative against cancer cell populations, other important factors should be also taken into account (e.g., information exchange for populations grouped in clusters, possible attacks from external entities like viruses and pathogens, impact of these micro-robots on human immune system). All in all, the radius and strength of interactions are constraints specific to synthetic biology design.

4.4.2 Local Repulsion

As previously emphasized, the emergent sporadic competitive behavior may prove beneficial for the micro-robotic swarm to explore more space and increase the chances of detecting abnormal cellular dynamics. Consequently, in what follows, we investigate the impact of local repulsion interaction on the overall swarm performance. In our simulations, we consider the interaction range to be 30 μm, while the radius of each micro-robot sphere is set to 2 μm. The parameters in Eq. (2) are as follows $C_a = 0.000001$, $e_r = 350$, $e_a = 350$.

Figure 10a shows the percentage of space covered as a function of the three repulsion strength coefficients (i.e., $C_r = 0.01$, $C_r = 0.05$, and $C_r = 0.5$). One can observe that increasing the strength of the repulsion interactions among micro-robots leads to an increased space coverage within the same time interval. In addition, the stronger the repulsion interaction among micro-robots is, the higher the deviation of the hitting time distribution is from the Gaussian law (see Fig. 10b). While the repulsion interaction seems to induce a similar behavior to the volume exclusion rules, from a qualitative perspective, its impact is less significant. One can find two plausible reasons for this phenomenon: (1) Local repulsion represents a soft interaction, while the volume exclusion rule leads to a hard interaction. In other words, under the local repulsion interaction, there is a non-zero probability that two micro-robots may get closer to each other than a predefined distance, while in volume exclusion rule this probability is strictly zero. (2) The interaction potential function used in our approach assumes that the repulsion force decays exponentially fast with distance. To remedy this situation, one can choose different interaction potential functions in Eq. (2) for various communication methods (e.g., chemical signaling, magnetic signaling) and strengthen the role of repulsions.

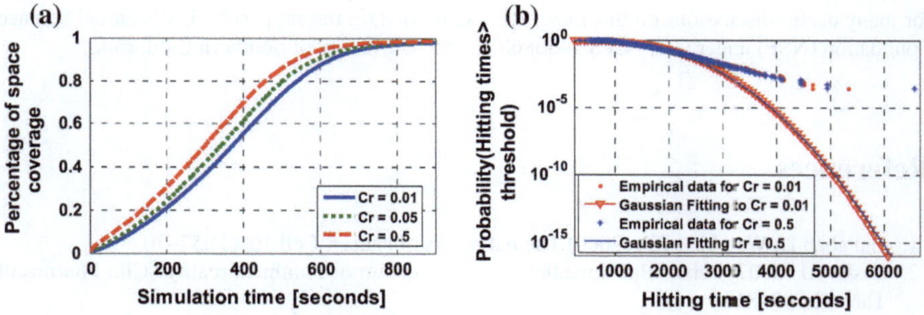

Fig. 10 **a** Micro-robots space coverage for different strengths of repulsion forces. Local repulsion interactions make micro-robots able to cover more space over the same amount of time. Note that repulsion interaction is conservatively assumed to be short-range chemical diffusion; mid-range repulsion can result in even more space covered over the same time period. **b** Probability of hitting times being larger than a certain threshold for different strengths of repulsion forces. The results show that the hitting time of micro-robots with repulsion interactions tends to deviate from Gaussian distribution, i.e., stronger interactions induce a longer tail in the distribution. (Figure reproduced with permission from [37])

5 Conclusions and Future Work

In this paper, we have proposed a non-equilibrium statistical physics description of the dynamics of dense networks of multicellular systems with application to abnormal cell detection and drug delivery by flagellated bacteria. Towards this end, we have developed a new mathematical formalism based on a new master equation which takes into consideration the volume exclusion principle and various rules of intercellular communication among cells.

Our results show that the volume exclusion rules and the attraction and repulsion interactions have significant implications on the probability of detecting abnormal cells, as well as the distribution of drug delivery (hitting) times both at short and long timescales. Consequently, our approach can become an essential tool when designing such dense networks of interacting micro-robots for healthcare applications. The framework is general enough that interested researchers can further modify it and adjust the action probabilities (or coefficients) to suit their interests for studying any cooperating swarm dynamics (e.g. micro-robots, cellular computing machines) and nanonetworks (e.g., passive diffusion-based molecular communication networks, active molecular communication network using flagellated bacteria).

In the future, we plan to model more realistic effects of micro-robots swimming in viscous fluids, such as the stochastic intracellular reactions, extracellular hydrodynamical interactions, and various molecular communication mechanisms among micro-robots and other cells. This will allow us to connect our theoretical model with wet-implementations for real diagnostic and drug delivery applications.

Acknowledgements This work was supported in part by the US National Science Foundation (NSF) under Grant CPS-1135850. Authors thank Prof. Metin Sitti and his collaborators at CMU for many useful discussions on this topic. P.B. acknowledges the support by US National Science Foundation (NSF) under Grant CPS 1453860 and the University of Southern California.

References

1. Hanahan D, Weinberg RA (2000) The hallmarks of cancer. Cell 100(1):57–70
2. Lesko LJ (2007) Personalized medicine: elusive dream or imminent reality? Clin Pharmacol Ther 81:807–816
3. Lesko LJ, Schmidt S (2012) Individualization of drug therapy: history, present state, and opportunities for the future. Clin Pharmacol Ther 92(4):458–466
4. Wei G, Marculescu R (2014) Don't Let history repeat itself: optimal multidrug quorum quenching of pathogens network. In: Proceedings of acm the first annual international conference on nanoscale computing and communicatio
5. Anderson J, Clarke E, Arkin A, Voigt C (2006) Environmentally controlled invasion of cancer cells by engineered bacteria. J Mol Biol 355(4):619–627
6. Behkam B, Sitti M (2009) Characterization of bacterial actuation of micro-objects. In: ICRA09 IEEE international conference on robotics and automation, pp 1022–1027
7. Saeidi N, Wong C, Lo T, Nguyen H, Ling H, Leong S, Poh C, Chang M (2011) Engineering microbes to sense and eradicate pseudomonas aeruginosa, a human pathogen. Mol Syst Biol 7 (1)
8. Arabagi V, Behkam B, Cheung E, Sitti M (2011) Modeling of stochastic motion of bacteria propelled spherical microbeads. J Appl Phy 109
9. Zhuang J, Wei G, Carlsen R, Edwards M, Marculescu R, Bogdan P, Sitti M (2014) Analytical modeling and experimental characterization of chemotaxis in serratia marcescens. Phys Rev E 89(5):052704
10. Berg H, Brown D (1972) Chemotaxis in escherichia coli analysed by three-dimensional tracking. Nature 239(5374):500–504
11. Berg H (2000) Motile behavior of bacteria. Phys Today 53(1):24–30
12. Axelrod R (1997) The complexity of cooperation: agent-based models of competition and collaboration. Princeton University Press
13. Gazi V, Passino K (2011) Swarm stability and optimization. Springer, New York
14. Naldi G (2010) Mathematical modeling of collective behavior in socio-economic and life sciences. Springer
15. Schweitzer F, Farmer J (2007) Brownian agents and active particles: collective dynamics in the natural and social sciences. Springer
16. Vicsek T, Zafeiris A (2012) Collective motion. Phys Rep 517(3–4):71–140
17. Gregori M, Akyildiz I (2010) A new nanonetwork architecture using flagellated bacteria and catalytic nanomotors. Sel Areas Commun IEEE J 28(4):612–619
18. Nakano T, Suda T, Okaie Y, Moore MJ, Vasilakos, AV (2014) Molecular communication among biological nanomachines: a layered architecture and research issues. IEEE Trans NanoBiosci 13(3):169–197
19. Nakano T, Kobayashi S, Suda T, Okaie Y, Hiraoka Y, Haraguchi T (2014) Externally controllable molecular communication. IEEE J on Sel Area Comm 32(12):2417–2431
20. Srinivas KV, Eckford AW, Adve RS (2012) Molecular communication in fluid media: the additive inverse Gaussian noise channel. IEEE T Inform Theory 58(7):4678–4692
21. Ruhi NA, Bogdan P (2015) Multiscale modeling of biological communication. In: IEEE International Conference on Communications (ICC), London, pp 1140–1145
22. Baker R, Yates C, Erban R (2010) From microscopic to macroscopic descriptions of cell migration on growing domains. Bull Math Biol 72(3):719–762
23. Xue C, Othmer H (2009) Multiscale models of taxis-driven patterning in bacterial populations. SIAM J Appl Math 70(1):133

24. Gueron S, Levin S (1995) The dynamics of group formation. math biosci 128(1–2):243–264
25. Vicsek T, Czirók A, Ben-Jacob E, Cohen I, Shochet O (1995) Novel type of phase transition in a system of self-driven particles. Phys Rev Lett 75(6):1226–1229
26. Yates C, Baker R, Erban R, Maini P (2009) Refining self-propelled particle models for collective behaviour
27. Bush SF (2010) Nanoscale communication networks. Artech House
28. Eckford AW (2007) Nanoscale communication with brownian motion. In: 41st Annual conference on information sciences and systems CISS07 IEEE, pp 160–165
29. Nakano T et al (2012) Molecular communication and networking: opportunities and challenges. Nanobiosci IEEE Trans 11(2):135–148
30. MacDonald JT, Gibbs JH, Pipkin AC (1968) Kinetics of biopolymerization on nucleic acid templates. biopolymers 6(1):1–25
31. Domb C, Lebowitz JL (2000) Phase transitions and critical phenomena. Academic Press
32. Grimmett G (2010) Probability on graphs: random processes on graphs and lattices. Cambridge University Press
33. Liggett T (1985) Interacting particle systems. Springer
34. Schadschneider A, Chowdhury D, Nishinari K (2010) Stochastic transport in complex systems: from molecules to vehicles. Elsevier
35. Spitzer F (1970) Interaction of markov processes. Adv Math 5:246–290
36. Bogdan P, Wei G, Marculescu R (2012) Modeling populations of micro-robots for biological applications. In: IEEE international conference on communications (ICC), 10–15 June 2012, pp 6188–6192
37. Wei G, Bogdan P, Marculescu R (2013) Bumpy rides: modeling the dynamics of chemotactic interacting bacteria. Sel Areas Commun IEEE J 31(12):879–890
38. Bassler B (2002) Small talk: cell-to-cell communication in bacteria. Cell 109(4):421–424
39. Einstein A (1956) Investigations on the theory of the brownian movement. Dover
40. Murray J (2002) Mathematical biology, vol 2. Springer
41. von Smoluchowski M (1906) Zur kinetischen theorie der brownschen molekularbewegung und der suspensionen. Ann Phys 326(14):756–780
42. Gillespie D (1977) Exact stochastic simulation of coupled chemical reactions. J Phys Chem 81(25):2340–2361
43. Ross S (1990) A course in simulation. Prentice
44. Wei G, Bogdan P, Marculescu R (2013) Efficient modeling and simulation of bacteria-based nanonetworks with BNSim. Sel Areas Commun IEEE J 31(12):868–878
45. Budrene E, Berg H et al (1991) Complex patterns formed by motile cells of escherichia coli. Nature 349(6310):630
46. Camazine S (2003) Self-organization in biological systems. Princeton University Press
47. Rappel W, Thomas P, Levine H, Loomis W (2002) Establishing direction during chemotaxis in eukaryotic cells. Biophys J 83(3):1361–1367
48. Li Z, Duan Z, Huang L (2009) Leader-follower consensus of multi-agent systems. In: American control conference ACC,09 IEEE, pp 3256–3261

Computational Biosensors: Molecules, Algorithms, and Detection Platforms

Elebeoba E. May, Jason C. Harper and Susan M. Brozik

Abstract Advanced nucleic acid-based sensor-applications require computationally intelligent biosensors that are able to concurrently perform complex detection and classification of samples within an in vitro platform. Realization of these cutting-edge computational biosensor systems necessitates innovation and integration of three key technologies: molecular probes with computational capabilities, algorithmic methods to enable in vitro computational post processing and classification, and immobilization and detection approaches that enable the realization of deployable computational biosensor platforms. We provide an overview of current technologies, including our contributions towards the development of computational biosensor systems.

Keywords Biomolecular logic systems · Digital biosensors · DNA computing · Computational biosensors · Intelligent biosensor · Application- and Substrate-specific algorithms · DNA detection · Deoxyribozymes · Enzymes

1 Introduction

The ability to detect and discriminate nucleic acid sequences is necessary for a wide variety of applications: high throughput screening, mutation tracking for disease emergence, genetically modified organism (GMO) monitoring, molecular computing, biometrics fingerprinting, and various genotype associated studies. Traditional sensor systems are multistep platforms that often times rely heavily on

In memory of Dr. Susan M. Brozik, deceased January 2014.

E.E. May (✉)
University of Huston, Huston, TX, USA
e-mail: eemay@uh.edu

J.C. Harper · S.M. Brozik
Sandia National Laboratories, Albuquerque, NM, USA

J. Suzuki et al. (eds.), *Modeling, Methodologies and Tools for Molecular and Nano-scale Communications*, Modeling and Optimization in Science and Technologies 9, DOI 10.1007/978-3-319-50688-3_23

off-platform post-processing to determine the outcome of detection or classify the biomolecule detected. Current high-throughput systems use traditional or silicon-based computing for interpretation of molecular recognition events. Complex bioinformatics algorithms are tasked with de-noising and processing sensor output signals. This approach can be error prone and not easily integrated into emerging microsystem technologies for hand held, lab-on-chip systems. Portable computational biosensor systems would be particularly useful for monitoring and diagnostic applications in resource-limited settings or situations, such as those encountered in developing countries and in military medical support applications.

The objective of this chapter is to provide an overview of capabilities needed for creating advanced, computationally intelligent, nucleic acid-based microsensor platforms that are reliable, deployable, and limits reliance on off-platform computational post-processing. In this chapter we review and discuss technologies and algorithmic advances necessary for the realization of integrated computational biosensor systems including: Nucleic acid based probes that enable detection and computation; Algorithm development for probe selection and target classification that transfers computational intelligence from silicon-based processing to nucleic acid based computational elements; Development of immobilization and detection approaches that permit the implementation and integration of computational biosensors on deployable platforms.

Nucleic acid based sensors use various types of probes, with single-stranded DNA, such as those used in microarrays, remaining the more widely employed. As an alternative to single stranded DNA technologies, we will discuss the use of molecular beacon, single-stranded oligonucleotide probes that incorporate structure to enhance functionalization [33, 118, 127]. Molecular beacons are able to detect mutations in target sequences and can be multiplexed; these properties make molecular beacons effective platforms for detecting genetically modified targets in various biosensor systems. Catalytic molecular beacons, such as deoxyribozymes, integrate enzymatic properties and modularity into standard beacons [98]. We will review characteristics of molecular beacon based probes and discuss their viability as molecular substrates for constructing computational biosensor platforms.

A major challenge in the development of computational biosensors is in the de novo development of application and substrate-specific algorithms that enable the transfer of computational intelligence from a silicon-based system to a nucleic acid based system. The algorithmic challenge is multifaceted and linked, with application-specific library design being the first aspect, and the second being the development of computational primitives that can systematically compute a classification outcome using the molecular substrate. We will discuss various approaches to address this challenge, including information-based methods for probe design and classification algorithm generation.

One of the drawbacks of optical-based detection platforms for molecular beacons is the reliance on laboratory-based equipment for analysis. Secondly, fluorescence detection may limit the integration of the computational sensor with traditional silicone-based microprocessors used in hand held, portable sensor systems. Electrochemical detection provides significant advantages over other common detection and signal transduction methods. Specifically, it is sensitive, rapid, and can provide greater specificity and sensitivity over optical detection methods, as interfering background fluorescence does not adversely affect the electrochemical signal. Electrochemical detection systems are also more amenable to miniaturization and integration given that the signal is already in a form that can be interpreted by silicone microprocessors and integrated circuit fabrication methods can be leveraged to produce the miniaturized components. We discuss the development of immobilization and detection technologies that enable the design and implementation of application specific integrated computational biosensor systems.

2 Computational Biosensors

Computational biosensors have roots in the emerging field of biocomputing. Increased research in biocomputing has resulted in significant advances in processing of chemical and biochemical information; this success is due, in part, to the innate high specificity and selectivity of biological molecules (e.g. enzymatic catalysis of substrate to product; nucleic acid binding of Watson-Crick base pairs). Further, the general compatibility of biomolecules allows for the intimate assembly of differing biomolecules within cascading networks capable of performing diverse reactions. Biocomputing systems have been reported that are assembled from proteins/enzymes [96, 113], RNA [124], DNA [101], and even whole cells [94]. Exploiting properties intrinsic to biomolecules for unconventional computing has resulted in greater success over chemical computing strategies alone [42].

Biocomputing shows perhaps the greatest promise for use with analytical systems, particularly for biomedical applications [60, 64, 115]. Use of biocomputing and logic operations could yield a novel class of computationally intelligent biosensors that can accept diverse input signals and systematically compute and output an actionable signal (Wang and Katz 2010). Additionally, signal output from the biosensors can be coupled to signal-responsive materials or processes that allow the system to sense, and then act [81]. Such 'intelligent' biosensors would offer significant advantages over traditional biosensors which typically accept only a single input, and then output a signal that must be post-processed and further analyzed by a well-trained operator for meaningful conclusions to be drawn.

Successful implementation of biocomputing principles with biosensing has led to significant and exciting advances in intelligent computational biosensors. Often referred to as 'digital biosensors,' or 'bio-logic' analysis, the vast majority of reported systems rely on either enzyme cascades or networks of DNA probes. Enzyme-based computational biosensors will be introduced and briefly discussed. A thorough review of enzyme logic systems lies outside the scope of this chapter which focuses on nucleic acid probe-based computational biosensors.

2.1 Enzyme-Based Computational Biosensors

Enzyme logic systems rely on multiple enzymes (at least 2, typically 3 or more) to develop a signaling cascade that results in the processing of chemical information in a manner that is comparable with Boolean logic operations (e.g. AND, OR, NAND, NOR, XOR, XNOR) [7, 82, 106, 132]. Combining the logic operations in a small logic network results in biosensor output in the form of a Yes/No response. Inputs to the sensor are typically chemicals and/or enzyme(s), and the Yes/No output is typically a change in solution pH, color, optical density, or electrochemical properties. This approach has been used to perform simple arithmetic functions [6]. More interestingly, enzyme logic systems have been coupled to signal-responsive materials forming switchable membranes [110], emulsions [70], nanoparticle assemblies [69], or modified electrode surfaces [83, 134].

Enzyme-based computational biosensors have been developed that focus on biomedically relevant markers/signaling molecules at their physically relevant concentrations. For example, an enzyme-based computational biosensor was developed to process biochemical information related to the pathophysiological conditions associated from traumatic brain injury and hemorrhagic shock [78]. Other biomedically relevant enzyme-based computation biosensors have been reported, including one capable of discriminating biomarkers characteristic of liver injury, soft tissue injury, and abdominal trauma [133]. Of note is the robustness of this system which was designed to operate in serum solutions spiked with injury biomarkers in order to mimic real medical samples.

Transitioning enzyme-based computational biosensors from in vitro to in vivo analysis is particularly challenging as operations are composed of several biocatalytic steps that must each be optimized to minimize signal noise, ensure adequate signal amplification, and reduce the impacts of many different interferants present in real biological fluids. The long-term instability of enzymes under ambient conditions further complicates development of practical, fieldable sensing systems relying on enzymes for signal processing. For further information of enzyme-based computational biosensors, the reader is referred to these excellent review articles [119, 120].

2.2 Nucleic Acid-Based Computational Biosensors

DNA-based biocomputing is a well-developed field, due in part to the high stability and rich information capacity of DNA molecules [101]. Typically, DNA biocomputing systems solve analytical problems using a combinatorial approach. DNA sequences are generated that encode all possible solutions to the problem. The correct solution is selected from the DNA library using polymerase chain reaction (PCR), affinity-purification and gel electrophoresis designed to only amplify/select DNA strands that meet the criteria for a correct solution. This strategy was first used by Adleman in 1994 [2] to solve a NP-complete problem (no efficient algorithms are known to solve NP-problems). Researchers have proposed several applications using a DNA computing framework including algorithms for breaking the Data Encryption Standard, DNA encryption methods, and techniques to investigate nature's cellular computing processes [3, 41, 46, 56]. More recently, Adleman and coworkers solved a 20 variable 3-SAT problem (a NP-complete problem requiring exponential time to solve, with 1,048,575 possible truth assignments) using DNA biocomputing [12].

DNA-based biocomputing has also been successfully employed for biomedically relevant applications. A stochastic computing system was reported in which a varying concentration of biomedical marker inputs provided competition between two alternative biochemical pathways that output a DNA molecule encoding the result [1]. This technique was subsequently used to analyze the concentration of cancer-related molecular indicators (i.e. mRNA). In the presence of the marker, the system output was a single stranded antisense DNA therapeutic against the cancer mRNA [8]. Such 'intelligent' systems combining sensing, computation, and action hold great promise for in situ medical diagnosis and treatment [93].

In nucleic acid-based computation and biosensing, the structure of the DNA probe(s) can have a profound impact on the functionality of the given system. The advantages and limitations of single-stranded DNA, DNA molecular beacons, and catalytic DNA molecular beacons are discussed below.

2.2.1 Single-Stranded Nucleic Acid Probes

The most widely employed DNA probe remains single stranded DNA. Used in DNA biocomputation systems, simple DNA-based biosensors, and DNA microarrays, single-stranded DNA probes are effective at detecting complementary DNA and RNA sequences due to strong Watson-Crick base pair binding [118]. Base pair binding results in exceptionally high specificity for the target sequence. Various methods are used to indicate that hybridization has occurred between the probe and the complementary target strand. These signal transduction methods are discussed in Sect. 4 of this chapter.

Despite the exceptionally high specificity intrinsic to DNA-based biosensors, the detection limit for most sensors is not sufficient to detect the target sequence at the very low concentrations typically found in real-world samples. This is partly due to the 1:1 target:signal stoichiometry which limits the number of signal events to the amount of target present in the sample. Thus, nearly all DNA-based biosensors rely on either (1) amplification of the target prior to introduction to the sensor, or (2) post-hybridization labeling that significantly amplifies the signal from each hybridization event. For example, a recent article reviewed detection technologies for identifying genetically modified organisms (GMO) in food and feed [5]. The authors stated that DNA-based biosensors are the 'leading edge' technology for simple, rapid and inexpensive testing for GMOs. However, they reported that for successful detection nearly all DNA-based GMO sensors required PCR amplification of samples extracted from raw ingredients and processed food prior to introduction to the sensor. The requirement of DNA sample pre-amplification via PCR makes these sensors difficult to implement as platforms for autonomous or field deployable biosensing.

Regarding post-hybridization amplification, a multitude of DNA-based sensors have been reported that use nanoparticles, dendrimers, supramolecular assemblies, and the like for labeling DNA hybridization events [79]. Detection of the label provides many orders of magnitude enhancement in the signal from each hybridization event. For example, Kawde and Wang used a triple amplification technique resulting in attomolar DNA detection limits [43]. As shown in Fig. 1, streptavidin coated polystyrene spheres (PS) decorated with biotinylated gold nanoparticles (AuNPs) were used as labels for DNA hybridization. Hybridization between the magnetic bead-bound DNA probe and the labeled target sequence (step a) was followed by further catalytic deposition of gold onto the AuNPs (step b). Detection of the binding event occurred by dissolution of the Au (step c) and stripping voltammetry in which the gold ions are accumulated onto a carbon electrode over many seconds, and then stripped from the electrode by reversing the applied potential. The multi-step strategy yielded a 300 attomolar (10^{-18} mol/L) DNA detection limit, eliminating the need for sample pre-amplification by PCR. However, the steps required for this, and other similar amplification strategies, significantly increase the complexity of the sensor making it impractical as a reliable and robust fieldable device.

Notwithstanding these limitations, the ability to multiplex DNA-based sensors into microarrays has resulted in the development of powerful analytical tools commonly referred to as gene chips, or biochips. DNA microarrays are widely used for high throughput analysis of DNA and RNA samples for differential gene expression levels, diagnosis of genetic diseases, detection and identification of infectious agents, determination of DNA-protein interactions, drug screening and forensic analysis [32, 118]. These hybridization chips are fabricated by

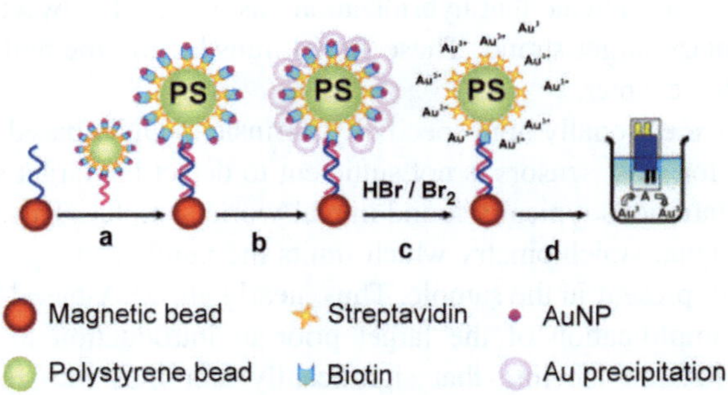

Fig. 1 Schematic illustration of amplified DNA detection employing probe DNA modified magnetic beads. **a** Treatment of probe with target DNA labeled with streptavidin conjugated polystyrene beads loaded with biotinlyated AuNPs, followed by **b** Au precipitation onto AuNP seeds, **c** dissolution of the Au and d) detection via electrochemical stripping. (Adapted with permission from [43]. Copyright 2004, Wiley.)

immobilization, or in situ synthesis, of thousands of single stranded DNA probes (up to 10^6 unique probes/cm^2) on a planar support (glass, silicon, or plastic) [118]. Target samples are introduced to the array and hybridization reactions are performed under conditions of high stringency such that mishybridizations are prevented. The vast majority of commercial gene chips utilize fluorescence for monitoring hybridization events using a bench-top fluorescent scanner, followed by software post-processing and expert analysis. Thus, these systems are not amenable for portable field use.

2.2.2 Molecular Beacon Nucleic Acid Probes

Molecular beacons are single-stranded oligonucleotide probes in which five to seven base pairs on either end of the strand are complementary to each other. The complementarity results in the strand ends hybridizing and forming a stem-loop structure [112], as shown in Fig. 2. The loop region of the strand contains a probe sequence that is complementary to the target sequence. Traditional molecular beacons contain a fluorophore and quencher on each arm of the stem. In the absence of target molecules, the stem-loop structure forms and the fluorophore is brought into close proximity to the quencher, suppressing fluorescence. When target DNA/RNA is present, hybridization occurs with the loop, breaking the stem region and separating the fluorophore and quencher. This results in a strong increase in probe fluorescence (see Fig. 2).

As opposed to traditional single-stranded DNA based biocomputing and biosensing systems, incorporation of molecular beacons that exploit structure, and changes in structure, adds another dimension of functionality to the system. Molecular beacons do not require additional labeling steps post-hybridization as the hybridization event itself results in an action (conformational change) that increases

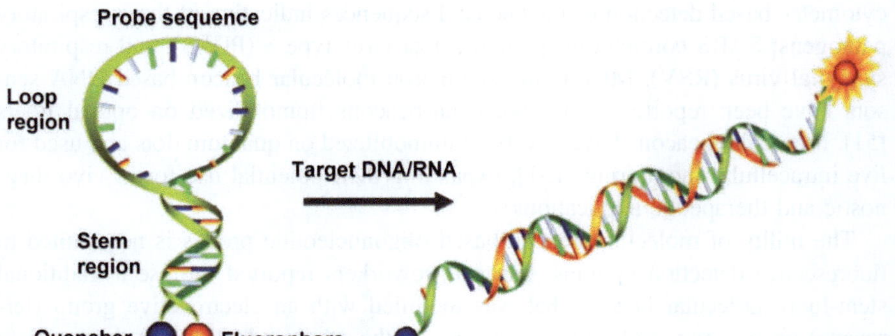

Fig. 2 Structural characteristics of a typical DNA molecular beacon probe showing formation of a stem-loop structure in the absence of target DNA/RNA. Breaking of the stem portion occurs when target is present and binds to the probe sequence loop region. (Adapted with permission from [107]. Copyright 2005, Royal Society of Chemistry.)

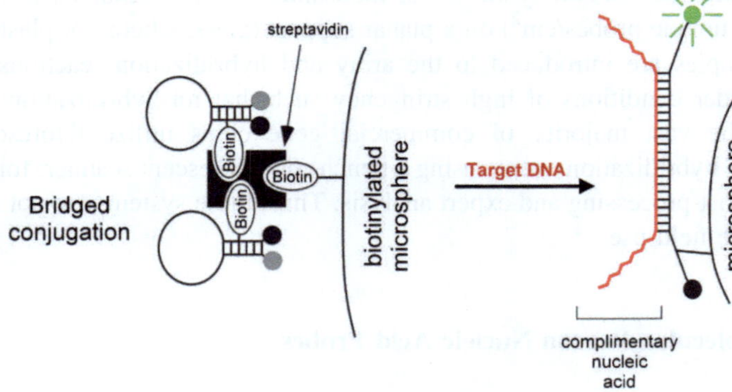

Fig. 3 Streptavidine bridged conjugation of biotinylated molecular beacons to a biotinylated microsphere. Introduction of target DNA and subsequent hybridization results in a conformational change, separating the fluorophore and quencher. (Adapted with permission from [33]. Copyright 2005, Oxford Journals.)

fluorescent output. This simplifies the system, increasing the likelihood that such a system could be developed into a deployable biosensor. Also, unlike microarray detection, molecular beacon probes can be used to detect the presence of target DNA or RNA sequences in complex mixtures [93].

Molecular beacons have found wide use in bioanalytical systems [107]. In one example, Horejsh et al. immobilized molecular beacons onto solid supports for multiplex detection of unlabeled nucleic acids in solution [33]. As shown in Fig. 3, biotinylated molecular beacons were conjugated to biotinylated microspheres using streptavidin. This streptavidin 'bridged' conjugation resulted in a nearly two fold improvement in signal to noise ratio over directly conjugating biotinylated molecular beacons to streptavidin coated microspheres. Using two different bead sizes and molecular beacons with two fluorophore colors allowed for flow cytometry based detection of nucleic acid sequences indicative of three respiratory pathogens: SARS coronavirus, parainfluenza virus type 3 (PIV-3), and respiratory syncytial virus (RSV). Micron and submicron molecular beacon-based DNA sensors have been reported using molecular-beacons immobilized on optical fibers [51]. Molecular beacons have also been immobilized on quantum dots and used for live intracellular monitoring [131], expanding their potential use for in vivo diagnostic and therapeutic applications.

The utility of molecular beacon-based oligonucleotide probes is not limited to fluorescence detection systems. Fan and coworkers reported the use a traditional stem-loop molecular beacon that was modified with an electroactive group (ferrocene) on one end, and a thiol molecule on the other end [26]. The thiol moiety facilitated self-assembly and immobilization of the molecular beacon onto a gold electrode surface, as shown in Fig. 4. In the absence of complementary DNA, the hairpin loop brings the ferrocene group in close proximity to the gold electrode surface, facilitating efficient redox reaction electron transfer. Upon hybridization the

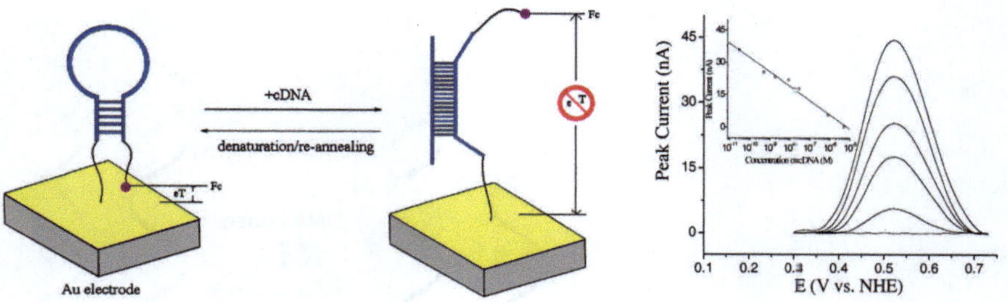

Fig. 4 Electrochemical detection using a molecular beacon oligonucleotide modified with a redox active ferrocene group on the 5′ end, and a thiol group on the 3′ end. The probe is immobilized via self-assembly between the thiol moiety and the Au electrode surface. Electron transfer (eT) is disrupted in the presence of target DNA. The *right* most panel shows background subtracted voltammograms with peak current decreasing as complementary DNA concentration increases. (Adapted with permission from [26]. Copyright 2003, PNAS USA.)

stem-loop structure is broken and the distance from the ferrocene tag to the electrode surface is increased. This reduced redox reaction associated electron transfer. This sensor is reusable and reagentless, making it well suited for device development and continuous monitoring of a given analyte. However, 'signal-off' sensors are more prone to false positive responses. Also, the differential in current response from no analyte present, to fully saturated with target analyte, was only 45 nA. Such a low signal differential can make analyzing the sensor output challenging.

Molecular beacon based sensors have also been reported with single nucleotide specificity, capable of identifying the presence of point mutations in the target sequence. For example, Wu and co-workers developed a 'signal-on' electrochemical DNA sensor based on DNA ligase and a 'reverse' molecular beacon [127]. In this approach, a single-stranded DNA capture probe was first immobilized onto a gold electrode surface via gold-thiol self-assembly. Introduction of the target DNA resulted in hybridization between the capture probe and a portion of the target DNA. The remaining portion of the target DNA is complementary to a detection DNA strand, and formed a 'sandwich.' As shown in Fig. 5, the detection DNA contains a phosphoryl group at the 5′ end, and a ferrocene tag at the free 3′ end. Addition of DNA ligase to the system joins the detection DNA to the surface immobilized capture DNA, forming a single DNA strand. Washing and thermal dehybridization removes the protein and target DNA. Incubation in blank hybridization buffer results in formation of a stem-loop structure as the detection DNA strand contains a 6-base sequence near the ferrocene group that is complementary to the 5′ end of the capture probe. This 'reverse' molecular beacon results in an increase in electron transfer when target DNA was present.

Wu and co-workers also reported that this system was capable of discriminating mutations in DNA sequences. Eight single-base mutations in the target sequence were evaluated. Point mutations at the nick location provided the lowest signal response, with G:T mismatches inducing slight higher currents. Relatively higher currents were observed when point mutations were on the 5′-side of the nick versus

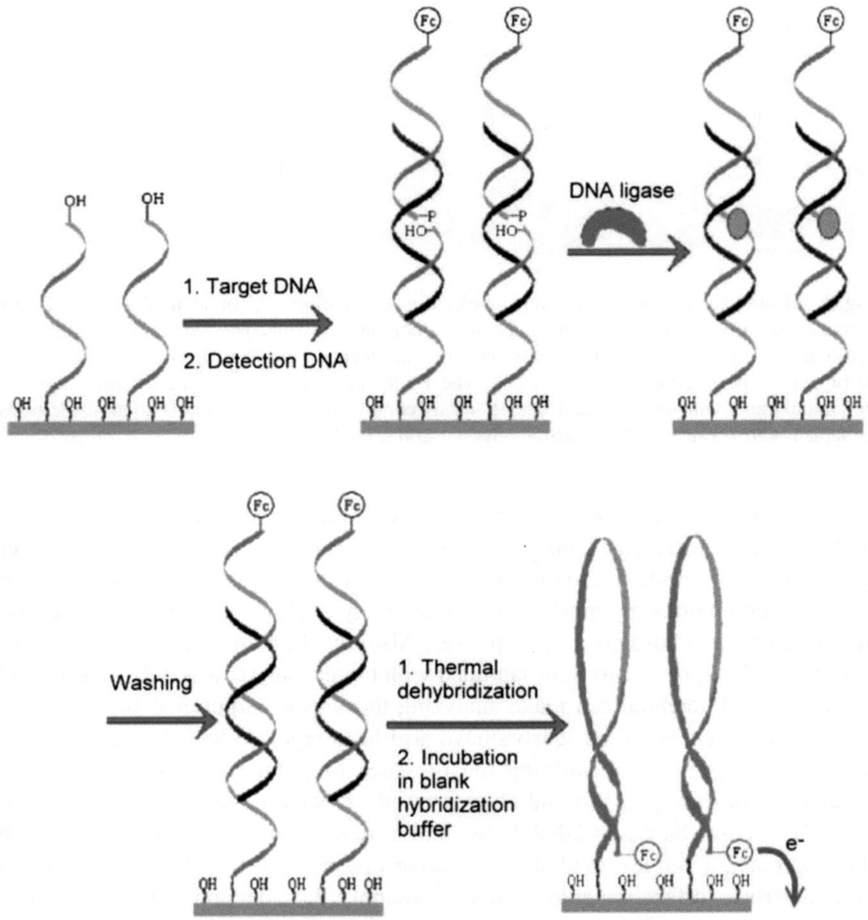

Fig. 5 Schematic illustration of a 'reverse' molecular beacon DNA detector with an electrochemical 'signal-on' mechanism. Target DNA is 'sandwiched' between the surface immobilized capture probe DNA and ferrecene label detection DNA. Enzymatic ligation joins the capture and detection sequences, resulting in a hairpin structure that positions the ferrocene label in close proximity to the electrode surface, permitting ready electron transfer. (Adapted with permission from [127]. Copyright 2007, Wiley.)

the 3′-side. Finally, the relatively highest currents from point mutation strands were measured when single-base mutations were at the third and fourth position from the ligation point. Still, the currents measured for these single-base pair mismatches were between only 0.6 and 5% of the signal obtained from the fully complementary target sequence which was approximately 70 nA at the highest target concentration. Using these subtle differences in the low nA current range to discriminate and identify point mutations would be challenging. Further, this detection scheme required several incubation/washing steps and addition of reagents, again making it difficult to employ in a setting outside the laboratory.

2.2.3 Aptamer Beacon Nucleic Acid Probes

Another important advance in structure-based nucleic acid biosensor systems came with the discovery of aptamers. First reported in 1990 by three independent groups [24, 86, 111], aptamers are single-stranded DNA or RNA molecules that fold into secondary and tertiary structures that bind to a target molecule with very high affinity (nanomolar to picomolar dissociation constants are typical). Often described as homologous to antibodies, aptamers can bind molecules from small inorganic ions, proteins, to even living cells [35]. Unlike antibodies, aptamers are more stable, simple to modify with functional groups, easy to purify and produce, can be developed against virtually any molecule, and do not require animals or cell lines to generate. A potential disadvantage of aptamers is that conditions under which the target molecule and aptamer are introduced (e.g. ionic strength, pH, temperature, protein content) cannot deviate significantly from those used during the SELEX selection process (systematic evolution of ligands by exponential amplification) or the aptamer will not fold and bind with high avidity.

'Aptamer beacons' [37] are similar to traditional beacons, typically contain a fluorophore and a quencher, and undergo a conformational change upon binding to the target that can be monitored by fluorescence signal readout. An early example is the aptamer beacon sensor for cocaine developed by Stojanovic et al. [98]. As shown in Fig. 6, a nucleic acid-based aptamer that recognizes cocaine was engineered with instability in one stem of a three-way junction that binds cocaine. The short stem was labeled with a fluorophore and a quencher that in the absence of cocaine remains open and emits fluorescence. Introduction of cocaine results in a conformational change that brings the fluorophore and quencher in close proximity, reducing fluorescence output (see Fig. 6, right panel). The aptamer beacon was selective over cocaine metabolites and could operate in serum, although it was

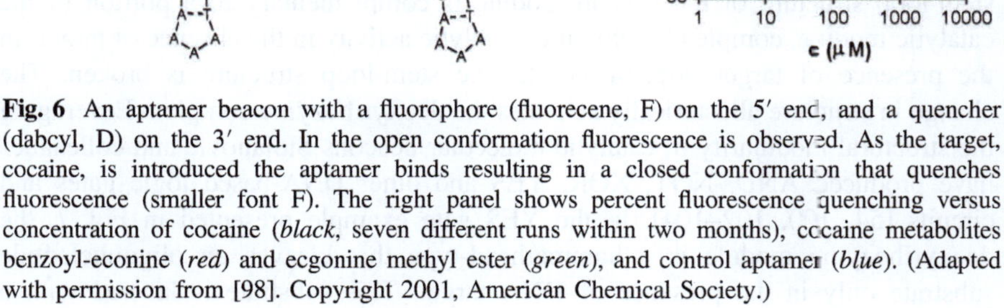

Fig. 6 An aptamer beacon with a fluorophore (fluorecene, F) on the 5′ end, and a quencher (dabcyl, D) on the 3′ end. In the open conformation fluorescence is observed. As the target, cocaine, is introduced the aptamer binds resulting in a closed conformation that quenches fluorescence (smaller font F). The right panel shows percent fluorescence quenching versus concentration of cocaine (*black*, seven different runs within two months), cocaine metabolites benzoyl-econine (*red*) and ecgonine methyl ester (*green*), and control aptamer (*blue*). (Adapted with permission from [98]. Copyright 2001, American Chemical Society.)

stable under only brief exposure to the serum, and the sensitivity (~10 μM detection limit) was not high enough for clinical or forensic applications.

Aptamer beacons have also been immobilized onto solid supports and modified with redox active groups for reagentless and label free electrochemical detection of small molecules [87] and proteins [52, 128]. Typically, the conformational change of the aptamer brings the electroactive group closer or farther away from the electrode surface in a manner similar to that shown in Fig. 4 of Sect. 2.2.2. The change in conformation results in an increase or decrease in redox current. Although these sensors provide a simple one-step protocol for detection with no labeling steps or addition of reagents, they again show small changes in current response between no analyte present, to fully saturated with target analyte (typically 10-200 nA differential) making analysis of the sensor output difficult.

2.2.4 Catalytic Molecular Beacon Nucleic Acid Probes

In the early 1980s it was discovered that naturally occurring single stranded ribonucleic acid molecules are capable of catalyzing certain biochemical reactions in a manner similar to enzymes. Termed 'ribozymes', or 'catalytic RNA,' these biological catalysts participate in RNA ligation, cleavage, synthesis, alkylation and acyl-transfer reactions, and n-glycosidic and petide bond formation [123]. A well-studied prototype catalytic RNA molecule is the hammerhead ribozyme which recognizes specific RNA sequences and catalyzes cleavage and ligation reactions at a specific site in the recognized region [34, 80]. Hammerhead ribozymes are also capable of self-cleavage.

Although DNA molecules are much more stable and serve as the primary carrier of genetic information in nature, naturally occurring catalytic DNA molecules have not yet been identified. However, single-stranded DNA molecules capable of performing catalytic reactions similar to RNA and enzymes can be engineered. Catalytic DNA, or deoxyribozymes, have been synthesized in the laboratory via an in vitro iterative selection process [14, 89]. Integrating this catalytic capability of single-stranded DNA with the structural dependent functionality of molecular beacons opens intriguing opportunities for development of complex, intelligent computational biosensors.

Like traditional molecular beacons, so called catalytic molecular beacons typically contain a catalytic module and a beacon module [99]. As shown in Fig. 7, the stem-loop structure of the beacon module is complementary to a portion of the catalytic module, completely inhibiting catalytic activity in the absence of target. In the presence of target oligonucleotide, the stem-loop structure is broken. The change in structure allosterically activates the deoxyribozyme complex. Leveraging the structural modularity of catalytic molecular beacons Stojanovic and colleagues have produced AND, NOT, XOR, YES and other DNA-based logic gates and circuits [54, 100, 102–104]. In the YES gate example presented in Fig. 7, the deoxyribozyme module is a hammerhead-type that cleaves an oligonucleotide substrate only in the presence of DNA target. This substrate is labeled with a

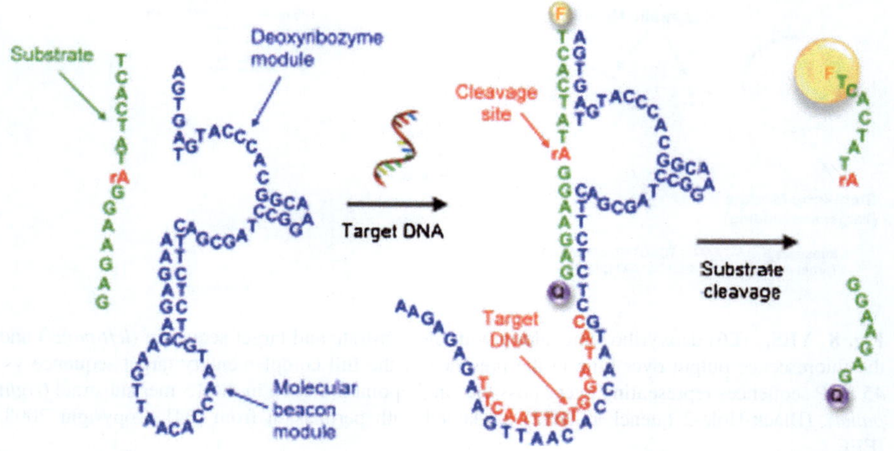

Fig. 7 A deoxyribozyme-based catalytic molecular beacon (*blue*) [99]. In the absence of target DNA (*red*), the stem-loop region blocks access of the substrate (*green*) to the catalytic region of the deoxyribozyme. In the presence of target DNA, hybridization occurs with the loop region of the molecular beacon. The conformational change exposes the catalytic region to the substrate, activating catalytic cleavage of the substrate. Labeling the substrate on either end with a fluorophore and a quencher allow catalytic activity and substrate cleavage to be monitored as an increase in fluorescence. (With permission from [64]. Copyright 2008, IEEE.)

fluorophore on the 5′ end, and a quencher on the 3′ end. Once the catalytic molecular beacon is activated by target DNA and substrate is cleaved, fluorescence output significantly increases.

Thus, in combination with DNA information processing techniques, DNA logic gates derived from catalytic molecular beacons are viable candidates for constructing intelligent computational biosensors. They can potentially integrate several molecular inputs, perform logical operations, and output an actionable signal and/or autonomously output the appropriate therapeutic. We have reported the development of a deoxyribozyme-based intelligent computational biosensor for the detection of genetic modifications in avian influenza [64, 65]. First, the deoxyribozyme YES_{iA} (E6) gate was constructed, based on a previously reported modular design [99], that appended a 15 nucleotide molecular beacon recognition loop into a hammerhead-type deoxyribozyme (Fig. 8, left panel). Using a substrate labeled with a fluorophore and quencher, the YES_{iA} (E6) deoxyribozyme exhibited a 14-fold increase in fluorescent signal intensity, resulting in a detection limit of 22.7 pM target DNA that was resolvable in only 3 min. This impressive detection limit and response time was a consequence of the catalytic behavior of the deoxyribozyme molecular beacon. Unlike traditional molecular beacons that are limited by a 1:1 target:signal stoichiometry, a single target DNA activates a given catalytic beacon, resulting in the cleaving of several substrate molecules. Thus, PCR amplification of target DNA prior to introduction to the sensor may not be

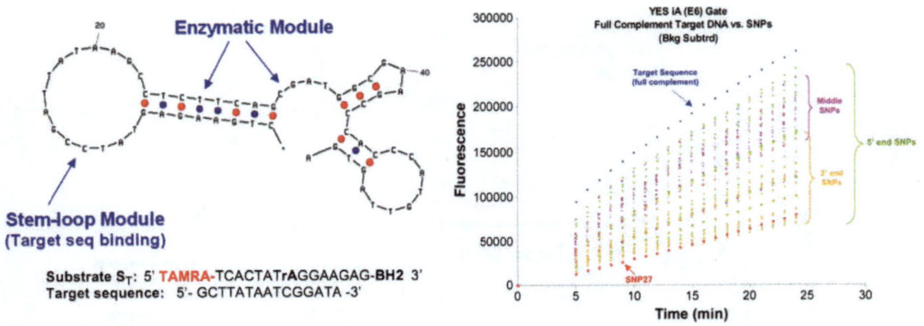

Fig. 8 YES_{iA} (E6) deoxyribozyme, along with the substrate and target sequence (*left panel*) and the fluorescence output over time in the presence of the full complementary target sequence vs. 45 SNP sequences representing every possible single point mutation in the 15-mer substrate (*right panel*). (Black-Hole-2 quencher, **BH2**). (Adapted with permission from [64]. Copyright 2008, IEEE.)

required for such systems, greatly simplifying development of a portable and robust diagnostic instrument.

We also showed that this system was very sensitive at discriminating the perfect complementary target input from sequences containing single base mutations. We compiled 45 single nucleotide polymorphism (SNP) containing target sequences, representing all possible single point mutations for each of the 15 positions in the target sequence. As shown in the right panel of Fig. 8, discrimination between the targets and SNP-containing sequences was possible in only 5 min. Further, the kinetic behavior of the YES_{iA} (E6) deoxyribozyme was impacted by the location and type of mutation in a quasi-predictable manner. This opened the exciting possibility of algorithm based analysis of the sensor output for predicting the exact location and composition of a given mutant [64]. This is potentially a powerful capability for DNA based sensors which typically cannot identify mismatches with high fidelity, and are not effective at detecting near neighbors for which the sensor was not designed. Development of a substrate-specific algorithm will be discussed more fully in Sect. 3.2.

In addition to the single input gate, a two input AND gate was developed for detecting H5N1 avian influenza virus DNA that was capable of differentiating point mutations that would increase human susceptibility. We designed the H5N1 $_{HA}AND_{PB2}$, gate to detect the mutated codon sequence in hemagglutinin (HA) protein that allows the virus to more readily bind to human cell receptors [129], and the mutated codon sequence in polymerase basic 2 protein (PB2) that allows for efficient replication in mammalian cells [91]. Early detection of viruses containing either or both mutations is critical to averting a pandemic. The $_{HA}AND_{PB2}$ gate showed the ability to discriminate each normal input target (HA only, PB2 only, HA + PB2) and each mutant input sequence (normal HA/mutated PB2, normal PB2/mutated HA, mutated HA/mutated PB2). The normalized rate of fluorescence output was inversely proportional to the number of total mismatches [64].

Other interesting intelligent biosensors exploiting the computational ability of deoxyribozyme probes include the work of Yashin et al. in which communication between deoxyribozyme gates immobilized on beads was demonstrated [130]. The detection of an orally administered drug inducing the release of a therapeutic peptide was reported by Taylor et al. [108]. For more information on deoxyribozymes for sensing and logic gate applications the reader is referred to the review by Willner et al. [122].

2.2.5 DNA Molecular Nanodevices

Aptamers, deoxyribozmes, and other physical properties of single stranded nucleic acids have been used to develop integrated intelligent biosystems. Capable of complex molecular detection and amplification schemes under autonomous and isothermal conditions, such systems are often described as 'molecular machines', 'nanodevices', or 'autonomous machines.' For example, by combining deoxyribozyme and DNA origami technologies, Lund et al. developed a DNA walker capable of autonomous behavior [53].

Detection of the small molecule, AMP, or the protein, lysozyme, was reported by Willner and co-workers using a aptamer-deoxyribozyme nanodevice [49]. The device consists of an aptamer sequence (region I) designed to target AMP or lysozyme. The aptamer recognition sequence is integrated with a deoxyribozyme sequence (region II) that in the presence of heme, will catalyze hydrogen peroxide reduction. The catalytic activity by the deoxyribozyme has been described as a horseradish peroxidase mimic. A second DNA strand that is complementary to portions of both the aptamer and deoxyribozyme regions is used to prevent catalytic activity except in the presence of the aptamer target and heme. Catalytic turnover of hydrogen peroxide was monitored colorimetrical using ABTS, or via chemiluminescence using luminol.

Other similar DNA molecular nanodevices employing amplification schemes have been reported by Willner's group for detection of single stranded DNA [126], cocaine [92], and even larger objects including viruses [9, 125]. For more information, the reader is referred to the recent and thorough review of nucleic acid based molecular devices by Krishnan and Simmel [44].

3 Application and Substrate-Specific Algorithm Development

The integration of DNA computing principles with recent advances in DNA-based catalytic molecular probes and nanodevices holds great promise for revolutionizing the field of bioanalytics and computational biosensors. However, application and substrate-specific algorithm development is key for integration within a platform viable for in-field detection.

Rapid development and deployment of application-specific microsensor detection systems requires algorithmic tools to map the biological detection challenge to the biosensor technology. Additionally computational biosensor-dependent technologies must incorporate algorithmic intelligence into the microsensor platform, enabling the execution of concurrent detection and classification tasks, such as systematic identification of nucleic acid sequences or polymorphisms that indicate the existence of microbial pathogens or disease-related genes in the presence of non-lethal agents.

We have outlined four general steps needed in the design of a computational biosensor (depicted in Fig. 9), where the first three steps focus on the development of computational algorithms for customizing the biological application to the sensor platform. Steps include: (1) Bioinformatics analysis to generate candidate probes; (2) Generation of probe-based classification algorithms; (3) Design and simulation of realizable biosensor and classification system; (4) System fabrication using computational biosensor-based substrate.

In this section we provide an overview of computational methods and algorithms used in the first three steps of the development process. We also provide an overview of methods used specifically for mapping biosensor applications to nucleic acid based computational sensor platforms.

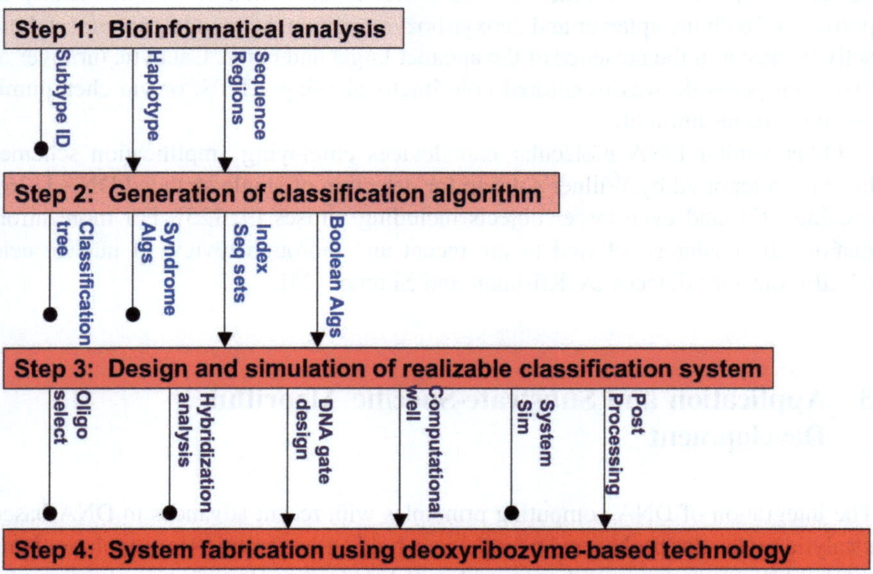

Fig. 9 Procedure for de novo generation of an application specific "intelligent" DNA sensor system

3.1 Bioinformatics Analysis and Subsequence Selection Algorithms

3.1.1 Methods for Subsequence Selection for Candidate Probes

Nucleic acid-based biosensor applications generally attempt to detect the presence of a subsequence target or set of subsequences that correspond to key components of a genome or transcription product. Identification of the most informative regions of a target that enable high detection of targets in a potentially noisy background and accurate classification of targets or variations in the target can be computationally challenging. Nucleic acid sensors and molecular beacons are viable options for genotyping, haplotyping and various gene and mRNA identification applications due to their relatively short target recognition site and their ability to discriminate polymorphisms in target sequences [17, 50, 67, 68, 77]. Various algorithmic methods such as those used in microarray design, haplotyping, and polymorphism detection have been explored for target identification and generation of candidate subsequences for use in nucleic acid based biosensors.

Subsequence selection is a multifaceted challenge requiring analysis and classification methods suited for the given biosensing application. Single nucleotide polymporphism (SNP) based applications have used information entropy to characterize and create initial sets of candidate SNPs for haplotyping and phenotype classification [47, 48]. Statistical and entropy based heuristics can yield subsequence groups with high correlation and minimal redundancy. Linear regression, Bayesian approaches, machine learning methods, genetic algorithms, and support vector machines are examples of methods used to search for the optimal SNP or subsequence subsets with maximal phenotype detection and classification performance [10, 30, 48].

3.1.2 Information Theory Based Subsequence Selection for Molecualr Beacon Probes

Molecular beacons and catalytic molecular beacons have short target recognition regions. Our method for finding highly informative subsequences was based on the integration of sequence alignment and information theory algorithms. As an example application we consider the design of a deoxyribozyme-based computational sensor for identification and monitoring of H5N1, the causative agent of influenza (May et al. [65]. Surveillance plus antigenic and genetic analysis have been used to monitor known H5N1 variants and identify emerging sublineages [97]. Large-scale sequencing efforts, such as the St. Jude Influenza Genome Project [71], provide genetic information for the design of computational biosensors that can aid in tracking and characterizing regional strains of H5N1. Since the genomic sequence of influenza viruses is highly mutable [27, 72] given a set of subtypes or strains, it was challenging to find several highly conserved regions that sufficiently distinguish the groups. In designing the FluChip diagnostic microarray, Mehlmann et al. manually inspected a phylogenetic tree of influenza sequences to identify clades that had sufficient sequence similarity [66]. These clades were then used in an automated probe extraction strategy. To encourage process uniformity and minimize design time, we developed extraction methods that minimize the manual intervention in the probe identification process.

Using 164 H5N1 hemagglutinin (HA) gene sequences (NCBI Influenza Virus Resource) from samples spanning five different geographic locales (Egypt, Hong Kong, Indonesia, Thailand, and Vietnam), we generated a multiple alignment and consensus sequence for each locale using MUSCLE [21]. We then created a multiple alignment of the five consensus sequences, in order to index the nucleotides across all sequences uniformly. To find regions of low variation within a locale, we used an information theory approach. We set a window size of $k = 5$ nucleotides, and for each index, we calculated the k-mer (words of length k) sequence entropy using the standard entropy equation [18]

$$E_i = - \sum_{w \in words_i} p_i(w) \log_2 p_i(w)$$

where the sum is over all k-length words occurring at position i in the alignment, and $p_i(w)$ is the proportion of words found at position i that are equal to w. Thus, positions with low k-mer sequence diversity—e.g., only one or two unique words occurring throughout the alignment at that position—will have low entropy, while positions that have high sequence diversity will have high entropy. We included only the ungapped words in the entropy calculation and introduce a uniform gap penalty to disfavor gapped positions in our calculation.

Once each alignment was annotated with the k-mer entropy per position, we extracted indices that had low sum-entropy spanning 15 nucleotides, the length of the target-binding region of the deoxyribozyme. Using an empirically determined entropic cutoff of $E^{15} < 8$ bits, we collated all of the unique 15- nucleotide words at each low sum-entropy position; a percentile cutoff could also be employed. We reduced the candidate probe set by removing the most infrequent words, such that at least 90% of the individuals in the alignment contained one of the words in the reduced set. While using a larger k-mer would seem the most direct method, we found that large k-mers decreased the resolution of the algorithm and the resulting candidate probe set (Fig. 10).

The initial probe set was subjected to additional heuristic filters. Candidate target subsequences were: (1) Checked to ensure that they were not cross reactive with non-HA H5N1 gene sequence; (2) Annotated according to their maximum classification sensitivity over all regions/locales (defined as $\frac{\text{\#individuals in a locale containing the word}}{\text{total \#individuals in a locale}}$) and a specificity measure equal to the entropy of the distribution of classification sensitivities. Candidate probes with sensitivity less than 0.75 and specificity scores less than 1 (maximum is $-\log 2\ \frac{1}{5}$) were removed. We subsequently performed a final filtering step to identify probe

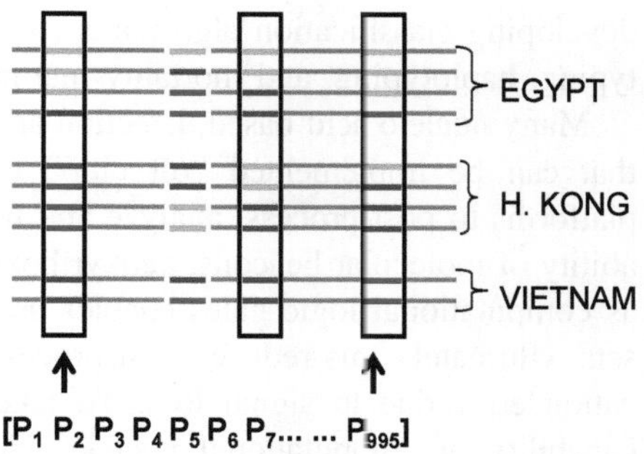

Fig. 10 Use of alignment and information algorithms for candidate probe selection for strain classification of regional H1N1 strains

sequences that had a high probability of structural incompatibility with the deoxyribozyme gate architecture (see Sect. 3.3). We applied our algorithmic pipeline to identify candidate probes for distinguishing influenza subtypes and were able to extract 392 candidate marker sequences on the HA gene spanning the five regional subtypes. Our results suggest that our method is robust to sequence groups with high variation.

3.2 Application and Substrate Specific Algorithm Generation

3.2.1 Classification Algorithms for Nucleic Acid Biosensor Applications

Nucleic acid based biodetection applications determine the presence or absence of a target using a subset of the target's genomic space. Depending on the sensor platform, the number of target probes incorporated in the system can vary in number and length. Consequently once an initial set of candidate probes are generated, further refinement of the target subsequence set may be required. Algorithmic or heuristic-based methods must generate a set of probes that when detected can be used to computationally classify the sample as one of multiple targets or a non-target, thus the probe selection and the sensor output classification algorithm are strongly correlated.

Use of genome wide association and polymorphism data to predict disease phenotypes or to identify drug resistant pathogens require the selection of optimal probe or SNP sets for maximal classification outcome. Statistical methods have been used to rank SNPs in order to optimize disease prediction outcome using single or ensemble classification algorithms [59]. Classification methods used include decision trees, logistic regression, support vector machines and Bayesian methods. In addition to statistical methods, regression methods, support vector machines, Mahalanobis distance, and machine learning methods have been used in developing classification algorithms for various application areas including genotyping, haplotyping, and mortality rate prediction [30, 47, 84].

Many nucleic acid based detection applications rely on classification algorithms that can be implemented "off chip" using traditional silicon-based computing platforms to post-process, analyze and interpret the detection event. However, the ability of molecular beacons, deoxyribozymes in particular, to functionally operate as computational logic gates enables us to concurrently detect and classify probe sets. Ultimately this reduces post-processing cost and potentially reduces classification error due to signal loss. To take advantage of the integrated computing capability of computational nucleic acid molecules [39, 105], we explore new algorithmic methods that couple detection and classification while considering potential computational and design limitations of the molecular beacon.

3.2.2 Algorithmic Design of Intelligent Deoxyribozyme Sensors

In order to demonstrate the capabilities of the deoxyribozyme platform for computational biosensing, we used the 392 candidate marker sequences generated from the HA gene spanning three of the five regional H5N1subtypes (Egypt, Hong Kong, Indonesia, Thailand, and Vietnam) to develop a prototype computational biosensor system [65]. Our system was designed de novo and was able to concurrently perform complex detection/classification tasks in the biological substrate, classifying a nucleic acid sample as belonging to one of the three regional subtypes based on the HA gene. The ability to compute differentiates this system from other molecular sensor systems. By increasing the complexity of the executed algorithm one can potentially increase the sensitivity and specificity of the biosensor, hence improving accuracy through computation.

We reduced the region classification problem to an experimentally tractable form: develop a deoxyribozyme-compliant algorithm that detects and classifies a candidate strain as originating from Egypt, Hong Kong, or Indonesia, using a reduced probe set of 75 candidate probes. The reduced probe set was selected from the candidate probe set based on their maximum classification sensitivity values. Using coding theoretic methods described in prior work [62–65] we formulated the classification and design problem as an inverse error control code (ECC), where we determine the optimal parity check matrix, H^T_{EgHkIn}, such that

$$H^T_{EgHkIn} * C_{Region_i} = S_{Region_i}$$

where C_{Region_i} is the set of all sample sequences from Region$_i$, for i = *{*Egypt, Hong Kong, Indonesia*}*; a sample sequence (a row in C_{Region_i}) is a 75-bit binary sequence, where a 1 at location p indicates that probe sequence p is present in the sample sequence. S_{Region_i} is the syndrome matrix for Region_$_i$ and contains the syndrome vector for each sequence in $C_{Region\_i.}$ The solution for the optimal H^T_{EgHkIn} concurrently maximized the inter-region hamming distance $d_{hamming}$ ($S_{Region\_i,}$

Table 1 Deoxyribozyme compliant computational wells for detection and classification of H1N1 regional strains

$CompWell_1$	$probe_{21} \oplus probe_{30} \oplus probe_{37} \oplus probe_{42} \oplus probe_{50} \oplus probe_{62} \oplus probe_{73}$
$CompWell_2$	$probe_{27} \oplus probe_{32} \oplus probe_{33} \oplus probe_{36} \oplus probe_{42} \oplus probe_{43} \oplus probe_{47} \oplus probe_{74}$
$CompWell_3$	$probe_{34} \oplus probe_{38} \oplus probe_{45} \oplus probe_{75}$

S_{Region_j}) of the regional syndrome matrix, minimized the intra-region hamming distance $d_{hamming}$ (S_{Region_i}, S_{Region_i}), and maximized the number of zeros in H^T_{EgHkIn}. These constraints were designed to produce efficient classification algorithms using a minimal set of probes.

We used our coding theory based method to generate an ($n = 75$, $k = 72$), deoxyribozyme compliant detection and classification algorithm. The algorithm resulted in a three column regional syndrome matrix, which corresponds to three computational wells, and used a genetic algorithm to search the space of feasible linear block codes that satisfied our design constraints. The resulting H^T_{EgHkIn} used 18 of the 75 probe sequences from the reduced set. Our detection and classification algorithm for each of our three computational wells (CompWell) was reported as follows (⊕ indicates exclusive-OR):

Table 1 lists the probe sequences used in each computational well. The algorithm produced three regionally distinct average syndrome patterns:

$$S^{Avg}_{Egypt} = (0, 0, 0) \quad S^{Avg}_{HongKong} = (0, 1, 1) \quad S^{Avg}_{Indonesia} = (1, 1, 1)$$

Based on the resulting syndrome patterns the computational biosensor system should be able to correctly distinguish the Egypt and Indonesian strains, but may not be as successful when distinguishing between the Hong Kong and Indonesian strains. A longer syndrome vector, hence additional computational wells, should result in a more accurate computational biosensor system. The algorithm in the computational wells can be used to design multiple deoxyribozyme gates to implement the prototype system.

3.3 Design of Nucleic-Acid Based Biosensor Substrates

The desired outcome of the algorithmic design process is the development of a realizable biosensor and classification system. A key to successful system fabrication is the design of structurally and functionally accurate molecular probes. Several computational tools, methods, and integrated pipelines for designing molecular beacons and various nucleic acid-based biosensor systems have been reported. Specific to molecular beacon probes, computational methods for ribozyme design using a random search method and partition function for calculating base-pair bindings was developed by Penchovsky and Breaker [74]. Building on public RNA secondary structure calculation tools, Hall et al. developed methods for computationally designing aptazymes based on free energy profiles [28]. Additional methods for molecular design and simulation of nucleic acid-based circuits composed of multiple interacting molecular probes have also been explored [31, 75, 85]. Drawing on existing tools and thermodynamic-based methods, we developed computational methods and metrics for evaluating candidate deoxyribozyme gates used to implement in vitro computational biosensors [64, 65].

3.3.1 Designing Realizable Deoxyribozyme Gates

As in all probe design steps, the influence of the target sequence on the probe structure can result in undesirable secondary structure. For deoxyribozymes this is particularly critical as the change in conformation is key to the enzymatic and Boolean functionality of the computational gate. Using the E6 catalytic core sequence [54] and the probe sequences generated by our algorithmic pipeline, we screened candidate probes to ensure that the target loop sequences had minimal complementarity to the catalytic portion of the deoxyribozyme, as well as minimal complementarity to other target sequences when multiple loops exist on the same gate.

Based on the form of the equations generated for the three computational wells, we required a series of probe sequences to be combined and incorporated into XOR functions. At the time of our prototype design there did not exist a deoxyribozyme gate that implemented the XOR function, therefore in lieu of an XOR gate we designed two complementary ANDNOT gates with the loops reversed to perform the same logic ($p \oplus q \Rightarrow (p \wedge \neg q) \vee (\neg p \wedge q)$). To determine which loops were pairwise compatible, we optimized selection based on string edit distance, calculated using Smith-Waterman local alignment [121], and hybridization energy calculated using the UNAFold software package. Using a graph theoretic approach we found the optimal probe-pairing using an implementation of Edmonds' Algorithm [22] written by Eppstein [25]. Unpaired probe sequences were incorporated into YES gates. We designed two complementary ANDNOT gates for each paired probe sequence using the E6 deoxyribozyme catalytic sequence. For each gate, we generated the minimum free energy structure using UNAFold and verified its structural similarity to the reported ANDNOT gate structure, however minor structural variations were permitted in the target recognition regions.

3.3.2 Computational Modeling and Simulation of Deoxyribozyme Gates

Due to the high cost and long turnaround times associated with experimental biosensor measurements, computational, or "in silico", prediction of biosensor performance is highly desirable prior to system fabrication. Computational predictions can be used to quickly and inexpensively screen proposed new biosensor designs and provide molecular-level rationale for experimentally observed phenomena. We have demonstrated the ability to computationally predict experimentally observed deoxyribozyme system performance using DNA hybridization thermodynamics information [64].

Ideally the use of molecular simulation could serve as a plausible method for investigation of the atomistic details of deoxyribozyme gate function using CHARMM [55] and AMBER [20] force fields. Both have been developed to include parameters for nucleic acid molecules and have been shown to be reasonably reliable models in representing the structural and dynamic properties of nucleic acids. These force field functional forms are included in massively parallel molecular dynamics (MD) software packages such as LAMMPS (Large-scale atomic/molecular massively parallel simulator; *lammps.sandia.gov*) and have been used in many biomolecular simulation studies. Unfortunately, complete observation of real gate dynamics would require orders of magnitude more simulation time than was feasible even given high performance computing resources. However valuable information can be extracted from short molecular scale simulations including insight into important 3D structural effects in the hybridization thermodynamics predictions.

Using DNA hybridization thermodynamics we can predict properties of candidate gates by accumulating contributions from base pair matches and nearest neighbor interactions. Such calculations are possible using assigned parameters that are fit from extensive experimental measurements and methods pioneered by the SantaLucia lab that has enabled nucleic acid hybridization thermodynamics prediction [76, 88]. Thermodynamics based methods developed by SantaLucia and colleagues have been implemented in two comparable software packages: the freely available HyTher web server and the commercial Oligonucleotide Modeling Platform (OMP) sold by DNA Software, Inc. Our initial prediction pipeline used HyTher, but ultimately we used OMP to develop a pipeline for deoxyribozyme gate and computational well design [65].

Using OMP to compute DNA hybridization thermodynamics, we developed a Python pipeline for computing ribozyme performance. Experimental conditions can serve as inputs to OMP, including primary sequences of the oligonucleotides, solution temperature, ionic concentrations, and buffer conditions. The output of OMP includes: melting temperature (Tm), hybridization free energy change (ΔG), and activated deoxyribozyme concentration ((active)). Although the three thermodynamic quantities correlated well with the experiments, we found that (active), the concentration of activated deoxyribozymes, produces the best agreement. We used the data to relate experimentally-measured changes in fluorescence with respect to time to the concentration of activated ribozymes as follows:

$$\frac{\delta fluorescence}{\delta t} = k[substrate][active]$$

where k is a rate constant and (substrate) is the concentration of substrate molecules cleaved by the deoxyribozyme gate. We set the concentration threshold for activated ribozyme to a 1 nM as a cutoff to distinguish between pairs producing a strong fluorescence signal (on or logic 1), and those pairs that produce no distinguishable signal (off or logic 0). Using this approach we were able to simulate individual gate performance and simulate system performance for our prototype system.

We used the algorithmic methods described in Sects. 3.1 through 3.3 to develop and test a simple prototype system and demonstrate the feasibility of using in silico design to build a deoxyribozyme-based computational biosensor. We devised a reduced (N = 52, K = 47) coding-based algorithm for subtype classification resulting in a three-bit classification algorithm:

$CompWell_1 =$	$probe_{23} \oplus probe_{48}$
$CompWell_2 =$	$probe_4 \oplus probe_{23} \oplus probe_{24} \oplus probe_{34}$
	$probe_{43} \oplus probe_{50}$
$CompWell_3 =$	$probe_{20} \oplus probe_{51}$

where CompWell represents the computational well, which is realized by combining the discretized binary output of the microplate wells containing each of the YES-gates in the classification algorithm. The YES-gate target probe sequences are listed in Table 2. We used our OMP-based pipeline to design, simulate, and generate a realizable.

YES-gate only system and a YES-gate/AND-NOT gate implementation. The AND-NOT gate implementation uses two AND-NOT gates to implement the exclusive-OR (⊕) function, which we performed in our post-processing step. Our computational biosensor system was implemented, in vitro and we tested the system's performance in H5N1 regional subtype classification using our fluorescence-based detection platform and synthesized input DNA sequences [64, 65]. The results of the experimental tests for the YES- gate configuration and the predicted system output is shown in Fig. 11. The binary microwell values were combined to compute the value of each computational well for a sample sequence

Table 2 Probe sequences for region identification for the (47, 52) code

Probe Id	Sequence
4	GATCCTCCTTTTTA
20	GCTATATCAAAACCC
23	CAGACAAGGCTATAT
24	AAGGCTATATCAAAA
34	ACAACATACACCCTC
43	CCAATCATGATGCCT
48	TACTAGACCCAAAGT
50	GCTACTAGACCCAAA
51	CTAGACCCAAAGTAA

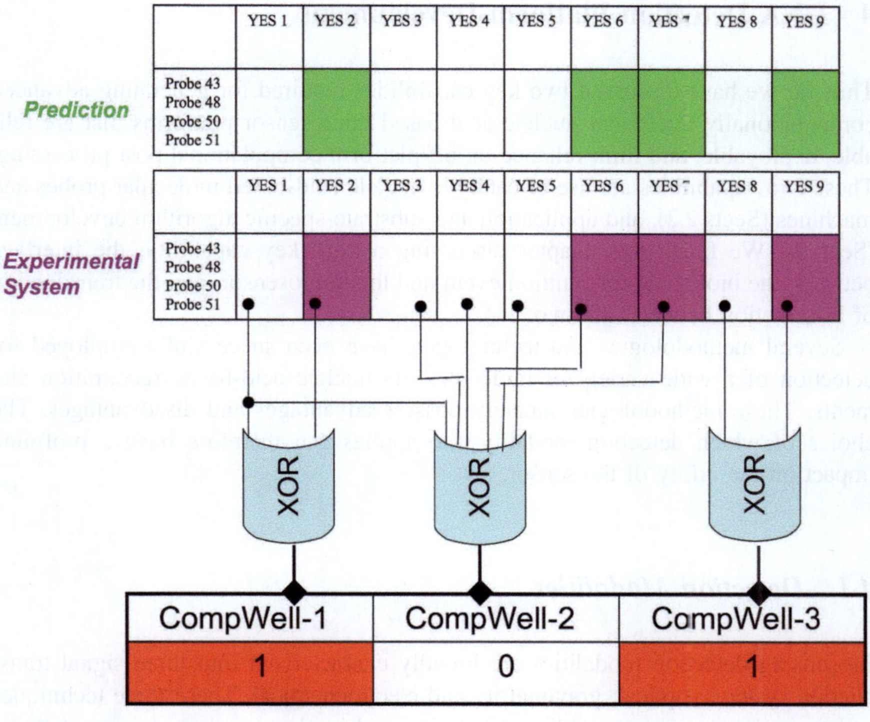

Fig. 11 Schematic for calculation of computational well values from experimental subwells. The probe sequences, representing the genomic components of the viral strain, are added to each well containing a specified YES-deoxyribozyme gate. The predicted (*green*) and experimentally determined (*purple*) outputs concur and are used to calculate the binary values of the computational wells. The three-bit binary syndrome pattern indicates viral strain association for the probe sequences

Table 3 Syndrome classification patterns for subtype classification

Subtype Class	$CompWell_1$	$CompWell_2$	$CompWell_3$
Egypt	0	1	0
Hong Kong	1	0	1
Indonesia	1	1	1

set as shown. The binary syndrome pattern of the computational wells determined which H5N1 strain the input sequence set belonged (Table 3). Using the binary patterns, we correctly identified the subtype of 97.3% of the simulated sample sequences.

4 DNA Detection Platform Development

Thus far we have discussed two key capabilities required for generating advanced computationally intelligent nucleic acid-based microsensor platforms that are reliable, deployable, and limit reliance on off-platform computational post-processing. These two capabilities are use of catalytic nucleic acid-based molecular probes and machines (Sect. 2.2), and application and substrate-specific algorithm development (Sect. 3). We finish this chapter discussing a third key capability, the interface between the biological recognition event and the microsensor, and the transduction of information between the two.

Several methodologies and technologies have been successfully employed for detection of a wide variety of molecules via nucleic acid-based recognition elements. These methodologies innately possess advantages and disadvantages. The choice of which detection modality one applies can therefore have a profound impact on the utility of the sensor.

4.1 Detection Modalities

Biosensing detection modalities are broadly characterized into three signal transduction systems: optical, gravimetric, and electrochemical. These three techniques and their characteristics enabling the design and implementation of computational biosensors will be briefly discussed.

4.1.1 Optical Detection

Optical nucleic acid detection methods encompass fluorescence, surface plasmon resonance (SPR), surface enhanced Raman spectroscopy (SERS), luminescence, colorimetric, and similar techniques for transducing the biorecognition event. The most widely used optical biodetection modality is fluorescence. This method relies on the use of fluorescent labels, or the generation of a fluorescent signal, for detection of the biorecognition event. This approach often yields simple to interpret, semi-quantitative data with very low detection limit. Use of fluorophores of differing colors also allows for multianalyte biodetection. However, fluorescence-based detection requires an excitation light source, optics to filter the excitation wavelength from the emitted fluorescence wavelength, and a photo detector sensitive to the emitted fluorescence. These components are typically integrated together into bench-top laboratory equipment (e.g. fluorescence microscope, fluorometer) that is expensive, not portable, and requires training and post-signal processing.

Developments in light emitting diode (LED) technology have led to advances in portable fluorescence-based DNA detection technologies. For example, Wand and

Trau recently reported a generic DNA bioassay using an LED/photodiode system that fits into an aluminum briefcase and can detect PCR products at a concentration as low as 97 fmol [116]. We anticipate continued progress in portable fluorescence-based DNA detection as these technologies are further refined.

Surface plasmon resonance (SPR) is a technique that monitors minute changes in refractive index occurring at the interface between materials with differing refractive indices. Commonly, a prism is coated with a thin gold layer that is functionalized with single-stranded DNA probes. A solution containing the analyte flows over this surface. Infrared or visible light at a single wavelength near the surface plasmon resonance condition is directed incident to the gold surface at the resonance angle, resulting in a minimum of the reflected light intensity. Upon binding of the target DNA to the probe surface, the surface resonance condition is altered, resulting in a change in the refractive index [16, 73]. Thus, a significant advantage of SPR-based DNA detection is that no label is required. However, false positives due to non-specific binding of non-target DNA and other components in the analyte solution are significant challenges. Non-specific binding is a common concern with nearly all label-free based biodetection systems.

Surface enhanced Raman spectroscopy (SERS) is a surface technique that has been used for DNA detection in which rough metallic nanostructures are used to substantially increase very weak Raman scattering. When molecules absorb onto the roughened nanostructures, incident light excites surface plasmons, or forms charge-transfer complexes, resulting in a significant enhancement in excitation intensity which can be as high as 10^{9-11} [45]. This allows for detection of molecules on the SERS substrate that would normally be undetectable using Raman alone. This phenomenon has resulted in the bourgeoning field of SERS-based detection of molecules of biomedical interest and the development of many SERS probes/substrates [90]. For example, the Vo-Dinh group has used SERS for detection of DNA hybridization to targets specific for breast cancer and the HIV-1 viral gene [114]. Nanomaterials modified with aptamers have also been widely used for SERS-based biosensing [117]. Although sensitive and specific, this technique can still require target amplification (typically PCR) as the detection limits are not sufficiently low for detection of DNA levels in real-world samples. Also, although portable Raman spectrometers are currently commercially available, these devices can be challenging to operate in the field, must be calibrated to the specific target, can be confounded by complex sample solutions, and are expensive [61].

We already discussed an aptamer-DNAzyme molecular nanodevice that produces a colorimetric signal, or chemiluminescent signal, in the presence of the target analyte and hemin [49], Sect. 2.2.5). Optical and chemiluminescence approaches are advantages in that the results can typically be read by eye, thus they do not require laboratory equipment or significant training to read. However, the detection limit of these systems is often poor as sufficient signal must be generated to be visually discernable. Better detection limits are possible using spectrometers for absorbance/light intensity measurements. However, using such a device negates the advantage of using a detection system that can be read by eye. In the case of the aptamer-DNAzyme system mentioned above, in which the optical signal is catalytically amplified beyond the 1:1 target to signal ratio, the detection limit of AMP was still only 4 μM. Thus, optical based systems for DNA detection have not yet found significant use outside the laboratory.

4.1.2 Gravimetric Detection

Gravimetric detection modalities rely on measuring small changes in mass upon target binding at the sensor surface. Quartz crystal micobalances (QCMs) are the most wide reported gravimetric biodetection devices. These devices are composed of a quartz wafer sandwiched between two electrodes (typically gold). One of the electrodes is then functionalized with a thin film containing the capture probe. An oscillating electric field is applied between the two electrodes, resulting in a mechanical resonance in the crystal that is very sensitive to mass [15]. The relation between the electrical and mechanical oscillations to the mass on the crystal is described by the Sauerbrey equation:

$$\Delta f = \frac{-2\Delta mnf_0^2}{A\sqrt{\mu\rho}}$$

where Δf is the oscillation frequency, Δm is the mass change, A is the piezoelectrically active area of the crystal, and μ and ρ are the shear modulus and density of the crystal, respectively.

Such sensors are capable of detecting nanogram changes in mass, in real-time. DNA-based QCM detectors measure the increase in mass due to hybridization of target to the capture probe, resulting in a decrease in resonance frequency. Another significant advantage to this system is the target does not need to be labeled. However, such devices suffer from the same non-specific binding limitation, resulting in false positive measurements, common to most label free sensors [36].

Surface acoustic wave (SAW) devices operate in a similar fashion. A thin piezoelectric crystal is modified with interdigitated electrodes. An oscillating electric field generates a surface acoustic wave, or a bulk acoustic wave, that is launched across the crystal. The crystal surface is functionalized with the capture probe. As the acoustic energy is confined to the thin surface region of the crystal substrate, binding of target molecules to the surface induces perturbations to the wave, altering its amplitude and velocity. Thus, the surface wave device can operate as a mass or viscosity sensor [13, 19].

Most reports of detection of DNA using these devices require PCR amplification. However, some sensors have been reported that detect target DNA in non-amplified samples. For example, Karamollaoglu et al. reported detection of a genetic insert in tobacco plants from PCR amplified DNA samples, and from samples fragmented by digestion and ultrasonication, using a QCM-based DNA biosensor [40].

The surface chemistry used to modify the crystal surface with the DNA capture probe can have a profound impact on the specificity, detection limit and dynamic range of the system. Several reports comparing gold-thiol to thiol-dextran [57], gold-thiol to biotin-streptavidin [58], and gold-thiol to amine-gluteraldehyde [40], all show statistically significant difference in sensitivity and dynamic ranges. Typically, the best results are obtained from the gold-thiol surface as these molecules assemble more densely (increased probe density, reduced sites for non-specific binding) and are comparatively lower in mass (surface chemistry does not contribute as significantly to frequency changes), allowing for more sensitive discrimination target binding induced changes in surface mass.

These devices are amenable for use outside the lab, with portable versions reported [4, 11]. However, they can be high cost, are non-trivial to construct, and require sensitive measurements and data post-processing. This makes production of simple to operate and low cost gravimetric-based DNA detection devices challenging.

4.1.3 Electrochemical Detection

Electrochemical detection methods utilize changes in reduction/oxidization (redox) reactions and electron transfer (including conductivity) properties associated with binding of the target molecule to a conducting substrate. Substrates are most often gold, platinum, or carbon. Conducting polymers and other metallic materials have also been successfully employed for electrochemical biosensing. The substrates can be either planar solids, or three dimensional with high surface area and porosity, or nanoscale in dimension. DNA capture probes are immobilized onto the conducting surface, and hybridization of target DNA to the capture probe is monitored directly or indirectly.

Direct methods typically measure changes in current directly from redox reactions with nucleic acid bases in the target DNA, most notably, the direct oxidation of guanine [38] which can be further amplified by using a catalytic mediator such as $(Ru(bpy)_3)^{2+}$ [95]. Background signal from the capture probe can be significantly reduced by replacing guanines with uracil, a nucleobase that also binds to cytosine, but oxidizes at a potential much higher than guanine. Thus, only when the target strand hybridizes to the surface is a current from guanine oxidation obtained. Direct detection can also occur by measuring changes in the electrode/electrolyte interface properties upon target hybridization typically by electrochemical impedance spectroscopy (EIS) or conductivity, although these methods are generally not very sensitive without addition of a redox mediator probe (e.g. ferricyanide, methylene blue).

Indirect methods make use of redox or electroactive mediators that can intercalate with hybridized double-stranded DNA. Examples of electroactive double stranded DNA intercalators include ethidium bromide and daunomycin. Alternatively, and most commonly, indirect methods employ labels such as enzymes, metallic or semiconductor nanoparticles, dendrimers, liposomes, carbon nanotubes, nanowires, etc., which enable a wide variety of electrochemical detection schemes and also significantly amplify the measured signal output [79]. In these schemes it is the label that is detected, or provides the electrochemical signal, indicating hybridization has occurred.

Electrochemical detection provides significant advantages over other common detection and signal transduction methods. Specifically, it can be very sensitive, with detection limits in the pico- and fempto- molar range, detecting DNA in non PCR amplified samples [109]. Also, electrochemical detection can be rapid and provide greater specificity and sensitivity over optical detection methods as interfering background fluorescence does not adversely affect the electrochemical signal. Unlike fluorescent or biological molecules which bleach or degrade with time, electroactive molecules are typically more stable and insensitive to the environment.

Further, multianalyte biodetection is often simpler with electrochemical schemes by using either multiple electroactive labels with differing redox potentials, or an array of individually addressable electrodes that are selectively modified with differing capture probes. For example, we reported the first simultaneous electrochemical detection of DNA and protein on a single platform by selectively immobilizing single stranded DNA probes on five individually addressable electrodes, and immobilizing antibody probes on another four individually addressable electrodes, in a nine element array [29].

Importantly, electrochemical detection systems are also much simpler to miniaturize as integrated circuit fabrication methods can be leveraged to produce the miniaturized components. These components have relatively low power requirements and low mass, making them idea for potable, handheld, or leave behind sensor modules. In another example, we recently reported the development of an inexpensive and low power electrochemical DNA detector that contained an array of nine individually addressable gold electrodes [23]. Using electroaddressable aryl diazonium chemistry, the electrodes were selectively patterned to detect DNA sequences specific to breast cancer, colorectal cancer, and provide information regarding non-specific binding on the array (negative control). The device also contained a resistive platinum heating element that permitted melting of double stranded DNA sample solution, allowing for multanalyte detection from double stranded DNA target on the array. Following detection, the array surface could be renewed via high temperature stripping utilizing the on-chip resistive heating element.

Such inexpensive and simple multanalyte DNA sensors show great potential for integration with nucleic acid probes that incorporate structure and catalytic function, in combination with DNA information processing techniques, resulting in intelligent computational biosensor platforms that can truly revolutionize DNA biodetection.

Acknowledgements We would like to thank our colleagues who collaborated on our original computational biosensor work: P. Dolan, P. Crozier, M. Lee, M. Manginell, and R. Polsky.

Sandia National Laboratories is a multi-program laboratory managed and operated by Sandia Corporation, a wholly owned subsidiary of Lockheed Martin Corporation, for the U.S. Department of Energy's National Nuclear Security Administration under contract DE-AC04-94AL85000.

References

1. Adar R, Benenson Y, Linshiz G, Rosner A, Tishby N, Shapiro E (2004) Stochastic computing with biomolecular automata. Proc Natl Acad Sci USA 101(27):9960–9965
2. Adleman LM (1994) Molecular computation of solutions to combinatorial problems. Science 266(5187):1021–1024
3. Adleman L, Rothemund P, Roweis S, Winfree E (1999) On applying molecular computation to the data encryption standard. J Comput Biol 6(1):53–63
4. Andle JC, Vetelino JF (1994) Acoustic wave biosensors. Sens Actuators A 44(3):167–176
5. Arugula MA, Zhang Y, Simonian AL (2014) Biosensors as 21st century technology for detecting genetically modified organisms in food and feed. Anal Chem 86(1):119–129
6. Baron R, Lioubashevski O, Katz E, Niazov T, Willner I (2006) Elementary arithmetic operations by enzymes: a model for metabolic pathway based computing. Angewandte Chemie-International Edition 45(10):1572–1576
7. Baron R, Lioubashevski O, Katz E, Niazov T, Willner I (2006) Logic gates and elementary computing by enzymes. J Phys Chem A 110(27):8548–8553
8. Benenson Y, Gil B, Ben-Dor U, Adar R, Shapiro E (2004) An autonomous molecular computer for logical control of gene expression. Nature 429(6990):423–429
9. Beissenhirtz MK, Elnathan R, Weizmann Y, Willner I (2007) The aggregation of Au nanoparticles by an autonomous DNA machine detects viruses. Small 3(3):375–397
10. Bevilaqua A, Rodrigues F (2011) L. do Amaral. SNPs classification: building biological high-level knowledge using genetic algorithms. Integrated computing technology communications in computer and information. Science 165:50–58
11. Bisoffi M, Hjelle B, Brown DC, Branch DW, Edwards TL, Brozik SM Bondu-Hawkins VS, Larson RS (2008) Detection of viral bioagents using a shear horizontal surface acoustic wave biosensor. Biosens Bioelectron 23(9):1397–1403
12. Braich RS, Chelyapov N, Johnson C, Rothemund PWK, Adleman L (2002) Solution of a 20-variable 3-SAT problem on a DNA computer. Science 296(5567):499–502
13. Branch DW, Brozik SM (2004) Low-level detection of Bacillus anthracis simulant using love-wave biosensors on 36°YX LiTaO3. Biosens Bioelectron 19(8):849–859
14. Breaker RR (1997) DNA enzymes. Nat Biotechnol 15(5):427–431
15. Bunde RL, Jarvi EJ, Rosentreter JJ (1998) Piezoelectric quartz crystal biosensors. Talanta 46 (6):1223–1236
16. Campbell CT, Kim G (2007) SPR microscopy and its applications to high-throughput analyses of biomolecular binding events and their kinetics. Biomaterials 28(15):2380–2392
17. Carter J, Balaraman V, Kucharski C, Fraser T, Fraser M (2013) A novel dengue virus detection method that couples DNAzyme and gold nanoparticle approaches. Virol J 10:201
18. Cover TM, Thomas JA (1991) Elements of information theory. Wiley, New York
19. Du J, Harding GL, Ogilvy JA, Dencher PR, Lake M (1996) A study of love-wave acoustic sensors. Sens Actuators, A 56(3):211–219
20. Duan Y, Wu C, Chowdhury S, Lee MC, Xiong G, Zhang W, Yang R, Cieplak P, Luo R, Lee T, Caldwell J, Wang J, Kollman P (2003) A point-charge force field for molecular mechanics simulations of proteins based on condensed-phase quantum mechanical calculations. J Comput Chem 24(16):1999–2012

21. Edgar RC (2004) MUSCLE: multiple sequence alignment with high accuracy and high throughput. Nucleic Acids Res 32(5):1792–1797
22. Edmonds J (1965) Paths, trees and flowers. Can J Math 17:449–467
23. Edwards TL, Harper JC, Polsky R, Lopez DM, Wheeler DR, Allen AC, Brozik SM (2011) A parallel microfluidic channel fixture fabricated using laser ablated plastic laminates for electrochemical and chemiluminescent biodetection of DNA. Biomicrofluidics 5:044115
24. Ellington AD, Szostak JW (1990) Invitro selection of RNA molecules that bind specific ligands. Nature 346(6287):818–822
25. Eppstein D (2003) Maximum cardinality matching in general graphs
26. Fan CH, Plaxco KW, Heeger AJ (2003) Electrochemical interrogation of conformational changes as a reagentless method for the sequence-specific detection of DNA. Proc Natl Acad Sci USA 100(16):9134–9137
27. Ghedin E, Sengamalay NA, Shumway M, Zaborsky J, Feldblyum T, Subbu V, Spiro DJ, Sitz J, Koo H, Bolotov P, Dernovoy D, Tatusova T, Bao Y, St George K, Taylor J, Lipman DJ, Fraser CM, Taubenberger JK, Salzberg SL (2005) Large-scale sequencing of human influenza reveals the dynamic nature of viral genome evolution. Nature 437 (7062):1162–1166
28. Hall B, Hesselberth J, Ellington A (2007) Computational selection of nucleic acid biosensors via a slip structure model. Biosens Bioelectron 22(9):1939–1947
29. Harper JC, Polsky R, Wheeler DW, Dirk SM, Brozik SM (2007) Selective immobilization of DNA and antibody probes on electrode arrays: Simultaneous electrochemical detection of DNA and protein on a single platform. Langmuir 23(16):8285–8287
30. He J, Zelikovsky A (2007) Informative SNP selection methods based on SNP prediction. IEEE Trans Nanobiosci 6(1):60–67
31. Hockenberry AJ, Jewett MC (2012) Synthetic in vitro circuits. Curr Opin Chem Biol 16(3–4):253–259
32. Hoheisel JD (2006) Microarray technology: Beyond transcript profiling and genotype analysis. Nat Rev Genet 7(3):200–210
33. Horejsh D, Martini F, Poccia F, Ippolito G, Di Caro A, Capobianchi MR (2005) A molecular beacon, bead-based assay for the detection of nucleic acids by flow cytometry. Nucleic Acids Res 33(2):e13
34. Hutchins CJ, Rathjen PD, Forster AC, Symons RH (1986) Self-cleavage of plus and minus RNA transcripts of avocado sunblotch viroid. Nucleic Acids Res 14(9):3627–3640
35. Iliuk AB, Hu L, Tao WA (2011) Aptamer in bioanalytical applications. Anal Chem 83 (12):4440–4452
36. Janshoff A, Galla HJ, Steinem C (2000) Piezoelectric mass-sensing devices as biosensors—An alternative to optical biosensors? Angewandte Chemie-International Edition 39 (22):4004–4032
37. Jhaveri S, Rajendran M, Ellington AD (2000) In vitro selection of signaling aptamers. Nat Biotechnol 18(12):1293–1297
38. Johnston DH, Glasgow KC, Thorp HH (1995) Electrochemical measurement of solvent accessibility of nucleobases using electron transfer between DNA and metal complexes. J Am Chem Soc 117(35):8933–8938
39. Jung C, Ellington AD (2014) Diagnostic applications of nucleic acid circuits. Acc Chem Res 47(6):1825–1835
40. Karamollaoglu I, Oktem HA, Mutlu M (2009) QCM-based DNA biosensor for detection of genetically modified organisms (GMOs). Biochem Eng J 44(2–3):142–150
41. Kari L, Kari J, Landweber LF (1999) Reversible molecular computation in ciliates. In: Karhumaki J, Maurer H, Paun G, Rozenberg G (eds) Jewels are Forever. Contributions on theoretical computer science in honor of Arto Salomaa. Springer, Berlin, pp 353–363
42. Katz E, Privman V (2010) Enzyme-based logic systems for information processing. Chem Soc Rev 39(5):1835–1857

43. Kawde AN, Wang J (2004) Amplified electrical transduction of DNA hybridization based on polymeric beads loaded with multiple gold nanoparticle tags Electroanalysis 16 (1–2):101–107
44. Krishnan Y, Simmel FC (2011) Nucleic acid based molecular devices. Angewandte Chemie-International Edition 50(14):3124–3156
45. Kudelski A (2008) Analytical applications of Raman spectroscopy. Talanta 76(1):1–8
46. Landweber LF, Kari L (1999) The evolution of cellular computing: nature's solution to a computational problem. Biosystems 52(1–3):3–13
47. Long N, Gianola D, Rosa G, Weigel K, Avendano S (2009) Comparison of classification methods for detecting associations between SNPs and chick mortality. Gen Sel Evol 41:18
48. Li X, Liao B, Cai L, Cao Z, Zhu W (2013) Informative SNPs Selection Based on Two-Locus and Multilocus Linkage Disequilibrium: Criteria of Max-Correlation and Min-Redundancy. IEEE/ACM Trans Comput Biol Bioinf 10(3):688–695
49. Li D, Shlyahovsky B, Elbaz J, Willner I (2007) Amplified analysis of low-molecular-weight substrates or proteins by the self-assembly of DNAzyme-aptamer conjugates. J Am Chem Soc 129(18):5804–5805
50. Lin SY, Probert W, Lo M, Desmond E (2004) Rapid detection of isoniazid and rifampin resistance mutations in Mycobacterium tuberculosis complex from cultures or smear-positive sputa by use of molecular beacons. J Clin Microbiol 42(9):4204–4208
51. Liu X, Farmerie W, Schuster S, Tan W (2000) Molecular beacons for DNA biosensors with micrometer to submicrometer dimensions. Anal Biochem 283:56–63
52. Liu Y, Tuleouva N, Ramanculov E, Revzin A (2010) Aptamer-based electrochemical biosensor for interferon gamma detection. Anal Chem 82(19):8131–8136
53. Lund K, Manzo A, Dabby N, Michelotti N, Johnson-Buck A, Nangreave J, Taylor S, Pei R, Stojanovic M, Walter N, Winfree E, Yan H (2010) Molecular robots guided by prescriptive landscapes. Nature 465:206–210
54. Macdonald J, Li Y, Sutovic M, Lederman H, Pendri K, Lu WH, Andrews BL, Stefanovic D, Stojanovic MN (2006) Medium scale integration of molecular logic gates in an automaton. Nano Lett 6(11):2598–2603
55. MacKerell AD Jr, Banavali N, Foloppe N (2001) Development and current status of the CHARMM force field for nucleic acids. Biopolymers 56(4):257–265
56. Maley CC (1998) DNA computation: theory, practice, and prospects. Evol Comput 6 (3):201–229
57. Mannelli I, Minunni M, Tombelli S, Mascini M (2003) Quartz crystal microbalance (QCM) affinity biosensor for genetically modified organisms (GMOs) detection. Biosens Bioelectron 18(2–3):129–140
58. Mannelli I, Minunni M, Tombelli S, Wang R, Spiriti MM, Mascini M (2005) Direct immobilization of DNA probes for the development of affinity biosensors. Bioelectrochemistry 66(1–2):129–138
59. Manor O, SegalPredicting E (2013) Disease Risk Using Bootstrap Ranking and Classification Algorithms. PLoS Comput Biol 9(8):e1003200
60. Margulies D, Hamilton AD (2009) Digital analysis of protein properties by an ensemble of DNA quadruplexes. J Am Chem Soc 131(26):9142–1943
61. Markert H, Ring J, Campbell N, Grates K (2011) A comparison of four commercially available portable Raman spectrometers. National Forensic Science Technology Center. http://www.nfstc.org/?dl_id=214
62. May E, Vouk M, Bitzer D, Rosnick D (2004) Coding theory based models for protein translation initiation in prokaryotic organisms. BioSyst J 76(1–3):249–260
63. May E, Vouk M, Bitzer D (2006) An Error-Control Coding Model For Classification of Escherichia coli K-12 ribosome binding sites. IEEE EMB Mag 25(1):90–97
64. May EE, Dolan PL, Crozier PS, Brozik SM, Manginell M (2008) Towards de novo design of deoxyribozyme biosensors for GMO detection. IEEE Sens J 8(6):1011–1019

65. May E, Lee M, Dolan P, Crozier P, Brozik S, Manginell M (2008b) Computational sensing and in vitro classification of GMOs and biomolecular events. In: Proceedings of the 26th army science conference
66. Mehlmann M, Dawson ED, Townsend MB, Smagala JA, Moore CL, Smith CB, Cox NJ, Kuchta RD, Rowlen KL (2006) Robust sequence selection method used to develop the FluChip diagnostic microarray for influenza virus. J Clin Microbiol 44(8):2857–2862
67. Mhlanga M, Malmberg L (2001) Using molecular beacons to detect single-nucleotide polymorphisms with real-time PCR. Methods 25(4):463–471
68. Monroy-Contreras R, Vaca L (2011) Molecular beacons: powerful tools for imaging RNA in living cells. J Nucleic Acids Article ID 741723, 15 pp
69. Motornov M, Zhou J, Pita M, Gopishetty V, Tokarev I, Katz E, Minko S (2008) "Chemical transformers" from nanoparticle ensembles operated with logic. Nano Lett 8(9):2993–2997
70. Motornov M, Zhou J, Pita M, Tokarev I, Gopishetty V, Katz E, Minko S (2009) An integrated multifunctional nanosystem from command nanoparticles and enzymes. Small 5 (7):817–820
71. Obenauer JC, Denson J, Mehta PK, Su X, Mukatira S, Finkelstein DB, Xu X, Wang J, Ma J, Fan Y, Rakestraw KM, Webster RG, Hoffmann E, Krauss S, Zheng J, Zhang Z, Naeve CW (2006) Large-scale sequence analysis of avian in- fluenza isolates. Science 311(5767):1576–1580
72. Palese P, Young JF (1982) Variation of influenza A, B, and C viruses. Science 215 (4539):1468–1474
73. Pattnaik P (2005) Surface plasmon resonance—applications in understanding receptor-ligand interaction. Appl Biochem Biotechnol 126(2):79–92
74. Penchovsky R, Breaker RR (2005) Computational design and experimental validation of oligonucleotide-sensing allosteric ribozymes. Nat Biotechnol 23(11):1424–1433
75. Penchovsky R (2012) Engineering integrated digital circuits with allosteric ribozymes for scaling up molecular computation and diagnostics. ACS Synthet Biol 1(10):471–482
76. Peyret N, Seneviratne PA, Allawi HT, SantaLucia J Jr (1999) Nearest-neighbor thermodynamics and NMR of DNA sequences with internal AA, CC, GG, and TT mismatches. Biochemistry 38(12):3468–3477
77. Piatek A, Tyagi S, Pol A, Telenti A, Miller L, Kramer F, Alland D (1998) Molecular beacon sequence analysis for detecting drug resistance in Mycobacterium tuberculosis. Nat Biotechnol 16(4):359–363
78. Pita M, Zhou J, Manesh KM, Halamek J, Katz E, Wang J (2009) Enzyme logic gates for assessing physiological conditions during an injury: towards digital sensors and actuators. Sens Actuators B-Chem 139(2):631–636
79. Polsky R, Harper JC, Brozik SM (2012) Nanomaterial-based electrochemical DNA detection. In: Ozsoz M (ed) Electrochemical DNA Biosensors. Pan Stanford Publishing, Singapore, pp 427–480
80. Prody GA, Bakos JT, Buzayan JM, Schneider IR, Bruening G (1986) Autolytic processing of dimeric plant-virus satellite RNA. Science 231(4745):1577–1580
81. Privman M, Tam TK, Bocharova V, Halamek J, Wang J, Katz E (2011) Responsive interface switchable by logically processed physiological signals: Toward "smart" actuators for signal amplification and drug delivery. ACS Appl Mater Interfaces 3(5):1620–1623
82. Privman V, Pedrosa V, Melnikov D, Pita M, Simonian A, Katz E (2009) Enzymatic AND-gate based on electrode-immobilized glucose-6-phosphate dehydrogenase: Towards digital biosensors and biochemical logic systems with low noise. Biosens Bioelectron 25 (4):695–701
83. Privman M, Tam TK, Pita M, Katz E (2009) Switchable electrode controlled by enzyme logic network system: approaching physiologically regulated bioelectronics. J Am Chem Soc 131(3):1314–1321
84. Rabbee N, Speed T (2006) A genotype calling algorithm for affymetrix SNP arrays. Bioinformatics 22(1):7–12

85. Ramlan E, Zauner K-P (2013) In-silico design of computational nucleic acids for molecular information processing. J Cheminform 5:22
86. Robertson DL, Joyce GF (1990) Selection in-vitro of an RNA enzyme that specifically cleaves single-stranded-DNA. Nature 344(6265):467–468
87. Rowe AA, Miller EA, Plaxco KW (2010) Reagent less measurement of aminoglycoside antibiotics in blood serum via an electrochemical, ribonucleic acid aptamer-based biosensor. Anal Chem 82(17):7090–7095
88. SantaLucia J Jr (1998) A unified view of polymer, dumbbell, and oligonucleotide DNA nearest-neighbor thermodynamics. Proc Natl Acad Sci USA 95(4):1460–1465
89. Santoro SW, Joyce GF (1997) A general purpose RNA-cleaving DNA enzyme. Proc Natl Acad Sci USA 94(9):4262–4266
90. Schluecker S (2009) SERS microscopy: nanoparticle probes and biomedical applications. ChemPhysChem 10(9–10):1344–1354
91. Shinya K, Hamm S, Hatta M, Ito H, Ito T, Kawaoka Y (2004) PB2 amino acid at position 627 affects, replicative efficiency, but not cell tropism, of Hong Kong H5N1 influenza A viruses in mice. Virology 320(2):258–266
92. Shlyahovsky B, Di L, Weizmann Y, Nowarski R, Kotler M, Willner I (2007) Spotlighting of cocaine by an autonomous aptamer-based machine. J Am Chem Soc 129(13):3814–3815
93. Simmel FC (2007) Towards biomedical applications for nucleic acid nanodevices. Nanomedicine 2(6):817–830
94. Simpson ML, Sayler GS, Fleming JT, Applegate B (2001) Whole-cell biocomputing. Trends Biotechnol 19(8):317–323
95. Sistare MF, Holmberg RC, Thorp HH (1999) Electrochemical studies of polynucleotide binding and oxidation by metal complexes: effects of scan rate, concentration, and sequence. J Phys Chem B 103(48):10718–10728
96. Sivan S, Tuchman S, Lotan N (2003) A biochemical logic gate using an enzyme and its inhibitor. Part II: The logic gate. Biosystems 70(1):21–33
97. Smith GJD, Fan XH, Wang J, Li KS, Qin K, Zhang JX, Vijaykrishna D, Cheung CL, Huang K, Rayner JM, Peiris JSM, Chen H, Webster RG, Guan Y (2006) Emergence and predominance of an H5N1 influenza variant in China. Proc Natl Acad Sci USA 103 (45):16936–16941
98. Stojanovic MN, de Prada P, Landry DW (2001) Aptamer-based folding fluorescent sensor for cocaine. J Am Chem Soc 123(21):4928–4931
99. Stojanovic MN, de Prada P (2001) D. W. Catalytic molecular beacons. ChemBioChem 2 (6):411–415
100. Stojanovic MN, Mitchell TE, Stefanovic D (2002) Deoxyribozyme-based logic gates. J Am Chem Soc 124(14):3555–3561
101. Stojanovic MN, Semova S, Kolpashchikov D, Macdonald J, Morgan C, Stefanovic D (2005) Deoxyribozyme-based ligase logic gates and their initial circuits. J Am Chem Soc 127 (19):6914–6915
102. Stojanovic MN, Stefanovic D (2003) A deoxyribozyme-based molecular automaton. Nat Biotechnol 21(9):1069–1074
103. Stojanovic MN, Stefanovic D (2003) Deoxyribozyme-based half-adder. J Am Chem Soc 125 (22):6673–6676
104. Stojanovic MN, Stefanovic D, LaBean T, Yan H (2005) In: Willner I, Katz E (eds) Bioelectronics: from theory to applications. Wiley-VCH, Weinheim, pp 427–455
105. Stojanovic MN, Stefanovic D, Rudchenko S (2014) Exercises in Molecular Computing. Acc Chem Res 47(6):1845–1852
106. Strack G, Pita M, Ornatska M, Katz E (2008) Boolean logic gates that use enzymes as input signals. ChemBioChem 9(8):1260–1266
107. Tan L, Li Y, Drake TJ, Moroz L, Wang K, Li J, Munteanu A, Yang CJ, Martinez K, Tan W (2005) Molecular beacons for bioanalytical applications. Analyst 130(7):1002–1005

108. Taylor SK, Pei R, Moon BC, Damera S, Shen A, Stojanovic MN (2009) Triggered release of an active peptide conjugate from a DNA device by an orally administrable small molecule. Angewandte Chemie-International Edition 48(24):4394–4397
109. Tichoniuk M, Ligaj M, Filipiak M (2008) Application of DNA hybridization biosensor as a screening method for the detection of genetically modified food components. Sensors 8 (4):2118–2135
110. Tokarev I, Gopishetty V, Zhou J, Pita M, Motornov M, Katz E, Minko S (2009) Stimuli-responsive hydrogel membranes coupled with biocatalytic processes. ACS Appl Mater Interfaces 1(3):532–536
111. Tuerk C, Gold L (1990) Systematic evolution of ligands by exponential enrichment—RNA ligands to bacteriophage-T4 DNA-polymerase. Science 249(4968):505–510
112. Tyagi S, Kramer FR (1996) Molecular beacons: probes that fluoresce upon hybridization. Nat Biotechnol 14(3):303–308
113. Unger R, Moult J (2006) Towards computing with proteins. Proteins-Structure Function and Bioinformatics 63(1):53–64
114. Vo-Dinh T, Wang H-N, Scaffidi J (2010) Plasmonic nanoprobes for SERS biosensing and bioimaging. J Biophotonics 3(1–2):89–102
115. Von Maltzahn G, Harris TJ, Park J-H, Min D-H, Schmidt AJ, Sailor MJ, Bhatia SN (2007) Nanoparticle self-assembly gated by logical proteolytic triggers. J Am Chem Soc 129 (19):6064–6065
116. Wang C, Trau D (2011) A portable generic DNA bioassay system based on in situ oligonucleotide synthesis and hybridization detection. Biosens Bioelectron 26(5):2436–2441
117. Wang G, Wang Y, Chen L, Choo J (2010) Nanomaterial-assisted aptamers for optical sensing. Biosens Bioelectron 25(8):1859–1868
118. Wang J (2000) From DNA biosensors to gene chips. Nucleic Acids Res 28(16):3011–3016
119. Wang J, Katz E (2011) Digital biosensors with built-in logic for biomedical applications. Isr J Chem 51(1):141–150
120. Wang J, Katz E (2010) Digital biosensors with built-in logic for biomedical applications-biosensors based on a biocomputing concept. Anal Bioanal Chem 398 (4):1591–1603
121. Waterman MS, Eggert M (1987) A new algorithm for best subsequence alignments with application to tRNA-rRNA comparisons. J Mol Biol 197(4):723–728
122. Willner I, Shlyahovsky B, Zayats M, Willner B (2008) DNAzymes for sensing, nanobiotechnology and logic gate applications. Chem Soc Rev 37(6):1153–1165
123. Wilson DS, Szostak JW (1999) In vitro selection of functional nucleic acids. Annu Rev Biochem 68:611–647
124. Win MN, Smolke CD (2008) Higher-order cellular information processing with synthetic RNA devices. Science 322(5900):456–460
125. Weizmann Y, Beissenhirtz MK, Cheglakov Z, Nowarski R, Kotler M, Willner I (2006) A virus spotlighted by an autonomous DNA machine. Angewandte Chemie-International Edition 45(44):7384–7388
126. Weizmann Y, Cheglakov Z, Pavlov V, Willner I (2006) Autonomous fueled mechanical replication of nucleic acid templates for the amplified optical detection of DNA. Angewandte Chemie-International Edition 45(14):2238–2242
127. Wu Z-S, Jiang J-H, Shen G-L, Yu R-Q (2007) Highly sensitive DNA detection and point mutation identification: An electrochemical approach based on the combined use of ligase and reverse molecular beacon. Hum Mutat 28(6):630–637
128. Xiao Y, Lubin AA, Heeger AJ, Plaxco KW (2005) Label-free electronic detection of thrombin in blood serum by using an aptamer-based sensor. Angewandte Chemie-International Edition 44(34):5456–5459
129. Yamada S, Suzuki Y, Suzuki T, Le MQ, Nidom CA, Sakai-Tagawa Y, Muramoto Y, Ito M, Kiso M, Horimoto T, Shinya K, Sawada T, Kiso M, Usui T, Murata T, Lin YP, Hay A, Haire LF, Stevens DJ, Russell RJ, Gamblin SJ, Skehel JJ, Kawaoka Y (2006)

Haemagglutinin mutations responsible for the binding of H5N1 influenza A viruses to human-type receptors. Nature 444(7117):378–382
130. Yashin R, Rudchenko S, Stojanovic MN (2007) Networking particles over distance using oligonucleotide-based devices. J Am Chem Soc 129(50):15581–15584
131. Yeh H-Y, Yates M, Mulchandania A, Chen W (2010) Molecular beacon–quantum dot–Au nanoparticle hybrid nanoprobes for visualizing virus replication in living cells. Chem Commun 46:3914–3916
132. Zhou J, Arugula MA, Halamek J, Pita M, Katz E (2009) Enzyme-based NAND and NOR logic gates with modular design. J Phys Chem B 113(49):16065–16070
133. Zhou J, Halamek J, Bocharova V, Wang J, Katz E (2011) Bio-logic analysis of injury biomarker patterns in human serum samples. Talanta 83(3):955–959
134. Zhou J, Tam TK, Pita M, Ornatska M, Minko S, Katz E (2009) Bioelectrocatalytic system coupled with enzyme-based biocomputing ensembles performing boolean logic operations: Approaching "smart" physiologically controlled biointerfaces. ACS Appl Mater Interfaces 1 (1):144–149

Digital Body

Aftab Ahmad

Nano-technology is rapidly maturing to help realize bio-chemical sensors that can be implanted in human body with integrated transceivers to share symptoms' data with each other internally to the body, and with outside world. Such implanted sensor networks can assist future medical practitioners to administer drug delivery remotely without interfering with patients' daily activities. However, in order to make this technology commercially viable, a digital computer model of human body is imperative. A model that can be customized to include any patient and is applicable locally for a single device as well as body-wide implant network. This chapter discusses some new approaches for creating such computer simulation models for human body.

1 Human Nano-engineering

The field of medicine is experiencing an interesting paradigm in terms of a marriage between life sciences and engineering. As reported in the Fourth Aspen Brain Forum (Aspen CO, September 18–20, 2013), brain implantation and deep brain stimulation (DBS) are slowly maturing. Along with cardiac implants in cardiology, implants for diabetes monitoring and drug administration, a host of neuromuscular implants for control (such as retina implants), and prosthetic limb design are working proficiently in a number of patients all over the world, especially among industrially advanced nations. Networking of these devices is the next step in this

A. Ahmad (✉)
City University of New York, John Jay College of Criminal Justice, New York, USA
e-mail: aahmad@jjay.cuny.edu

J. Suzuki et al. (eds.), *Modeling, Methodologies and Tools for Molecular and Nano-scale Communications*, Modeling and Optimization in Science and Technologies 9, DOI 10.1007/978-3-319-50688-3_24

wave of human engineering. Implant networking can achieve several objectives. Examples are, monitoring of the correlation between various pathological prognoses, such as nephrological and cardiac disorders, brain and artificial limb communications, and automatically informing the primary care or first responder from within the body. Human body offers a very complex networking environment, which is wireless in nature but does not have many signal propagation mechanisms like the wireless environment for a cell phone. Therefore, the channel model for human body finds it hard to inherit the research carried out for wireless environment, and needs a new-look approach. This chapter provides such an approach and outlines a computer simulation technique that can be used by future medical professionals in getting assistance to implant in situ devices that require minimum power to provide prolonged functionality. One of the distinguishing characteristics of the proposed modeling approach is that it can be employed in ensuring that the implant is in an area of the body where it should provide the best usage scenario, and the definition of required minimum power is that if there is a choice of location for the implant area, the exact location is determined by the minimum power path between the implant under consideration and the implant with which it will network.

1.1 The Human Body Channel

Human body is a complex system of materials with dynamic propagation characteristics. Determining the power budget for a communications signal path within the body depends on a number of factors, such as, the composition of the materials on the path, whether the clearest path will be external to the body or internal, the variance of composition for a given subject, the transmitted power, sensitivity of the receivers, whether the subject is healthy or sick, in motion or static, and the wireless spectrum, to name a few. In order to identify the best location for a communications path between multiple implants, a medical professional (MP) will have to consider what body volume is available as a choice for each implant, what is the signal attenuation profile of each location pair for transceivers within the available body volumes, the energy mode and recharging or power source replacement mechanisms.

The proposed modeling technique in this chapter will help the MPs determine the path loss profile within the human body. It will have two major components: a general component that applies to all humans and a reconfigurable component that will be derived by the MP with the help of the subject's body material composition and characteristics of those materials on the communications path within the body. Depending upon the path profile thus obtained, the power source employed, the emitted power and the receiver sensitivity, the MP will be able to determine how frequently the power source should be re-energized or replenished.

1.2 Methodology

The technique is originally proposed in [1]. Some terms in this section therefore may appear for the first time to the reader who has not seen [1], but they will become clear later in this chapter. In the original paper, an element by element mechanism for determining signal loss profile has been proposed. The size of the 3-D cube shaped element should be configurable depending on the heterogeneity of the body area under-consideration, available power sources and the required accuracy of modeling. Computing power is easily available in today's desktop machines to consult a body composition database, use the readily determinable *coarse* model from this database for rough power estimate, and determine the optimal locations for a given subject by incorporating the available choices of paths and other factors.

2 Propagation Models

Extensive empirical research has been carried out in modeling propagation conditions in wireless and dielectric media. Usually, the goal of these models is to determine the resources needed to provide a communications capability between two devices when the channel is stochastic in nature. One of the resources is the transmit power required for given environment & receiver sensitivity, which is the minimum value of the receive signal power that a device can detect. For most applications, the environments provide extremely random conditions. The approach taken to resolve this variability is by defining certain types of environments [2, 3], such as urban, sub-urban, rural, space, inside buildings, in busy square, etc., so as to develop models for each environment. Empirical methods are employed to determine the average signal attenuation and distribution of randomness in the signal, characterized as path loss and fading respectively. For most applications of wireless communications, delay spread and Doppler's bandwidth are also important. For more accurate measurements, and especially in antenna design, the electromagnetic wave propagation equations provide a better solution [4].

2.1 Limitations of Body Modeling

Human body departs, in many ways, from most environments studied for wireless communications [5]. The intractability of implanted systems perhaps stands out among all others factors. There are critical issues relating to supplying and continuation of power as well [6]. The impact of radiation on tissues is yet another debate that seems to be never-ending and asks for lowest possible power levels. The variety of materials in various parts of body and variety of bodies are other issues.

Since this is an area where gamble is not an option, precision and accuracy are the only choice while designing a channel model for body for intra-body communications. The IEEE 802.15.6 project is set up to come up with physical layer and medium access control (MAC) sub-layer specifications for implantable devices that will use the 402 MHz frequency band. Such a system is called MICS (medically implantable communications system).

2.2 Existing Work

Efforts have been reported in modeling human body and head. In [7], the authors have used tissue and geometrical variations properties in a galvanic coupling description of body. The model employs numerical simulations and measurements to characterize transmission of signals in individuals. In this quite relevant work, the authors model the human body for intra-body communications. They have plotted E-field around the body and measured and simulated various characteristics including power spectrum density (PSD). In [7, 8], the frequency ranges used are below 100 MHz. However, IEEE 802.15.6 devices will use the FCC allocated band at 400 MHz [9], and there is enough interest in UWB to warrant a modeling study at this spectrum as well. In one of the latest works [10], the authors have used simulations to characterize the human body as a communications medium for implants network at higher frequency bands, including MICS (400 MHz), ISM (2.4 GHz) and UWB (3.1 GHZ+). The finite difference time domain (FDTD) mechanism used in this paper is the one researchers are looking up to as computational power is being harnessed in desktop and smaller computing machinery. A frequency dependent FDTD has been employed in [11] along with Matlab® for determining the body signal propagation characteristics at the UWB band. In other related work on different frequency bands, one can find reference to ultrasonic frequencies [12].

The NIST model is an attempt to create a 3D immersive model specifically for the on-body and in-body sensor networks. A modelling technique, closer to the one we discuss here in the sense of being empirical, is discussed in [13]. It is based on measurements conducted in an anechoic chamber in the Ultrawide Band (UWB) frequency range. The frequency range is the limitation in this case. The modelling technique proposed in thus chapter can be made independent of the frequency spectrum and used for any frequency range for which propagation characteristics can be available by plugging them in the simulation program. In other applications, such as sports and entertainment, there is a need of body modeling but with different goals. For example, in [14], statistical modeling of human body share and pose is discussed from a transformation in poses point of view. By adding the propagation properties such as with the help of proposed work, such models can be made universal in usage, by applying them for body contour simulation, as well as implantable sensors. The IEEE 802.15 group attracted several suggestions, such as [15] that can give the propagation characteristics. However, the problem with these models is that they are either of inflexible nature or will require the help of a computer database for accommodating differences in individual bodies. Such models can be the candidates for the proposed work as can be seen in the next section Eq. (1).

Modeling techniques could depend on environment in which the signal propagates, as well as the targeted accuracy, and customization for each transaction or application. The idiosyncrasy of modeling human body can be understood from the following example in terms of the same for a cell phone. *It is like modeling a cell phone channel that can be applied to individual phones, for each instance of mobility*. Calculation of a precise power budget for this modeling is certain to require unreasonably large amount of resources, but cell phones have a big advantage that they have cell phone provider's tower with a base station on one end of communications that has no dearth of power. As a result of this asymmetric need, power control mechanisms are employed in cell phone communications to optimize the usage of power on a per transaction basis from the side of phone. In the absence of such power control mechanism, as is the case for intra-body communications, either a precise model based on electro-magnetic wave propagation through the body or a simulation model that gets live and constant feedback from the implants' communications data is perhaps the best approach, if not the only approach. Since the material composition of the body changes substantially from point to point, the former can only provide accuracy for a small section. Therefore, we resort to the later approach, that is, have a propagation model that can be tuned in precision as more data becomes available. The parameters that can be tuned in our proposed technique are the path-loss exponent and the variance of fading. It is reasonable then to start with some well-known values of the propagation characteristics of body materials and employ the standard path loss model, and then adjust the model parameters based on real data for each person. Since not all materials in the body are expected to follow the same profile of change with time, each material has to be dealt with individually. With these assumption, we are ready to describe our proposed model in the next section. Before ending this section, however, we will list modeling approaches in Table 1. The table is based on [2].

Table 1 Modeling approaches

Modeling approach	Examples	Applicability to body modeling
Path loss models	Okumura, Hata	Accuracy issue
Small scale fading	Rayleigh, Ricean	Applicable with a good path-loss model
Impulse response	Statistical Time Spread, SIRCIM	Measurement equipment issue
Joint angle-delay estimation (JADE)	JADE-MUSIC, SI-JADE	Body movements issue

3 The Proposed Modeling Approach

In the proposed model, we depart from the traditional approaches in many ways:

(i) We assume that the body consists of cubes, called *elements*, of a constant side d_o. Value of d_o can be changed depending on the application of using the model. For example for areas that are easier to approach surgically, with relatively homogeneous body composition, can have relatively large d_o—in general. That is also the difference between this assumption and FDTD, that is, d_o can be large depending on the body section, type of implant and its compatibility.

(ii) We assume that the propagation characteristics of each material type are unique enough to warrant its own model. Transitions between materials are modelled separately from in-material propagation. Thus, the model will require two path loss exponents for muscle and bone and one for transition from one to the other, assuming the transition is symmetric. If γ_{kl} is the path loss exponent for propagation from material type k to l and vice versa, then γ_k is the same for within the material type k. In this way, the path loss exponent in our model consists of an $n \times n$ matrix γ.

(iii) The second assumption allows us to model fading individually for each material type, which makes sense even if the distribution of fading is the same for all material types. Thus, $\boldsymbol{X}$ be the vector with components X_k ($k = 1, 2, \ldots n$) for the fading factor for material k with a standard deviation of $\sigma_{\boldsymbol{k}}$.

(iv) We assume that the path loss for the first d_o is known, not a deviation from other approaches.

Figure 1 shows how the body element is viewed.

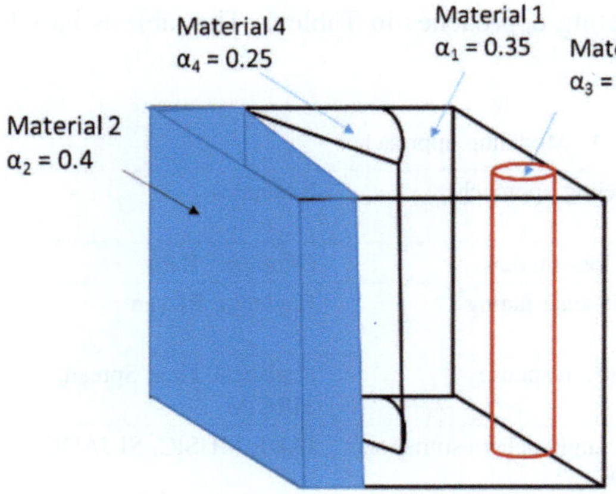

Fig. 1 Body element as depicted by the proposed model

In view of the above assumptions, the general model of the path loss is of the following form.

$$\boldsymbol{PL}(d,\lambda) = \alpha \cdot [\boldsymbol{A}(d_o) + \boldsymbol{B}(\lambda)] + 10 \cdot \boldsymbol{Y} \cdot log(R) + \boldsymbol{X} \tag{1}$$

In Eq. (1),

$\mathbf{A}(d_o)$ is the attenuation constant vector that depends on d_o and has components for different material types.
$\mathbf{B}(\lambda)$ is the vector for attenuation factor that depends on frequency of the signal, α is the composition vector for various materials such that $\Sigma\alpha_k = 1$. As a special case when the transitions between materials are abrupt, the $\gamma_{ij} = 0$ for $i \neq j$, and Eq. (1) becomes a set of linearly independent equations.

4 Computer Simulation Scenarios

In this section, we will give two examples; one for coarse grain implementation and the other for fine grain implementation. In this Chapter, our main focus is on fine grain model.

4.1 Course Grained Model

The coarse grain model is for implants that are approachable to simple surgical procedures, or by considering each major limb as a propagation unit and applying to the limb as having uniform characteristics. Figure 2 shows a breakdown of human body into such propagation areas.

Example—Hip Prosthetics

An example of such implants is a network of coordinating hip implants for a person with artificial legs. The implants are in an area that are relatively easier to operate upon and are composed of a few material types. Depending on the location of sensors, the intervening material can be a single muscle. In this case, the matrix γ is a 2×2 or 3×3 matrix with only diagonal non-zeros for abrupt material boundaries. In some cases, such as brain implants, or implants for monitoring pregnancy, the interconnection will be within the same limb and a coarse grain model is good even with high accuracy. The grain in this case is defined by the scope of modeling, which is individual limb.

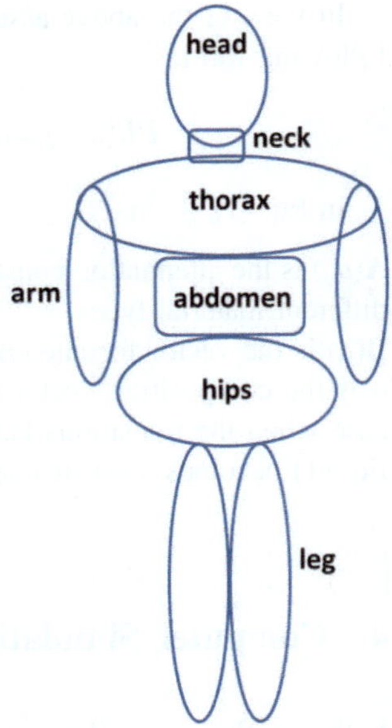

Fig. 2 Coarse grain implementation example

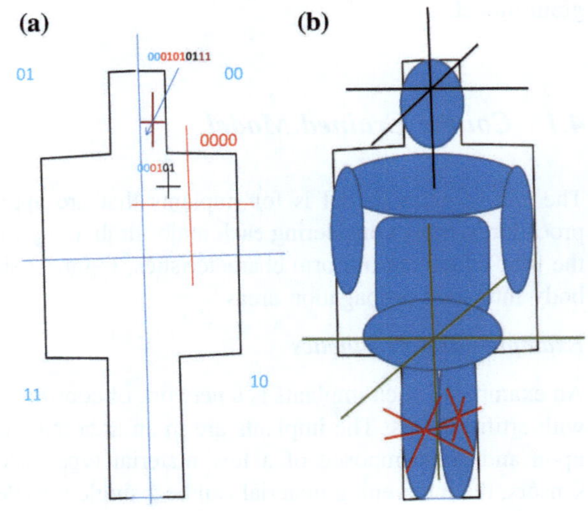

Fig. 3 **a** Fine grain and **b** coarse grain comparison

4.2 Fine Grain Model

We elaborate this model in greater detail and use the bulk of discussion for the fine grain model. For areas that have more complex composition, are hard to reach and operate upon, a fine grain model can be adopted from Eq. (1). Figure 3 shows how we scale this model as compared to the coarse grain. For ease of demonstration we have drawn Fig. 3a in 2 dimensions. The third dimension can be incorporated easily as will be shown later in this section. Each element is designated by a bit pattern that also defines the resolution and location of the element. The model resolution is controlled by hierarchical layering of resolution reference axes. As shown in the figure, the first level divides the body into four parts, requiring two bits for designation of each element (quadrant). Therefore, if an implant instance can tolerate a power variance given in one of these quadrants, a two-bit resolution can be used to identify the location of an element where the implant is located and propagation properties for that quadrant can be employed. The second layer of reference axes can narrow down the location to one-fourth of each quadrant area, further narrowing down the implant location and its neighborhood.

Figure 3a shows four hierarchies. To extend this to 3-D, one bit will be added to each layer, making it 12-bit for a four layer model. Each element in the fourth hierarchy is specified using 8 bits. In an 3 *M*-bit body, there are $2^{(3M)}$ elements. For $M = 0$, the whole body is considered as one element and for $M > 0$, there are *M* hierarchies of grains or sections. For this reason, we called this model the *M-grain model.*

4.2.1 How to Create an M-Grain Model

It follows from the discussion above that an M-grain model views body as consisting of $n = 2^{3M}$ 3-D elements, each consisting of a composition of various materials. The following Lemma describes an important property of this model.

Lemma 1 *If we know the material composition of all body elements for a given value of* $M = \mu$*, then we also know the composition of materials for all elements for values of* $m < \mu$.

Lemma 1 can be explained by taking an actual value with reference to Fig. 3. We stick to 2-D example for ease of understanding. For the 2-D case, the number of elements will be modified to 2^{2M} for *M*-grain model. In Fig. 3a, if we know material composition of all sections for $M = 3$, then each element needs 6 bits (b_0, b_1, b_2, b_3, b_4, b_5) to be identified. Out of these 6 bits the left 4 bits constitute all the section IDs for $m = 2$. Thus, (0, 0, 0, 1, b_4, b_5) is the second from top-right elements out of the 16 (red) elements for all values of b_4 and b_5. Figure 4 illustrates this.

Lemma 1 leads to the conclusion that the resolution of the *M*-grain model can be scaled back to a lower resolution model by masking groups of *d* bits from the right side of the location IDs of the elements in a *d*-dimension. The following theorem is based on Lemma 1.

Theorem 1 *In d-Dimension implementation of the M-grain model, the material properties of resolution layer m < M are obtained by masking the right (M-m).d bits of each element.*

Proof The proof of Theorem 1 follows from Lemma 1.

It may be noted that there is a difference between the coarse grain model and the low-resolution-layers fine grain model. In the coarse grain model we assume that

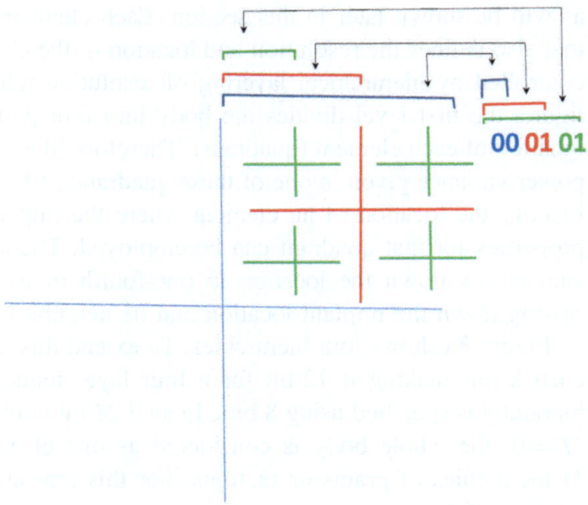

Fig. 4 Lemma 1 demonstration for mapping from $M = 3$ to $m = 2$

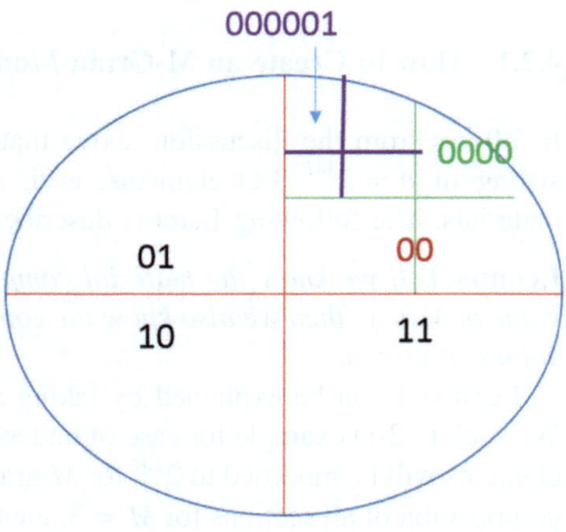

Fig. 5 Coarse and fine grained models are easily applicable to human head

each limb is designed as one or a few elements, thus requiring a different reference axis for each limb or organ according to convenience. A fine grain model can be applied to each limb of a coarse grain model. In Sect. 5, we will discuss how to store the material properties in a computer to be used to create individual models.

4.3 Modeling the Human Head

Figure 5 shows how the fine grain model can be applied to head as a limb.

In this figure, a 2D shape of head has been shown. The head is divided into a flexible number of *elements*. Each element identifies a certain part of the head and consists of a composition of materials.

5 Computer Simulation of Proposed Model

Body is a complex system of systems, in every sense of the word 'system'. To add to it, propagation of signals is a stochastic mechanism in any system and depends on the electromagnetic properties of the material including permeability, permittivity and conductivity, among others. However, technology is available [16] for measuring these properties of body materials and they have been measured to large extent. Incorporating these properties into propagation models to evaluate the path loss exponents is a subject of extensive research [8]. Additionally, the fading properties need to be incorporated in the model as well.

Another difficulty in simulating such a model for medical applications arises from the versatility of human subjects. Even if we have separate simulation systems for the male and female bodies, there is sufficient variety within each sex to warrant a case-by-case simulation. Given the criticality of implants when they are needed, this is a requirement that needs to be addressed anyway. Therefore, we contend that the strengths of the M-grain model will build slowly as the technology advances for measuring material properties of parts of bodies of individuals. One such strength is that it allows for defining the body element as an object in an object-oriented programming (OOP) language, such as C++. Thus, in an actual implementation

Using an object-oriented language, an element can be defined as a class, and then inherited in the class that will be body. A database consisting of general values of propagation exponents can store a general body that can be loaded by default. For people with variation from the general characteristics, the actual values of the propagation constants can be changed in the desired implant communications path to customize only the required area and save processing by keeping the rest of this default model (*virtual* body) the same.

5.1 Simulation Usage

In a typical application of the model, following is a sequence of steps:

Step 1 The physician has a computer program that is based on Eq. (1). The program employs, from a database, the material properties of various body parts that have default values

Step 2 The physician enters the two body areas that are to host the implants in the computer. A general composition of the materials is available in the program. These can be manually changed by the physician depending on a physical exam

Step 3 The physician enters the value for d_o in a unit of length and the longest length L_M to be considered to determine M. This satisfies an equation $(L_M/d_o)^3 = 2^{3M}$. From this, the value of M is given by:

$$M = log_2(L_M/d_o)\text{bits} \tag{2}$$

The value of L_M is influenced by whether a limb or a part is considered in isolation or the whole body is considered. For some applications such as interconnection between heart sensor and diabetes sensor, multiple parts or whole body may be a better consideration for determining M. As a result of Eq. (2), the total number of elements in the L_M length is 2^{3M}. The implicit assumption in deriving Eq. (2) is that the body section of length L_M is a perfect cube, which is not the case for most part. However, the real purpose of this way of modeling is not be accurate about the location of boundaries, but to be accurate in modeling signal propagation within each body element through the use of proper material composition, and to be accurate in determining signal power received. For isotropic antennas propagation, the assumption of cubic element has the additional benefit that the model can be tested empirically by measuring power received at the body boundaries that occur at distances smaller than L_M and comparing it with the model. This is shown in Fig. 6.

Step 4 Solve Eq. (1) typically employing numerical methods available extensively these days.

Step 5 The physician can determine the required implant power source based on the solution of Eq. (1), criticality of the situation, sensitivity of the receiver and experience. At this point a decision can be made and surgical operation can be planned.

It must be mentioned here that more research on this modeling technique is needed and is expected to result in performance evaluation studies in future.

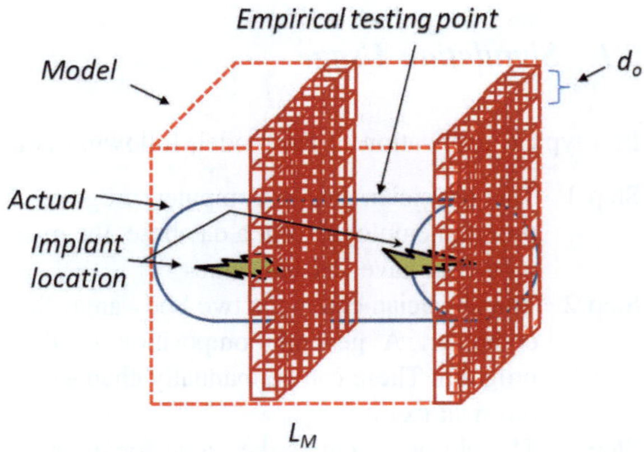

Fig. 6 Difference between modeling and actual shape of body area can be used for empirical model verification

6 Conclusion

We have presented a modeling approach to determine propagation channel characteristics of human body. The proposed method makes use of several available techniques with a new look to accommodate the body characteristics. These include:

- The use of path loss formula along with fading component.
- Considering the body as constituent of elements of a size that can be used to make the model scalable and flexible as per need.
- Considering each element as composed of multiple body materials and employing material characteristics in a matrix of path loss exponents.
- Using simulation for an initial general implementation based on physician experience and testing, and later refinement based on collected data from implanted transceivers.

More work is in progress in relating the propagation characteristics to path loss models, defining computer program for simulation and testing of the model for real life situations.

One of the important aspect of the proposed modeling technique is its relation to other proposals of body modeling for simulation. The proposed technique will create a database of human body consisting of 3-D elements. The size of the elements in general is a choice depending on the accuracy required for modeling. In a scenario where the elements are so small that there is one type of material, then using the information of organ and limb dimensions for a subject, it can be accurately applied to any variation of body. The simulation performance, however, will degrade as the element size reduces. Therefore, the decision of element size should be made based on the factors such the material composition on the signal path, how often and how much data the implant needs to process and the recharging mechanisms available.

References

1. Ahmad A (2013) A scalable human body modeling technique for networkable implants. In: Proceedings of the 9th international conference on body area networks, Boston MA, September–October 2013
2. Sarkar TK, Ji Z, Kim K, Medouri A, Salazar-Palma M (2003) A survey of various propagation models for mobile communication. Antennas Propag Mag IEEE 45(3):51–82
3. Chen AM, Rao RR (1998) On tractable wireless channel models. *In:* The ninth IEEE international symposium on personal, indoor and mobile radio communications, 1998, vol 2, pp 825–830
4. Jensen MA, Wallace JW (2004) A review of antennas and propagation for MIMO wireless communications. IEEE Trans Antennas Propag 52(11)

5. Schwiebert L, Gupta SKS, Weinmann J (2001) Research challenges in wireless networks of biomedical sensors. In: Proceedings of the 7th annual international conference on Mobile computing and networking, July 2001, Rome, Italy, pp 151–165
6. Heetderks WJ (1988) RF powering of millimeter- and submillimeter-sized neural prosthetic implants. IEEE Trans Biomed Eng 35(5):323–327
7. Wegmueller MS, Kuhn A, Froehlich J, Oberle M, Felber N, Kuster N, Wolfgang F (2007) An attempt to model the human body as a communication channel. IEEE Trans Biomed Eng 54 (10):1851–1857
8. Gupta SKS, Lalwani S, Prakash Y, Elsharawy E, Schwiebert L (2003) Towards a propagation model for wireless biomedical applications. IEEE international conference on communications, 2003. ICC '03, vol 3, 11–15 May 2003, pp 1993–1997
9. Kwak K-S, Ullah S, Ullah N (2010) An overview of IEEE 802.15.6 standard. In: 2010 3rd international symposium on applied sciences in biomedical and communication technologies (ISABEL), 7–10 November 2010, pp 1–6
10. De Santis V, Feliziani M (2011) Intra-body channel characterization of medical implant devices. In: EMC Europe 2011 York, 26–30 September 2011, pp 816–819
11. Tayamachi T, Wang Q, Wang J (2007) Transmission characteristic analysis for UWB body area communications. In: International symposium on electromagnetic compatibility, 2007. EMC 2007, 23–26 October 2007, pp 75–78
12. Galluccio L, Melodiay T, Palazzo S, Santagati GE (2012) Challenges and implications of using ultrasonic communications in intra-body area networks. In: Proceedings of the 9th annual conference on wireless on-demand network systems and services (WONS), pp 182–189
13. Geng Y, Wan Y, He J, Pahlavan K (2013) An empirical channel model for the effect of human body on ray tracing. In: 2013 IEEE 24th international symposium on personal indoor and mobile radio communications (PIMRC). IEEE, pp 47–52
14. Hasler N, Stoll C, Sunkel M, Rosenhahn B, Seidel H-P (2009) A statistical model of human pose and body shape. In: EUROGRAPHICS 2009, vol 28, no 2
15. Yazdandoost KY (2009) Channel model for body area network (BAN). In: IEEE P802.15-08-0780-09-0006, April 27, 2009
16. Gabriel C, Gabriely S, Corthout E (1996) The dielectric properties of biological tissues: I. Literature survey. Phys Med Biol 41:2231–2249
17. Paradiso JA, Starner T (2005) Energy scavenging for mobile and wireless electronics. Pervasive Comput IEEE 4(1):18–27. doi:10.1109/MPRV.2005.9
18. Wu X, Ma L, Huang KS, Gao Y, Chen Z (2005) Generic-model based human-body modeling. In: Entertainment computing ICEC 2005, Lecture notes in computer science, vol 3711, pp 203–214
19. Hagedorn J, Sayrafian-Pour K, Yang W-B, Terill JE (2015) (NIST staff), Visualization of body area networks. http://www.nist.gov/itl/math/hpcvg/ban.cfm. Accessed 12 Feb 2015

FSC
www.fsc.org
MIX
Papier aus verantwortungsvollen Quellen
Paper from responsible sources
FSC® C105338

If you have any concerns about our products,
you can contact us on
ProductSafety@springernature.com

In case Publisher is established outside the EU,
the EU authorized representative is:
Springer Nature Customer Service Center GmbH
Europaplatz 3, 69115 Heidelberg, Germany

Printed by Libri Plureos GmbH
in Hamburg, Germany